D0882611

ASTRONOMY AND
ASTROPHYSICS LIBRARY

Springer
Berlin
Heidelberg
New York
Hong Kong
London
Milan
Paris
Tokyo

dents. Consequently, subjects such as practical alignment and test techniques, as well as maintenance aspects, occupy a significant part. Nevertheless, there are inevitably major overlap areas with both Bahner's and Schroeder's books which the informed reader will recognise. This overlap, involving repetitions in a somewhat different context, is unavoidable for a complete presentation.

Bahner's book included sections on achromatic objectives for refracting telescopes, astrographic objectives and oculars. No such material is included in this book. The refractor as such and the optical design of oculars are only of historical interest in large telescope optics and are only mentioned in this context. Of course, refracting *elements* still play an important role in wide-field telescopes, field correctors and focal reducers, and these are dealt with in Chapters 3 and 4. In general, mirrors supply the optical power while refracting elements have only the subordinate but important role of improving the imagery.

I favour the morphological approach with a strong emphasis on the historical background of the subject. In this sense, Chapter 5 is to be seen as essential background for understanding the current situation in telescope optics. For the background of the general theory of optical aberrations and diffraction, the reader is referred to specialist books in the field of optics. Only the essential consequences of Gaussian optics, third order theory and diffraction theory are given: the emphasis is on a complete treatment of the *application* to reflecting telescope optics.

At the suggestion of the publisher, the work has been split into two volumes. The first volume deals with the historical development (but there is no claim to completeness as a history of telescope optics - that would be a separate work) and the theory of reflecting telescope optics, including that of the refracting corrector elements. The second volume deals with technical aspects and modern developments in general. Although there is considerable cross-referencing between the volumes, the split is a logical one, since each volume has its own entity.

Every attempt has been made to give complete references to the international literature. It is hoped that the work will be useful, apart from its own content, as a "source book" of the subject.

While I was writing the book, three further works on the subject were published: "Telescope Optics" by Rutten and van Venrooij (1988), "Astrooptik" by Laux (1993) and "Reflective Optics" by Korsch (1991). The first two are primarily destined for amateurs, but have equally great value for professionals. As with the works of Bahner and Schroeder, there is considerable overlap with my material and I have referred to them liberally in my text. I only became aware of Korsch's work when my own text was finished, but again there is inevitably considerable overlap of treatment. However, not only the content and aim of these five works, all admirable, are very different, but also their styles. In this sense, I feel confirmed in my own enterprise.

Preface

The development of telescope optics is a fascinating story. Until this century, the optical theory of reflecting telescopes was essentially limited to the Cartesian treatment of axial imagery. In 1905, Karl Schwarzschild initiated a revolution by applying third order (Seidel) theory to the field imagery of 2-mirror telescopes. Since then, the whole gamut of possible telescope systems has been invented, analysed and, in many cases, tried out in practice.

Over all its history, the optical development of the telescope has also depended on *technical* inventions, above all in mirror materials, glasses, support systems and means of achieving high reflectivity. Over the last 30 years, developments have been particularly spectacular, above all in manufacture and test techniques and generally in enhancing the image quality available.

When I started this work in 1988 there was little literature in book form available on telescope optics. Two of the best were in German: "Die Fernrohre und Entfernungsmesser" by König-Köhler (1959) and the monograph on "Teleskope" by K. Bahner in "Handbuch der Physik", Vol. XXIX, which appeared in 1967. A major part of this latter work was devoted to a condensed, but excellent exposition of the theory of telescope optics. Inevitably, more modern technical developments which have since assumed great importance could not be included; furthermore, the fact that it was written in German has reduced its impact and dissemination to a limited section of the interested community.

In 1987, "Astronomical Optics" by D. J. Schroeder appeared. Harland Epps kindly drew my attention to this excellent book in 1988 and I reflected then whether scope for a further work on the subject still existed. I finally concluded that it did: Schroeder's book covers a much wider field, since "astronomical" optics includes the broad subject of astronomical instruments, whereas my intention was (and remains) only the comprehensive coverage of the optics of the *reflecting* telescope, in the broadest interpretation of that term. Furthermore, Schroeder's work emerged more from the university orbit and includes much basic optical theory addressed to graduate students who need, and can profit from, the whole physics background.

The aim of the present book is different from Schroeder's. It is addressed primarily to specialists in the field, both in the astronomical community itself and in the industries concerned, although I hope it may also be useful to stu-

60 cm telescope and pages 500–501 with the historic group photo of the ISU Meeting at Mt. Wilson in 1910. The latter two photos, previously unknown in the astronomical community, were kindly supplied by Dr. Don Osterbrock, to whom I express my grateful thanks. I consider the above three historic print/photos to be a major enrichment of the book.

A change hopefully made correctly throughout the book is the spelling of the name "Abbe" *without* an acute accent on the "e". This error was pointed out by a reviewer in an Irish journal and has been confirmed by a former colleague of mine at Carl Zeiss. Abbe himself *pronounced* his name in later life (as virtually everyone does today, also in Germany) as though there were an accent, but he apparently never wrote it that way!

Apart from the valuable help from outside sources, acknowledged above or in the text, I also owe a great debt of gratitude to a number of ESO colleagues: Uta Grothkopf and Angelika Treumann of the ESO library, for their admirable service in literature procurement, particularly concerning historical aspects of Petzval's work; Bernard Delabre for information on his P.F. corrector; Gero Rupprecht for information on the ESO Linear ADC system; Philippe Dierickx for help with the setting-up of the new Spot Diagram (Fig. 4.18); Ed Janssen for completing this Spot Diagram, for the new figure of the LADC (Fig. 4.36) and corrections to two existing figures (see above); Stephane Guisard for pointing out two errors; and, above all, Lothar Noethe for many discussions and suggestions arising from his detailed knowledge of the book – I sometimes think he knows parts of it better than I know them myself! My deep thanks are again due to my wife Anne, who always serves as my "Delphic Oracle" on the English language; to Springer-Verlag for the admirable cooperation in all respects, also to Uwe Matrisch of the firm LE-TeX in Leipzig for excellent work in setting up the final form of this edition; and to the management of ESO for their invaluable general support.

I hope, of course, that this second edition will be often reprinted. Minor corrections, hopefully few, can then be made. But I am virtually certain that I shall not produce a third edition. Unlike the matter of RTO II, the theory of RTO I is, I believe, largely complete. Innovations may be made with minor modifications of known solutions for centered systems or with new "Schiefspiegler" with several mirrors. But radically new optical design solutions seem unlikely. Revolutionary progress is more likely in the domain of RTO II or in aspects not treated there, such as interferometry.

July 2003   *Ray N. Wilson*

Preface to the 2nd edition

A corrected reprint of the first edition appeared in 2000. It was a requirement that the pagination remain unaltered, but nevertheless, apart from minor and format corrections on 17 pages, a number of corrections or additions of substance could be incorporated. These included minor corrections to Figs. 1.3 b) and 2.8 and to the text of Fig. 5.18. The most important change of all was probably the complete revision of the historical treatment of Cassegrain in the Portrait Gallery, due to the superb research of Baranne and Launay on his identity, published in 1997. Additions of substance were text on pages 21, 323 and 487 (Portrait Gallery – Mersenne) and corrections on pages 117 (y^2 to y^4), 174 (concerning the scale of Fig. 3.37), 263 (Fig. 3.96 instead of 3.97 in the text), 341 (sign in the text equation below Eq. (4.36)), Table 5.2 (concerning UKIRT), Table A.15 (first symbol), pages 505 (Ref. 3.71) and 531 (Brown and Cassegrain). Several of these errors were pointed out by readers, to whom I express my gratitude.

The present 2nd edition contains all the material of the first edition, unchanged apart from some further corrections, but with 25 pages more of additional explanations or new material, including 5 new figures (2 in Chap. 4, 2 in Chap. 5 and 1 following the Portrait Gallery). Significant text additions are on pages 1–2, 22, 43–46, 63, 85–86, 86–87, 117, 120–122, 126, 129–130, 131–132, 214, 222–223, 232–233, 262, 269, 278–279, 281, 324, 328, 370–372, 402–404, 426–429, 433–435, 500–501. The most important of these are the following: pages 43–46 with the correction to Eq. (2.53) and the extensive justification of the definition of m_2 in Eq. (2.55) – due to a most fruitful correspondence with Dr. Dan Schroeder, for which I express here my grateful thanks; pages 120–122 where the mathematical argument has been completely reformulated; pages 232–233 and 278–279 where the remarkable new analytical procedure of Rakich and Rumsey for setting up 3- or 4-mirror telescope solutions is briefly discussed; pages 370–372 with the new prime-focus corrector due to Bernard Delabre of ESO including a new "Spot-Diagram" in the same standard format used in the first edition; pages 402–404 with a description of the new Linear ADC-corrector used in the ESO VLT; pages 426–429 with the historical print of the casting of the blank of the Melbourne reflector – kindly supplied by Peter Hingley, librarian of the RAS, to whom my grateful thanks; pages 433–435 with a historic photo of Ritchey at his

Karl Schwarzschild ca. 1908 (courtesy Martin Schwarzschild)

To the memory of

Karl Schwarzschild

(1873–1916)

who developed the first complete aberration theory
of reflecting telescopes

To the memory of

Harold Hopkins

(1918–1994)

a great physicist, teacher and friend,
who revealed to me the beauty and power
of aberration theory

Raymond N. Wilson
Waaler Str. 29
85296 Rohrbach, Germany

Cover picture: Karl Schwarzschild, surrounded by, from left to right, Marin Mersenne, René Descartes, James Gregory, William Herschel, Ludwig von Seidel, George Ritchey, Henri Chrétien, Bernhard Schmidt.
(From various sources acknowledged in the book.)

Libary of Congress Cataloging-in-Publication Data

Wilson, R. N. (Ray N.)
Reflecting telescope optics / R.N. Wilson.– 2nd ed.
p. cm. – (Astronomy and astrophysics library, ISSN 0941-7834)
Includes biblographical references and index.
Contents: 1. Basic design theory and its historical development
ISBN 3-540-40106-7 (acid-free paper)
1. Reflecting telescopes. 2. Reflection (Optics) 3. Reflecting telescopes–Design and construction. I. Title. II. Series.
QB88.W55 2004
522'.2–dc22   2004041321

ISSN 0941-7834
ISBN 3-540-40106-7 Second Edition Springer-Verlag Berlin Heidelberg New York
ISBN 3-540-58964-3 First Edition Springer-Verlag Berlin Heidelberg New York

Springer-Verlag is a part of Springer Science+Business Media

springeronline.com

© Springer-Verlag Berlin Heidelberg 1996, 2004
Printed in Germany

Typesetting and production: LE-TEX Jelonek, Schmidt & Vöckler GbR, Leipzig
Cover design: design & production GmbH, Heidelberg

Printed on acid-free paper   SPIN: 10920343   55/3141/YL - 5 4 3 2 1 0

R. N. Wilson

Reflecting Telescope Optics I

Basic Design Theory
and its Historical Development

Second Edition
With 237 Figures

 Springer

Chapter 3 of Vol. I, dealing with the aberration theory of reflecting telescopes, is the longest and certainly one of the most important in the whole work. It is in this area that there is the greatest overlap with the above books. However, an illustration of the major, and legitimate, differences in presentation is the data given on the optical quality of systems discussed. Spot-diagrams are the commonest way of representing the quality according to geometrical optics. Rutten-van Venrooij and Laux give virtually complete spot-diagram analyses of the systems they discuss, a very valuable feature. To keep Vol. I within reasonable bounds, I have preferred to limit myself to chosen examples, intended to illustrate with spot-diagrams the key points of the development. Some of these are taken from the literature; but most of those in Chapter 3 (and a few in Chapter 4) have been optimized by Bernard Delabre of ESO from starting systems I set up from the basic theory, or with minor modifications emerging from the calculations. I am deeply grateful for this major contribution to the work.

I owe a great debt of gratitude to many specialist members of the astronomical community and associated industrial concerns, particularly Carl Zeiss (Oberkochen) and REOSC (Paris), who have generously supplied information. This debt extends, too, to many ESO colleagues. Above all, I am grateful to the ESO management for supporting the project and for extensive help in establishing the final text. In the detailed work, I wish to thank specifically, as well as Bernard Delabre mentioned above, Marion Beelen, Samantha Milligan, Baxter Aitken (who has not only played a major role in the text-processing but also kindly read through the entire work), Ed Janssen (who drew and formatted the figures) and Hans-Hermann Heyer for much hard work and enthusiastic support. My gratitude is also due to Richard West for general encouragement and support. Finally, I thank the publisher, Springer-Verlag, for excellent cooperation, and, last but by no means least, my wife Anne, for much help with the text and, above all, for patience throughout the whole task.

D-85296 Rohrbach   Ray N. Wilson
January 1996

Contents

1 Historical introduction 1
- 1.1 Period 1608–1672 1
- 1.2 Period 1672–1840 11
- 1.3 William Herschel's telescopes 15

2 Basic (Gaussian) optical theory of telescopes 21
- 2.1 Basic function of a telescope 21
- 2.2 The ideal optical system, geometrical optics
  and Gaussian optics 23
  - 2.2.1 The ideal optical system and Gaussian concept 23
  - 2.2.2 Geometrical optics and geometrical wavefronts 26
  - 2.2.3 The Gaussian optics approximation 27
  - 2.2.4 The conventional telescope with an ocular 36
  - 2.2.5 Basic forms of reflecting telescope 40
  - 2.2.6 The scale of astronomical telescopes and the
    magnification in afocal use of compound telescopes ... 54
  - 2.2.7 "Wide-field" telescopes and multi-element forms 55

3 Aberration theory of telescopes 57
- 3.1 Definition of the third order approximation 57
- 3.2 Characteristic Function and Seidel (3rd order) aberrations:
  aberration theory of basic telescope forms 59
  - 3.2.1 The Characteristic Function of Hamilton 59
  - 3.2.2 The Seidel approximation:
    third order aberration coefficients 63
  - 3.2.3 Seidel coefficients
    of some basic reflecting telescope systems 65
  - 3.2.4 Analytical (third order) theory for 1-mirror
    and 2-mirror telescopes 69
  - 3.2.5 Higher order aberrations and system evaluation 82
  - 3.2.6 Analytical expressions for a 1-mirror telescope
    and various forms of 2-mirror telescopes (Classical,
    Ritchey-Chrétien, Dall-Kirkham, Spherical Primary) .. 88
  - 3.2.7 Other forms of aplanatic 2-mirror telescopes
    (Schwarzschild, Couder) 111

3.2.8 Scaling laws from normalized systems
to real apertures and focal lengths 126
3.3 Nature of third order aberrations and conversion formulae
from wavefront aberration to other forms 128
3.3.1 Spherical aberration (S_I) 128
3.3.2 Coma (S_{II}) 131
3.3.3 Astigmatism (S_{III}) and field curvature (S_{IV}) 135
3.3.4 Distortion (S_V) 138
3.3.5 Examples of conversions 139
3.3.6 Conversions for Gaussian aberrations 139
3.4 The theory of aspheric plates 140
3.5 The role of refracting elements in modern telescopes:
chromatic variations of first order and third order aberrations 146
3.6 Wide-field telescopes 148
3.6.1 The symmetrical stop position: the Bouwers telescope . 148
3.6.2 The Schmidt telescope 151
3.6.3 The Maksutov telescope 165
3.6.4 More complex variants of telescopes
derived from the principles of the Schmidt,
Bouwers and Maksutov systems 174
3.6.5 Three- or multi-mirror telescopes (centered) 223
3.7 Off-axis (Schiefspiegler) and decentered telescopes 255
3.7.1 Two- and three-mirror Schiefspiegler 255
3.7.2 The significance of Schiefspiegler theory
in the centering of normal telescopes:
formulae for the effects of decentering of 2-mirror
telescopes 261
3.8 Despace effects in 2-mirror telescopes 279
3.8.1 Axial despace effects 279
3.8.2 Transverse despace effects 287
3.9 Zernike polynomials 288
3.10 Diffraction theory and its relation to aberrations 293
3.10.1 The Point Spread Function (PSF) due to diffraction
at a rectangular aperture 293
3.10.2 Coherence 297
3.10.3 The Point Spread Function (PSF) due to diffraction
at a circular aperture 298
3.10.4 The Point Spread Function (PSF) due to diffraction
at an annular aperture 302
3.10.5 The diffraction PSF in the presence
of small aberrations............................... 304
3.10.6 The diffraction PSF in the presence
of small aberrations and an annular aperture......... 310

3.10.7 The diffraction PSF in the presence of larger
aberrations: the Optical Transfer Function (OTF) 312
3.10.8 Diffraction effects at obstructions in the pupil
other than axial central obstruction 322

4 Field correctors and focal reducers or extenders 325
4.1 Introduction 325
4.2 Aspheric plate correctors 327
4.2.1 Prime focus (PF) correctors using aspheric plates 327
4.2.2 Cassegrain or Gregory focus correctors
using aspheric plates 340
4.3 Correctors using lenses 348
4.3.1 Prime focus (PF) correctors using lenses 348
4.3.2 Secondary focus correctors using lenses 372
4.4 Atmospheric Dispersion Correctors (ADC) 392
4.5 Focal reducers and extenders.......................... 404
4.5.1 Simple reducers and extenders in front of the image... 404
4.5.2 Wide-field focal reducers (FR) as a substitute
for a prime focus 406
4.5.3 Other Cassegrain focal reducers.................... 414

5 Major telescopes from Lord Rosse to about 1980 419
5.1 Major telescopes in the speculum mirror epoch to 1865 419
5.2 Glass optics telescopes up to the Palomar 200-inch.......... 431
5.3 Reflectors after the 200-inch Palomar Telescope
up to about 1980 449

Appendices 465
A. List of mathematical symbols 467
B. Portrait gallery...................................... 487

References 502

List of Figures 513

List of Tables 527

Name Index 531

Subject Index 536

1 Historical introduction

The history of the telescope for astronomical purposes is certainly one of the most inspiring aspects of the development of science and technology. Many excellent accounts exist as chapters in more general works, in biographies or accounts of specific projects, but only a few books are devoted solely to the historical development. The works of King (1955) [1.1] and Riekher (1957, 1990) [1.2] are notable. The recent second edition of Riekher has been heavily modified and includes excellent new material which partially displaces interesting older material of the first edition. An updating of the work of King would be an enrichment of the subject but, failing that, the author hopes the present work will, with this chapter and with Chap. 5, to some extent fill this gap, at least on the optics side.

1.1 Period 1608–1672

The purpose of this introduction, as well as the content of Chap. 5, is to illustrate the course of development of telescope *optics* as a consistent logical process, an understanding of which is essential to a clear appreciation of the current situation and the way to the future. Precisely the early history, from 1608 up to 1672, is particularly instructive, since not only the basic theory of the *reflecting* telescope was completely and correctly expounded, but also all its basic forms. The theory of the *refractor* evolved more slowly but it remained for at least 100 years the more powerful practical tool purely for *manufacturing* reasons which will be considered in detail in RTO II, Chap. 1. This early history is treated particularly well in the classic work on telescopes by Danjon and Couder [1.3]. Another interesting account from an earlier epoch was given by Grant [1.4].

Above all in the Anglo-Saxon literature, Isaac Newton is often represented as the inventor of the reflecting telescope. But this is far from the truth: Newton's great merit was that he was the first person who *made* a reflecting telescope which could rival the refractors of the time. But the idea of a reflecting telescope goes right back to the origins of the refractor which emerged in a practical form between 1608 and 1610 in Holland and Italy. The first genuinely scientific analysis of the optical function of the refractor was given by Kepler in 1611 [1.2]. However, Galileo understood the basic imaging properties of his first telescope (1610) far better than most of his contemporaries and recognised immediately that the convex lens objective could, in principle, be replaced by a concave mirror. This was clear from his contact with Sagredo, Marsili and Caravaggi,[1] in which not only concepts but also attempts to make a reflecting telescope were documented [1.3]. One of the first serious attempts at manufacture was by Zucchi in 1616 [1.1] [1.3]. He states he procured a bronze concave mirror "executed by an experienced and careful artist of the trade" and used it directly with a negative Galilean eyepiece. This implies a "front-view" reflector of the type later introduced by William Herschel. Since a "front-view" telescope must necessarily have used an inclined beam to avoid obstruction by the observer's head (Fig. 1.1), it may be that field coma and astigmatism were also a significant factor in the poor image produced (see Chap. 3). But, in all probability, the manufacturing

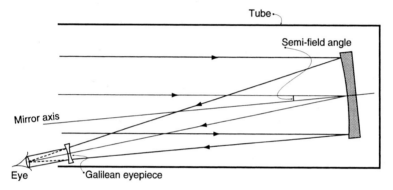

Fig. 1.1. Zucchi's attempt at a Herschel-type front-view reflecting telescope, 1616

---

[1] Cesare Caravaggi of Bologna should not be confused with the well-known painter Michelangelo Merisi da Caravaggio who lived in Lombardy and died in 1610. (According to Hockney [1.10], Caravaggio was one of the first painters who most successfully used a concave mirror to project the visual scene on to a screen and thereby achieve amazingly accurate perspective effects). I am most grateful to Dr. S. D'Odorico of ESO and Prof. F. Bertola of Padua University for copies of original letters of B. Imperiali to Galileo (21 March, 1626), C. Marsili to Galileo (7 July, 1626) and Galileo to Marsili (17 July, 1626), referring to the construction of a reflecting telescope by Caravaggi. These make it clear that Caravaggi, named Caravagio (sic) by Imperiali, was certainly already dead in 1626 when the letters were written. Riekher [1.2] states that he *died* in that year, but this is not clear from the above letters. Marsili refers to an attached drawing of the telescope, but this was subsequently lost. The comments in the letters do not make its construction clear, nor when it was made. Probably it had the same form as the Zucchi telescope of 1616. Unless further evidence emerges, which seems unlikely, we must conclude that Zucchi was the first person actually to construct a reflecting telescope.

quality of the mirror was too poor to produce a useful image on axis. Mirrors are more critical than lenses: the reason why surface accuracy requirements for mirrors are higher than for lens surfaces is given below and in RTO II, Chap. 1.

Not long after, soon after 1630, Descartes invented analytical geometry, a tool which enabled him to confirm not only the *paraboloid* as the necessary form for a concave telescope mirror, which was already well-known, but also the aspheric forms necessary for theoretically perfect imagery on the axis of a lens system. (Kepler had anticipated this work to a limited extent with his pioneer views on vision and the correction of aberration in the human eye in "Dioptrice", 1611 [1.1]). With this brilliant analysis, Descartes already laid down in his works "Traité du monde ou de la lumière" (1634) and "Dioptrique" (1637) the complete theory for the elimination of *spherical aberration* (Fig. 1.2) by suitable aspheric surfaces on mirrors or lenses. Descartes' theory

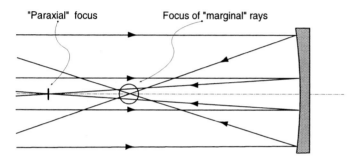

Fig. 1.2. Spherical aberration of a spherical concave telescope mirror. "Paraxial" rays are nominally at a negligible height from the axis

recognized the fundamental role of the conic sections in general as defined by second degree equations. For lenses free of spherical aberration, he even established the curve of fourth degree (Cartesian oval) required for imagery with two finite conjugates [1.5]. A spherical mirror (or *single* lens surface) has no axis, but an aspheric mirror has an axis uniquely defined by its equation. Thus Descartes' theory enabled a complete theoretical prescription for perfect *axial* imagery of mirror and lens systems in the sense that each element was itself free from spherical aberration. It is remarkable that no further advance was made in the basic theory of mirror forms of reflecting telescopes until Schwarzschild [1.6] in 1905, 270 years later! Schwarzschild was concerned to optimize imagery in the field as well as on axis (see Chap. 3), whereas at the time of Descartes the problem was far more fundamental: how to *make* the form prescribed by Descartes to get adequate axial imagery. Improvement of field imagery relative to axial imagery would have been a problem of no practical meaning at that time.

The significance of Descartes' work for reflecting surfaces was well understood by another contemporary Frenchman, Mersenne [1.1] [1.3], with whom Descartes was in close contact. Mersenne proposed the forms shown in Fig. 1.3, published in his work "L'Harmonie Universelle" in 1636. Although Mersenne's work is often referred to, its full significance is rarely appreciated – and certainly could not be appreciated by Mersenne or his contemporaries. His proposals had the following novel and remarkable features:

a) The invention of the *compound* reflecting telescope comprising two curved mirrors, in both the Gregory and Cassegrain forms.
b) The use of the second mirror as *ocular*.
c) As a consequence of b), Mersenne was the first to propose the *afocal* reflecting telescope with a parallel beam entering and leaving the mirror system, i.e. a beam compressor consisting of two mirrors in the same sense that the refracting telescopes of the time were beam compressors consisting of a large positive and small negative (eyepiece) lens (see Fig. 1.3).

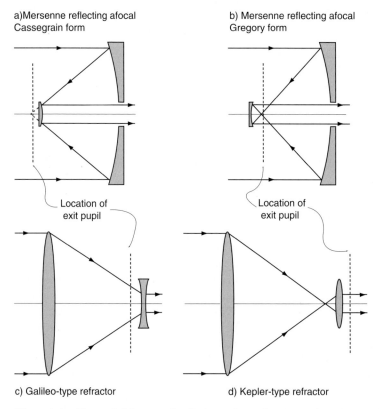

a)Mersenne reflecting afocal Cassegrain form

b) Mersenne reflecting afocal Gregory form

Location of exit pupil

Location of exit pupil

c) Galileo-type refractor

d) Kepler-type refractor

Fig. 1.3. Two of Mersenne's designs for reflecting telescopes, adapted from "L'Harmonie Universelle", 1636, and King [1.1], compared with Galileo-type and Kepler-type refracting telescopes

d) Because of his acquaintance with Cartesian theory, Mersenne was fully aware that the correction of spherical aberration for each mirror required confocal paraboloids for their forms.

e) Without knowing it, Mersenne invented, above all with his Cassegrain[2] form using a convex secondary, the basis of the first *mirror telephoto system* with strong telephoto effect. The properties of the lens telephoto system were first recognized by Kepler in 1611 but it was forgotten and re-invented by Barlow in 1834, then as a photographic objective by Dallmeyer in 1891 [1.7]. This property was of central importance for the future development of the reflecting telescope (see Chap. 2).

f) Also without knowing it, Mersenne invented the *aplanatic* (and *anastigmatic*) reflecting telescope in a limit case where the "aplanatic" and "classical" forms converge. This will be discussed in detail in Chap. 3. An "aplanatic" solution also corrects field coma for modest field angles, a property which would have been too sophisticated for Mersenne, or even Descartes, to understand. Its significance would only have become clear after the work of Fraunhofer, or of Petzval and Seidel in aberration theory (see Chap. 3). In fact, it was only with Schwarzschild's work in 1905 [1.6] that it would have been possible to fully understand this property of the Mersenne telescopes; and much later still that the clear recognition took place with the Wolter telescopes for X-ray astronomy [1.8] which are simply extreme forms of Mersenne telescopes with small modifications.

Ironically, Descartes himself completely failed to recognize the interest of Mersenne's proposals. He raised a number of objections [1.3], of which only one had real validity: that the exit pupil of the Mersenne forms using the smaller mirror as an eyepiece was inaccessible to the eye pupil giving severe field limitation. In fact, the Galileo-type refracting telescope current at the time had the same disadvantage, although to a lesser extent (see Fig. 1.3 and Chap. 2). Descartes also raised the interesting objection that the Mersenne telescopes "should not be less long than equivalent refractors if one wished for the same effect, so the construction would be no easier", a remark which showed he had failed to grasp the potential telephoto property. In the Mersenne afocal form, Descartes was right that there was little effective gain in compactness over the equivalent Galileo- or Kepler-type refracting telescopes. Only with the work of Gregory (see below), with his proposal to form a *real image* with a 2-mirror system, was the enormous telephoto advantage compared with a primary image telescope revealed. However, this was not generally understood until much later.

In fact, Descartes wished only to apply his theory to *lens* forms, not to mirrors, since he believed the failure or inability to produce the theoretical aspheric required to eliminate spherical aberration was responsible for

---

[2] The term "Cassegrain form" is used here because it is normal terminology. However, a further 36 years were to pass before the proposal of Cassegrain was made – see below.

the image defects of the refracting telescopes of the time. This major error came from Descartes' theory of the nature of light and colour and hence of chromatic aberration [1.1] which was completely inadequate to explain scientifically the real nature of the problem of the colour aberration of refracting telescopes. With the authority of Descartes behind it, this erroneous interpretation led to a major effort by the opticians of the time to manufacture aspheric forms on objective lenses, a problem which was not only completely insoluble with the technical means available but also futile, since spherical aberration was negligible with the low relative apertures which makers of refractors were forced to use because of chromatic aberration. In general, the manufacturing problem, also for primary mirrors for reflectors, was not the inability to produce the correct aspheric but the inability to produce an adequate *sphere* (see Chapters 2, 3, 5 and RTO II, Chap. 1).

No doubt discouraged by the negative reaction of Descartes, Mersenne apparently abandoned any attempts to actually *make* an afocal reflector. In any event, as we shall see, the manufacture of a relatively steep secondary operating as an eyepiece would, indeed, have required an aspheric form as Descartes insisted, and this would anyway have failed at that time. Nevertheless, Mersenne must be accorded the credit for inventing, on paper, the definitive basic geometrical form of the modern telescope. One of the principal motivations for Mersenne was certainly to solve the problem of the observer head obstruction in Zucchi's earlier direct front-view (Herschel) arrangement of Fig. 1.1. This problem was clearly seen as a fundamental disadvantage of the concave mirror as a replacement for the refracting convex objective.

While the simple refractor made progress – mainly through Huygens – by increased length to minimize chromatic (and also spherical) aberration and produced notable astronomical discoveries, the reflector had still not been realized in practice when Gregory made a further major advance in its theoretical form. In 1663, he proposed the Gregory form [1.3], apparently without knowledge of the work of Mersenne [1.7], as shown in Fig. 1.4.

Gregory analysed three possibilities with considerable scientific rigour: a telescope comprising only lenses (refractors as then used); a telescope comprising only mirrors (one of the Mersenne telescopes, Fig. 1.3(b)), and a telescope combining mirrors and a lens or lenses. He recommended the last, an important step from the basic afocal Mersenne concept to a modern form in which a compound (2-mirror) reflecting telescope forms a real image. He also showed a positive eyepiece (Kepler, 1611, in "Dioptrice"). This had two advantages: the exit pupil of the telescope was accessible to the eye, thereby markedly increasing the field over that of the Galilean telescope; a real intermediate image was provided enabling the use of a reticle. But Kepler's work was not realized in practice, and little known till about 1650. Thus, the Gregory proposal was the first reflecting telescope form that had, in principle, all the advantages: back view (no obstruction by the observer's head), moderate telephoto effect (reduced length – see Chap. 2), a reasonable field,

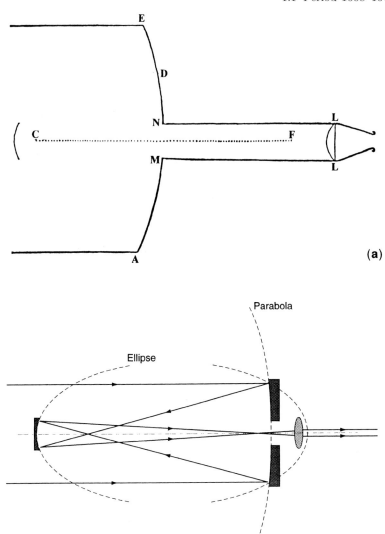

Fig. 1.4. (a) Facsimile of the Gregory telescope from "Optica Promota", 1663 (after Danjon and Couder [1.3]). (b) Raypath of the Gregory form (after King [1.1])

and an upright image, a useful advantage at that time. Gregory not only gave the correct geometrical forms for the two mirrors – parabola and ellipse – but also provided a stray-light baffle through the long tube separating the primary from the eyepiece (Fig. 1.4(a)). He also gave the correct formula for the equivalent focal length of the mirror combination. Gregory attempted to convert his admirable concept into practice with the help of the London opticians Reeves and Cox whom he commissioned to make a telescope of 6-foot focus [1.1][1.3]. Inevitably, if only because of the elliptical secondary, their efforts were a complete failure, and Gregory abandoned the attempt.

Newton was well informed of Gregory's proposals and attempts at manufacture. His determination to attempt to make a reflecting telescope was a result of his classic experiment in 1666 demonstrating that white light was composed of the colours of the spectrum which were refracted differently by a prism. This was, of course, not only a radical advance in the theory of light and vision but also provided the correct explanation of the chromatic aberration which had limited the performance of refractors since their invention nearly 60 years earlier. Unfortunately, Newton arrived too hastily at the conclusion that refraction and chromatic dispersion were linked by the same linear function for all refracting materials. Backed by his authority, this error delayed the invention of the achromatic objective for over 50 years. Nevertheless, the scientifically correct explanation of chromatic aberration was a huge step forward. Newton also showed [1.1] that the effect of spherical aberration in typical refracting telescopes was only about 1/1000 of that of chromatic aberration, thus correcting the illusion generated by Descartes concerning the need for aspheric lens surfaces to correct the observed defects. Furthermore, Newton's error led to the construction of the first reflecting telescope which could rival the better refractors.

Newton solved the front-view obstruction problem of the observer's head by the elegant device of his plane mirror at 45° (Fig. 1.5). This invention, which seems simple to us today, had eluded all his predecessors with the possible exception of Zucchi, whose ideas were too vaguely expressed to warrant

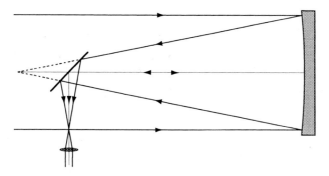

Fig. 1.5. The Newton reflecting telescope, 1668

a claim to its invention [1.3]. In fact, the Newton plane mirror was even more important than it seemed; for it was the *only* form proposed which had any chance of manufacturing success, since the manufacturing problems of the *secondary mirrors* of the Mersenne, Gregory or (later)Cassegrain solutions were insuperable for a considerable time to come (see Chap. 5 and RTO II, Chap. 1). Forewarned by the manufacturing problems of Gregory's over-ambitious plan, Newton himself made his first telescope in 1668 with a primary of only about 3.4 cm aperture and a focal length of about 16 cm. He used a plano-convex (Kepler) eyepiece giving a magnification of about 35 times. The "turned-down edge" of the primary was masked off by a diaphragm near the exit pupil behind the eyepiece. Newton claimed the performance was comparable with, or better than that of a Galileo-type refractor of 4-foot focal length, although light losses were inevitably much higher. No doubt the success of Newton's manufacture of the optics was due to his more scientific approach compared with the "spectacle lens quality" delivered by most opticians (he invented polishing on a pitch lap or was, at least, the first to publish it), and to the more modest size of the telescope. A duplicate instrument, apparently of better quality, was the one that became famous through its demonstration at the Royal Society.

This first phase in the development of the reflecting telescope was completed by the announcement by de Bercé in 1672 of the invention by Cassegrain of the form that bears his name (Fig. 1.6). De Bercé's description was very poor, in no way comparable with the excellent exposition by Gregory of his form 9 years earlier. Newton was forthright in his condemnations, above all in order to refute claims of priority on Cassegrain's part, which were anyway weakly based. His criticisms of the need to make a hyperboloidal secondary and the difficulties of achieving that were justified; but his other criticisms had no validity. Both he and his contemporaries in England and elsewhere (e.g. Huygens) completely failed to appreciate the huge telephoto advantages; also apparently its links with the proposals of Mersenne. The Cassegrain form simply completed the theoretical possibilities of the 2-mirror compound reflector as expounded by Mersenne and elaborated by Gregory, by using the convex secondary instead of the concave one in the Gregory-type "mixed" telescope of mirrors forming a real image observed by a refracting eyepiece.

The year 1672 marked the end of the first phase of telescope development and was followed by a period of consolidation. This first phase, lasting only some 64 years, had produced remarkable results. It was totally dominated by *optics*, the problem of getting a good image. Since planetary observation dominated the astronomical research of the time, the demands were high. *Mechanical* problems were also serious, but mainly on account of the length of telescopes required to reduce chromatic aberration to acceptable levels. The situation during this first phase regarding the optical development of refractors and reflectors was remarkable and can be summed up as follows:

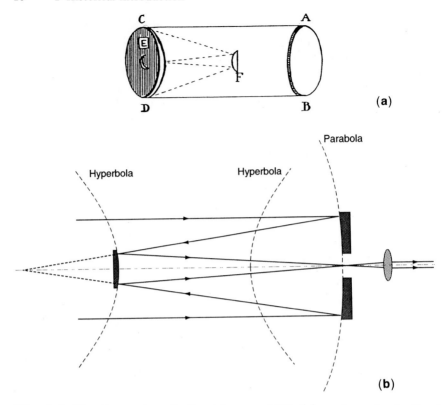

Fig. 1.6. The Cassegrain reflecting telescope, 1672, (**a**) as drawn by de Bercé. (**b**) Raypath of the Cassegrain form (after King [1.1])

Refractors: The optical theory of the refractor made no advance after Galileo up to the time of Newton, except for the positive Kepler eyepiece, because the physical origin of the dominant colour aberration was not understood. Newton explained chromatic aberration and laid the basis for the scientific theory of light propagation and colour vision, but – by a serious error of interpretation – *blocked* further development of the refractor for over 50 years. In spite of these failures to advance the theory, the refractor remained the only practical form available *because it could be manufactured with sufficient precision to achieve imagery of the quality limit set by the chromatic aberration of long telescopes.*

Reflectors: The situation here was exactly the opposite of that of refractors. The optical theory developed rapidly through the work, above all, of Descartes, Mersenne, Gregory, Newton and Cassegrain. It was so complete that no real further advance was made till 1905! However, even for minimum quality requirements, manufacture had proved impossible until Newton's first reflector in 1668. The reason for this is quite simple and lies in elementary optical theory (see RTO II, Chap. 1): *the precision required for the same performance of a reflecting surface is about 4 times higher than that for a refracting surface.* This increase of manufacturing precision compared with refracting surfaces was not possible even for spherical surfaces and out of the question for *secondaries* of compound telescopes, which required an aspheric form. *Primaries* could remain spherical if the relative apertures were reasonably low. *This, combined with the centering problem of 2-mirror telescopes (Chapters 3 and 5), is the reason why the Newton (or Herschel) forms were the only ones feasible in practice for about 180 years!* A further central problem in the early development of the reflector was the poor reflectivity, and hence poor efficiency, of the speculum metal used and, of course, its further reduction due to tarnishing. For a given aperture, therefore, refractors had a huge advantage in light-gathering power.

1.2 Period 1672–1840

For the reflecting telescope to attempt to rival the long refractors of the day, the theoretical work described above had to be complemented by major improvement in practical manufacture. Realistically, the Newton optical geometry was pursued. In 1721, a Newton reflector of 6 inches aperture and 62 inches focus (f/10.3) (Fig. 1.7) was presented by John Hadley to the Royal Society of London [1.1]. This telescope, only about 6 feet in length, was tested and compared with the 123-foot focus refractor of Huygens with a comparable effective aperture of the order of 6 inches [1.3]. Inevitably, the refractor gave the brighter image, but Hadley's reflector gave comparable definition. The ease of manipulation greatly impressed Bradley and Pound during their comparative tests.

Hadley's work was the birth of the practical reflecting telescope. His workmanship, as shown in Fig. 1.7, was so beautiful that it would do credit to a modern amateur even if the design is not what one would attempt to produce today. He invented the autocollimation pinhole test at the centre of curvature [1.1], the first scientific test during manufacture of a primary mirror. Using this test, Hadley was able to estimate in a qualitative way the errors from a *true sphere* of the mirror during working. This was his great advance over his predecessors, who had effectively been "working blind" using methods associated more with spectacle lens manufacture, for which the quality requirements were much lower. Following Cartesian theory, Hadley attempted to "flatten" the outer parts of the mirror to produce a paraboloid. It is doubtful whether this attempt produced any improvement, since the difference between the spherical and paraboloidal form for a 6-inch aperture working at f/10.3 was negligible. Although, as has been stated above, the basic theory for perfect axial imagery in all the fundamental forms of the reflecting telescopewas already laid down by 1672, *this was not the case for a correct analysis of manufacturing tolerances.* What was needed for this

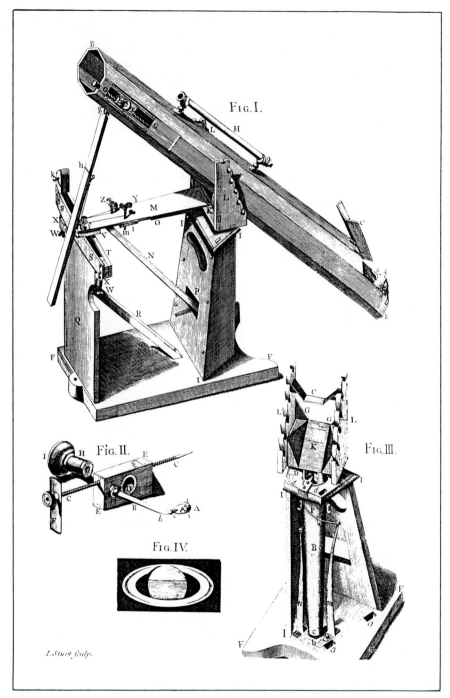

Fig. 1.7. John Hadley's 6-inch, f/10.3 Newton reflector, 1721 (courtesy Royal Astronomical Society, through Peter Hingley)

in the case of primary mirrors was the series expansion of the circle which gives directly the difference from the parabola with the same vertex curvature (see § 3.1), together with an idea of what surface accuracy was required to give good imagery. In fact, with Descartes' analytical geometry combined with his own binomial theorem and "method of fluxions" (calculus), Newton possessed, in principle, the mathematical tools not only to derive the series expansion of the circle but also to deduce, by differentiation, the transverse image aberration corresponding to the difference between the sphere and the paraboloid. If Hadley had known this, he could have compared (perhaps very crudely) the values with the transverse errors shown by his pinhole test, although the interpretation would have been strictly geometrical, without knowledge of the decisive influence of the wave theory of light which was only clarified by Rayleigh about 150 years later. In any event, there is no evidence that an analysis of tolerances using a series expansion of the circle took place till long after Newton. The full understanding of the convergence limitations of the binomial theorem was published in 1715 but not really noticed or understood till about 1772.

The more generalized series expansion of a conic section, had it been known, would also have scientifically validated the concentration on the Newton form of the reflector: it would have shown (see Chap. 3) that the aspheric secondary mirrors in the form proposed by Cassegrain required a far greater difference from the sphere than the primaries.

A further notable advance in the manufacturing quality of reflectors was made by James Short, who dominated the scene both for quality and volume of output between 1733 and 1768. It was Short who raised the level of the reflector to that of a practical instrument surpassing the old "long" refractors. Most of his telescopes were in the *Gregory form* using a concave elliptical secondary. Short was therefore the first optician who mastered the manufacture of appreciably aspheric *concave* mirrors to adequate quality for the astronomical observations of the time; aspherics being essential for his telescopes because he used much steeper primaries than his predecessors, giving relative apertures between f/3 and f/8 at the Gregory focus. It should be remembered that, even at that time, the demands on resolution were high because of the emphasis on planetary observation. It is very important to note that Short nearly always made the Gregory and not the Cassegrain form. It is far easier to make a concave than a convex mirror (see Chap. 3 and RTO II, Chap. 1) because of the test procedures involved. As was unfortunately common at the time, Short gave no account of his manufacturing methods. It seems possible he invented the "overhang" method of aspherising [1.1]. In 1752 he made an 18-inch Gregory telescope for the King of Spain (price £1200) which remained for many years the largest reflector in the world. In 1749 he introduced a "universal" equatorial mount, but not in a very stable form. In all, according to Baxandall [1.1], he made well over a thousand telescopes.

As was so often the case in the fascinating history of the development of the telescope, the advance of the reflector to surpass the single lens objective refractor (above all because of compactness) was accompanied by a major breakthrough in the refractor. The development of the achromatic objective was delayed (see above) by Newton's false conclusion about the link between refraction and dispersion. This was first challenged in 1695 by David Gregory, Savilian professor of astronomy at Cambridge and nephew of James Gregory, on the basis of the supposed achromatism of the eye. In 1729, Chester Moor Hall worked out the basic theory of an achromatic doublet and had such objectives made by opticians in London, but they apparently failed to understand their full significance. About 1750, John Dollond became aware of this and also of a paper by Euler (1747, in ignorance of Hall's previous work) expounding the theory of an achromatic doublet. Dollond had previously believed Newton's statement but then convinced himself by experiments that an achromat was possible. A letter to Dollond from Klingenstierna in Uppsala confirmed that Newton's conclusions had been wrong. Dollond then succeeded in making his first achromatic objectives but suppressed all reference or credit to Hall, Euler or Klingenstierna. John Dollond's behaviour in this respect was far from admirable, his son Peter Dollond's subsequent behaviour was even worse (he not only denied all credit to Hall, but also blocked the rights of the opticians with whom Hall had worked – Bass, Bird and Ayscough – by a successful patent action) [1.1]. In 1790, sixty years after Hall's original invention, it became clear that Hall had established the theory not only of correction of primary chromatism but also of spherical aberration. It was Clairaut, finally, who put the theory into modern form (1761–1764) and established the normal doublet form (still known in France as "un Clairaut") also corrected for spherical aberration for one wavelength [1.1]. Clairaut also pointed out the inevitable residual "secondary spectrum". Furthermore, he investigated by ray-tracing the imagery of such objectives *off-axis*, the first time imagery in the field was considered, effectively discovering the aberrations of coma, astigmatism and field curvature (see Chap. 3).

A major limitation in the optimum manufacture of achromats at that time was the impossibility of correctly measuring the refractive indices for a specified colour. Another problem was the diameter limitation of glass of good quality, above all flint glass. Four inches was already a large blank. In the constant battle for supremacy between refractor and reflector, this was the principal factor limiting the development of the achromat before the improvements in glass manufacture by Fraunhofer and Guinand. In spite of this, the transmission of achromats was so much higher than the reflectivity of reflectors with two speculum mirrors that some of the best achromats of Peter Dollond (triplets giving reduced secondary spectrum) with apertures of 3.8 inches rivalled appreciably larger reflectors. In 1777, tests by Maskelyne with such a Dollond telescope resolved η Coronae Borealis with a separation of *0.9 arcsec.*

In this period of intensive development of the refractor, two important advances were made by Jesse Ramsden, an excellent instrument maker. In 1775 he discovered what was later known as the "Ramsden disk", i.e. the exit pupil of a telescope (Fig. 1.8). Vague ideas of the exit pupil were certainly current after the introduction of the Kepler (positive) eyepiece, but Ramsden was the first to explain it correctly. He was also, perhaps, the first to make a Cassegrain telescope of "reasonable" quality. In fact, Ramsden's Cassegrain telescope was a special form which he called a catoptric (i.e. all reflecting) micrometer which used a split convex secondary whose tiltable halves formed the micrometer. However, the instrument was not successful, probably confirming that the manufacture of the Cassegrain secondary was beyond the technology of the time, above all from a test point of view (RTO II, Chap. 1).

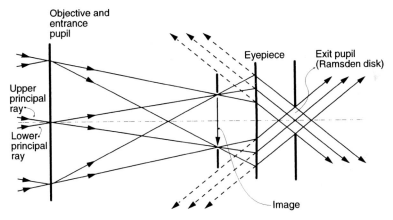

Fig. 1.8. The "Ramsden disk" (exit pupil) explained by Ramsden in 1775 (after King [1.1])

1.3 William Herschel's telescopes

It is my feeling that William Herschel was the greatest astronomer of all time. But I would assert with more confidence that Herschel was the greatest *telescope maker* of all time. His work [1.1] produced the biggest quantum jump in light-gathering power, while maintaining or even improving resolution, that has occurred since Galileo. Figure 1.9 shows his portrait at the age of 46 in 1785. His achievements seem all the greater if one reflects that he only turned his attention to astronomy in a practical way at the age of 35.

Fig. 1.9. Sir William Herschel (1738–1822) painted by L.T. Abbot in 1785 (courtesy Deutsches Museum, Munich)

Fig. 1.10. William Herschel's "large" 20-foot focus telescope, aperture 18.8 inches (f/12.8), completed in 1784, (reproduced from an engraving of 1794, courtesy Science History Publications Ltd., Cambridge, England)

Herschel started off as an amateur in Bath with refractors, going up to 30 feet length with purchased objectives. Because of the inconvenience of the long, thin tubes, he borrowed a Gregory-type reflector. He found this so much better that he tried to purchase a 5 or 6-foot[3] reflector, but its price was beyond his means. So he started to make reflectors himself and had made a $5\frac{1}{2}$-foot Gregory telescope before the end of 1773. However, *alignment* of this form gave him such trouble that he switched to the Newton form. He made a 7-foot telescope giving excellent performance on Saturn. By May 1776 he

---

[3] At this stage of development, telescopes were still normally characterized by their focal lengths rather than their apertures.

was working on a 20-foot telescope. Herschel was the first to realize that the key to success was the production of a high quality *spherical* surface and that the difference between the sphere and parabola was minimal for his primaries which normally had relative apertures of f/12 – f/13. Apparently he was also the first to use, systematically, pitch polishers cut into squares. His discovery of Uranus in 1781 was made with an excellent 7-foot telescope with 6.2 inches aperture made in 1778.

As his projects became more ambitious in size, Herschel was forced to use a higher copper content in casting his speculum metal blanks to avoid fracture. Typically, for large blanks he used 73% copper and 27% tin. The lower tin content made the mirror more prone to tarnish.

His work became known to the Royal Society and the Greenwich Observatory. Comparisons with the best telescopes otherwise available were made in 1782 and revealed the marked superiority of Herschel's instruments. He then received an official salary as "Royal Astronomer" but augmented this, after the move to Datchet in 1782, by making and selling reflecting telescopes. In three years he made and sold about 60 telescopes, most of 7 to 10 feet in focal length. This was a prodigious feat in view of the extensive telescope development for his own work and the astronomical observations performed with them.

For his work on nebulae, starting about 1784 from Messier's catalogue, Herschel had available his "large" 20-foot telescope with an aperture of 18.8 inches (f/12.8). This Newton-type telescope (Fig. 1.10) was to prove to be his greatest achievement from the point of view of practical astronomical results. However, he had plans for an even larger telescope of similar altazimuth mounting, with an aperture of 48 inches. A blank cast in London in 1785 was polished and gave reasonable results. A second blank cracked in cooling (Herschel did not understand the necessity of slow cooling in the casting furnace) and a third was given an even higher proportion of copper, a guarantee of rapid tarnishing. The focal length was 40 feet (f/10), a formidable mechanical undertaking. The giant telescope (Fig. 1.11) went into operation in 1789 and gave "pretty sharp images". The extra light-gathering power enabled Herschel to discover two further satellites of Saturn, Enceladus and Mimas. But, compared with the 20-foot telescope, the 40-foot was a relative failure. The high content of copper in the mirror caused rapid tarnishing; but, above all, Herschel had reached a telescope size where *mechanical problems* became the limiting factor, rather than the problems of optical figuring, though these were also formidable. His simple support system, a radial iron ring, gave rise to considerable flexure problems when the telescope was used at appreciable zenith distance. This was exacerbated by the relatively thin mirror, although this had the advantage of reducing the thermal sensitivity. Herschel used the "front-view" form (Fig. 1.1) which bears his name, having already experimented with it with the 20-foot.

Fig. 1.11. William Herschel's largest telescope: 4 feet in aperture, 40-foot focus (f/10), completed in 1789 (courtesy Deutsches Museum, Munich)

For freshly polished mirrors Herschel determined a reflectivity of 67%, justifying his use of the Herschel focus. The optical quality of his smaller telescopes must have been extraordinarily good. In well-conceived experiments using terrestrial objects he established that a telescope of *8.8 inches* would show a "real disc", as distinct from the "spurious disc", of *0.25 arcsec*. This suggests clearly that the telescope was *diffraction limited*, a conclusion supported by his observations when the telescope was stopped down. His observational achievements indicate that his best telescopes, including the 20-foot, were "seeing limited" for the seeing at Slough in England at that time, a quality of resolution not significantly improved till the 20th century.

A plausible measure of the optical quality of Herschel's two largest telescopes can be derived from his use of the Herschel "front-view" focus position (Fig. 1.1). With a reasonable estimate of the eyepiece position relative to the tube axis, the tangential field coma (see Chap. 3) of the 20-foot was about 0.28 arcsec, that of the 40-foot about 0.58 arcsec. These amounts of coma were presumably smaller than, or at most comparable with, the best atmospheric seeing and other telescope errors, implying excellent quality even by modern standards (see RTO II, Chap. 3 and 4).

In a classic paper [1.9] presented to the Royal Society in 1799, Herschel analysed with great rigour "the power of penetration into space" of his telescopes, as distinct from their formal magnifying power. We shall consider his criterion further in RTO II, Chap. 4.

Herschel was so much in advance of his colleagues, both astronomically and technically, that a further 50–60 years were to pass before his achievements could be surpassed. To understand the further advances in the reflecting telescope, the theory of Chapters 2 and 3 is required. The modern historical development from about 1840 will be taken up again in Chap. 5.

2 Basic (Gaussian) optical theory of telescopes

"God invented the number, the Devil invented the sign" – remark by Christoph Kühne at Carl Zeiss, Oberkochen, about 1969, concerning the problems caused by the sign in geometrical and technical optics.

2.1 Basic function of a telescope

A telescope is a device whose basic technical purpose is very simple: it should provide a high quality image of distant objects, which may be point sources or extended objects. The telescope must be constructed in such a way that it can be pointed at the desired object field. Furthermore, if the object is moving, as in the astronomical case as a result of the earth's rotation, the telescope must provide the means to compensate this movement. So an astronomical telescope has three basic functions:

- High quality imaging
- Pointing
- Tracking

The first requirement is the main subject of this book, but the other two aspects are closely related to it and cannot be achieved without the telescope's image-forming property. Adequate tracking is also essential to high image quality and represents one of the most difficult technical requirements in modern telescopes. Until the invention of photography and its first applications to astronomy in the mid-nineteenth century, the only *detector* available to investigate the information contained in the image was the human eye. The normal device for doing this was the ocular, and the combined form of imaging device known as the refracting telescope (objective and ocular) is that normally explained in elementary textbooks. However, the basic form of a telescope is simply a device producing a real image for some detector, used either directly in the image plane or indirectly by a transfer system such as an ocular to the human eye or an instrument to a modern detector. The basic function is therefore that of a *photographic camera*, a device which produces a real image of distant objects for detection by a photographic emulsion. A modern astronomical telescope is simply a photographic camera with a huge aperture and focal length, but working with a modern detector and an angular field which is small compared with those of photographic objectives, even for cases of so-called "wide-field telescopes".

Even before the invention of photography, the use of the telescope in this basic mode was known in the form of the "camera obscura", in which the real image was projected on to a screen. In its original form, probably invented by della Porta [2.1][2.2] before the invention of the telescope, the image was produced by a pinhole camera. The system was used later, above all for solar projection, the pinhole being replaced by a telescope objective functioning exactly as a camera objective. The "camera obscura" had the disadvantages of light and resolution losses due to the remittance properties of the screen and, above all, the fact that the image size was too small because the focal length of the early objectives was fairly short. The combination forming the refracting telescope by the addition of an ocular was therefore a revolutionary advance as it provided substantial magnification, i.e. the equivalent of a "camera obscura" of much longer focal length. This can best be understood from the fundamental concept for the human eye of the *minimum distance of distinct vision* (d_{min}) or *near point*. This varies strongly with age, but the standard average value is normally taken to be 25 cm. This means that the minimum distance for observation of the image in the "camera obscura" will be about 25 cm, and may be more if the linear size of the image gives an angular field at this distance which is too large for vision in comfort, for example with a wide field, long focus objective. Considered as a telescope, the magnification of the "camera obscura" is then with the objective focal length f'_{co}

$$f'_{co}/d_{min} \; ,$$

which expresses the ratio of the angular size of the object or image, as seen from the objective, relative to that of the image on the screen seen by the observing eye. If the screen and eye are replaced by an ocular and eye to give a normal afocal refractor (see § 2.2.4), the angular magnification is the ratio of the focal lengths

$$f'_{co}/f'_{ocular} \; ,$$

whereby f'_{ocular} can easily be made as short as 10 mm or even less. With $f'_{ocular} = 10$ mm , the length of the refractor for the same magnification is only about 1/25 of that of the equivalent "camera obscura". Furthermore, the eye behind the ocular is in a relaxed state focused on infinity rather than the state of maximum accommodation (ca. 25 cm) required for the "camera obscura". The role of the screen in the "camera obscura" is to scatter remitted light in such a way that sufficient light intensity reaches the eye pupil for the maximum semi-field angle observed. It plays the role of a light-inefficient "field lens", but with the advantage that the scattering properties of the screen are the same over a large area so that the position of the image on it is uncritical. A field lens would image the objective (entrance pupil) on to the eye pupil. This function is shown for a Kepler-type refractor in Figs. 1.8 and 2.8 and plays an important role in many telescope designs.

Before we consider the precise function of such telescope systems, it is necessary to review some of the basic properties of optical imaging systems in general.

2.2 The ideal optical system, geometrical optics and Gaussian optics

2.2.1 The ideal optical system and Gaussian concept

An excellent treatment of the basic theory of optical systems is given by Welford [2.3], to which the reader is referred for a full account. Here we will give a condensed version of the essential points.

An ideal optical system is assumed to have a unique axis of symmetry, building a so-called *centered optical system*. Most practical optical systems are conceptually centered systems, including telescope optical systems. However, there are important exceptions. The general theory of non-centered systems is very much more complex because of the vast increase in parametric freedom.

The essential theory of imagery by an ideal optical system was first laid down by Gauss in his classic work of 1841 [2.4]. He introduced the concept of principal planes as the equivalent of any centered optical system and the strict definition of focal length.

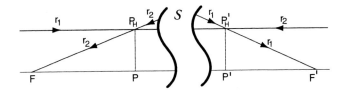

Fig. 2.1. The ideal optical system: the principal planes and unit magnification between them

In Fig. 2.1, let F P P' F' represent the axis of an ideal centered system S which may consist of any number and arrangement of centered elements. The word "centered" implies symmetry of all elements to the axis so that the axis cuts all elements normally. Let the ray r_1 enter from the left at any arbitrary height above the axis and parallel to it, i.e. from the axial point of an infinitely distant object. This ray emerges at some height on the image side of the system S and crosses the axis at the point F'. Let r_2 be a ray at identical height, also parallel to the axis, but entering in reverse direction from the right. This ray emerges at some height on the left hand side of S and cuts the axis at F. Let r_1 cut r_2 (its own projection with reversed direction) at P'_H, where the suffix H simply implies a higher point than P' in the plane of the figure. Similarly, r_2 cuts r_1 at P_H. The points P and P' are then constructed by dropping perpendiculars from P_H and P'_H to the axis.

F and F' are the *focal points* of the system; P and P' are the *principal points*, the planes through PP_H and $P'P'_H$ perpendicular to the axis the *principal planes*. Since r_1 and r_2 both pass through P_H and P'_H, these points must form object and image to each other, i.e. they are *conjugate points*.

Similarly, PP_H and $P'P'_H$ are *conjugate planes*. Since, by the nature of the construction, $PP_H = P'P'_H$, it follows that the magnification from one plane to the other is unity. Although, by definition, F and P are always in the "object space" and F' and P' are always in the "image space", the points F' and P' may lie to the left of the system; or F' may lie to the right and P' to the left. An analogous situation obtains for F and P.

The distance $P'F'$ is the *image-side focal length*, denoted by f', while PF is the *object-side focal length*, denoted by f. We shall see that $|f'| = |f|$ if the media (i.e. refractive indices) are the same in both object and image spaces.

The four points F, F', P, P' along the axis completely define the properties of the ideal optical system. They lead at once to a construction enabling the complete geometrical determination of image formation, as shown in Fig. 2.2.

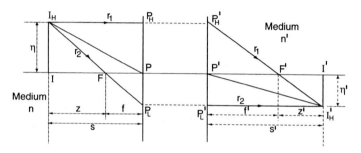

Fig. 2.2. Geometrical construction of ideal image formation

The rays r_1 and r_2 are constructed from the object point I_H above the axis and placed at a distance PI to the left of the object principal plane $P_H P P_L$, where the suffix L refers to the lower half of the principal plane. r_1 is parallel to the axis, passes from P_H to P'_H with unit magnification and crosses the axis at F'. The ray r_2, now proceeding in the *incident direction from left to right, which is defined as the normal direction of incident light*, is drawn from I_H through F to P_L. After unit magnification transfer to P'_L, r_2 must proceed parallel to the axis, since it passes through F, meeting r_1 at I'_H and thereby defining the position and size of the image $I'I'_H$. For generality, $|P'F'|$ has been made *not* equal to $|PF|$, implying that the object and image media are different, e.g. air and water. In telescope systems, this is very rarely the case.

A decision must now be taken on the *signs* of the geometrical distances involved. This is a matter of fundamental importance in geometrical optics and many sign conventions have been proposed and used in its long history. In accordance with Welford [2.3] and the modern trend in optical design in general, *we shall use throughout this book a strict Cartesian system*. Only in this way can complete generality of formulation be achieved, even though the price may be an apparent complication of certain simple cases. This is a price well worth paying to avoid ambiguities and errors.

Let the axial distances z and z' be measured from F and F', so that z is negative, z' positive. Similarly, measuring f and f' from P and P' respectively, f is negative and f' is positive. Also η is positive, η' is negative. Then

$$\eta'/\eta = -f/z = m \tag{2.1}$$

and

$$\eta'/\eta = -z'/f' = m \tag{2.2}$$

Combining these two, we have the well-known formula

$$zz' = ff', \tag{2.3}$$

Newton's conjugate distance equation.

Equations (2.1) and (2.2) also define the *magnification* m of the system. However, these simple relationships in z and z' do not usually represent the most convenient form, since P and P' are usually more convenient reference points for the object and image positions than F and F'. Let $PI = s$, $P'I' = s'$. As with z and z', s is negative in Fig. 2.2 and s' is positive. Then

$$s = z + f, \quad s' = z' + f' \tag{2.4}$$

From (2.1) and (2.2):

$$s = f\frac{(m-1)}{m}, \quad s' = -f'(m-1) \tag{2.5}$$

Eliminating m gives

$$s = \frac{-fs'}{f'-s'}, \tag{2.6}$$

which reduces to

$$\frac{f'}{s'} + \frac{f}{s} = 1 \tag{2.7}$$

If the media in object and image space are identical, the usual case, then it will be shown that $f = -f'$ and (2.7) reduces to the well-known "lens formula"

$$\frac{1}{s'} - \frac{1}{s} = \frac{1}{f'} \tag{2.8}$$

From (2.5) we have for the magnification m:

$$m = -\frac{s'f}{sf'}, \tag{2.9}$$

or, for a system in one medium

$$m = \frac{s'}{s} \tag{2.10}$$

The marked difference in form between Eqs. (2.3) and (2.7) arises from the fact that P and P' are conjugate points whereas F and F' are *not* conjugates.

Telescopes are usually used on very distant, or effectively infinitely distant objects. In this case $s \to \infty$ and the above formulae for the magnification m become zero and hence meaningless if the system terminates at the first image plane $I' I'_H$ in Fig. 2.2, the case of the photographic camera or normal modern telescope. We shall see in § 2.2.6 that the concept of *magnification* is replaced by the concept of *scale*.

Finally, we must introduce the concept of *nodal points*. In Fig. 2.2, the rays $I_H P$ and $P' I'_H$ do *not* have the same angle to the axis because, for generality, the object and image spaces have been shown with different media whose refractive indices are n and n' respectively. The consequence is that $|f|$ and $|f'|$ are not numerically identical but are related by Eq. (2.20) below. If the media are identical, then any ray cutting the principal point P will leave P' with the same angle to the axis. This is the general property of *nodal points*. They are the same as the principal points unless $n' \neq n$. Since this case is very unusual in telescope optics, it will not be considered further here. But it should always be borne in mind for exceptional cases where an image is not formed in air.

2.2.2 Geometrical optics and geometrical wavefronts

We have used above the concept of a *ray* of light without defining its exact meaning. Everyday experience with shadows makes the ray concept plausible as the straight line light path from a point source to any point it illuminates in the same medium. The definition of a ray is closely linked to that of a *geometrical wavefront*. The geometrical wavefront is the surface of constant phase, or optical path, which was introduced by Fermat, and published in 1667. The wavefront is orthogonal to the rays which, to the approximation of *geometrical optics*, are the paths along which the radiation is propagated. The limitations of this interpretation, arising from the wave nature of light at boundaries, are the subject of *physical optics* and *diffraction theory*. Much of telescope theory can be handled by geometrical optics, though diffraction limitations are of great importance in defining the theoretical limits of resolution and the tolerancing of optical errors.

In Fig. 2.3, the point source I sends out the wavefront W in object space. This object geometrical wavefront is, by definition, exactly spherical with its centre at I. It advances with the speed of light in the object medium and strikes some centered optical system with principal planes at P and P'. Suppose this system has sufficient optical power to form a point image I' of the point object I. Then the optical system has transformed the incident

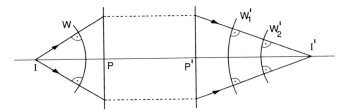

Fig. 2.3. Geometrical wavefronts and rays

wavefront W into the emergent wavefront W_1' centered on I'. If the wavefront W_1', and equally the wavefront W_2' at a later instant, are strictly spherical, then the image I' is perfect from the point of view of geometrical optics. If, however, the wavefronts are *not* exactly spherical, the phase error involved is a measure of the geometrical aberrations of the optical system. This concept will be developed in Chap. 3 in detail. At this stage, it is important to note that, in a given medium, the phase error remains constant with the advance of the wavefront: any phase error in W_1' will be identical in W_2'. The time $t_{II'}$ the light takes to travel from I to I' is

$$t_{II'} = \frac{1}{c_l} \int_I^{I'} n \, ds \, , \tag{2.11}$$

where c_l is the velocity of light, n the *refractive index* of the medium and ds an element of the path. The quantity $\int n \, ds$ (which reduces to $\sum n \Delta s$ for practical systems with a few finite jumps in the refractive index n) is called the *optical path length*. The wavefronts are surfaces of constant optical path length.

Equation (2.11) defines the refractive index n completely but it is better known as a consequence of *Snell's law of refraction* at a surface separating two different optical media, which states that the incident and refracted rays are coplanar with the normal and that

$$n' \sin i' = n \sin i \, , \tag{2.12}$$

where i and i' are the angles of incidence and refraction to the normal. We shall see in the next section that this relation is even more powerful and general than the word "refraction" implies, because "reflection" can be handled as a special case of refraction by the same equation.

A more general approach to the geometrical optics of the phenomenon of refraction is provided by *Fermat's principle*. This approach is dealt with exhaustively in such works as Born-Wolf [2.5] or Schroeder [2.6], to which the reader is referred.

2.2.3 The Gaussian optics approximation

Figs. 2.1 and 2.2 represent refraction effects in a plane, the plane of the paper or *principal section*, as a two-dimensional process. The most important aspects of the theory of centered optical systems can be dealt with in this section: the y, z plane, where z is the direction of the optical (system) axis and y the height in this section above the axis. Any light ray not lying in this section, i.e. for which the x coordinate is not always zero, is called a *skew ray*. We shall refer to these in Chap. 3.

In the principal section, any surface of a centered optical system can be defined, with origin at its intersection with the axis, by the equation (see §3.1)

$$z = \frac{c}{2}y^2 + a_1 y^4 + a_2 y^6 + \cdots , \tag{2.13}$$

only even powers appearing because of axial symmetry. The quantity c is the curvature, the reciprocal of r, the radius of curvature, while $a_1, a_2...$ are constants. Wavefronts emerging from axially placed object points and passing through the system would also be expressed by Eq. (2.13). For object points in the principal section which are not on the axis ($y \neq 0$), the central ray of the beam (the *principal ray* to be defined in connection with pupils, below) will define an oblique line in the principal section as the equivalent of the z-axis, to which a polynomial can also be applied for wavefronts centered on such a principal ray. Because of the asymmetry of such an object point to the axis, this polynomial will be an extended form of Eq. (2.13) which includes odd terms as well as even ones. It should be noted, however, that the only odd term of lower order than the first term of Eq. (2.13) is a linear term, implying a tilt of the wavefront. But this implies an incorrect image height: if this height is chosen correctly, the linear term is zero and the first quadratic term of (2.13) also remains the first for off-axis object points.

Gauss [2.4] first developed a complete theory of image formation based only on the approximation of the first, quadratic, term in Eq. (2.13). This is also called the parabolic term since it defines a parabola if all other terms are zero. Further properties of Eq. (2.13) will be discussed in §3.1: at this stage we are concerned with the significance of the Gaussian approximation, whereby only the first term is considered.

Higher order terms than the first can only be negligible if the height y above the axis is very small compared with the radius of curvature r. This domain of validity is called the *paraxial region*. By definition, the difference between a parabolaand a sphere, or any other conic section (see Chap. 3), is negligible in the paraxial region. The definition therefore excludes all qualitative influences of the precise form of refracting or reflecting surfaces, since the wavefronts are also defined in a way that ignores the effects of differing forms. In other words, the Gaussian or paraxial region is concerned, by definition, with ideal image formation. It is solely concerned with the *position* and *nominal size* of an image, *not* with its *quality*.

Since the heights y of the ray intersections are small, it follows that the angles u, u' of the rays with the axis, and the angles of incidence and refraction i, i' with the refracting (or reflecting) surfaces are small. This leads to another important interpretation of the paraxial region. Snell's law (Eq. (2.12)) can be written, expanding the sines by Maclaurin's theorem

$$n'\left(i' - \frac{i'^3}{3!} + \frac{i'^5}{5!}......\right) = n\left(i - \frac{i^3}{3!} + \frac{i^5}{5!}.......\right) \tag{2.14}$$

To the Gaussian approximation all terms after the first are negligible, so that Snell's law for the paraxial region becomes simply

$$n'i' = ni, \tag{2.15}$$

a simple *linear* equation.

The basic tool of optical design is the exact tracing of the path of optical rays through the system. The general formulation for doing this is given in specialist works on optics, e.g. Welford [2.3], Herzberger [2.7]. Although it is normally expressed in a vectorial, algebraic form adapted to modern computers, the exact formulation corresponds to the exact form of Snell's law of Eq. (2.12). "Paraxial rays" are traced with a reduced, simplified formulation based on the linear equation (2.15). All the equations involved are also *linear* and are given below. This linearity has the following important consequence: although *in physical reality* the heights y and angles u and i of a paraxial ray (or its diagrammatic representation) must be very small if the results are to be valid, the calculated results for the position and size of paraxial images are independent of the input conditions. Thus, if for a real physical lens surface whose diameter far exceeds the paraxial region, a ray is traced from an axial object point to the aperture edge with the exact formulae and then for the same aperture height with the paraxial formulae, the difference gives directly the effect of the terms in Eq. (2.13) beyond the first. These terms determine the *quality* of the image, in other words, the extent to which it is affected by *aberrations*. This is the subject of Chap. 3.

The laws of Gaussian optics completely define the position and size of the image in any telescope system, irrespective of the number and nature of its elements. The paraxial form of Snell's law is the basic law. We shall now deduce some other relationships.

In Fig. 2.4 the object medium has index n, the image medium n'. In the paraxial sense, the principal planes PP_H and $P'P'_H$ can be considered as wavefronts of the rays r_1 and r_2. For r_1, the optical path $FP_HP'_HF'_H$ must be equal to $FPP'F'$ or

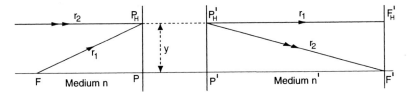

Fig. 2.4. The relationship between the focal lengths

$$nFP_H + [P_H P_H'] + n'f' = -nf + [PP'] + n'f' \tag{2.16}$$

Now from the paraxial approximation of Eq. (2.13) for the wavefront we have with $PP_H = y$

$$nFP_H = -nf - \frac{ny^2}{2f} , \tag{2.17}$$

giving from (2.16)

$$-n\frac{y^2}{2f} = [PP'] - [P_H P_H'] \tag{2.18}$$

For ray r_2, with the same height y, we have

$$n'\frac{y^2}{2f'} = [PP'] - [P_H P_H'] \tag{2.19}$$

Eqs. (2.18) and (2.19) give

$$\frac{n'}{f'} = -\frac{n}{f} , \tag{2.20}$$

or, for a system in a vacuum (or air, as a normal approximation)

$$f = -f' , \tag{2.21}$$

as used in Eq. (2.8).

The quantity n'/f' in Eq. (2.20) is called the *power* of the optical system and is often denoted by K. The reciprocal of K is the *equivalent focal length* and is, in a vacuum (or air as approximation), given by $f' = -f$. Eq. (2.7) can then, from Eq. (2.20), be written in the general form

$$\frac{n'}{s'} - \frac{n}{s} = K = \frac{n'}{f'} \tag{2.22}$$

Figure 2.5 illustrates the derivation of what is probably the most fundamental law of Gaussian optics, in that it is related to the first law of thermodynamics because it governs the energy throughput of optical systems. This is the *Lagrange Invariant* (also known, variously, as the Smith, Lagrange, Helmholtz relationship).

The transverse magnification m of the system is given by η'/η where the object and image points I and I' on axis are conjugate and I_H and I_H' at height η and η' respectively. From Eq. (2.9)

$$\frac{\eta'}{\eta} = m = -\frac{s'f}{sf'} , \tag{2.23}$$

while from Eq. (2.22) and Fig. 2.5 with optical distances s'/n' and s/n

$$\frac{\eta'}{\eta} = \frac{ns'}{n's} , \tag{2.24}$$

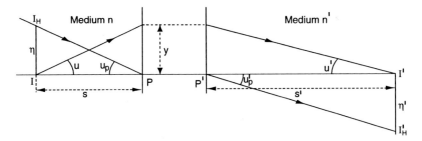

Fig. 2.5. Derivation of the Lagrange Invariant

or

$$\frac{n'\eta'}{s'} = \frac{n\eta}{s} \tag{2.25}$$

Let the height of the ray defining the aperture be y. Because they are conjugate planes of unit magnification, the height y is the same at both principal planes. Then

$$u = -\frac{y}{s} \quad \text{and} \quad u' = -\frac{y}{s'} , \tag{2.26}$$

bearing in mind the Cartesian sign convention for the angles. Combining with Eq. (2.25) we obtain

$$n'u'\eta' = nu\eta = H , \tag{2.27}$$

H being the Lagrange Invariant. In a system in a vacuum (or air) with $n' = n$, the invariant reduces to

$$u'\eta' = u\eta \tag{2.28}$$

This relation is of great importance in practical work with telescopes. If, for example, a "focal reducer" is put into the emerging beam of a telescope to reduce the emerging f/no from f/8 to f/4, then u' in (2.28) is doubled and the image size η' *must* be halved. So long as normal (centered) optical systems are used, the Lagrange Invariant will always apply. The connection with thermodynamics is clear from the fact that the total flux collected by an optical system from a uniformly radiating object surface is proportional to H^2, as we shall see below.

   In telescope optics, *afocal systems* have great significance. An afocal system has an image or object, or both, at infinity. In the case of telescopes, the term normally implies that both are at infinity. If we use the forms of Eq. (2.26) for u and u' to give y and y', and the equivalent forms

$$\eta = u_{pr}s \quad \text{and} \quad \eta' = u'_{pr}s' \tag{2.29}$$

for η and η', where u_{pr} and u'_{pr} are the angles subtended by the object and image over the distances s and s' respectively (see Fig. 2.6, where $|s|$ and $|s'|$

tend to infinity for the afocal case, and also Eq. (2.50)), then substitution in (2.27) gives the afocal form of the Lagrange Invariant:

$$n'y'u'_{pr} = nyu_{pr} \qquad (2.30)$$

This will be demonstrated in the next section in the specific case of an afocal telescope. We shall see that an afocal telescope has its principal planes at infinity.

The suffix (pr) of u_{pr} and u'_{pr} in Eqs. (2.29) and (2.30) actually refers to the *principal ray*, which is defined relative to the *entrance* and *exit pupils* of a system. Only in the afocal case, where all rays of the entrance and emergent beams are parallel, is it possible to ascribe the field angles u_{pr} and u'_{pr} to the rays incident on P and emerging from P', except in the rare cases where the entrance and exit pupils are coincident with the principal planes. With the general definition of u_{pr} and u'_{pr} related to the pupils, Eq. (2.29) is only true in the afocal case.

The *entrance pupil* of an optical system is some aperture which limits the diameter of the light beam entering the system for all field angles which the *field stop* allows to pass. The field stop limits the area of the object or image. Because of the axially symmetrical nature of centered optical systems, the limiting aperture is normally circular or an approximation thereto. Figure 2.6 shows this schematically. A is the *aperture stop*, E the *entrance pupil*,

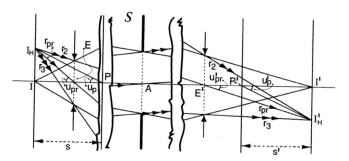

Fig. 2.6. Aperture stop, entrance and exit pupils

E' the *exit pupil*. The aperture stop limits the beam width such that beams from all object points which pass through the *field stop*, of radius II_H, pass centrally through it. This implies that all the other elements in the system S are *sufficiently large* to allow the passage of the extreme rays r_2 and r_3 from the upper edge I_H of the object through the system. In other words, all the other elements in the system must have a free diameter *larger* than the axial beam emanating from I as it passes through them. The light beam cannot think! If no clear aperture stop is defined in a system, the light will be limited by that aperture in the system which is effectively the smallest. But, in such a case, aperture limitations at other elements may obstruct one side of oblique beams. This phenomenon is called *vignetting*. Sometimes an aperture stop is deliberately made large enough to allow vignetting by other elements. This may be done, for example in certain photographic systems, to remove rays from the oblique image-forming beams which would otherwise cause unacceptable aberrations.

In Fig. 2.6, the paraxial image of A traced backwards towards the front of the system, is formed at E *in the object space*. This is the *entrance pupil*. The *principal ray* is, by definition, that ray that passes through the centre of the aperture stop A. Again, by definition in the paraxial sense, it must also pass through the centre of E. The real principal ray, entering the system at an angle which exceeds the paraxial region, may depart slightly from this theoretical path: this is the effect of *pupil aberration*. The entrance pupil E is therefore the paraxial image of the aperture stop A formed by backward imagery through all optical elements in front of the aperture stop. E may be real or virtual. For an aperture stop inside a relatively compact, complex system, E is usually virtual. In such cases, although the entrance pupil must, by definition, be mathematically in the "object space", it may physically be well to the right of the aperture stop or, indeed, of the whole physical system. In Fig. 2.6 it is shown to the left, physically in the object space, for the sake of clarity. This implies that the total optical power of that part of the optical system to the left of A must be positive (i.e. like a convex lens) and sufficient to form a real image of A. Such cases are uncommon.

Similarly, the *exit pupil E'* is the paraxial image of A formed in the forward direction of the light by all those elements which are to the right of A. Again, E' may be real but is more often virtual. If one places one's eye at the image plane I' of an optical system and moves it over the plane, the exit pupil is that aperture out of which the light appears to come.

The angle $I_H EI$ in Fig. 2.6 is the angle to the axis u_{pr} of the incident principal ray, angle $I'_H E' I'$ the angle u'_{pr} of the emergent principal ray. In general $|u'_{pr}| \neq |u_{pr}|$. In telescope systems with telephoto characteristics (see § 2.2.5), there is normally a large difference between $|u'_{pr}|$ and $|u_{pr}|$. But in an optical system placed in one medium, the ray angles u'_p and u_p are the same because of the nodal point property of the principal planes P and P' in this case. As mentioned above, in the afocal case all rays from the infinitely distant object (or image) point are parallel so that $u_p = u_{pr}$ (or $u'_p = u'_{pr}$). This means that, in the normal telescope case *in one medium and with the object at infinity*, the condition $u_p = u'_p = u_{pr}$ obtains. In this case

$$\eta' = u_{pr} f' , \tag{2.31}$$

a very convenient relation.

We shall see below in Chap. 3 the role that pupil position can play in various telescope systems. An important special case remains to be dealt with: *telecentric systems*.

Figure 2.7 shows a *telecentric aperture stop*, an aperture stop A placed at the focal point F of an imaging system. A is also the entrance pupil E since there is no element to the left of it. The axial image of the object point I is formed at I', the image of the point I_H in the object field at I'_H. The principal ray r_{pr} passes through the centre of the stop A (and E) and is refracted by the system parallel to the axis so that *the exit pupil is at infinity.*

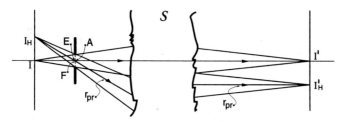

Fig. 2.7. Telecentric aperture stop

Telecentric pupils are commonly used in mechanical measuring systems. For example, a graticule placed in the image at $I'I'_H$ will give measurements of the object size which are independent of focus error.

Before leaving the Gaussian properties of optical systems in general, we must consider the application of the general formula of Eq. (2.22) to the specific case of a refraction or reflection at a *single surface*. It is easily shown [2.3] that

$$f = \frac{-nr}{n' - n} \quad \text{and} \quad f' = \frac{n'r}{n' - n} \,, \tag{2.32}$$

where n and n' are the refractive indices in the object and image spaces, f and f' the corresponding focal lengths and $r = 1/c$ is the radius of curvature of the surface. Then Eq. (2.22) becomes

$$\frac{n'}{s'} - \frac{n}{s} = K = -\frac{n}{f} = \frac{n'}{f'} = \frac{n' - n}{r} \,, \tag{2.33}$$

again following the strict Cartesian system, measured from the refracting surface, for the distances s, s', f, f' and r. The radius r is therefore positive for a surface which is convex to the incident light from the left.

It can also be shown [2.3] that the general case of refraction, as expressed by Snell's law of Eq. (2.12), can be extended to *reflection* at a surface if we write $n' = -n$ and maintain the same strict Cartesian system of signs. This implies that distances, after reflection, to a succeeding refracting or reflecting surface are also reversed in sign because the light direction is reversed. The general formula for reflection at a surface with optical power, i.e. finite curvature c, is then obtained by inserting $n = +n$, $n' = -n$ (or equally $n = -n$, $n' = +n$) in (2.33) giving

$$\frac{n}{s'} + \frac{n}{s} = \frac{n}{f'} = \frac{2n}{r} , \tag{2.34}$$

in which n becomes simply a uniform scaling factor reflecting the optical path laws expressed in Eqs. (2.11) and (2.20). Eq. (2.34) reduces, irrespective of the value of n, to the simple *reflection equation*

$$\frac{1}{s'} + \frac{1}{s} = \frac{1}{f'} = \frac{2}{r} \tag{2.35}$$

Finally, in order to fully understand the consequences of the sign convention for the signs of the various paraxial quantities defining the image forming process, it is a valuable exercise to consider the *paraxial ray tracing* formulae. These are discussed in detail by Welford [2.3] and various formulations are possible. As stated above, the equations are fundamentally linear and, in the general case for following the path of a ray through successive refracting or reflecting surfaces, consist of a refraction equation and a transfer equation to the next surface. Here, we follow the old procedure with individual steps, because the signs of intermediate quantities are of fundamental importance in understanding the numerous relationships of paraxial quantities in the basic telescope forms discussed in § 2.2.5 below. The complete paraxial ray trace for refraction (or reflection) at a surface and transfer to the next is then written:

a) Refraction at surface ν

$$\left.\begin{aligned}
i_\nu &= -\frac{(s_\nu - r_\nu)}{r_\nu} u_\nu \ \ \text{or} \ \ i_1 = \frac{y_1}{r_1} \ \text{if} \ u_1 = 0 \\[2mm]
i'_\nu &= \frac{n_\nu}{n'_\nu} i_\nu \\[2mm]
u'_\nu &= i'_\nu - i_\nu + u_\nu \\[2mm]
s'_\nu &= r_\nu \left(1 - \frac{i'_\nu}{u'_\nu}\right)
\end{aligned}\right\} \tag{2.36}$$

b) Transfer from surface ν to surface $(\nu + 1)$

$$\left.\begin{aligned}
s_{(\nu+1)} &= s'_\nu - d_\nu \\
u_{(\nu+1)} &= u'_\nu \\
y_{(\nu+1)} &= y_\nu + d_\nu u'_\nu
\end{aligned}\right\} \tag{2.37}$$

c) Transfer from the last surface l to the image

$$\left.\begin{aligned}
\text{Back focal distance} \qquad s'_l &= -\frac{y_l}{u'_l} \\[2mm]
\text{Equivalent focal length (efl)} \ f' &= -\frac{y_1}{u'_l}
\end{aligned}\right\} , \tag{2.38}$$

if the paraxial ray is traced for an object at infinity. The symbols are defined in Figs. 2.2 and 2.5 and the equations above, apart from d_ν which defines the axial distance from surface ν to surface $(\nu + 1)$.

2.2.4 The conventional telescope with an ocular

Although we shall afterwards be concerned only with the reflecting telescope in its various forms, it is instructive to consider the Gaussian optics of the conventional refractor with an ocular, for comparison. This is shown in Fig. 2.8. In normal use with an ocular, such a telescope is *afocal* both in the object and image spaces. Both these spaces are in the same medium, normally air.

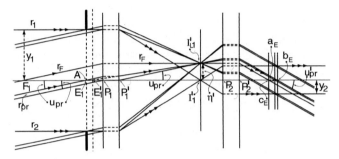

Fig. 2.8. Gaussian optics of a conventional refracting telescope with ocular (afocal in both object and image spaces)

The objective is represented by its two principal planes P_1 and P_1'. For a classical "thin" doublet objective, P_1 and P_1' will be quite close together and the entrance pupil is also close to P_1. For generality, we show it at E_1, in front of the objective, which is also the aperture stop A. The field stop has the radius $I_1' I_{L1}'$ and defines the image size. Since the object is at infinity, we have a parallel incident beam, so that all incident rays for a given object field angle have the same angle u_{pr} to the axis, the angle of the principal ray through A. The image size (radius), $I_1' I_{L1}' = \eta'$ is given by (2.31) as $\eta' = u_{pr} f_1'$, where f_1' is the focal length of the objective and u_{pr} is in radians.

P_2 and P_2' are the principal planes of the ocular. In normal adjustment, with a relaxed eye, the ocular will be focused such that the eye focuses on infinity, giving an afocal emergent beam which implies that all rays associated with a given image direction are parallel.

From the construction of Fig. 2.1 for the principal planes of a total optical system, we see that, in this afocal case, the rays r_1 and r_2, which are incident parallel to the axis, also emerge parallel to the axis. The principal planes of the total afocal system are therefore at infinity.

It is important to remember that, in the Gaussian (paraxial) region, all angles are defined by the first term of the Maclaurin expansion of (2.14). There is then, by definition, no difference between the natural angle, the sine and the tangent. The principal planes are correctly represented as planes to this approximation. As soon as the next term in (2.13) or (2.14) is considered, this is no longer true. This will be encountered with the *sine condition* in Chap. 3.

The *exit pupil* of the afocal telescope is at a_E, the paraxial image of A. If A were placed at P_1, the exit pupil would shift slightly to b_E. If A were placed at F_1, the objective would be telecentric and the exit pupil would shift slightly in the other direction to c_E. However, because of the afocal exit beam, the diameter of the exit pupil is independent of its position and is solely determined by the diameter of the axial exit beam. This is the property that was recognized by Ramsden, giving the name *Ramsden disk* to the exit pupil – see Fig. 1.8.

The essential property of such an afocal telescope is therefore *beam compression*: the incident parallel beam of diameter $2y_1$ is compressed into an emergent parallel beam of diameter $2y_2$. From Fig. 2.8 it is clear that the *beam compression factor C* is given by

$$C = y_1/y_2 , \tag{2.39}$$

or

$$|C| = |E_1|/|E_2'| , \tag{2.40}$$

if E_1 and E_2' express the diameters of entrance and exit pupils respectively, E_2' being the diameter of the Ramsden disk.

In such a conventional telescope with an ocular, another essential property is the *magnification m*. For an infinitely distant object and image, m is defined simply as the viewing angle under which the image is viewed relative to the object:

$$m = u_{pr}'/u_{pr} \tag{2.41}$$

Consider now the ray r_F in Fig. 2.8 as one of the rays with angle u_{pr} of the incident beam forming the image point I_{L1}' and also passing through the object focal point F_1 of the objective. After refraction, r_F must pass through I_{L1}' parallel to the axis and is refracted by the ocular through the axial point c_E, also with angle u_{pr}' because of beam parallelism. Now the heights of r_F at the objective and ocular are equal, so we have

$$-f_1 u_{pr} = f_1' u_{pr} = -f_2' u_{pr}' , \tag{2.42}$$

since $P_2' c_E = f_2'$, the focal length of the ocular. From (2.41) and (2.42)

$$m = u_{pr}'/u_{pr} = -f_1'/f_2' , \tag{2.43}$$

showing that the magnification is negative and an image inversion has taken place in such a conventional, Kepler-type telescope with a positive ocular.

Also

$$C = y_1/y_2 = P_1' I_1'/I_1' P_2 = f_1'/f_2 = -f_1'/f_2' , \tag{2.44}$$

or

$$C = m \tag{2.45}$$

The beam compression ratio is therefore identical to the magnification and can, in practice, by measurement of the diameter of the Ramsden disk, give an accurate determination of m. This has important practical consequences in the visual use of telescopes in general. For example, an expensive 10×50 binocular designed for low light levels and bird-watching will have $m = 10$ and an aperture $|E_1| = 50$ mm, giving an exit pupil diameter $|E_2'|$ of 5 mm according to (2.40). If it is used in bright sunlight, the observer's eye pupil will shrink to 2 mm (or less). This effective exit pupil is projected back to the entrance pupil and diaphragms the objective to 20 mm (or less), 16% of its area, thereby wasting an expensive instrument.

The same danger exists in visual astronomical observation. With a field of faint stars, an eye pupil of 5 mm is quite possible; but if a bright object such as Jupiter or the moon is observed, the pupil will probably hardly exceed 1 mm diameter. If an ocular is used with a modern 3.5 m telescope on such an object, a magnification of 3500 is required to compress the beam down to 1 mm diameter for passage into the eye. Such a magnification is higher than is normally acceptable for atmospheric seeing conditions, since it increases the normal resolution of the eye of about 1 arcmin to about 1/60 arcsec, so a lower value is preferred. This gives better views but diaphragms off the telescope. William Herschel was well aware of these problems (see RTO II, Chap. 4) and accordingly used magnifications as high as possible with his larger telescopes.

The eye pupil problem is one of the main reasons why visual judgement of larger telescopes is notoriously dangerous and usually too optimistic. Oculars are no longer used for professional astronomical observation, but are often applied for technical purposes, above all during set-up. The focal length of the ocular to avoid diaphragming must be

$$f_2' \leq E_2' N_1 \ , \tag{2.46}$$

where E_2' is the estimated eye pupil diameter and N_1 the f/no of the objective or primary image-forming system.

The magnification and beam compression laws of (2.44) and (2.45) are related to thermodynamics in that they are concerned with the energy transfer conditions. We mentioned above in § 2.2.3 that the energy transferred is proportional to H^2, where H is the Lagrange Invariant. Ignoring absorption or reflection losses, the capacity of a telescope to transfer energy is defined by the "Throughput" or "Light Transmission Power" as defined by Hansen [2.8][2.9][2.10]:

$$LTP = kn^2 \frac{y^2 \eta^2}{(f_1')^2} \tag{2.47}$$

Here, n is the refractive index, y^2 a measure of the area of the pupil, η^2 a measure of the area of the image, f_1' the distance between them, the focal length, and k a constant. The LTP has the dimensions $[L]^2$.

The Lagrange Invariant for the case of an afocal telescope can be deduced at once from (2.43) and (2.44) since

$$m = \frac{u_{pr}'}{u_{pr}} = C = \frac{y_1}{y_2} \ , \tag{2.48}$$

or

$$u_{pr}'y_2 = u_{pr}y_1 \ , \tag{2.49}$$

for a telescope in air. In the general case, with object and image media n and n', we have

$$n'u_{pr}'y_2 = nu_{pr}y_1 \ , \tag{2.50}$$

in agreement with Eq. (2.30), there given in notation for a single-element transfer for object and image distances tending to infinity.

Above we have considered the normal case of the afocal telescope with the eye focused on infinity. Of course, it is possible, though more tiring for the observer, to focus the ocular further in, giving a diverging emergent beam. The final (virtual) image can then be observed at any desired distance down to the minimum distance of distinct vision, normally 25 cm. The angle u_F' of the ray r_F after refraction by the ocular remains unchanged, but the other angles, including u_{pr}' will change somewhat because of the divergence of the beam.

Alternatively, the ocular can be focused further out, giving a converging emergent beam. This can no longer be focused by the eye, but the technique is widely used by amateurs in order to project a real image, such as that of the sun, on to a screen. This case is instructive in comparison with the reflecting telescope forms treated in § 2.2.5 below. In Fig. 2.9, let the plane P_1 represent an objective so thin that its principal planes fall together, and P_2 a similar plane for the ocular. Let the latter be defocused outwards to produce a converging beam and a real image at I_2'. If the emergent ray r is projected to meet the projection of the incident ray r, parallel to the axis, the plane P' where they intersect is the image-side principal plane of the total system, which is no longer at infinity. The effective focal length is f'. Since it is measured from P' to I_2', it is negative. The focal length f' will determine

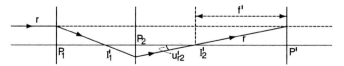

Fig. 2.9. Image principal plane in the defocused telescope of Fig. 2.8, producing a real image at I_2'

the *scale* (see § 2.2.6) of the real image I_2'. The smaller the defocus, the smaller will be u_{r2}' and the larger the value of f', which can be made far larger than $f_1' = P_1 I_1'$, the focal length of the objective. In other words, the defocused telescope becomes a strong *telephoto objective*, a property identical with that of reflecting telescopes of the Gregory and Cassegrain forms. We shall see that the case of Fig. 2.9 corresponds exactly to the Gregory telescope.

The telephoto effect of Fig. 2.9 arises from the fact that the defocused ocular largely neutralises the power of the objective. It is easily shown [2.3] that the total power K of two separated thin lenses in air is given by

$$K = K_1 + K_2 - dK_1 K_2 , \tag{2.51}$$

where K_1 and K_2 are their powers and d the separation. In air, therefore

$$\frac{1}{f'} = \frac{1}{f_1'} + \frac{1}{f_2'} - \frac{d}{f_1' f_2'} , \tag{2.52}$$

giving $1/f' = 0$ if $d = f_1' + f_2'$, the afocal case of Fig. 2.8. If d is slightly increased, a small negative balance of power remains giving a long negative focal length. In the next section, we shall consider the more general form of Eq. (2.52), applicable to mirrors.

The object space principal plane P, corresponding to P' for the image space in Fig. 2.9, may be constructed by tracing a ray parallel to the axis backwards into the ocular end of the system. P will lie to the left of P_1 giving a positive value of f so that, in air, $f = -f'$, as required by (2.21).

Finally, it should be mentioned that the defocused Galileo-type telescope, using a negative ocular inside the objective focus, is the similar equivalent of the Cassegrain reflector. In this case, f_2' in (2.52) is negative. The properties of the Galileo-type refractor can be demonstrated by the analogous construction to Fig. 2.8. The length is more compact with $d = f_1' - f_2'$ for the afocal case, but the exit pupil is virtual and on the opposite (objective) side of the ocular from the image space where the eye must be placed (see also Fig 1.3). It is therefore impossible to place the eye pupil at the exit pupil, giving a severe field limitation. The Kepler-type, giving an accessible exit pupil at the Ramsden disk, is therefore the definitive form for most purposes, though the Galileo-type still survives, because of its compactness, in opera glasses.

2.2.5 Basic forms of reflecting telescope

The modern astronomical telescope is a reflector, in its normal function a large photographic objective forming a real image for some detector. We are concerned here with the Gaussian properties of the basic forms of reflecting telescope.

2.2.5.1 Prime focus telescopes with a single powered mirror. The simplest forms are the Herschel and Newton forms, using only the *prime focus* of a concave primary mirror (Fig. 2.10). From the point of view of

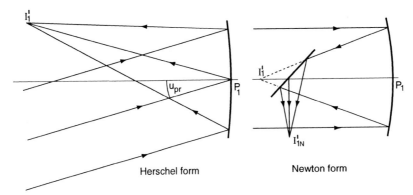

Fig. 2.10. Prime focus forms of reflecting telescope

Gaussian optics, these forms are identical. The optical power provided by the "thin" concave primary is the equivalent of a "thin" convex lens as objective. The pole of the primary embodies both principal points and is normally the aperture diaphragm and both entrance and exit pupils. In the Newton case, this implies that the flat must be large enough to accept the defined angular field of the detector. The sag of the mirror surface is meaningless in paraxial terms. In large modern telescopes the direct axial focus point I_1' can be used directly without the Newton flat since obstruction problems are no longer serious. The inclination of the central incident beam in the Herschel case has no significance in Gaussian optics, although it is very important if the higher terms (aberrations) are considered.

Consider now (Table 2.1) the *signs* of the basic paraxial parameters in the ray-tracing scheme of Eqs. (2.36) to (2.38) for direct imagery by reflection of an object at infinity. Strict adherence to this sign convention is essential for the understanding and correct use not only of the Gaussian relationships but also the entire aberration theory of Chap. 3.

Table 2.1. Gaussian optics of a prime focus reflecting telescope with a single powered mirror: sign of the paraxial parameters

Surface	Positive quantities	Negative quantities
1	y_1 n_1 i_1' u_1'	r_1 i_1 n_1' s_1'
Image Space		s_l' $f'(N_1)$

2.2.5.2 Basic compound telescopes with two powered mirrors. The two most important telescope forms with two powered mirrors are those invented early in the history of the reflector, the Gregory and Cassegrain forms, shown in Figs. 2.11 and 2.12 respectively. We shall establish a number of paraxial relationships which are common to both these forms and are of great importance, not only for the Gaussian layout of telescopes but also in the general aberration theory of Chap. 3.

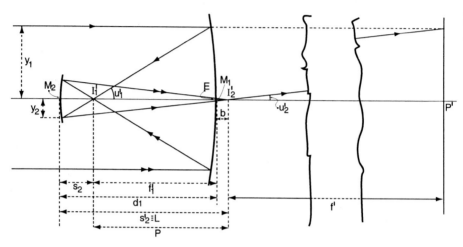

Fig. 2.11. Gaussian optics of a Gregory telescope

As stated in § 2.2.4 above, the Gregory form is the equivalent of a defocused Kepler-type refractor, the Cassegrain of a defocused Galileo-type. The image side principal planes P' are constructed as shown, the Gregory form having then a negative focal length f', the Cassegrain form a positive f'. The object side principal planes P, constructed by tracing a parallel beam backwards into the system from the right, lie well to the left of the system and to the left of the image principal planes P' in both cases. If the telescopes are made afocal, with I_1' at the focal point of M_2, the principal planes are at infinity.

Although the operation may at first sight seem trivial, it is very instructive to trace the paraxial ray through the system using Eqs. (2.36) - (2.38) step by step, for example with the normalized data of Cases 3 and 7 of Table 3.3 in Chap. 3 where the paraxial parameters are used for calculating the third order aberrations. The ray trace for the first surface (prime focus) is common to both Gregory and Cassegrain. Table 2.2 shows the signs of the paraxial quantities, the most important aspect being the sign reversals in certain cases between the Gregory and Cassegrain forms.

The strong shifts of the principal planes with the analogy of the defocused refracting telescopes indicate the most important property of the Gregory and Cassegrain forms: they are strong *telephoto systems* because the power of the

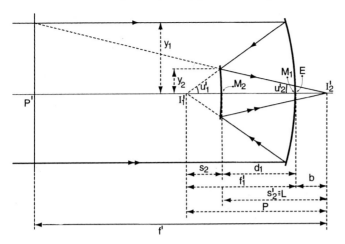

Fig. 2.12. Gaussian optics of a Cassegrain telescope

primary is largely annulled by the secondary. This situation is expressed by the equivalent relationship to Eq. (2.52) for the total residual power of two separated mirrors, which can be written [2.3]:

$$\frac{n'}{f'} = \frac{n'_1}{f'_1} + \frac{n'_2}{f'_2} - \frac{d_1}{n'_1}\frac{n'_1 n'_2}{f'_1 f'_2}, \tag{2.53}$$

where d_1/n'_1 is the effective optical separation of the mirrors. If all the refractive indices are set to $+1$, this reduces to the lens formula of (2.52). For our 2-mirror case and with our sign convention of Table 2.2, $n_1 = n'_2 = n' = +1$ and $n'_1 = -1$ in the normal case in air. The separation d_1 is negative in both cases as is $f'_1 = r_1/2$; while $f'_2 = r_2/2$ is positive in the Gregory and negative in the Cassegrain case. Eq. (2.53) reduces to

$$\frac{1}{f'} = -\frac{1}{f'_1} + \frac{1}{f'_2} - \frac{d_1}{f'_1 f'_2} = \frac{-f'_2 + f'_1 - d_1}{f'_1 f'_2} \tag{2.54}$$

In both the Gregory and the Cassegrain case, $|d_1|$ is slightly greater than for the afocal case, giving an $|f'|$ far larger than $|f'_1|$. f' is then negative for Gregory and positive for Cassegrain. The strong telephoto effect is evident from the construction of Figs. 2.11 and 2.12.

Another obvious way of interpreting the telephoto effect is to consider the secondary as a magnifier increasing the effective focal length without increasing the length of the system. This brings up the important question of the *definition* of the magnification m_2 of the secondary.[1] For an optical

---

[1] I am deeply grateful to Dr. Daniel Schroeder for querying this matter, leading to an excellent discussion by correspondence. As a result, a more detailed justification of the definition I chose in this book is now given here for the 2nd. edition.

Table 2.2. Gregory and Cassegrain telescope forms: sign of paraxial ray trace quantities (*denotes sign inversion between Gregory and Cassegrain)

Surface		Gregory form		Cassegrain form	
		Positive quantities	Negative quantities	Positive quantities	Negative quantities
1		y_1 n_1 i_1' u_1'	$r_1 (f_1')$ i_1 n_1' s_1' d_1	y_1 n_1 i_1' u_1'	$r_1 (f_1')$ i_1 n_1' s_1' d_1
2	* * * *	s_2 u_2 $r_2 (f_2')$ i_2 n_2' u_2' $s_2' (= L)$	y_2 n_2 i_2'	s_2 u_2 y_2 i_2 n_2' $s_2' (= L)$	$r_2 (f_2')$ n_2 i_2' u_2'
Image Space	*		$f'(N)$	$f'(N)$	

system with finite conjugates on both sides (Fig. 2.5), the normal definition is the *transverse magnification* as defined in Eq. (2.23). A compound telescope (Gregory or Cassegrain) is *afocal* on the object side but usually gives images at a finite distance both from the primary and from the secondary. In the Gregory case, the corresponding image heights η_1' and η_2' are negative and positive respectively, giving a negative transverse magnification. In the Cassegrain case, both heights are negative, giving a positive transverse magnification. This is the definition used in Schroeder's book [2.6]. However, both forms – and in modern times particularly the Cassegrain because of its predominant use – can be used in the *afocal* mode invented by Mersenne (Fig. 1.3). We shall see in Chap. 3, from their remarkable aberrational properties as both aplanatic and anastigmatic telescopes, that the Mersenne telescopes may be seen as the *fundamental forms* of the reflecting telescope. We recall, too, that the original refracting telescope of Fig. 2.8, with ocular, was also afocal. Above all, since modern Cassegrain telescopes have relatively small secondaries with high telephoto effects, the aplanatic Ritchey-Chrétien form only departs minimally from the Mersenne afocal form: the primaries of such telescopes have a hyperbolic form of eccentricity only slightly higher than the Mersenne parabola. For generality, therefore, it seems desirable to define the magnification of the secondary in a way that is also valid in the *afocal case*. The obvious parameter to use would be the ratio of the *field angles*, as in Eq. (2.41) for the afocal refractor. Unfortunately, in the *focal case*, the angular magnification of the secondary defined as u'_{pr2}/u'_{pr1} deviates from the transverse magnification η'_2/η'_1 because of the discrepancy between the exit pupil (and maybe the entrance pupil) and the principal planes. However, the *signs* emerging from this definition are instructive: with our Cartesian ray trace system of Table 2.2, the angular magnification of the secondary is *negative* in the Cassegrain telescope. But since the angular magnification of the primary is also *negative* (normally -1), the total angular magnification is *positive*. This is why, if we were to look at the moon through an afocal Mersenne Cassegrain telescope, using the secondary as an ocular as Mersenne proposed, we would see the moon *upright*, albeit with minute angular field because of the unfavourable exit pupil position (Fig. 1.3).

We see, therefore, that the *sign* of the magnification of an optical arrangement in general or a Cassegrain secondary in particular, can depend on the measuring parameter chosen. Nevertheless, the discrepancy of the values given by the definitions in the focal case between transverse image heights and field angles is not acceptable. Fortunately, there is an excellent alternative which removes this discrepancy and is also applicable to the afocal case. This definition of the magnification of the secondary m_2 is simply the ratio of the final to the primary *focal length*, a ratio which is anyway fundamental to telescope designers. This also gives a direct measure of the telephoto effect. With this definition, we have for the focal case from Eq. (2.24) and the Lagrange Invariant of Eq. (2.27)

$$m_2 = \frac{f'}{f'_1} = \frac{n'_2\eta'_2}{n'_1\eta'_1} = \frac{u'_1}{u'_2} = \frac{s'_2}{s_2} = \frac{L}{s_2} \, , \qquad (2.55)$$

in which the normal transverse magnification ratio η'_2/η'_1 is multiplied by n'_2/n'_1. Now in Table 2.2 for the Cassegrain case n'_2 is positive and n'_1 is negative while, as mentioned above, η'_2 and η'_1 are both negative. It follows also that f'_1 is negative, with f' positive. The definition of Eq. (2.55) therefore gives the same value for m_2 as the transverse magnification, but with reversed, that is *negative* sign in the Cassegrain case. Bahner [2.11] also defined m_2 as f'/f'_1, but he used a non-Cartesian sign convention of special form, not suitable for general application, giving f'_1 as positive. But other non-Cartesian ray tracing systems, such as those of Conrady [2.12] or Hopkins [2.13], give f'_1 as *negative*, in agreement with the strict Cartesian system. Since the definition of Eq. (2.55) can be applied perfectly generally to both focal and afocal cases, this has been adopted for this book. The resulting *negative* value for m_2 in the Cassegrain case then inevitably produces discrepancies in the equations for the aberration theory compared with other treatments such as that of Schroeder. It is not a question of right or wrong, but a question of preference. The choice according to Eq. (2.55) has been deliberately made with the aim of following an absolutely consistent Cartesian scheme throughout this book, also giving preference to the focal lengths as fundamental parameters rather than the transverse image heights. Correspondingly, of course, m_2 is *positive* for the Gregory telescope.

One could define the telephoto effect relative to the primary simply as

$$T_p = m_2 \ , \tag{2.56}$$

but the *true telephoto effect* T in the normal photographic sense of the equivalent focal length relative to the constructional length L is given by

$$T = \frac{f'}{L} \ , \tag{2.57}$$

where $L = M_2 I_2'$ in Figs. 2.11 and 2.12. In terms of ray trace parameters $L \equiv s_2'$. Note that Eq. (2.57) is only valid if $|L| \geq |d_1|$, i.e. if $b \geq 0$, since otherwise the true length of the system is given by $|d_1|$, not $|L|$.

Another fundamental parameter of these compound telescope forms is the *axial obstruction ratio* R_A. This is given by

$$R_A = \frac{y_2}{y_1} = \frac{s_2}{f_1'} = \frac{L}{f'} \ , \tag{2.58}$$

from the geometry of Figs. 2.11 and 2.12. Eqs. (2.57) and (2.58) give the simple relationship

$$T = 1/R_A \tag{2.59}$$

This result essentially explains why the compound (2-mirror) telescope has triumphed over the single mirror form and is the standard solution for modern astronomy. Historically it was desirable to have a large scale (see § 2.2.6 below) and therefore a long focal length f'. A solution with a large T giving reduced length is exactly what is needed. The fact that a large value of T gives a small axial obstructionfrom (2.59) is a marvellous added attraction. Of course, we have said nothing about the difficulties of manufacture and test which, as we shall see in Chap. 5 and in RTO II, Chap. 1 and 3, have been formidable.

Formally, from (2.57) and (2.59), there is no difference between the Gregory and Cassegrain solutions. However, there is a further parameter to be introduced which makes the Cassegrain solution much superior for normal purposes, i.e. where a real primary image is of no consequence. The notable exception is the Gregory form for solar telescopes, in which the intense heat of the bulk of the solar image can be absorbed at the real prime focus. The parameter favouring the Cassegrain form is the position of the final image I_2'. For a fixed, convenient position behind the primary and a given value of

R_A, the Cassegrain allows a larger relative aperture u_2' or a shorter value of f' because L is shorter. This is evident from Figs. 2.11 and 2.12 but will be proven by the formulae below.

Applying (2.34) and (2.35) to the secondary, we have

$$\frac{1}{s_2'} + \frac{1}{s_2} = \frac{1}{f_2'} , \tag{2.60}$$

giving with $L \equiv s_2'$

$$\frac{s_2'}{s_2} = \frac{L}{f_2'} - 1 \tag{2.61}$$

From (2.55) and (2.57)

$$T = \frac{f'}{L} = \frac{f_1'}{f_2'} = \frac{f_1'}{L} \tag{2.62}$$

Now from (2.58) and (2.61) we have with $L \equiv s_2'$

$$s_2 = R_A f_1' = \frac{f_2' L}{L - f_2'} , \tag{2.63}$$

giving

$$f_2' = \frac{R_A L f_1'}{L + R_A f_1'} \tag{2.64}$$

The image position is defined by the parameter b, where

$$b = L + f_1' - s_2 = L + f_1'(1 - R_A) , \tag{2.65}$$

from (2.63).

The relative aperture of the emergent beam is defined by u_2' from Figs. 2.11 and 2.12 as

$$u_2' = -\frac{y_2}{L} = -\frac{R_A y_1}{L} , \tag{2.66}$$

from (2.58). Eq. (2.66) is the relationship which proves the advantage of the Cassegrain form over the Gregorian. If y_1 and R_A are predefined, then the angle u_2' depends only on L. Now L is given for a predefined image position b from (2.65) as

$$L = b - f_1'(1 - R_A) , \tag{2.67}$$

in which f_1' is negative in both cases and R_A is, from (2.58), positive in the Cassegrain form and negative in the Gregory form. Since L is a positive quantity in both cases, it is always larger in the Gregory form for a given R_A. Therefore, u_2' must be smaller for the Gregory form from (2.66) unless $|R_A|$ is zero, a limit case of no practical significance. Typically $|R_A| = \frac{1}{3}$. If $b = 0$, Eq. (2.67) shows that L is twice as large for the Gregory form. T is the same in both cases, so $|f'|$ is twice as large for the Gregory form from

(2.57). From (2.55) m_2 is also twice as large, making the solution technically more extreme and sensitive. But, above all, the "speed" of the telescope in the photographic sense is reduced to a quarter by the halving of u_2' in (2.66). In other words, the *Light Transmission Power* defined in Eq. (2.47) is much higher in the Cassegrain case for given values of R_A, T, y_1 and b.

Combining (2.66) and (2.67) gives

$$u_2' = \frac{R_A y_1}{f_1'(1 - R_A) - b} = \frac{R_A y_1}{f_1' - s_2 - b} \tag{2.68}$$

In § 2.2.4 we introduced the basic parameter C, the compression ratio, for an *afocal* telescope, defined from (2.48) by

$$C = y_1/y_2$$

If either the Gregory or Cassegrain form is made *afocal*, it becomes a Mersenne telescope as discussed in Chap. 1 (see Fig. 1.3). In this case

$$C = 1/R_A , \tag{2.69}$$

from (2.58). As in the refracting case, the Lagrange Invariant of (2.50) also gives

$$C = m , \tag{2.70}$$

where m is the magnification of the total afocal system.

We return now to the normal case with a real image. So far, the parameter d_1 has hardly appeared in our paraxial relationships. It is nevertheless an important parameter, defined by

$$s_2 = f_1' - d_1 \tag{2.71}$$

From (2.58) we have

$$R_A = \frac{y_2}{y_1} = \frac{L}{f'} = \frac{L}{m_2 f_1'} = \frac{s_2}{f_1'} = 1 - \frac{d_1}{f_1'} = 1 - \frac{m_2 d_1}{f'} , \tag{2.72}$$

relations of great importance in the aberration theory of Chap. 3.

Further useful relations are

$$s_2' = L = \frac{f_2' s_2}{s_2 - f_2'} , \tag{2.73}$$

from (2.61), and

$$d_1 = b - s_2' = b - L \tag{2.74}$$

Also, from (2.71) and (2.72)

$$L = m_2 s_2 = m_2(f_1' - d_1) = f' - m_2 d_1 \tag{2.75}$$

From (2.54)

$$f' = \frac{f_1' f_2'}{f_1' - f_2' - d_1} \tag{2.76}$$

Eliminating f' from (2.75) and (2.76):

$$L = \frac{f_2'(f_1' - d_1)}{f_1' - f_2' - d_1} \tag{2.77}$$

Transposing (2.72):

$$f' = d_1 \left(\frac{m_2}{1 - R_A} \right) \tag{2.78}$$

Eliminating s_2 and f_1' from (2.71), (2.72) and (2.76):

$$f' = L - d_1 + \frac{L d_1}{f_2'} \tag{2.79}$$

Eliminating L from (2.74) and (2.79) gives

$$f_2' = \frac{d_1(b - d_1)}{f' - b + 2d_1} \tag{2.80}$$

We now introduce a further important constructional parameter P in Figs. 2.11 and 2.12, representing the distance between the primary and secondary images. This is a positive quantity in both the Gregory and Cassegrain forms and is defined by

$$P = L - s_2 = b - f_1' = b - s_2 - d_1 , \tag{2.81}$$

from (2.71). Now we can write, using (2.55)

$$s_2 = s_2 \left(\frac{L - s_2}{L - s_2} \right) = \frac{P}{\left(\frac{L}{s_2} - 1 \right)} = \frac{P}{m_2 - 1} \tag{2.82}$$

Transposing (2.63) gives

$$f_2' \left(\frac{L}{s_2} + 1 \right) = L = m_2 s_2 , \tag{2.83}$$

from (2.55). Combining (2.83) with (2.82) gives finally with $L/s_2 = m_2$

$$f_2' = P \left[\frac{m_2}{(m_2 + 1)(m_2 - 1)} \right] = P \left(\frac{m_2}{m_2^2 - 1} \right) \tag{2.84}$$

This is probably the most useful of all the paraxial formulae for Gregory and Cassegrain telescopes, for the following reason. In setting up the telescope design parameters it is normal to start with the diameter D_1 and the f/no of the primary, thereby defining f_1'. Also, the position of the final image I_2' behind the primary is defined by technical considerations. This defines P. Usually a final f/no for the emergent beam is envisaged, which defines m_2 from (2.55). Then Eq. (2.84) gives $f_2' = r_2/2$. The fundamental parameter remaining, which is dependent, is R_A. From (2.58) and (2.82), we have

$$R_A = \frac{s_2}{f_1'} = \frac{P}{f_1'(m_2 - 1)} = \frac{P}{f' - f_1'} = \frac{m_2 P}{f'(m_2 - 1)} \tag{2.85}$$

These forms are useful, but a clearer indication of the driving parameters is given by substituting for P from (2.81) to give

$$R_A = \frac{b - f_1'}{f_1'(m_2 - 1)} = \frac{\bar{b} - 1}{m_2 - 1},$$

(2.86)

where

$$\bar{b} = \frac{b}{f_1'} = \frac{L}{f_1'} + (1 - R_A),$$

(2.87)

from (2.65).

Equation (2.86) makes it clear what must be changed if R_A is too high (assuming $b \ll |f_1'|$). m_2 *must be increased*. So either f' must be increased or f_1' reduced. If the latter is preferred, to maintain the Light Transmission Power, the primary will be steeper and more difficult to make, but T will be increased giving a more compact solution. The denominator of (2.86) reveals the superiority of the Cassegrain form again since it is larger, m_2 being negative in the Cassegrain form, if $|m_2|$ is the same for Gregory and Cassegrain solutions.

If f_2' and P are given parameters, Eq. (2.84) gives for m_2:

$$m_2 = \frac{1 + \left[1 + 4\left(\dfrac{f_2'}{P}\right)^2\right]^{1/2}}{2f_2'/P}$$

(2.88)

Finally, there are useful relations linking f_2', L and m_2. From (2.55) and (2.81) we have

$$P = L - s_2 = L - \frac{L}{m_2} = L\left(\frac{m_2 - 1}{m_2}\right)$$

(2.89)

Substitution in (2.84) gives the simple relation

$$f_2' = \frac{L}{m_2 + 1} = s_2\left(\frac{m_2}{m_2 + 1}\right),$$

(2.90)

from (2.55), a result which is also given directly by the reflection equation (2.35). We shall see in Chap. 3 that this is one of the most important paraxial relations in the aberration theory of 2-mirror telescopes.

The *position of the entrance pupil* has no influence on the Gaussian parameters of these telescope systems, but it does affect the diameters of the mirrors required. Conventionally, the entrance pupil is placed *at the primary* for the simple reason that the primary is the most difficult and expensive element in the system and is most efficiently used in this way. In this case, a supplement in diameter at the secondary is required if vignetting in the field is to be avoided. For the axial beam at the secondary, we have from (2.72)

$$y_2 = y_1 \frac{L}{f'} = y_1 \frac{L}{m_2 f_1'} \,, \tag{2.91}$$

while, for the field, the supplement $(y_{pr})_2$ given by

$$(y_{pr})_2 = (u_{pr}')_1 d_1 = -(u_{pr})_1 d_1 \tag{2.92}$$

is required, for an object at infinity. The total diameter $(D_{TOT})_2$ is then given by

$$(D_{TOT})_2 = 2\left[\left|y_2\right| + \left|(y_{pr})_2\right|\right] = (D_{AX})_2 + 2\left|(u_{pr}')_1 d_1\right| \tag{2.93}$$

The position of the *exit pupil* in the normal case of entrance pupil at the primary is determined by tracing a paraxial ray from the primary to the secondary. If this is turned round and the ray traced from left to right in the normal way, Eq. (2.35) gives

$$(s_2')_E = \frac{(r_2)_E (s_2)_E}{2(s_2)_E - (r_2)_E} = \frac{(r_2)_E (-|d_1|)}{-2|d_1| - (r_2)_E} \,, \tag{2.94}$$

in which $(s_2')_E$ is the exit pupil distance from the secondary, $(s_2)_E$ the entrance pupil distance and $(r_2)_E$ the radius of curvature of the secondary in the reversed ray trace. Now $(r_2)_E$ is *positive* in the Cassegrain form (Fig. 2.13) and $(s_2)_E$ is negative, i.e. $-|d_1|$. Therefore, in the Cassegrain form, $(s_2')_E$ is always positive, i.e. the exit pupil always lies *behind* the secondary, somewhat nearer than the prime focus I_1' if $b > 0$. In the Gregory form, $(s_2')_E$ is always *negative*, so the exit pupil lies in front of the secondary, again somewhat nearer than the prime focus if $b > 0$.

In spite of the advantages of efficient use of the primary in the conventional choice of the entrance pupil E at M_1, the increasing importance of infra-red observation in astronomy often forces the decision to place the aperture stop of the telescope at the secondary. Since there is no subsequent element with

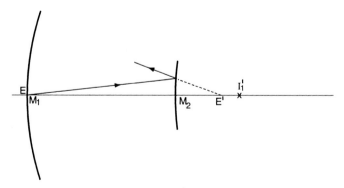

Fig. 2.13. Exit pupil position E' in the Cassegrain form with the entrance pupil E at the primary

optical power, the *secondary* then becomes the *exit pupil* E'. The paraxial image of the secondary in the primary is then the entrance pupil E. Tracing from E' at M_2 to M_1 gives from Fig. 2.14

$$(s_1')_E = \frac{(r_1)_E (s_1)_E}{2(s_1)_E - (r_1)_E} = \frac{(r_1)_E (-|d_1|)}{-2|d_1| - (r_1)_E} \tag{2.95}$$

Now $(r_1)_E$ and $(s_1)_E$ are both negative. In the Cassegrain form $|(r_1)_E| > 2|d_1|$, so the denominator is always positive, as is also the numerator. Therefore, $(s_1')_E$ is always positive, giving a virtual entrance pupil to the right of the primary. With the geometry of Case 3 in Table 3.3 (Chap. 3), the distance $M_1 E$ is $+3.44|f_1'|$. In the Gregory form $2|d_1| > |(r_1)_E|$, so that the denominator of (2.95) is always negative, the numerator always positive. The entrance pupil E is therefore real and in front of the primary. With the geometry of Case 7 of Table 3.3, $M_1 E = -5.44|f_1'|$. In both Cassegrain and Gregory forms, a supplement will have to be added to the diameter of M_1 if vignetting of the field is to be avoided. The supplement is $2|(y_{pr})_1|$ where $(y_{pr})_1$ is given by

$$(y_{pr})_1 = u_{pr} M_1 E \tag{2.96}$$

The reader is reminded of the remarks in §2.2.3 concerning the means of fixing the aperture stop in an optical system. If, in a telescope laid out to have its stop at the primary without vignetting at the secondary, a stop is laid over the secondary such that the axial beam at the primary is reduced to the extent required to accommodate the required field surplus according to (2.96) with its actual diameter, then the exit pupil is at the secondary and no vignetting takes place. This conversion process represents an elegant solution to the problem of switching between the aperture stop at the primary and the secondary. However, the diaphragm over the secondary has to be designed to give an acceptable level of IR emissivity.

In Tables 2.1 and 2.2 above, we gave a list of the paraxial ray trace quantities to indicate the signs for use in the Gregory and Cassegrain forms in the numerous relations derived. These relations also include some derived quantities. Since correct use of the signs is essential, these are given in Table 2.3.

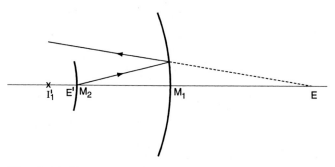

Fig. 2.14. Entrance pupil position E in the Cassegrain form with the exit pupil E' at the secondary

Table 2.3. Signs of derived quantities from the paraxial ray trace for the Gregory and Cassegrain forms

* Sign	Gregory form		Cassegrain form	
inver-sion	Positive quantities	Negative quantities	Positive quantities	Negative quantities
	L		L	
*	m_2			m_2
*		T	T	
*		R_A	R_A	
*		C	C	
	P		P	

An important limit case remains to be considered. If, in Eq. (2.84), m_2 becomes either $+1$ or -1, the denominator becomes zero. The case $m_2 = +1$ corresponds to a real limit case of the *Gregory telescope* in which the secondary has its centre of curvature at the primary image I_1' in Fig. 2.11. The final image is then returned to I_1' and $P = 0$, so that f_2' is indeterminate from (2.84). In fact, it would be equal to $s_2/2$. The case is of no practical interest since it represents an autocollimator returning the incident parallel beam into space.

The case $m_2 = -1$, on the other hand, has considerable practical interest as the *limit case* of a *Cassegrain telescope* in which the secondary has zero power and becomes a plane mirror (Fig. 2.15). Since P is real and positive, the numerator of (2.84) remains finite and $f_2' = \infty$. For this limit case, we can write, using $\sim$ to denote the plane mirror solution,

$$\tilde{f}' = -\tilde{f}_1' , \tag{2.97}$$

and from (2.58)

$$\tilde{L} = -\tilde{R}_A \tilde{f}_1' , \tag{2.98}$$

giving correctly $f_2' = \infty$ from (2.64). Now, from (2.65)

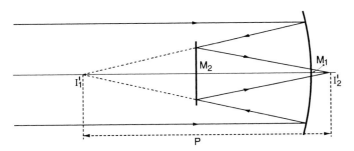

Fig. 2.15. Limit case of a Cassegrain telescope with a plane secondary mirror

$$\tilde{L} = \tilde{b} - \tilde{f}_1'(1 - \tilde{R}_A) \tag{2.99}$$

Eliminating $\tilde{L}$ from (2.98) and (2.99) gives

$$\tilde{R}_A = \frac{\tilde{f}_1' - \tilde{b}}{2\tilde{f}_1'} \tag{2.100}$$

If $\tilde{b} > 0$, then $\tilde{R}_A > 0.5$. With $\tilde{b} = 0$, $\tilde{R}_A = 0.5$.

The folding flat reflector thus has a very high axial obstruction ratio $\tilde{R}_A$, whereby at least a quarter of the light is obstructed at centre field. If $\tilde{b} = 0$, the telephoto effect $\tilde{T}$ is still 2 from (2.59), even though the secondary has no optical power. This therefore gives a major reduction in length compared with a simple prime focus. For this reason the solution has been used in special cases such as the astrometric reflector at the U.S. Naval Observatory, where image stability considerations ruled out a powered secondary because of possible image variations from decentering (see Chap. 3).

2.2.6 The scale of astronomical telescopes and the magnification in afocal use of compound telescopes

As the equivalent of a large-scale photographic objective, the modern astronomical telescope is essentially defined by its free diameter D, its effective focal length f' and its semi-angular field u_{pr} in object space. The parameter magnification of a conventional refractor using an ocular in the afocal mode is replaced by the *scale* of the telescope image. The scale is usually expressed, because of the traditional use of arcsec as the angular measure of fine structures, in arcsec/mm or its inverse function mm/arcsec. From Eq. (2.31) we have

$$u_{pr} = \frac{\eta'}{f'} \text{ rad} = 206\,265\frac{\eta'}{f'} \text{ arcsec} \tag{2.101}$$

for an object at infinity with $u_p' = u_{pr}$. Setting $\eta' = 1$ mm gives the scale S as

$$S = \frac{206\,265}{f'(\text{mm})} \text{ arcsec/mm} \tag{2.102}$$

and the inverse scale $\overline{S}$ as

$$\overline{S} = \frac{f'(\text{mm})}{206\,265} \text{ mm/arcsec} \tag{2.103}$$

As a typical example of practical values, the ESO New Technology Telescope has $D = 3500$ mm and an f/11 focus, giving $f' = 38500$ mm. Then $S = 5.36$ arcsec/mm and $\overline{S} = 0.187$ mm/arcsec.

The f/no, denoted by N for the final image of a compound telescope and N_1 for the primary image, is given by

$$N = \frac{f'}{D_1} = \frac{f'}{2y_1} = -\frac{1}{2u_2'} = m_2 N_1 = m_2 \frac{r_1}{4y_1} \qquad (2.104)$$

According to our sign convention, N is negative for a Gregory telescope and positive for a Cassegrain. N_1 is a negative quantity. See Tables 2.1 and 2.2.

In §2.2.5 above, the use of Gregory or Cassegrain systems in the afocal (Mersenne) form (Fig. 1.3) was mentioned. Such systems obey the same magnification laws as the conventional afocal refractor, given in §2.2.4, whereby the compression ratio C is equal to the angular magnification m from Eq. (2.45). The principal use of such afocal forms is as a supplementary feeder system equivalent to afocal supplementary lens systems placed before an objective in photography. These increase the effective focal length. If the magnification of the afocal system is $m = C$ from (2.45), then the emergent field angle $u_{pr}' = mu_{pr}$ and the objective forms an image of height $m\eta'$ instead of η'. Now from (2.101) $m\eta' = mu_{pr}f'$, so for the real object field u_{pr} the effective focal length has become mf'. In the astronomical telescope afocal feed system, the purpose may also be to increase the effective aperture of the system being fed. A limitation of such afocal Gregory or Cassegrain feed systems is the unfavourable exit pupil position near the secondary combined with the field angle magnification.

For amateur use on a small scale, a conventional positive ocular may be used to observe the real image of the reflecting telescope. This will again follow the laws of the afocal refractor. Since the exit pupil of the Gregory or Cassegrain reflector is near the secondary, the ocular is working under similar conditions to those of a refractor of modest focal length.

2.2.7 "Wide-field" telescopes and multi-element forms

In this chapter, we have considered only the basic forms of telescope, all invented in the seventeenth century. Later forms with more elements and "wide field" forms also have design aspects requiring Gaussian optics. Above all, the theory of pupils is essential. From §2.2.5, it is clear that there is already a major complication in the general Gaussian properties in advancing from one powered element to two separated powered elements. More complex forms are best treated by tracing a paraxial aperture ray and a paraxial principal ray through the system and deriving from these general formulae on the basis of aberration theory. (Alternatively, recursion formulae are given in §3.6.5.2 which enable the determination of the paraxial parameters required for any number of centered reflecting surfaces). This is the subject of Chap. 3.

3 Aberration theory of telescopes

3.1 Definition of the third order approximation

The limits of the theory of Gaussian optics were defined in Chap. 2 by the expansions of Eqs. (2.13) and (2.14):

$$z = \frac{c}{2}y^2 + a_1 y^4 + a_2 y^6 \dots$$

$$n'\left(i' - \frac{i'^3}{3!} + \frac{i'^5}{5!} \dots \cdot\right) = n\left(i - \frac{i^3}{3!} + \frac{i^5}{5!} \dots\right)$$

All terms above the first were neglected. Snell's law of refraction reduces to a linear law in the Gaussian region. This gives, then, the theory of centered systems to the *first order*. If the next term is considered, we have *third order theory*. This is concerned with the lowest order terms which affect the *quality* of the image, whereas Gaussian optics is only concerned with its position and size. The aberrations affecting image quality are generally most important in the third order approximation, so the aberration theory of telescopes is mainly concerned with this approximation. The theory of higher order aberrations is extremely complex. It has not played any appreciable role in telescope development, since the exact total effect of all orders can be easily calculated by *ray tracing* using modern computers. But this in no way reduces the value of third order theory which remains essential for a correct understanding of the properties of different telescope forms.

In Eq. (2.13), the polynomial defines the form either of a refracting or reflecting *surface*, or of a *wavefront* defining the quality of an axial image point. We saw in §2.2.2 (Fig. 2.3) that a perfect image point is associated with a perfectly spherical wavefront. Aberration theory is concerned with phase errors from this perfect spherical form which fall within the third order region. The rays are the normals to the wavefront, as shown in Fig. 3.1.

$W' = AA'$ is the wavefront (phase) aberration of the ray ABC passing, if $W' = 0$, through the image point I_o' on the principal ray $E'OI_o'$. The ray actually cuts the principal ray at B and the image plane at C. Its *longitudinal aberration* is BI_o' and its *lateral aberration* CI_o'. All three forms have their uses but W' is, physically, the most meaningful. The lateral aberration is related to the *angular aberration* $\delta u_p'$. Interpreted, using the magnification

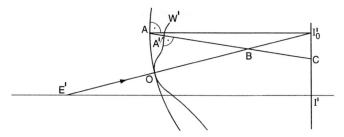

Fig. 3.1. Wavefront, longitudinal and lateral aberration

laws, in object space, this is particularly important in telescope systems. We shall give conversion formulae between W' and $\delta u'_p$.

Let us consider now the precise forms that Eq. (2.13) assumes for the sphere with circular section and for other conic sections, all with axial symmetry. Let the z-axis define the direction of axial symmetry and the ordinate y the height in the principal section.

The equation of a circle referred to its pole is

$$z = r - r \left(1 - \frac{y^2}{r^2}\right)^{1/2} \tag{3.1}$$

Expanding this:

$$z = \frac{y^2}{2r} + \frac{1}{8}\frac{y^4}{r^3} + \frac{1}{16}\frac{y^6}{r^5} + \dots \tag{3.2}$$

or

$$z_{Sph} = \frac{c}{2}y^2 + \frac{c^3}{8}y^4 + \frac{c^5}{16}y^6 + \dots \,, \tag{3.3}$$

with $c = 1/r$. For a spherical surface, therefore, the constants of its section a_1 and a_2 in (2.13) are:

$a_1 = c^3/8$ (third order term)

$a_2 = c^5/16$ (fifth order term)

The first (Gaussian) term gives the equation of a parabola, but the nature of the conic section is only defined by the second term. If it is zero, the surface section really is a parabola with vertex curvature c.

For an *ellipse*, the equivalent equation to (3.1) is

$$z = a - a \left(1 - \frac{y^2}{b^2}\right)^{1/2} \tag{3.4}$$

where a and b are the semi-axes. The expansion gives

$$z = \frac{1}{2}\left(\frac{a}{b^2}\right)y^2 + \frac{1}{8}\left(\frac{a}{b^2}\right)^2\frac{y^4}{a} + \frac{1}{16}\left(\frac{a}{b^2}\right)^3\frac{y^6}{a^2} + \dots \tag{3.5}$$

Now it can easily be shown that the vertex curvature is

$$c = \frac{a}{b^2} = \frac{1}{a(1 - \varepsilon^2)} \, ,$$ (3.6)

where ε is the eccentricity. Then

$$z_{Ell} = \frac{c}{2}y^2 + \frac{c^2}{8a}y^4 + \frac{c^3}{16a^2}y^6 + \dots$$ (3.7)

A hyperbola gives the same result except for alternating negative signs:

$$z_{Hyp} = \frac{c}{2}y^2 - \frac{c^2}{8a}y^4 + \frac{c^3}{16a^2}y^6 - \dots$$ (3.8)

Considering now the three equations (3.3), (3.7) and (3.8) and following Schwarzschild [3.1], they can all be written from (3.6) in the form:

$$z = \frac{c}{2}y^2 + \frac{c^3}{8}(1 - \varepsilon^2)y^4 + \frac{c^5}{16}(1 - \varepsilon^2)^2 y^6 + \dots$$ (3.9)

or, introducing the Schwarzschild (conic) constant b_s [3.1] as

$$b_s = -\varepsilon^2$$ (3.10)

$$z = \frac{c}{2}y^2 + \frac{c^3}{8}(1 + b_s)y^4 + \frac{c^5}{16}(1 + b_s)^2 y^6 + \dots$$ (3.11)

To the third order (second term), all surfaces have conic sections, uniquely defined by b_s as follows:

$$\left.\begin{array}{llc} b_s = 0, & \varepsilon = 0, & \text{circle (sphere)} \\ b_s = -1, & \varepsilon = 1, & \text{parabola} \\ -1 < b_s < 0, \, 0 < \varepsilon < 1, & \text{ellipse} \\ b_s < -1, & \varepsilon > 1, & \text{hyperbola} \end{array}\right\}$$ (3.12)

A further case is of interest, as it can be useful in optical systems:

$$b_s > 0, \quad \varepsilon \text{ imaginary, oblate spheroid (minor axis of an ellipse)}$$ (3.13)

The formulation of Eqs. (3.11) and (3.12) is the essential basis of aberration theory for telescopes.

3.2 Characteristic Function and Seidel (3rd order) aberrations: aberration theory of basic telescope forms

3.2.1 The Characteristic Function of Hamilton

In 1833, Hamilton [3.2] published one of the most profound and elegant analyses in the history of geometrical optics: the Characteristic Function. Based solely on the property of symmetry of a centered optical system about its axis, he deduced the general form of the aberration function in terms of three fundamental parameters: the aperture radius (ρ) (normalized to 1 at the edge), the field radius (σ) (normalized to 1 at the edge), and the azimuth angle of the plane containing the ray and principal ray in the image forming wavefront. Following Hopkins [3.3], we will define these normalized parameters as

$$\rho, \sigma \text{ and } \phi,$$

the dashes of ρ', σ' and ϕ', denoting the image space, being omitted for simplicity. Hamilton showed that, because of symmetry, these parameters can only appear in the general aberration function in the forms

$$\rho^2, \quad \sigma^2, \quad \rho\,\sigma\,\cos\phi$$

The aberration function must take the form, expressed as *wavefront aberration*:

$$
\begin{aligned}
W(\sigma,\rho,\phi) \;=\; {}_0k_{00} &+ ({}_0k_{20}\rho^2 + {}_1k_{11}\sigma\rho\cos\phi + {}_2k_{00}\sigma^2) \\
&+ ({}_0k_{40}\rho^4 + {}_1k_{31}\sigma\rho^3\cos\phi + {}_2k_{22}\sigma^2\rho^2\cos^2\phi \\
&+ {}_2k_{20}\sigma^2\rho^2 + {}_3k_{11}\sigma^3\rho\cos\phi + {}_4k_{00}\sigma^4) + \ldots
\end{aligned}
\tag{3.14}
$$

It is easily shown that some of these terms must be zero because of the definitions of Gaussian optics. All terms ${}_2k_{00}\sigma^2$, ${}_4k_{00}\sigma^4 \ldots$ are zero; also the constant term ${}_0k_{00}$ must be zero with the normal definition of the wavefront. The first terms that remain are then

$$
\begin{aligned}
W(\sigma,\rho,\phi) \;=\; {}_0k_{20}\rho^2 &+ {}_1k_{11}\sigma\rho\cos\phi + {}_0k_{40}\rho^4 + {}_1k_{31}\sigma\rho^3\cos\phi \\
&+ {}_2k_{22}\sigma^2\rho^2\cos^2\phi + {}_2k_{20}\sigma^2\rho^2 + {}_3k_{11}\sigma^3\rho\cos\phi + \ldots
\end{aligned}
\tag{3.15}
$$

This can be expressed as the general function

$$W(\sigma,\rho,\phi) = \sum {}_{(l+n)}k_{(m+n),n}\sigma^{(l+n)}\rho^{(m+n)}\cos^n\phi,
\tag{3.16}$$

in which l, m are even positive integers or zero, and n is any positive integer or zero. However, the terms with

$$
\begin{aligned}
m &= n = 0 \\
l &= n = 0 \text{ and } m = 2 \\
l &= m = 0 \text{ and } n = 1
\end{aligned}
$$

are excluded if the Gaussian conditions are defined as zero. In telescope optics, this is not, in general, the case for the latter two of these conditions.

The type of aberration depends on the functions of ρ and $\cos\phi$, whereas the function of σ shows how the effect varies in the field.

Table 3.1 shows the first, third and fifth order aberrations of this general function.

Table 3.1. Aberration types from the Characteristic Function

m	$n = 0$	$n = 1$	$n = 2$	$n = 3$	Order
0		$_1k_{11}\sigma\rho\cos\phi$			First:
2	$_0k_{20}\rho^2$				$N_H = 1$
					$(l + m$
					$+2n = 2)$
0		$_3k_{11}\sigma^3\rho\cos\phi$	$_2k_{22}\sigma^2\rho^2\cos^2\phi$		Third:
2	$_2k_{20}\sigma^2\rho^2$	$_1k_{31}\sigma\rho^3\cos\phi$			$N_H = 3$
4	$_0k_{40}\rho^4$				$(l + m$
					$+2n = 4)$
0		$_5k_{11}\sigma^5\rho\cos\phi$	$_4k_{22}\sigma^4\rho^2\cos^2\phi$	$_3k_{33}\sigma^3\rho^3\cos^3\phi$	Fifth:
2	$_4k_{20}\sigma^4\rho^2$	$_3k_{31}\sigma^3\rho^3\cos\phi$	$_2k_{42}\sigma^2\rho^4\cos^2\phi$		$N_H = 5$
4	$_2k_{40}\sigma^2\rho^4$	$_1k_{51}\sigma\rho^5\cos\phi$			$(l + m$
6	$_0k_{60}\rho^6$				$+2n = 6)$

Each aberration type, defined by a column with a given value of n, represents an infinite series in ascending powers of ρ. The order of an aberration term is defined from Eq. (3.16) as

$$N_H = (\text{Sum of powers of } \sigma \text{ and } \rho) - 1 \qquad (3.17)$$

The two aberrations in the first group are *first order aberrations*, i.e. errors in the region of Gaussian optics. The term $_1k_{11}\sigma\rho\cos\phi$ denotes a lateral shift of each image point from its nominal Gaussian position, the error being proportional to the nominal image height, i.e. a scaling error. In wavefront terms, this error is of great practical importance in telescope optics as it represents *pointing, tracking or guiding error*. It is best understood as a tilt of the imaging wavefront by a shift of its centre of curvature from the nominal position. The other Gaussian term $_0k_{20}\rho^2$ is also of great practical importance as it represents *defocus* from the correct nominal image plane. Defocus therefore gives a parabolic departure from the perfect reference sphere, which, in the Gaussian region, simply means a change of radius of the wavefront. This Gaussian concept of defocus can, of course, be extended at once to cover all orders from Eq. (3.11), as a change of radius of a *spherical* wavefront.

The five aberrations in the second group are the *monochromatic third order aberrations* or *Seidel aberrations* [3.4] and are, with the first order effects above, the dominant errors of telescope images and optical systems in general.

The column with $n = 0$ represents a family with no dependence on ϕ, so the effect is symmetrical to the axis. The terms

$_0k_{40}\rho^4$, $_0k_{60}\rho^6$, $\ldots$

represent *spherical aberration*, a symmetrical fourth power phase error of the wavefront in the third order case. The term

$$_2 k_{20} \sigma^2 \rho^2$$

is a defocus error which also depends on the square of the field: it is therefore *field curvature*.

The column with $n = 1$ has two third order terms. The term $_1 k_{31} \sigma \rho^3 \cos \phi$ is one of the most important in practical telescope optics: *third order coma*. The other term $_3 k_{11} \sigma^3 \rho \cos \phi$ has the same aperture function $m + n = 1$, $n = 1$ as the equivalent first order (tilt) term and is therefore also a lateral shift of the image from its nominal position. But it is no longer simply an error of scale since it varies with the cube of the field size. It is the aberration known as *distortion*. Distortion does not reduce the quality of a point image, but displaces it from its nominal position in a non-linear manner. In most basic telescope systems, distortion is small and of little significance. However, as soon as field correctors or other elements are introduced fairly near the image, distortion can become important.

The last third order term is in the column with $n = 2$. This is $_2 k_{22} \sigma^2 \rho^2 \cos^2 \phi$ and represents *astigmatism*, again extremely important in practical telescope optics.

The fifth order terms can be interpreted in a similar way as higher order effects, combining the basic types of aberration. The column $n = 0$ gives all axisymmetric combinations of spherical aberration and field curvature. The nomenclature of the other columns is somewhat arbitrary. There is general agreement that the columns with even values of n are "astigmatic" types, in combination with various orders of spherical aberration and field curvature. For example, $_2 k_{42} \sigma^2 \rho^4 \cos^2 \phi$ is normal fifth order astigmatism.

Some authors also call the types with odd values of n greater than 1 astigmatic. Personally, I prefer the term *"comatic"* for all odd values of n. Certainly, $n = 1$ is a comatic family, the term $_1 k_{51} \sigma \rho^5 \cos \phi$ being normal fifth order coma. But $_3 k_{31} \sigma^3 \rho^3 \cos \phi$ is also a form of fifth order coma. $_5 k_{11} \sigma^5 \rho \cos \phi$ is fifth order distortion. The fifth order term with $n = 3$, $_3 k_{33} \sigma^3 \rho^3 \cos^3 \phi$, is also a term with considerable importance in practical telescope optics, as we shall see in later chapters. We shall call it "triangular coma".

The field dependence in the aberrations, the power of σ, is of no consequence for the degrading effect on the image. Nevertheless, it is of great importance in telescope optics to know that, for third order aberrations, spherical aberration is independent of the field, coma linearly dependent, astigmatism dependent on the square, and distortion on the cube.

We shall return to these formulations later in connection with telescope testing and active control of telescope optics. It will also be necessary to compare the properties of the above Hamilton formulation with those of Zernike polynomials and "natural modes". But optical design procedures are still largely based on the classical Hamilton/Seidel formulation which has many virtues of simplicity and clear physical interpretation. For this reason we shall use it in the treatment of telescope systems in this chapter.

3.2.2 The Seidel approximation: third order aberration coefficients

The Hamilton formulation of Table 3.1 is of little *direct* use in optical design because the coefficients k are not known in terms of the constructional parameters of the optical system. This was the problem first solved by Seidel [3.4], and possibly, to some extent, earlier by Petzval.[1] An excellent summary is given by Bahner [3.5] which we take over here in an expanded form. However, the reader familiar with Bahner's admirable book should note that there are differences of sign arising from our use of the strict Cartesian sign convention of Welford [3.6]. The formulation is instructive, and nothing reveals better how a telescope form achieves its correction, and what its limitations are, than an analysis of the third order aberrations. However, such calculations are rarely performed by hand in an age when powerful optical design programs, often working with PC's, can calculate these aberrations reliably and

---

[1] The extent to which Petzval possessed a *practical* version of third order theory, as so admirably formulated by Seidel, has been hotly debated for over a hundred years, even before Petzval's death in 1891. His first, and only seriously scientific paper in 1843 [3.151], reveals a clear understanding of the Hamilton (Characteristic Function) basis, almost certainly independently developed, but gives no *explicit* formulae. Subsequent publications of Petzval are semi-popular and scientifically trivial in comparison. The best classical historical analysis of Petzval's work was given by von Rohr in 1899 [3.152]. He points out the excellent quality of Petzval's famous portrait objective, including balance of higher order spherical aberration at the remarkable relative aperture for that time of $f/3.4$. He also states that Petzval did *not* use, or indeed believe in, iterative trigonometrical ray tracing, although this had been first published in complete form in 1778 in the (then) well-known book by Klügel [3.153]. This is an astonishing and important piece of information: for if Petzval used no ray tracing, the only way he could have calculated such an excellent objective would have been by calculation of the aberration coefficients to the third order and, at least for spherical aberration, probably to the fifth order as well! Without explicit formulae such as those of Seidel, this would have been impossible. But von Rohr confirms the general view that Petzval never published such formulae or his means of deriving them, a tragedy for his historical reputation.

New light has recently been thrown on the subject through the admirable thesis of Rakich (see §3.6.5.1 and §3.7.2.4). Rakich quotes a paper given in 1900 [3.164] by the notable designer of photographic objectives, H.L. Aldis, which has remained unmentioned in the classical literature. Aldis gives equations for third order aberrations *which he attributes directly to Petzval*, without any mention of Seidel! However, he gives no reference to written work by Petzval, nor to any meeting with him with verbal information – Petzval had died nine years earlier in Vienna. After further study, I hope to publish an analysis of this Aldis paper, regarding Petzval's work, in 2004.

effortlessly. As in all calculations in geometrical optics, great care with the sign convention is always required.

A basic aspect of third order calculations must be emphasized from the start: the linear superposition of the effects of successive surfaces in the optical system. In other words, the Seidel surface coefficients $(S_q)_\nu$ are calculated from the formulae below to give by algebraic addition the Seidel sum for each aberration:

$$\sum S_q = \sum_\nu (S_q)_\nu , \tag{3.18}$$

where $\sum S_q$ is the total aberration for one of the five monochromatic Seidel aberrations and $(S_q)_\nu$ the contribution of surface ν.

Because the physical interpretation is the most direct, and because wave-front aberration (in general, not just for third order) can be added up algebraically through an optical system, we have given the formulation of the Characteristic Function above in terms of wavefront aberration. In telescope optics, both wavefront and lateral (angular) aberration (see Fig. 3.1) are essential: we shall constantly be making use of both systems and switching from one to the other. It is useful first to consider the *wavefront formulation* of the Seidel contributions. The derivation may be found in the works of Hopkins [3.3] and Welford [3.6]. The coefficients S_I, S_{II}, ... correspond to the third order constants $_0k_{40}$, $_1k_{31}$, ... of the Characteristic Function of Table 3.1. They are expressed in terms of the following paraxial parameters deduced from paraxial ray traces at each surface ν through the system:

y_ν: the paraxial ray height (often, but not necessarily, set equal to the actual marginal ray height of the beam)

A_ν: $n_\nu i_\nu = n'_\nu i'_\nu$, the Snell Invariant for the paraxial ray

$\overline{A}_\nu$: $n_\nu i_{pr\nu} = n'_\nu i'_{pr\nu}$, the Snell Invariant for the paraxial principal ray

$\Delta\left(\frac{u}{n}\right)_\nu$: $\left(\frac{u'}{n'} - \frac{u}{n}\right)_\nu$, the so-called "aplanatic" parameter

H_ν: $n_\nu u_\nu \eta_\nu = n'_\nu u'_\nu \eta'_\nu$, the Lagrange Invariant

$(P_c)_\nu$: $c_\nu \left(\frac{1}{n'} - \frac{1}{n}\right)_\nu = c_\nu \Delta\left(\frac{1}{n}\right)_\nu$, the Petzval sum of surface ν, where c_ν is the curvature $1/r_\nu$ of the surface

τ_ν: $c_\nu^3 (n'_\nu - n_\nu) b_{s\nu} y_\nu^4$, the influence of the aspheric form defined by the Schwarzschild constant $b_{s\nu}$

$(HE)_\nu$: $\left(\frac{y_{pr}}{y}\right)_\nu$, defining the effect of the pupil position relative to surface ν.

Then the wavefront aberration

$$W' = f(\sigma, \rho, \phi) \tag{3.19}$$

can be expressed in terms of the five Seidel coefficients defined by:

$$\sum S_I = -\left[\sum_\nu A^2 y \Delta\left(\tfrac{u}{n}\right)\right] + \sum_\nu \tau \qquad \text{--spherical aberration}$$

$$\sum S_{II} = -\left[\sum_\nu A\bar{A} y \Delta\left(\tfrac{u}{n}\right)\right] + \sum_\nu (HE)\tau \qquad \text{--coma}$$

$$\sum S_{III} = -\left[\sum_\nu \bar{A}^2 y \Delta\left(\tfrac{u}{n}\right)\right] + \sum_\nu (HE)^2 \tau \qquad \text{--astigmatism}$$

$$\sum S_{IV} = -\left[\sum_\nu H^2 P_c\right] \qquad \text{--field curvature}$$

$$\sum S_V = -\left[\sum_\nu \tfrac{\bar{A}}{A}\left(H^2 P_c + \bar{A}^2 y \Delta\left(\tfrac{u}{n}\right)\right)\right]$$
$$\qquad + \sum_\nu (HE)^3 \tau \qquad \text{--distortion}$$

$$(3.20)$$

This formulation will enable us to transform Eq. (3.19) into an explicit form (Eq. (3.21) giving the corresponding coefficients of the Characteristic Function by tracing a paraxial ray and a paraxial principal ray through the system). Only in specific, simple cases is it feasible to express the Seidel coefficients in terms of the basic constructional parameters of the system, rather than the ray-trace derived parameters used in Eq. (3.20). Reflecting telescopes consisting of one or two mirrors are a case where explicit formulation is feasible.

The formulation of the Seidel coefficients of Eqs. (3.20) and the definition of the parameters involved given above, are identical with those of Welford [3.6]. Conventionally, field curvature S_{IV} is defined as positive for a thin positive (biconvex) lens giving a real image. However, the curved image surface is concave to the incident light, i.e. negative in the Cartesian sense. The same property will hold for telescope systems with a positive focal length. Because of the close interplay between field curvature and astigmatism, we shall retain the Welford sign convention of S_{IV} to maintain consistency in the total formulation of the aberrations. The sign convention of Eqs. (3.20) is based on the consistent principle that the *wavefront* aberration be positive if the coefficients are positive, i.e. a positive phase shift in the direction of the light as shown in Fig. 3.1. It is not possible to have a consistent Cartesian regime of aberrations in all three systems of aberration definition: wavefront, lateral, longitudinal. Of these, priority is given to the wavefront definition since it is physically the most meaningful.

3.2.3 Seidel coefficients of some basic reflecting telescope systems

The paraxial aperture ray and paraxial principal ray are traced according to Eqs. (2.36) to (2.38). Conventionally, the starting values for such ray traces are the same values as for the traces of the real rays according to the exact formulae [3.6] [3.7], as discussed in § 2.2.3, where it was shown that the linear nature of the paraxial equations makes it immaterial for the result what starting values are taken. Following Bahner [3.5] and others, we shall therefore derive the Seidel coefficients for some *normalized* telescope systems, in air, for which

$$
\begin{aligned}
f' &= \pm 1 \\
y_1 &= +1 \text{ (object at } \infty) \\
u_{pr1} &= +1 \\
\eta' &= \pm 1 \text{ (since } \eta' = f'u_{pr1} \text{ in air)} \\
H &= n'u'\eta' = -1
\end{aligned}
$$

In these examples, the entrance pupil is taken to be at the conventional position, the primary mirror. This normalization is practical since it allows direct comparison of aberration values for different systems.

The normalized starting values for the paraxial aperture ray (y_1) and the paraxial principal ray (u_{pr1}) given above correspond to a semi-aperture of $f/0.5$ and a semi-field of 1 radian, values far exceeding those of most real, practical systems. The resulting coefficients can easily be converted into the real coefficients of given systems by using the aperture and field dependencies of the third order aberrations as given by the Characteristic Function of Eq. (3.15) and Table 3.1. The conversion of the coefficients of Eq. (3.20) to wavefront aberration for third order aberrations W_3' is then given from the normalized parameters y_{m1}, η_m' to the real parameters y_1, η' by (see [3.3] [3.6]):

$$
\begin{aligned}
W_3'(y_1, \eta') = &\frac{1}{8}\left(\frac{y_1}{y_{m1}}\right)^4 \sum S_I + \frac{1}{2}\left(\frac{y_1}{y_{m1}}\right)^3 \left(\frac{\eta'}{\eta_m'}\right)\sum S_{II}\cos\phi \\
&+\frac{1}{4}\left(\frac{y_1}{y_{m1}}\right)^2\left(\frac{\eta'}{\eta_m'}\right)^2\left[(3\sum S_{III} + \sum S_{IV})\cos^2\phi\right. \\
&\qquad\qquad\left. + (\sum S_{III} + \sum S_{IV})\sin^2\phi\right] \\
&+\frac{1}{2}\left(\frac{y_1}{y_{m1}}\right)\left(\frac{\eta'}{\eta_m'}\right)^3\sum S_V\cos\phi
\end{aligned}
\right\} \quad (3.21)
$$

Tables 3.2 and 3.3 give, respectively, the paraxial data and third order aberration coefficients for the nine cases shown. The significance of the results will be discussed in the next sections in connection with the analytical formulations. The values in the tables can be used as a reference for checking the correct use of these formulations. Distortion $(\sum S_V)$ is not given as it is normally of little significance, but it can easily be derived from the values given and from Eqs. (3.20).

Equations (3.20) are the fundamental formulae of the aberration theory of telescopes. We saw in the definitions of the quantities involved that the parameter $(HE)_\nu$ defines the effect of the pupil position relative to the surface ν, and is determined from the paraxial principal ray trace, giving $(y_{pr})_\nu$, and the paraxial aperture ray trace, giving y_ν. This concept leads to an important set of equations known as the *stop-shift formulae* (see [3.3] [3.6]). These define the effect of a stop shift, or pupil shift, in any centered optical system. In these formulae, $\sum S_I$, $\sum S_{II}$, etc. are the values for the original stop position

Table 3.2. Paraxial values for deriving the Seidel coefficients (Table 3.3) for some basic telescope systems

Case	Surface ν	r_ν	$(b_s)_\nu$	d_ν	y_ν	A_ν	$\bar{A}_\nu$	$\left(\frac{A}{\bar{A}}\right)_\nu$	$\Delta\left(\frac{u}{n}\right)_\nu$	τ_ν	$(HE)_\nu$
1. Spherical mirror (EP at primary)	1	-2.0	0	–	+1.0	-0.5	+1.0	-2.0	-1.0	0	0
2. Parabolic mirror (EP at primary)	1	-2.0	-1.0	–	+1.0	-0.5	+1.0	-2.0	-1.0	-0.25	0
3. Classical Cassegrain telescope (EP at primary) $m_2 = -4$	1	-0.5	-1.0	-0.19375	+1.0	-2.0	+1.0	-0.50	-4.0	-16.0	0
	2	-0.150	-2.77778	–	+0.22500	-2.50	+2.29167	-0.916668	+3.0	+4.21875	+0.86111
4. Ritchey–Chrétien (RC) Cassegrain telescope (EP at primary) $m_2 = -4$	1	-0.5	-1.03629	-0.19375	+1.0	-2.0	+1.0	-0.50	-4.0	-16.58064	0
	2	-0.150	-3.16010	–	+0.22500	-2.5	+2.29167	-0.916668	+3.0	+4.79940	+0.86111
5. Dall–Kirkham (DK) Cassegrain telescope (EP at primary) $m_2 = -4$	1	-0.5	-0.73633	-0.19375	+1.0	-2.0	+1.0	-0.50	-4.0	-11.78125	0
	2	-0.150	0	–	+0.22500	-2.5	+2.29167	-0.916668	+3.0	0	+0.86111
6. Spherical Primary (SP) Cassegrain telescope (EP at primary) $m_2 = -4$	1	-0.5	0	-0.19375	+1.0	-2.0	+1.0	-0.50	-4.0	0	0
	2	-0.150	+7.75719	–	+0.225	-2.5	+2.29167	-0.916668	+3.0	-11.78125	+0.86111
7. Classical Gregory telescope (EP at primary) $m_2 = +4$	1	-0.5	-1.0	-0.30625	+1.0	-2.0	+1.0	-0.50	-4.0	-16.0	0
	2	+0.090	-0.36000	–	-0.225	-1.50	-2.40278	+1.60185	+5.0	-2.53125	-1.36111
8. Aplanatic Gregory telescope (EP at primary) $m_2 = +4$	1	-0.5	-0.97704	-0.30625	+1.0	-2.0	+1.0	-0.50	-4.0	-15.63265	0
	2	+0.090	-0.41224	–	-0.225	-1.50	-2.40278	+1.60185	+5.0	-2.89860	-1.36111
9. 3-mirror system of Korsch–Design I (EP at primary) $m_2 = -10$ $m_3 = -0.15$	1	-1.333333	-1.262939	-0.500000	+1.0	-0.750000	+1.0	-1.333333	-1.50	-1.065604	0
	2	-0.370370	-2.843216	+0.213675	+0.250000	-0.825000	+2.350000	-2.848485	+1.350000	+0.437211	+2.0
	3	-0.512821	-1.401477	(-0.217949 to image)	+0.217949	-0.575000	+1.183334	-2.057973	-0.850000	-0.048962	+5.921566

Table 3.3. Seidel coefficients for some basic telescope systems. The asterisk denotes the aspheric contribution

(In each asterisked surface row the upper number is the spherical contribution and the lower number is the aspheric contribution.)

Case	Surface ν	$(S_I)_\nu$	$(S_{II})_\nu$	$(S_{III})_\nu$	$(S_{IV})_\nu = -(P_c)_\nu$	Effective field curvature $2(S_{III})_\nu + (S_{IV})_\nu$
1. Spherical mirror (EP at primary)	1	+0.25	−0.5	+1.0	−1.0	+1.0
2. Parabolic mirror (EP at primary)	1*	+0.25 −0.25	−0.5 0	+1.0 0	−1.0 0	+1.0 0
	Sum	0	−0.5	+1.0	−1.0	+1.0
3. Classical Cassegrain telescope (EP at primary) $m_2 = -4$	1*	+16.0 −16.0	−8.0 0	+4.0 0	−4.0 0	+4.0 0
	2	−4.21875	+3.86719	−3.54493	+13.33333	+6.24347
	2*	+4.21875	+3.63281	+3.12825	0	+6.25650
	Sum	0	−0.50000	+3.58332	+9.33333	+16.49997
4. Ritchey-Chrétien (RC) Cassegrain telescope (EP at primary) $m_2 = -4$	1*	+16.0 −16.58064	−8.0 0	+4.0 0	−4.0 0	+4.0 0
	2	−4.21875	+3.86719	−3.54493	+13.33333	+6.24347
	2*	+4.79940	+4.13281	+3.55881	0	+7.11762
	Sum	0	0	+4.01388	+9.33333	+17.36109
5. Dall-Kirkham (DK) Cassegrain telescope (EP at primary) $m_2 = -4$	1*	+16.0 −11.78125	−8.0 0	+4.0 0	−4.0 0	+4.0 0
	2	−4.21875	+3.86719	−3.54493	+13.33333	+6.24347
	Sum	0	−4.13281	+0.45507	+9.33333	+10.24347
6. Spherical primary (SP) Cassegrain telescope (EP at primary) $m_2 = -4$	1	+16.0	−8.0	+4.0	−4.0	+4.0
	2	−4.21875	+3.86719	−3.54493	+13.33333	+6.24347
	2*	−11.78125	−10.14495	−8.73592	0	−17.47184
	Sum	0	−14.27776	−8.28085	+9.33333	−7.22837
7. Classical Gregory telescope (EP at primary) $m_2 = +4$	1*	+16.0 −16.0	−8.0 0	+4.0 0	−4.0 0	+4.0 0
	2	+2.53125	+4.05469	+6.49502	−22.22222	−9.23218
	2*	−2.53125	+3.44531	−4.68945	0	−9.37890
	Sum	0	−0.50000	+5.80557	−26.22222	−14.61108
8. Aplanatic Gregory telescope (EP at primary) $m_2 = +4$	1*	+16.0 −15.63265	−8.0 0	+4.0 0	−4.0 0	+4.0 0
	2	+2.53125	+4.05469	+6.49502	−22.22222	−9.23218
	2*	−2.89860	+3.94531	−5.37000	0	−10.74000
	Sum	0	0	+5.12502	−26.22222	−15.97218
9. 3-mirror system of Korsch–Design I (EP at primary) $m_2 = -10$ $m_3 = -0.15$	1*	+0.843750 −1.065604	−1.125000 0	+1.500000 0	−1.500000 0	+1.500000 0
	2*	−0.229711 +0.437211	+0.654328 +0.874422	−1.863844 +1.748844	+5.400000 0	+1.672312 +3.497688
	3*	+0.061250 −0.046896	−0.126052 −0.277699	+0.259411 −1.644412	−3.900000 0	−3.381178 −3.288824
	Sum	0	0	0	0	0

and $\sum S_I^+$, $\sum S_{II}^+$, etc. are the values resulting after the stop shift ∂E. Since, from the definitions above Eq. (3.19)

$$(HE)_\nu = \left(\frac{y_{pr}}{y}\right)_\nu ,$$

it follows that the factor $(H\partial E)_\nu$ is simply $\left[\frac{(\partial y_{pr})}{y}\right]_\nu$ because y_ν is independent of the stop position. The stop-shift formulae are given by:

$$\left.\begin{array}{rcl}
\sum_\nu S_I^+ & = & \sum_\nu S_I \\
\sum_\nu S_{II}^+ & = & \sum_\nu S_{II} + \sum_\nu (H\partial E)S_I \\
\sum_\nu S_{III}^+ & = & \sum_\nu S_{III} + \sum_\nu 2(H\partial E)S_{II} + \sum_\nu (H\partial E)^2 S_I \\
\sum_\nu S_{IV}^+ & = & \sum_\nu S_{IV} \\
\sum_\nu S_V^+ & = & \sum_\nu S_V + \sum_\nu (H\partial E)(S_{IV} + 3S_{III}) \\
& & + \sum_\nu 3(H\partial E)^2 S_{II} + \sum_\nu (H\partial E)^3 S_I
\end{array}\right\} \quad (3.22)$$

Thus, to third order accuracy, the coma coefficient $\sum S_{II}^+$ is independent of the stop position if the spherical aberration coefficient $\sum S_I$ is zero; and the astigmatism coefficient $\sum S_{III}^+$ is independent of the stop position if both $\sum S_I$ and $\sum S_{II}$ are zero. These principles are of great significance in understanding the effect of different stop positions in various telescope forms treated below.

In telescope optics, the *first order aberrations* of Table 3.1 are also of great practical significance for the reasons given in § 3.2.1 above. It can easily be shown [3.3] [3.6] that these can be expressed as wavefront aberrations in terms of the lateral and longitudinal focus shifts $\delta\eta'$ and δz of the centre of curvature of the image forming wavefront by

$$W_1'(y_1, \eta') = \left(\frac{y_1}{y_{m1}}\right)(n'u'\delta\eta')\cos\phi + \left(\frac{y_1}{y_{m1}}\right)^2 \left(-\frac{1}{2}n'u'^2\delta z\right) \quad (3.23)$$

The normalizing functions of $\frac{y_1}{y_{m1}}$ are given again from Eq. (3.15) and Table 3.1.

These first order aberrations play no part in the formal aberration theory of telescopes, which we now develop in detail below, since they are theoretically defined as zero. Their significance will be clear in RTO II, Chap. 2 and 3 where we are concerned with the adjustment and active control of telescope optics.

3.2.4 Analytical (third order) theory for 1-mirror and 2-mirror telescopes

3.2.4.1 General definitions. The results of Table 3.3 are derived directly from the values of the paraxial ray traces. Before commenting on the results, it is better to develop the general analytical theory of such telescope systems, giving the third order aberrations in terms of basic system parameters without the need to trace paraxial rays.

Table 3.3 normalized the system data with

$$y_1 = +1, \quad u_{pr1} = +1, \quad f' = \pm 1, \quad \eta' = \pm 1, \quad H = n'u'\eta' = -1$$

Bahner [3.5] derives the analytical formulae using this normalization, a practice which has frequently been followed and produces a simplification in the derivation. However, by omitting the parameters y_1 and f', defined as 1, this simplified formulation is no longer dimensionally correct. It also leads to confusion between the parameters y_2 and L, which have the same numerical value with the normalization from Eq. (2.58). Furthermore, if the paraxial relations of Chap. 2 are applied appropriately, the general (i.e. non-normalized) procedure becomes quite reasonable. This is now given, the form of the relations being similar to that of Bahner, but in generalised form, and with sign changes arising from our use of the strict Cartesian sign convention of modern optical design. The relations are thus perfectly general if the sign convention of Eqs. (2.36) to (2.38) and of Tables 2.1 and 2.2 is rigorously followed.

Equations (3.20) can be re-written in the form

$$\left. \begin{aligned}
\sum S_I &= \sum_\nu S_I^0 + \sum_\nu S_I^* = -\left[\sum_\nu A^2 y \Delta\left(\tfrac{u}{n}\right)\right] + \sum_\nu \tau \\
\sum S_{II} &= \sum_\nu (\tfrac{\overline{A}}{A}) S_I^0 + \sum_\nu (HE) S_I^* = \sum_\nu S_{II}^0 + \sum_\nu S_{II}^* \\
\sum S_{III} &= \sum_\nu (\tfrac{\overline{A}}{A}) S_{II}^0 + \sum_\nu (HE) S_{II}^* = \sum_\nu S_{III}^0 + \sum_\nu S_{III}^* \\
\sum S_{IV} &= -\sum_\nu H^2 P_c \\
\sum S_V &= \sum_\nu (\tfrac{\overline{A}}{A})(S_{III}^0 - H^2 P_c) + \sum_\nu (HE) S_{III}^*
\end{aligned} \right\} , \quad (3.24)$$

in which S_I^0, S_{II}^0, ... give the contributions due to a spherical surface, S_I^*, S_{II}^*, ... those due to the aspheric form. The quantities A, y and $\Delta(\tfrac{u}{n})$ for S_I^0 are derived from the paraxial aperture ray with Eqs. (2.36) to (2.38), while the multipliers $\tfrac{\overline{A}}{A}$ and HE require $\overline{A}$ and y_{pr} from the paraxial principal ray. In Tables 3.2 and 3.3, the entrance pupil was defined for the normal case as being at the primary mirror, i.e. $s_{pr1} = 0$ in Eq. (2.36). However, this limitation is not acceptable for a general formulation, for which the entrance pupil is at s_{pr1} from the first surface (primary). If the entrance pupil is to the left of the primary (i.e. in front of it), s_{pr1} is negative.

The other quantities required for the evaluation of (3.24) are the aspheric parameter τ, the Petzval curvature P_c and the Lagrange Invariant H.

Because practical use of telescopes is almost always confined to objects effectively at infinity, we retain this limitation here. However, the informed reader may prefer the general recursion formulae given in §3.6.5.2, below, for calculating the aberrations of any system of $\tilde{n}$ mirrors under general conditions.

3.2.4.2   1-mirror telescopes: single concave primary. The term "1-mirror telescopes" implies, of course, a single mirror with optical power and covers all folded forms with plane mirrors.

The paraxial ray trace equations (2.36) to (2.38) give with the definitions above Eq. (3.19):

$$
\left.
\begin{aligned}
A_1 &= \frac{1}{2}\frac{y_1}{f_1'}, & \Delta\left(\frac{u}{n}\right)_1 &= \frac{y_1}{f_1'} \\[2mm]
\overline{A}_1 &= -\frac{(s_{pr1}-2f_1')}{2f_1'}u_{pr1}, & y_{pr1} &= -s_{pr1}u_{pr1} \\[2mm]
\left(\frac{\overline{A}}{A}\right)_1 &= -\frac{(s_{pr1}-2f_1')}{y_1}u_{pr1}, & (HE)_1 &= -\frac{s_{pr1}}{y_1}u_{pr1} \\[2mm]
\tau_1 &= -\frac{1}{4}b_{s1}\left(\frac{y_1}{f_1'}\right)^4 f_1' \\[2mm]
P_{c1} &= -\frac{1}{f_1'}
\end{aligned}
\right\} \quad (3.25)
$$

If the above quantities are introduced into Eqs. (3.24), we can derive with minimal reduction:

$$
\left.
\begin{aligned}
(S_I)_1 &= -\left(\frac{y_1}{f_1'}\right)^4 \frac{f_1'}{4}(1+b_{s1}) \\[2mm]
(S_{II})_1 &= -\left(\frac{y_1}{f_1'}\right)^3 \frac{1}{4}\left[2f_1'-s_{pr1}(1+b_{s1})\right]u_{pr1} \\[2mm]
(S_{III})_1 &= -\left(\frac{y_1}{f_1'}\right)^2 \frac{1}{4f_1'}\left[4f_1'(f_1'-s_{pr1})+s_{pr1}^2(1+b_{s1})\right]u_{pr1}^2 \\[2mm]
(S_{IV})_1 &= +\frac{H^2}{f_1'}
\end{aligned}
\right\} \quad (3.26)
$$

The dimensions are length throughout, since the coefficients each represent a linear wavefront shift in the z-axis direction or along the principal ray. The various powers of (y_1/f_1') are the same as the aperture factors in Eq. (3.21). Similarly, the various powers of u_{pr1} correspond in angular measure to the image height factors in (3.21).

Another important property is that the terms in s_{pr1} and s_{pr1}^2 in the equations for $(S_{II})_1$ and $(S_{III})_1$ are simply the "stop-shift terms" containing corresponding powers of ∂E in Eq. (3.22), since the "stop shift" from the primary is expressed simply by $-s_{pr1}/y_1$ from (3.25).

In § 3.2.6, we shall discuss the consequences of Eqs. (3.26) for the design of 1-mirror telescopes.

3.2.4.3   2-mirror telescopes. *First surface (primary mirror):* We can take over the results of Eqs. (3.25) directly for the contributions of the primary to the total aberration. However, it is usually convenient to normalize to the final focal length f' of the complete 2-mirror system rather than to retain the primary focal length f_1'. From (2.55) $f_1' = \frac{f'}{m_2}$, giving

$$
\left.
\begin{aligned}
A_1 &= \frac{1}{2} m_2 \frac{y_1}{f'}, & \Delta\left(\frac{u}{n}\right)_1 &= m_2 \frac{y_1}{f'} \\[2mm]
\overline{A}_1 &= \left(1 - \frac{m_2 s_{pr1}}{2f'}\right) u_{pr1}, & y_{pr1} &= -s_{pr1} u_{pr1} \\[2mm]
\left(\frac{\overline{A}}{A}\right)_1 &= \left(\frac{f'}{y_1}\right)\left(\frac{2}{m_2} - \frac{s_{pr1}}{f'}\right) u_{pr1}, & (HE)_1 &= -\left(\frac{f'}{y_1}\right)\frac{s_{pr1}}{f'} u_{pr1} \\[2mm]
\tau_1 &= -\frac{1}{4} b_{s1} m_2^3 \left(\frac{y_1}{f'}\right)^4 f' \\[2mm]
P_{c1} &= -\frac{1}{f_1'} = -\frac{m_2}{f'}
\end{aligned}
\right\}
$$

$$(3.27)$$

From (3.26), we have for $(S_I)_1$

$$
(S_I)_1 = -\left(\frac{y_1}{f'}\right)^4 f' \frac{m_2^3}{4}(1 + b_{s1}) = (S_I)_1^0 + (S_I)_1^*, \tag{3.28}
$$

with

$$
\left.
\begin{aligned}
(S_I)_1^0 &= -\left(\frac{y_1}{f'}\right)^4 f' \zeta^0 \\[2mm]
(S_I)_1^* &= -\left(\frac{y_1}{f'}\right)^4 f' \zeta^*
\end{aligned}
\right\}
\tag{3.29}
$$

if we define, following Bahner [3.5] and others,

$$
\zeta = \zeta^0 + \zeta^* = \frac{m_2^3}{4} + \frac{m_2^3}{4} b_{s1} = \frac{m_2^3}{4}(1 + b_{s1}) \tag{3.30}
$$

Then from (3.28)

$$
(S_I)_1 = -\left(\frac{y_1}{f'}\right)^4 f' \zeta \tag{3.31}
$$

If the parameter ζ is introduced in the equations for $(S_{II})_1$ and $(S_{III})_1$ in (3.26) and f_1' replaced by $\frac{f'}{m_2}$, we deduce at once the set of aberration coefficients for the primary as

$$(S_I)_1 \quad = \quad -\left(\frac{y_1}{f'}\right)^4 f'\zeta$$

$$(S_{II})_1 \quad = \quad -\left(\frac{y_1}{f'}\right)^3 \left[m_2^2 \frac{f'}{2} - s_{pr1}\zeta\right]u_{pr1}$$

$$(S_{III})_1 \quad = \quad -\left(\frac{y_1}{f'}\right)^2 \left[m_2 f' - m_2^2 s_{pr1} + \frac{s_{pr1}^2}{f'}\zeta\right]u_{pr1}^2$$

$$(S_{IV})_1 \quad = \quad +\frac{H^2}{f_1'} \quad = \quad +H^2\frac{m_2}{f'}$$

(3.32)

Second surface (secondary mirror): The analytical expressions resulting from the paraxial aperture ray trace become appreciably more complex when extended with Eqs. (2.36) to (2.38) to the second surface. This complexity is even more marked for the paraxial principal ray. We give here only the basic steps in the reduction. The first essential parameters are:

$$A_2 \quad = \quad -\left(\frac{y_1}{f_1'}\right)\left(\frac{f_1' - d_1 - 2f_2'}{2f_2'}\right), \quad y_2 = \left(\frac{y_1}{f_1'}\right)(f_1' - d_1)$$

$$\Delta\left(\frac{u}{n}\right)_2 \quad = \quad -\left(\frac{y_1}{f_1'}\right)\left(\frac{f_1' - d_1}{f_2'}\right), \quad \tau_2 = \left(\frac{y_1}{f_1'}\right)^4 \frac{(f_1' - d_1)^4}{4f_2'^3}b_{s2}$$

(3.33)

The spherical aberration contribution $(S_I)_2$ is then calculated from (3.24) as a function of the quantities appearing on the right sides of Eqs. (3.33). This form can be greatly simplified by introducing from Eq. (2.72)

$$f_1' - d_1 = \frac{L}{m_2}$$

(3.34)

and from Eq. (2.90)

$$f_2' = \frac{L}{m_2 + 1}$$

(3.35)

The parameter L, the back focal distance from the secondary, is a very convenient parameter for the final expressions, together with the system focal length f' and the secondary magnification m_2. The reduction gives

$$(S_I)_2 = (S_I)_2^0 + (S_I)_2^*$$

(3.36)

with

$$(S_I)_2^0 = \left(\frac{y_1}{f'}\right)^4 L\frac{(m_2 + 1)^3}{4}\left(\frac{m_2 - 1}{m_2 + 1}\right)^2$$

(3.37)

$$(S_I)_2^* = \left(\frac{y_1}{f'}\right)^4 L\frac{(m_2 + 1)^3}{4}b_{s2}$$

(3.38)

Combining these gives

$$(S_I)_2 = \left(\frac{y_1}{f'}\right)^4 L\frac{(m_2+1)^3}{4}\left[\left(\frac{m_2-1}{m_2+1}\right)^2 + b_{s2}\right] \tag{3.39}$$

By analogy with the definition of ζ above for the primary mirror in Eq. (3.30), we now define

$$\xi = \xi^0 + \xi^* = \frac{(m_2+1)^3}{4}\left(\frac{m_2-1}{m_2+1}\right)^2 + \frac{(m_2+1)^3}{4}b_{s2} \,, \tag{3.40}$$

giving

$$\xi = \frac{(m_2+1)^3}{4}\left[\left(\frac{m_2-1}{m_2+1}\right)^2 + b_{s2}\right] \tag{3.41}$$

and

$$(S_I)_2^0 = \left(\frac{y_1}{f'}\right)^4 L\xi^0, \quad (S_I)_2^* = \left(\frac{y_1}{f'}\right)^4 L\xi^* \tag{3.42}$$

We have, finally

$$(S_I)_2 = \left(\frac{y_1}{f'}\right)^4 L\xi \tag{3.43}$$

In order to determine $(S_{II})_2$ and $(S_{III})_2$ from Eq. (3.24), we still require the parameters $\left(\frac{\overline{A}}{A}\right)_2$ and $(HE)_2$ from the paraxial principal ray. Eqs. (2.36) to (2.38) give

$$y_{pr2} = \frac{\left[s_{pr1}(d_1 - f_1') - d_1 f_1'\right]}{f_1'}u_{pr1} \tag{3.44}$$

and

$$\overline{A}_2 = \frac{\left[s_{pr1}(f_1' - d_1) + d_1 f_1' - 2f_2'(s_{pr1} - f_1')\right]}{2f_1'f_2'}u_{pr1} \tag{3.45}$$

Now substituting (3.34) and (3.35) in the expression for A_2 in (3.33), this reduces to

$$A_2 = \left(\frac{y_1}{f'}\right)\left(\frac{m_2-1}{2}\right) \,, \tag{3.46}$$

if also f_1' is replaced by $\frac{f'}{m_2}$ from (2.55). Similar substitutions in (3.45) lead to

$$\overline{A}_2 = \frac{\left[-s_{pr1}L(m_2-1) + 2f'^2 - d_1 f'(m_2-1)\right]}{2Lf'}u_{pr1} \tag{3.47}$$

Then from (3.46) and (3.47), we have finally

$$\left(\overline{\frac{A}{A}}\right)_2 = \left(\frac{f'}{y_1}\right)\left[-\frac{d_1}{L} + \frac{2f'}{L(m_2 - 1)} - \frac{s_{pr1}}{f'}\right]u_{pr1} \tag{3.48}$$

From the definitions above Eq. (3.19), we have

$$(HE)_2 = \frac{y_{pr2}}{y_2} \tag{3.49}$$

y_{pr2} and y_2 are given in Eqs. (3.44) and (3.33) respectively. Substituting again from (3.34) and (2.55) for f'_1, these equations reduce easily to

$$(HE)_2 = \left(\frac{f'}{y_1}\right)\left[-\frac{d_1}{L} - \frac{s_{pr1}}{f'}\right]u_{pr1} \tag{3.50}$$

For the field curvature $(S_{IV})_2$, we require $(P_c)_2$. From the definitions above Eq. (3.19), this is simply

$$(P_c)_2 = +\frac{1}{f'_2} = \frac{m_2 + 1}{L} = \frac{m_2 + 1}{f' - m_2 d_1} \tag{3.51}$$

from (2.55), (3.34) and (3.35).

The second equation of (3.24) can now be applied to the calculation of the coma contribution $(S_{II})_2$ using Eqs. (3.42) for $(S_I)_2^0$ and $(S_I)_2^*$ and (3.48) and (3.50) for $\left(\overline{\frac{A}{A}}\right)_2$ and $(HE)_2$. The reduction is made quite simple by the fact that the form of the equation (3.48) is the same as (3.50) except for an additional term, and that ξ^0 and ξ^* have a common factor. The result is

$$(S_{II})_2 = \left(\frac{y_1}{f'}\right)^3\left[-d_1\xi + \frac{f'}{2}(m_2^2 - 1) - s_{pr1}\frac{L}{f'}\xi\right]u_{pr1}, \tag{3.52}$$

in which the components $(S_{II})_2^0$ and $(S_{II})_2^*$ are given by

$$(S_{II})_2^0 = \left(\frac{y_1}{f'}\right)^3\xi^0\left[-d_1 - \frac{L}{f'}s_{pr1} + \frac{2f'}{(m_2 - 1)}\right]u_{pr1} \tag{3.53}$$

and

$$(S_{II})_2^* = \left(\frac{y_1}{f'}\right)^3\xi^*\left[-d_1 - \frac{L}{f'}s_{pr1}\right]u_{pr1} \tag{3.54}$$

These are applied in the third equation of (3.24), combined with (3.48) and (3.50) to determine $(S_{III})_2$. This leads to

$$(S_{III})_2^0 = \left(\frac{y_1}{f'}\right)^2\xi^0 L\left[\left(\frac{-d_1}{L} - \frac{s_{pr1}}{f'}\right) + \frac{2f'}{L(m_2 - 1)}\right]^2 u_{pr1}^2 \tag{3.55}$$

$$(S_{III})_2^* = \left(\frac{y_1}{f'}\right)^2\xi^* L\left(\frac{-d_1}{L} - \frac{s_{pr1}}{f'}\right)^2 u_{pr1}^2 \tag{3.56}$$

The sum of these, giving $(S_{III})_2$, involves reduction using (3.40) for ξ^0 and ξ^*, leading to

$$
(S_{III})_2 = \left(\frac{y_1}{f'}\right)^2 L \left\{ \left(\frac{-d_1}{L} - \frac{s_{pr1}}{f'}\right)^2 \xi + \left[\frac{-f'd_1}{L^2}(m_2^2 - 1)\right. \right.
$$
$$
\left. \left. - \frac{s_{pr1}}{L}(m_2^2 - 1) + \left(\frac{f'}{L}\right)^2 (m_2 + 1)\right] \right\} u_{pr1}^2
\tag{3.57}
$$

The final surface contribution is $(S_{IV})_2$, which is given directly by (3.51) from (3.24) as

$$
\left.
\begin{aligned}
(S_{IV})_2 &= -H^2 \left(\frac{1}{f_2'}\right) & = -H^2 \left(\frac{m_2 + 1}{L}\right) \\
&= -H^2 \left(\frac{m_2 + 1}{f' - m_2 d_1}\right) & = -H^2 \left(\frac{m_2^2 - 1}{m_2 P}\right)
\end{aligned}
\right\} ,
\tag{3.58}
$$

the last form being derived from Eq. (2.84) and giving the curvature contribution in terms of P, the separation of the primary and secondary images (Figs. 2.11 and 2.12), a parameter frequently pre-defined in a telescope system.

It remains to determine the total aberrations of the 2-mirror telescope by adding the contributions of the first and second surfaces.

From (3.32) and (3.43), the spherical aberration $\sum S_I$ is given by

$$
\sum S_I = (S_I)_1 + (S_I)_2 = \left(\frac{y_1}{f'}\right)^4 \left[-f'\zeta + L\xi\right]
\tag{3.59}
$$

The coma $\sum S_{II}$ is given directly from (3.32) and (3.52) as

$$
\sum S_{II} = (S_{II})_1 + (S_{II})_2 = \left(\frac{y_1}{f'}\right)^3 \left[-d_1\xi - \frac{f'}{2} - \frac{s_{pr1}}{f'}(-f'\zeta + L\xi)\right] u_{pr1}
\tag{3.60}
$$

The astigmatism $\sum S_{III}$ is given from (3.32) and (3.57). If, in the reduction, the relation from (3.34) and (2.55)

$$
f' = L + m_2 d_1
$$

is used, then all the terms in m_2 vanish and we have finally

$$
\left.
\begin{aligned}
\sum S_{III} &= (S_{III})_1 + (S_{III})_2 \\
&= \left(\frac{y_1}{f'}\right)^2 \left[\frac{f'}{L}(f' + d_1) + \frac{d_1^2}{L}\xi\right. \\
&\quad \left. + s_{pr1}\left(1 + \frac{2d_1}{f'}\xi\right) + \left(\frac{s_{pr1}}{f'}\right)^2 (-f'\zeta + L\xi)\right] u_{pr1}^2
\end{aligned}
\right\}
\tag{3.61}
$$

For the field curvature $\sum S_{IV}$, we have from (3.32) and (3.58):

$$
\begin{aligned}
\sum S_{IV} &= H^2 \left(\frac{1}{f_1'} - \frac{1}{f_2'} \right) = H^2 \left[\left(\frac{m_2}{f'} \right) - \left(\frac{m_2 + 1}{L} \right) \right] \\
&= H^2 \left[\left(\frac{m_2}{f'} \right) - \left(\frac{m_2 + 1}{f' - m_2 d_1} \right) \right] = H^2 \left[\left(\frac{m_2}{f'} \right) - \left(\frac{m_2^2 - 1}{m_2 P} \right) \right]
\end{aligned}
$$
(3.62)

If Eqs. (3.59), (3.60) and (3.61) are compared with the stop-shift formulae of (3.22), it will be seen that there is exact equivalence, the quantities S_I and S_{II} in (3.22) being the normalized quantities $(-f'\zeta + L\xi)$ and $(-d_1\xi - f'/2)$ in (3.59) and (3.60) respectively and $(H\partial E)$ being the normalized stop shift (s_{pr1}/f').

The form of the above equations (3.32) for the primary mirror and (3.59) to (3.62) for a 2-mirror telescope is essentially that given by Bahner [3.5] except that the formulae given here are general and not limited to normalized parameters as used in Table 3.3. Also, the sign convention differs with our use of the general Cartesian system of optical design defined in Tables 2.1, 2.2 and 2.3. Equations of this sort were first set up by K. Schwarzschild [3.1] in his classic paper of 1905, arguably the most fundamental and important paper ever written in telescope optics. Schwarzschild's formulation retained parameters equivalent to A and $\overline{A}$, so that certain properties were less evident than they are in the form given in Eqs. (3.59) to (3.62) above. We shall return to this matter in § 3.2.6, where individual solutions arising from the equations are considered.

In § 2.2.5.2 it was pointed out that the initial layout of a 2-mirror telescope is usually established with the parameters

y_1, f_1', m_2 (giving f') and P,

whereby the resulting axial obstruction ratio R_A must also be acceptable. The parameters d_1 and L used in Eqs. (3.59) to (3.62) are therefore *derived* parameters resulting from the basic ones. The conversion from d_1 and L to the basic parameters above, if these are preferred, is easily performed from two relations derived from Eq. (2.72):

$$
L = R_A f' = R_A m_2 f_1'
$$
(3.63)

$$
d_1 = f_1'(1 - R_A) = \frac{f'}{m_2}(1 - R_A)
$$
(3.64)

If P, the distance between the primary and secondary images, is preferred to the parameter R_A, the conversion is made from Eq. (2.85):

$$
R_A = \frac{P}{f' - f_1'} = \frac{m_2 P}{f'(m_2 - 1)}
$$
(3.65)

In Chap. 2, Fig. 2.15, we showed that the limit case with $m_2 = -1$ has practical significance as the *folded Cassegrain,* the secondary being a flat mirror; whereas the equivalent Gregory case with $m_2 = +1$ is of no practical interest, as there is no real image. The same result applies to the 2-mirror aberration formulae of Eqs. (3.59) to (3.62). If we set $m_2 = -1$ for the folded Cassegrain, then $f' = -f_1'$ and $\xi = 0$ from (3.41). The 2-mirror formulae then reduce to the formulae of Eqs. (3.32) for the primary alone. The folded Gregory case with $m_2 = +1$ gives the same result for the virtual image.

There is, however, a much more important limit case than the folded Cassegrain. This is the *afocal case,* whereby not only the object, but also the image is at infinity. Such a telescope is a 2-mirror *beam compressor,* as discussed in § 2.2.4 for conventional visual refractors. In the reflecting form as a Cassegrain or Gregory afocal system, it is a *Mersenne telescope* (Fig. 1.3). We shall see in § 3.2.6 that this form has profound significance in the theory of 2-mirror telescopes. It is therefore important to express the aberration formulae in terms valid for this limit case. Eqs. (3.59) to (3.62) are convenient for the normal focal telescope but not for the afocal case because the parameters f', L, m_2, ζ and ξ become infinite and the forms are indeterminate. We therefore set up equivalent equations using only finite parameters such as f_1', y_1, and d_1. The parameter m_2 is also used but may have no power higher than zero in the numerator. It should be noted that the parameter f' in the numerator is automatically replaced by y_1 if we reduce the power n of the generalizing factor $(\frac{y_1}{f'})^n$ to $(\frac{y_1}{f'})^{n-1}$ and multiply through by $(\frac{y_1}{f'})$. If ζ and ξ are expanded, f' is replaced by $m_2 f_1'$ and

$$y_2 = y_1 R_A = L\frac{y_1}{f'} \tag{3.66}$$

inserted from (2.58) to eliminate L, then (3.59) can be written as

$$\sum S_I = \left(\frac{y_1}{f_1'}\right)^3 \left\{-\frac{y_1}{4}(1+b_{s1}) + \frac{y_2}{4}\left(\frac{m_2+1}{m_2}\right)^3\left[\left(\frac{m_2-1}{m_2+1}\right)^2 + b_{s2}\right]\right\} \tag{3.67}$$

In the afocal case, $m_2 \to \pm\infty$ and the quantities $\left[\frac{(m_2+1)}{m_2}\right]^3$ and $\left[\frac{(m_2-1)}{(m_2+1)}\right]^2$ both $\to 1$, giving

$$\sum_{Afoc} S_I = \left(\frac{y_1}{f_1'}\right)^3\left[-\frac{y_1}{4}(1+b_{s1}) + \frac{y_2}{4}(1+b_{s2})\right] \tag{3.67a}$$

Similarly, the coma given by (3.60) becomes

$$\sum S_{II} = \left(\frac{y_1}{f_1'}\right)^3 \left\{ -\frac{d_1}{4}\left(\frac{m_2+1}{m_2}\right)^3 \left[\left(\frac{m_2-1}{m_2+1}\right)^2 + b_{s2}\right] - \frac{f_1'}{2m_2^2}\right\} u_{pr1} \\ - \left[\frac{s_{pr1}}{y_1}\sum S_I\right] u_{pr1} \Bigg\},$$

$$(3.68)$$

giving for the afocal case

$$\sum_{Afoc} S_{II} = \left(\frac{y_1}{f_1'}\right)^3 \left[-\frac{d_1}{4}(1+b_{s2})\right] u_{pr1} - \left[\frac{s_{pr1}}{y_1}\sum S_I\right] u_{pr1} \qquad (3.68a)$$

The expression (3.61) for the astigmatism leads, if L is eliminated from (2.72) by

$$L = m_2(f_1' - d_1)$$

to the unwieldy form:

$$\sum S_{III} = \left[\frac{y_1^2}{m_2(f_1'-d_1)}\right]\left[1 + \frac{d_1}{m_2 f_1'}\right] u_{pr1}^2 \\ + \left[\frac{y_1^2 d_1^2}{4f_1'^2}\left(\frac{1}{f_1'-d_1}\right)\left(\frac{m_2+1}{m_2}\right)^3 \left\{\left(\frac{m_2-1}{m_2+1}\right)^2 + b_{s2}\right\}\right] u_{pr1}^2 \\ + s_{pr1}\left[\frac{y_1^2}{m_2^2 f_1'^2} + \frac{y_1^2 d_1}{2f_1'^3}\left(\frac{m_2+1}{m_2}\right)^3 \left\{\left(\frac{m_2-1}{m_2+1}\right)^2 + b_{s2}\right\}\right] u_{pr1}^2 \\ + \left(\frac{s_{pr1}}{y_1}\right)^2 \sum S_I u_{pr1}^2 \Bigg\}$$

$$(3.69)$$

For the afocal case, this simplifies to

$$\sum_{Afoc} S_{III} = \left[\frac{y_1^2 d_1^2}{4f_1'^2(f_1'-d_1)}(1+b_{s2}) \\ +s_{pr1}\frac{y_1^2 d_1}{2f_1'^3}(1+b_{s2}) + \left(\frac{s_{pr1}}{y_1}\right)^2 \sum S_I\right] u_{pr1}^2 \Bigg\}$$

$$(3.69a)$$

Finally (3.62) for the field curvature becomes, again eliminating L from (2.72):

$$\sum S_{IV} = H^2 \left[\frac{1}{f_1'} - \left(\frac{m_2+1}{m_2}\right)\left(\frac{1}{f_1'-d_1}\right)\right] \qquad (3.70)$$

The afocal form is then

$$\sum_{Afoc} S_{IV} = H^2 \left[\frac{1}{f_1'} - \left(\frac{1}{f_1'-d_1}\right)\right] \qquad (3.70a)$$

Tables 3.4 to 3.6 give in resumé the four third order aberration contributions: Table 3.4 for a 1-mirror telescope (Eqs. (3.26)); Table 3.5 for a 2-mirror
telescope in *focal* form (Eqs. (3.59) to (3.62)); Table 3.6 for a 2-mirror telescope in *afocal* form (Eqs. (3.67a) to (3.70a)).

Table 3.4. Third order aberrations for a 1-mirror telescope (concave primary)

$$(S_I)_1 \quad = \quad -\left(\frac{y_1}{f_1'}\right)^4 \frac{f_1'}{4}(1 + b_{s1})$$

$$(S_{II})_1 \quad = \quad -\left(\frac{y_1}{f_1'}\right)^3 \frac{1}{4}[2f_1' - s_{pr1}(1 + b_{s1})]u_{pr1}$$

$$(S_{III})_1 \quad = \quad -\left(\frac{y_1}{f_1'}\right)^2 \frac{1}{4f_1'}\left[4f_1'(f_1' - s_{pr1}) + s_{pr1}^2(1 + b_{s1})\right]u_{pr1}^2$$

$$(S_{IV})_1 \quad = \quad +H^2\left(\frac{1}{f_1'}\right)$$

Table 3.5. Third order aberrations and associated relations for a 2-mirror telescope in *focal* form

$$\zeta \quad = \quad \zeta^0 + \zeta^* = \frac{m_2^3}{4}(1 + b_{s1})$$

$$\xi \quad = \quad \xi^0 + \xi^* = \frac{(m_2+1)^3}{4}\left[\left(\frac{m_2-1}{m_2+1}\right)^2 + b_{s2}\right]$$

$$\sum S_I \quad = \quad \left(\frac{y_1}{f'}\right)^4 (-f'\zeta + L\xi)$$

$$\sum S_{II} \quad = \quad \left(\frac{y_1}{f'}\right)^3 \left[-d_1\xi - \frac{f'}{2} - \frac{s_{pr1}}{f'}(-f'\zeta + L\xi)\right] u_{pr1}$$

$$\sum S_{III} \quad = \quad \left(\frac{y_1}{f'}\right)^2 \left[\frac{f'}{L}(f' + d_1) + \frac{d_1^2}{L}\xi + s_{pr1}\left(1 + \frac{2d_1}{f'}\xi\right)\right.$$

$$\left. + \left(\frac{s_{pr1}}{f'}\right)^2 (-f'\zeta + L\xi)\right] u_{pr1}^2$$

$$\sum S_{IV} \quad = \quad -\sum H^2 P_c = H^2\left(\frac{1}{f_1'} - \frac{1}{f_2'}\right)$$

$$= \quad H^2\left[\left(\frac{m_2}{f'}\right) - \left(\frac{m_2+1}{L}\right)\right]$$

$$= \quad H^2\left[\left(\frac{m_2}{f'}\right) - \left(\frac{m_2+1}{f'-m_2d_1}\right)\right] = H^2\left[\left(\frac{m_2}{f'}\right) - \left(\frac{m_2^2-1}{m_2 P}\right)\right]$$

Some conversion formulae from Chap. 2 (see Figs. 2.11 and 2.12):

$$L \quad = \quad f' - m_2 d_1 = f' R_A$$
$$f' \quad = \quad m_2 f_1'$$
$$f_2' \quad = \quad \frac{L}{m_2+1} = P\left(\frac{m_2}{m_2^2-1}\right)$$

Table 3.6. Third order aberrations and associated relations for a 2-mirror telescope in *afocal* form

$$\sum_{\text{Afoc}} S_I = \left(\tfrac{y_1}{f_1'}\right)^3 \left[-\tfrac{y_1}{4}(1+b_{s1}) + \tfrac{y_2}{4}(1+b_{s2})\right]$$

$$\sum_{\text{Afoc}} S_{II} = \left(\tfrac{y_1}{f_1'}\right)^3 \left[-\tfrac{d_1}{4}(1+b_{s2})\right]u_{pr1} - \left[\tfrac{s_{pr1}}{y_1}\sum S_I\right]u_{pr1}$$

$$\sum_{\text{Afoc}} S_{III} = \left(\tfrac{y_1}{f_1'}\right)^2 \left[\tfrac{d_1^2}{4(f_1'-d_1)} + s_{pr1}\tfrac{d_1}{2f_1'}\right](1+b_{s2})u_{pr1}^2$$

$$+ \left[\left(\tfrac{s_{pr1}}{y_1}\right)^2 \sum S_I\right]u_{pr1}^2$$

$$\sum_{\text{Afoc}} S_{IV} = H^2\left[\left(\tfrac{1}{f_1'}\right) - \left(\tfrac{1}{f_1'-d_1}\right)\right]$$

The conversion formula for y_2 is, from Chap. 2:

$$y_2 = y_1 R_A = y_1\left(\tfrac{f_2'}{f_1'}\right)_{\text{Afoc}}$$

Before considering the application of the above third order theory to various forms of 2-mirror telescope, we must deal briefly with the effects of higher order aberrations and the practical evaluation of systems with ray-trace programs.

3.2.5 Higher order aberrations and system evaluation

3.2.5.1 General definition of aspheric surfaces. In § 3.1 we gave the general definition of conic sections in terms of the Schwarzschild (conic) constant b_s, as represented by Eq. (3.11). The theory of § 3.2.4 above is limited to the second term of this equation, known as the third order, or Seidel, approximation. With steep apertures or significant field sizes, the third order approximation remains of immense value for understanding the basic correction potential of a given design, but is no longer adequate to describe the final image quality. Furthermore, departures may be necessary from the strict conic sections defined by (3.11) in order to compensate higher order aberrations. This requires a more general definition of aspheric surfaces.

From Eq. (3.4) we had for the general equation of an ellipse

$$z = a - a\left(1 - \frac{y^2}{b^2}\right)^{1/2} \tag{3.71}$$

The longer semi-axis a can be replaced by the vertex curvature c from (3.6) and (3.10):

$$c = \frac{a}{b^2} = \frac{1}{a(1 - \varepsilon^2)} = \frac{1}{a(1 + b_s)} \tag{3.72}$$

From Eq. (3.12), this expression is general for any conic section. If, now, both sides of (3.71) are multiplied by

$$a + a\left(1 - \frac{y^2}{b^2}\right)^{1/2} \, ,$$

we have the form

$$z = \frac{\frac{a^2}{b^2}y^2}{a + a\left(1 - \frac{y^2}{b^2}\right)^{1/2}} \tag{3.73}$$

From (3.72) and

$$b^2 = a^2(1 + b_s) \, , \tag{3.74}$$

this reduces to

$$z = \frac{cy^2}{1 + [1 - (1 + b_s)c^2y^2]^{1/2}} \tag{3.75}$$

This form still describes a strict conic section and can be expanded into the polynomial of Eq. (3.11). If $b_s = -1$, Eq. (3.75) reduces to the parabolic first term of (3.11):

$$z = \frac{c}{2}y^2 \tag{3.76}$$

Most optical design programs allow a modification of any desired conic section defined by b_s in (3.75) by adding aspheric terms of higher orders whose coefficients A_s, B_s, C_s, D_s can be chosen at will, giving the general form:

$$z = \frac{cy^2}{1 + [1 - (1 + b_s)c^2y^2]^{1/2}} + A_sy^4 + B_sy^6 + C_sy^8 + D_sy^{10} + \ldots \tag{3.77}$$

Frequently, the designer will find it convenient to define the conic section as a parabola with $b_s = -1$, so that the supplementary coefficients A_s, B_s, ... give the supplement to the parabola of (3.76). They can then be compared with the coefficients of (3.11) for any other conic defined by b_s. For the third order and fifth order terms, the comparison gives

$$b_s = \left(\frac{8A_s}{c^3}\right) - 1 = \left(\frac{16B_s}{c^5}\right)^{1/2} - 1 \tag{3.78}$$

If the coefficients A_s, B_s, ... are referred to the sphere, then, of course, the terms -1 in Eq. (3.78) vanish. In this way, it is a simple matter to determine the departure of a given surface from a conic form b_s defined by the third order coefficient A_s. From (3.11) it is clear that strict conics have only positive terms, but it should be borne in mind that both c and $(1 + b_s)$ can have positive or negative values. Sign reversals of terms compared with Eq. (3.11) will imply, if the contributions to z are significant, major departures in the higher orders from strict conic sections.

Another form, used by Laux [3.8(a)], defines the conic parameter as $\bar{b}_s$ by

$$\bar{b}_s = \frac{c}{2}(1 + b_s) \tag{3.79}$$

Since, from (3.72)

$$a = \frac{1}{c(1 + b_s)},$$

we have from (3.79)

$$a = \frac{1}{2\bar{b}_s} \tag{3.80}$$

From (3.71), this gives the general form for a strict conic section as

$$z = \frac{1}{2\bar{b}_s}\left[1 - (1 - 2\bar{b}_s cy^2)^{1/2}\right] \tag{3.81}$$

This can be compared with the form of Eq. (3.75). The form (3.81) is singular in the case of the parabola.

It is a matter of individual preference, but I personally prefer the basic form of Eq. (3.11), as derived by Schwarzschild [3.1], and the derived form using b_s of Eq. (3.75) which leads naturally to the general aspheric form of (3.77). The only weakness of the conventional (Schwarzschild) definition of b_s is revealed by Eq. (3.11): the actual amount of the aspheric third order contribution Δz to the sagitta z is given by

$$\Delta z = \frac{c^3}{8}b_s y^4 = \frac{1}{512N_s^3}b_s y , \tag{3.82}$$

where $N_s = \frac{f'}{D} = \frac{r}{2D}$ is the aperture number if the surface is used as a telescope mirror with an infinite object. Δz is therefore proportional to the inverse cube of the aperture number N_s. This means, for a given aperture height y, that the aspheric contribution Δz for an f/1 parabola with $b_s = -1$ is the same as for an f/10 hyperbola with $b_s = -1000$. It is important to bear the aperture number of the surface in mind when interpreting the practical significance of a given value of b_s.

3.2.5.2 Higher order aberrations, the Abbe sine condition and Eikonals.

In §3.2.1, the "orders" of the monochromatic aberrations arising out of Hamilton's Characteristic Function were discussed and the terms of the first three orders (first, third and fifth) were given in Table 3.1. The number of terms in an order increases quite rapidly as the order increases. Ray tracing embraces, by definition, all the orders, within the accuracy given by the computer. With modern computers, therefore, the effect of all the higher orders is given at once from the difference between ray tracing results and third order calculations, the latter being converted to lateral aberrations for comparison by the formulae of §3.3.

After the formulation by Seidel [3.4] for third order aberrations in 1856, there was considerable interest in developing the theory of higher orders, particularly to allow the explicit calculation of fifth order effects. Another area of intensive research was concerned with formulations for the total optical path through an optical system. In 1873, Abbe [3.9] discovered what is perhaps the most important relation of this type covering all the orders in *aperture* of coma (but only the third order in the field): the Abbe *sine condition*. Abbe's work was concerned with imagery in microscopes, but the condition had been discovered in another form by Clausius [3.10] in connection with energy transfer from thermodynamic considerations. The sine condition for freedom from coma is expressed in the normal (focal) case by

$$\frac{\sin U'}{u'} = \frac{\sin U}{u} , \qquad (3.83)$$

in which U and U' are the aperture ray angles to the axis of finite rays in the object and image space respectively and u and u' the paraxial equivalents. In the afocal case, the equivalent condition is

$$\frac{Y'}{y'} = \frac{Y}{y} , \qquad (3.84)$$

in which the aperture ray angles are replaced by the aperture ray heights. For the normal telescope case of an object at infinity and a finite image distance, the sine condition is simply

$$\frac{Y}{\sin U'} = \frac{y}{u'} = \text{const} = -f' , \qquad (3.85)$$

from (2.38).

The Abbe sine condition refers to the so-called linear coma, i.e. the sum of all coma terms in Table 3.1 which depend *linearly* on the field parameter:

$$Coma_{Lin} =_1 k_{31}\sigma\rho^3 \cos\phi +_1 k_{51}\sigma\rho^5 \cos\phi +_1 k_{71}\sigma\rho^7 \cos\phi + \dots$$

It is not trivial to prove that $Coma_{Lin}$, so defined and when set to zero for its correction, is the equivalent of the Abbe sine condition of Eq. (3.83). The proof is given by Welford [3.6] and includes the derivation of the Staeble–Lihotzky condition. This condition expresses the requirement for zero linear coma in the presence of uncorrected spherical aberration. Only if the latter is zero does the Staeble–Lihotzky condition reduce to the Abbe form of Eq. (3.83), which therefore applies only to *aplanatic* systems corrected for both spherical aberration and linear coma.

Equation (3.83) shows that the concept of *principal planes*, introduced in § 2.2.1 (Fig. 2.1) in the treatment of Gaussian optics for the paraxial domain, is no longer valid if coma is to be corrected with finite apertures. Eq. (3.83) requires that the planes be replaced by spheres, concentric to the object and image, as shown in Fig. 3.2. The spheres are, of course, simply wavefronts defined as passing through the principal points P and P'. The sine condition confirms what one would intuitively expect from the wavefront concept of image formation.

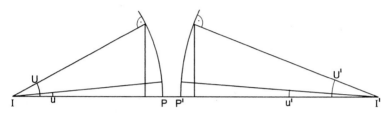

Fig. 3.2. The Abbe sine condition

Following Abbe, up to the effective implementation of computers, much elegant mathematical work was carried out in the hope of developing higher order aberration theories and techniques of calculating optical path lengths by extensions of the theorems of Fermat and Malus (see § 2.2.2). This led to the "Eikonal" of Bruns in 1895 and the modification (Winkeleikonal) used by Schwarzschild [3.1]. Accounts of this work and complete literature references are given by Czapski-Eppenstein [3.11(a)] and by Herzberger [3.7]. Modern computers, with their efficiency in ray tracing, have reduced the significance of this theory. Nevertheless, for a deeper understanding of the function of optical systems, many aspects retain great value, but it cannot rival third order theory in its significance and simplicity.

Modern developments in telescope optics involving active optics, as described in Vol. II of this work (RTO II, § 3.5) and relating particularly to the ESO New Technology Telescope (NTT) and Very Large Telescope (VLT), also make extensive use of aberration theory. However, in the basic layout of the active support systems for thin primary mirrors, higher order aberration theory is a useful rather than essential tool, since Zernike polynomials (§ 3.9) or natural vibration modes (RTO II, § 3.5.4) provide the information required. But classical higher order theory has assumed considerable significance in the general theory of decentering and in practical techniques for the alignment of the elements of modern telescopes (RTO II, § 2.2.1).

Of the earlier work on higher order aberration theory given by Herzberger [3.7], the most notable in the modern context is that of Buchdahl [3.154]. Buchdahl developed systematically the concept of quasi-stable parameters for a given order, equivalent to the use of paraxial parameters in third order theory. Probably the definitive treatment of this subject was given by Focke in 1965 [3.155], although its assimilation requires much hard work. Focke's aim is a modernised treatment unifying the theories of Schwarzschild, T. Smith [3.156] and Herzberger. T. Smith's classic paper investigated the dependence of aberration coefficients *of any order* on the object and stop positions. Focke gives explicit formulae, based on the Schwarzschild angle Eikonal and the Smith stop-shift formulae, for refraction at a surface which can be spherical or aspheric to the fifth order. The formulae cover all third order and fifth order terms, separated into spherical and aspheric parts. He also derives the Herzberger formulae, with the link to the Schwarzschild coefficients, and shows that they are in exact agreement with those of Buchdahl and have a simple form for *spherical* surfaces. This simplicity is lost for *aspheric* surfaces and Focke does not develop the Herzberger treatment for this case.

More recently de Meijère and Velzel [3.157] have analysed the dependence of third and fifth order aberration coefficients on the *definition* of the pupil coordinates. They show that the normal definition leads to 9 independent fifth order terms, in agreement with Table 3.1, whereas two other definitions lead to 12 independent fifth order terms. This is of considerable historical interest, because Petzval [3.11] [3.151] stated clearly already in 1843 that there were 12 such terms (see Footnote in § 3.2.2).

3.2.5.3 System evaluation according to geometrical optics from ray tracing.

Traditional representations of system quality according to geometrical optics were based on limited information from hand-traced rays, usually as transverse aberrations (intercept differences in the image plane). Such representations can still be very revealing and are still used [3.8]. However, fast computers have led to the universal use of *spot-diagrams* as a measure of geometrical image quality. They were introduced by Herzberger [3.7] [3.37]. To generate a spot-diagram, the entrance pupil is (usually) divided up into a rectangular grid with square mesh and a finite ray is traced through the system for each mesh intersection point [3.12]. Sometimes, other pupil divisions are used, such as concentric circles, but this gives less complete sampling of the pupil. The number of rays required will depend on the number of aberration orders present in the system and usually lies in the range 50 - 300. The usual spot-diagram (SD) representation is a matrix of points, the columns representing the different field heights with the SD for the axis at the bottom, while the rows represent different wavelengths (Fig. 3.3). Each matrix represents an image plane at a chosen axial focus point. Full understanding of the image potential often requires several such SD-matrices for different focus points.

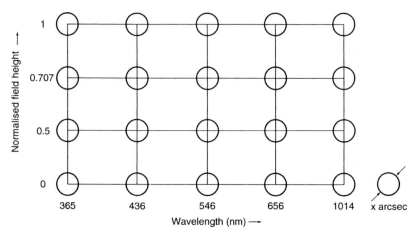

Fig. 3.3. Normal representation of spot-diagrams

For the given focus, each matrix represents the field plane for the various wavelengths on the chosen (arbitrary) scale, on which the SD give the geometrical point-spread-functions (PSF) in a vastly magnified form. Their scale is normally indicated by a circle whose diameter is given in arcsec for telescope systems, to permit direct comparison with the seeing. Catoptric systems, with only reflecting elements, will have a single-column spot-diagram matrix for a given focus. In such catoptric systems, we shall show a "through-focus" matrix in which the rows represent different focus shifts from the nominal focus.

In this book, the primary purpose is to present the theory and properties of various telescope forms in a coherent way. Examples of achievable quality will be given in typical cases with spot-diagrams. For more complete comparisons with SD, the reader is referred to Rutten and van Venrooij [3.12] and Laux [3.8].

3.2.6 Analytical expressions for a 1-mirror telescope and various forms of 2-mirror telescopes (Classical, Ritchey-Chrétien, Dall-Kirkham, Spherical Primary)

3.2.6.1 Introduction. The results derived in § 3.2.4 will now be used to derive the basic properties of 1-mirror telescopes and various common forms (Cassegrain and Gregory) of 2-mirror telescopes. The theory is given in wavefront aberration, whose conversion to other forms is given in § 3.3.

It should be noted that the classical forms of reflecting telescope only corrected *one* aberration: spherical aberration, giving image correction on the axis. The necessary theory was given by Descartes in 1634. This theory was sufficient for the long focal lengths and small visual fields for nearly three centuries! The problems lay not in lack of theory, but in manufacture, test methods, mirror support, the mechanics of mountings and operation. Only in 1905, impelled by the field demands of the slow, but increasingly dominant photographic emulsions and the dramatic advances of the reflecting telescope due to Ritchey (see Chap. 5), came the revolutionary work of Schwarzschild [3.1] [3.13]. This introduced third order theory into reflecting telescope design, in a remarkably complete form, and laid the way for all modern designs taking field correction into account.

Schwarzschild proved, using his definition b_s of the asphericity of Eq. (3.11), that two aspheric mirrors *in any geometry with sufficient space between them* could correct two Seidel aberrations. I call this the "*Schwarzschild theorem*". The two lowest terms from the point of view of field dependence (σ) in Table 3.1 are $S_I(\sigma^0)$ and $S_{II}(\sigma^1)$. Schwarzschild's work thereby opened the way to *aplanatic* telescopes, a term introduced by Abbe [3.9] for optical systems corrected for S_I and S_{II}.

We shall see in §§3.4 and 3.6.5 that the generalization of the above Schwarzschild theorem [3.13] is of central importance in assessing the possibilities of telescopes containing more elements than two aspheric mirrors and is, indeed, the basis of all modern telescope solutions.

3.2.6.2 1-mirror telescopes. Since plane mirrors introduce no aberrations, the single *centered* concave mirror (i.e. one mirror with optical power, or "powered" mirror) covers all the normal *prime focus (PF)* cases: direct PF, Newton focus and folding-flat Cassegrain (Fig. 2.15). The aberrations are given in Table 3.4.

Since any telescope must be free from $(S_I)_1$, the first equation gives the requirement $b_{s1} = -1$, the parabolic form. A *spherical* primary gives

$$(S_I)_1 \atop sph = - \left(\frac{y_1}{f_1'}\right)^4 \frac{f_1'}{4} \tag{3.86}$$

The second term in the bracket

$$s_{pr1}(1 + b_{s1})$$

for the coma $(S_{II})_1$ is also zero, which shows that the coma is independent of the stop position s_{pr1} because $(S_I)_1$ is zero, in accordance with the stop-shift formulae of Eqs. (3.22). The coma is then given by

$$(S_{II})_1 \atop parab = - \left(\frac{y_1}{f_1'}\right)^3 \frac{f_1'}{2} u_{pr1} \tag{3.87}$$

For a spherical mirror, or any non-parabolic form, the result of Eq. (3.87) only applies if the stop is at the primary; otherwise, the general form of Table 3.4 must be used.

For the astigmatism, the third formula reduces, for a parabolic form, to

$$(S_{III})_1 \atop parab = -\left(\frac{y_1}{f_1'}\right)^2 (f_1' - s_{pr1})u_{pr1}^2 \tag{3.88}$$

This is not independent of the stop position s_{pr1} because the coma is not zero: only the last term of the stop-shift formula in (3.22) vanishes because of zero $(S_I)_1$. As with the coma, if the stop is at the primary, the astigmatism is independent of its form.

If the normalization of Table 3.3 is used with $y_1 = +1$, $f_1' = -1$, $u_{pr1} = +1$ and the stop is placed at the primary with $s_{pr1} = 0$, then $(S_I)_1 = 0$, $(S_{II})_1 = -0.5$ and $S_{III} = +1.0$ as shown, in agreement with Eqs. (3.87) and (3.88).

Historically, there is another form of 1-mirror telescope which is not axially centered: the Herschel form (Figs. 1.1 and 2.10). The central part of the observed image then corresponds to the field angle u_{pr1} of the mirror tilt. This leads to the coma and astigmatism of Eqs. (3.87) and (3.88). Because of the dependence on the first and second power of u_{pr1} respectively, the astigmatism is negligible for small tilts and the coma is dominant.

3.2.6.3 2-mirror telescopes in normal Cassegrain or Gregory geometry

The results of §3.2.4.3 enable us now to derive analytical expressions for the third order aberrations of various 2-mirror solutions. In this section, we shall confine ourselves to those solutions based on normal Cassegrain or Gregory geometry: other 2-mirror systems will be treated in the following section. The theory given here will enable us to understand the (normalized) third order values of Table 3.3, derived from Eqs. (3.20) and paraxial ray traces. The conversion to lateral or angular aberration is given in §3.3. At this stage, we are only concerned with the *relative* values of the third order coefficients arising from different telescope solutions.

It should be remembered that the formulae we have derived above are completely general for 2-mirror telescope forms. It follows that the derivations below apply to both Cassegrain and Gregory forms. Which form applies will depend on the signs of the parameters m_2 and f' (negative and positive respectively for Cassegrain; positive and negative respectively for Gregory – see Tables 2.2 and 2.3), and on the values of d_1 and L which are negative and positive respectively in both cases.

a) Classical telescope: classical Cassegrain and Gregory. These forms are defined by a primary mirror of *parabolic* form ($b_{s1} = -1$) giving a prime focus corrected for $(S_I)_1$. From Eqs. (3.30) and (3.31) it follows that

$$\zeta = \frac{m_2^3}{4}(1 + b_{s1}) = 0 \tag{3.89}$$

$$(S_I)_1 = -\left(\frac{y_1}{f'}\right)^4 f'\zeta = 0 \tag{3.90}$$

Similarly, the secondary mirror must also contribute zero spherical aberration $(S_I)_2$. From Eqs. (3.41) and (3.43), we have

$$\xi = \frac{(m_2 + 1)^3}{4} \left[\left(\frac{m_2 - 1}{m_2 + 1} \right)^2 + b_{s2} \right] = 0 \tag{3.91}$$

$$(S_I)_2 = \left(\frac{y_1}{f'} \right)^4 L\xi = 0 , \tag{3.92}$$

from which it follows that

$$(b_{s2})_{cl} = - \left(\frac{m_2 - 1}{m_2 + 1} \right)^2 \tag{3.93}$$

For the classical Cassegrain given in Case 3 of Table 3.3 with $m_2 = -4$, Eq. (3.93) gives for the form of the secondary $b_{s2} = -2.778$, a strong hyperbola. In the classical Cassegrain, the foci of this hyperbola obviously lie at the primary and secondary images I_1' and I_2' of Fig. 2.12. Since, for a given radius r_2 and diameter $2y_2$, b_{s2} is a direct measure of the asphericity from Eq. (3.11), it is clear why the manufacture of Cassegrain secondaries, combined with the test problems resulting from its convex form (see RTO II, Chap. 1), remained unsolved for 180 years after the original proposal by Cassegrain.

Similarly, Eq. (3.93) gives for the classical Gregory of Case 7 in Table 3.3, with $m_2 = +4$, an elliptical form for the secondary with $b_{s2} = -0.360$. However, the asphericity on the surface is not necessarily smaller than that on the Cassegrain secondary, since the radius r_2 is steeper for the same value of m_2 because the quantity P in Eq. (2.84) is less in the Gregory solution. The manufacture of the Gregory secondary is easier because its concave form is more convenient for testing – see RTO II, Chap. 1.

If $m_2 \rightarrow \pm\infty$, Eq. (3.93) gives $b_{s2} \rightarrow -1$, giving the Mersenne telescopes with confocal paraboloids (Table 3.6). For a given diameter and curvature, therefore, the manufacture of the Cassegrain secondary will ease with increase of $|m_2|$. However, if increase of $|m_2|$ implies reduction in diameter (obstruction ratio R_A), as for a fixed final image position in the focal case according to Eq. (2.72), then this is not necessarily true, since the asphericity must be achieved over a smaller diameter – the aspheric function becomes "steeper".

The field coma $\sum S_{II}$ for the classical Cassegrain or Gregory is given from Eq. (3.60) and Table 3.5. The last term

$$\frac{s_{pr1}}{f'}(-f'\zeta + L\xi)$$

is in all cases zero if the stop is placed at the primary, giving $s_{pr1} = 0$. But it is anyway zero for the classical Cassegrain because $\sum S_I = 0$ by virtue of $\zeta = \xi = 0$. This is, again, a statement of the stop-shift formulae of Eq. (3.22). Since any useful telescope must be corrected for spherical aberration, the field coma is always independent of the position of the stop.

For the classical telescope forms, $\xi = 0$ from Eq. (3.91), so that the field coma is given from Eq. (3.60) by

$$\sum S_{II} = -\left(\frac{y_1}{f'}\right)^3 \frac{f'}{2} u_{pr1} \tag{3.94}$$

This is identical with the result of Eq. (3.87) for a 1-mirror telescope if we replace f_1' by f', the focal length of the 2-mirror combination. For the cases of Table 3.3 with y_1 and the final focal lengths normalized to unity, the field coma is identical for the 1-mirror and classical Cassegrain and Gregory cases, an important result. However, it must be remembered that the classical 2-mirror telescopes will give a lower value in practical cases if their relative aperture is smaller.

The field astigmatism is given by Eq. (3.61). With $\sum S_I = 0$, the last term is again zero. The other stop-shift term in the bracket can be written

$$\frac{s_{pr1}}{f'}\left(2d_1\xi + f'\right) ,$$

which contains in the bracket *twice* the stop-independent coma of Eq. (3.60). In the general case, then, the classical telescope has the astigmatism

$$\sum S_{III} = \left(\frac{y_1}{f'}\right)^2 \left[\frac{f'}{L}(f' + d_1) + s_{pr1}\right] u_{pr1}^2 , \tag{3.95}$$

since $\xi = 0$. If the stop is at the primary, this reduces to

$$\left(\sum S_{III}\right)_{s_{pr1}=0} = \left(\frac{y_1}{f'}\right)^2 \left[\frac{f'}{L}(f' + d_1)\right] u_{pr1}^2 \\
= \left(\frac{y_1}{f'}\right)^2 f'\left[\frac{f' + d_1}{f' - m_2 d_1}\right] u_{pr1}^2 \tag{3.96}$$

from (2.75). Since $L \ll 1$ in the normalized system of Table 3.3, the astigmatism of the classical telescope is markedly higher for the same final focal length than that of the 1-mirror telescope given by Eq. (3.88). If we set $m_2 = -1$ in Eq. (3.96), we have the limit case of a folding flat without optical power for the Cassegrain secondary and the astigmatism is the same as for the 1-mirror telescope with the stop at the primary.

In the *afocal* case, we have the limit case of the *Mersenne telescope* (Fig. 1.3 (a) and (b)) consisting of two confocal paraboloidal mirrors. Since f', L and m_2 are ∞, the focal forms (3.94) and (3.96) would be singular in the normalized forms of Table 3.3 without their initial factors $(y_1/f')^3$ and $(y_1/f')^2$ respectively, but reduce to zero with these factors. This is confirmed from the general afocal formulation of Table 3.6. From this, it is clear that, if

$$b_{s1} = b_{s2} = -1 , \tag{3.97}$$

then

$$\sum_{Afoc} S_I = \sum_{Afoc} S_{II} = \sum_{Afoc} S_{III} = 0 \tag{3.98}$$

Thus, in 1634, without knowing it and without the necessary theory of field aberrations to be able to understand its significance, Mersenne invented not only the first aplanatic telescope form, but also the first anastigmatic form [3.13]. We shall see that these properties of an afocal telescope are of great significance for the further development of modern forms of reflecting telescope.

The field curvature $\sum S_{IV}$ of the classical telescope is given in various forms by Eq. (3.62) and in Table 3.5 for the focal case; in Table 3.6 for the afocal case. In the Cassegrain case, if the telephoto effect T is large according to Eq. (2.57), then $L \ll f'$ and the second form of Eq. (3.62) shows that the field curvature from the secondary will be far higher than that of the primary. This gives the high values of the Cassegrain solutions of Table 3.3 compared with the 1-mirror telescope. This Cassegrain field curvature is *concave* towards the incident light. The last column of Table 3.3 gives the effective field curvature for best imagery in the presence of $\sum S_{III}$ and $\sum S_{IV}$. We see that the astigmatism increases the effective field curvature.

If the curvatures of M_1 and M_2 in a Cassegrain telescope are made equal, the system will yield a flat field with $\sum S_{IV} = 0$. However, this leads to an optical geometry which is rarely practicable from the point of view of the final image position or obstruction. This is easily demonstrated from Eq. (2.64):

$$f_2' = \frac{R_A L f_1'}{L + R_A f_1'} \tag{3.99}$$

For $\sum S_{IV} = 0$, we must set $f_2' = f_1'$. Then

$$L + R_A f_1' = R_A L ,$$

giving

$$f_1' = -L \left(\frac{1 - R_A}{R_A} \right) \tag{3.100}$$

Setting $R_A = 1/3$, a typical value, gives

$$f_1' = -2L \tag{3.101}$$

From Eq. (3.62):

$$\sum S_{IV} = H^2 \left[\frac{m_2}{f'} - \left(\frac{m_2 + 1}{L} \right) \right] = H^2 \left[\frac{1}{f_1'} - \left(\frac{m_2 + 1}{L} \right) \right] \tag{3.102}$$

If $\sum S_{IV} = 0$, then (3.101) and (3.102) give

$$m_2 = -1.5 \tag{3.103}$$

for a flat field with $R_A = 1/3$. This is the inevitably low value of m_2 associated with a weakly curved secondary. The consequence for the position of the image b relative to the primary is given by Eq. (2.65):

$$b = L + f_1'(1 - R_A) \tag{3.104}$$

With $R_A = 1/3$ and $f_1' = -2L$ from (3.101), we have

$$b = -L/3 \tag{3.105}$$

This means that the Cassegrain image is well to the left of the primary in Fig. 2.12, normally an unacceptable position.

If $R_A = 1/2$ is accepted, then

$$\begin{aligned}
f_1' &= -L \\
m_2 &= -2 \\
b &= +L/2 \,,
\end{aligned}$$

a normally acceptable image position; but $R_A = 1/2$ is normally too high. We see that the image position b is a very sensitive function in the flat field condition arising out of Eq. (3.62).

Normal modern Cassegrain telescopes rarely have $|m_2| < 3$ or $R_A > 1/3$. Indeed, the modern trend is to make $|m_2| \geq 4$, implying that field curvature is accepted as the lesser evil. If necessary, field-flattening lenses can be used, but one must beware of transverse chromatic aberration (see §3.5). Modern array detectors, such as CCDs, can compensate modest field curvature.

In Table 3.3, it is clear that the primary compensates some of the field curvature of the secondary, whereas in the Gregory case (Case 7) the effects are additive. However, the *effective* field curvatures with astigmatism are *numerically* quite similar. A very important advantage of the Gregory form is the fact that the *sign* of both $\sum S_{IV}$ and the effective field curvature is the opposite of that of the Cassegrain. This means that these field curvatures in the Gregory form are *convex* to the incident light. This is precisely the overcorrection required to compensate the natural field curvature of almost all instruments.

b) The aplanatic telescope and its Cassegrain (Ritchey-Chrétien) form. The Ritchey-Chrétien (RC) form of the Cassegrain telescope is, in practice, by far the most important modification of the classical Cassegrain telescope. It was originally proposed by Chrétien in 1922 [3.14], who was inspired by Schwarzschild's pioneer work in 1905 [3.1]. However, there is good evidence – see §5.2 – that Chrétien, at Ritchey's suggestion, had already established the aplanatic basis of his design by 1910, before he met Schwarzschild or became aware of his work. Schwarzschild established the fundamental design principles for field correction (see §3.2.6.1), but sought a faster system than that provided by the normal Cassegrain form. The Schwarzschild form also aimed for a remarkable angular field extension: it will be treated in §3.2.7. Schwarzschild, therefore, did not discover the RC telescope, although it was implicit in his theory [3.13]. It corrects the two conditions S_I and S_{II} to give an aplanatic telescope without any change to the desired Cassegrain geometry. Ritchey recognised the importance of this development for field extension in a very fast (for that time) Cassegrain with f/6.5. He then made the first such major RC telescope, with 1 m aperture, for the US Naval Observatory in the 1930s, a remarkable achievement with the test techniques available at that time. Without wishing to detract from the achievements of Ritchey, it might seem more logical to call this system the Schwarzschild-Chrétien, a name which would reflect more justly its origins in optical design theory. However, the name Ritchey-Chrétien is now firmly established.

The RC solution is defined by Eq. (3.60) for the field coma of a 2-mirror telescope, which must be set to zero for an aplanatic solution:

$$\left(\sum S_{II}\right)_{Aplan} = \left(\frac{y_1}{f'}\right)^3 \left[\left(-d_1\xi - \frac{f'}{2}\right) - \frac{s_{pr1}}{f'}(-f'\zeta + L\xi)\right] u_{pr1} = 0 \qquad (3.106)$$

For correction of $\sum S_I$, the second term in the square bracket is zero irrespective of the pupil position s_{pr1}. The condition for an aplanatic telescope is then

$$d_1\xi_{Aplan} + \frac{f'}{2} = 0 , \qquad (3.107)$$

giving

$$\xi_{Aplan} = -\frac{f'}{2d_1} \qquad (3.108)$$

Combining this with (3.41) gives

$$(b_{s2})_{Aplan} = -\left[\left(\frac{m_2 - 1}{m_2 + 1}\right)^2 + \frac{2f'}{d_1(m_2 + 1)^3}\right] \qquad (3.109)$$

for the general condition of an aplanatic, 2-mirror telescope.

Whether the telescope is of Cassegrain or Gregory form will again depend on the signs of m_2 and f' according to Tables 2.2 and 2.3 and the numerical value of d_1. Table 3.2 gives the values of b_{s2} for the RC and aplanatic Gregory cases with $m_2 = -4$ and $+4$ respectively. We see that there is a modest increase in eccentricity in both cases compared with the classical case. Eq. (3.109) can be written from (3.93)

$$(b_{s2})_{Aplan} = (b_{s2})_{cl} - \frac{2f'}{d_1(m_2 + 1)^3} , \qquad (3.110)$$

in which the second term gives the supplement required for the aplanatic solution. Since $(b_{s2})_{cl}$ is always negative, an increase of eccentricity always takes place because of the signs of the quantities in the Cassegrain and Gregory cases, giving a negative supplement.

Equation (3.109) completely defines the solution for the 2-mirror aplanatic telescope: the condition $\sum S_{II} = 0$ is achieved solely by the form of the secondary. The non-zero value of ξ of Eq. (3.108) corrects the field coma term of a primary mirror with the same final focal length f'.

Equation (3.110) confirms that the classical and aplanatic forms are identical in the afocal case of the Mersenne telescope. If we replace f' by $m_2 f_1'$ from Eq. (2.55), we can write

$$(b_{s2})_{\substack{Aplan \\ Afoc}} = (b_{s2})_{cl} - \frac{2f_1'}{d_1} \frac{m_2}{(m_2 + 1)^3} , \tag{3.111}$$

in which the supplementary term is zero if $m_2 \to \infty$.

The second condition for aplanatism, the correction $\sum S_I = 0$, must be established from the correct form of the primary. From Eq. (3.59), the condition is

$$-f' \zeta_{Aplan} + L\xi_{Aplan} = 0 , \tag{3.112}$$

giving from (3.108)

$$\zeta_{Aplan} = -\frac{L}{2d_1} \tag{3.113}$$

From (3.30), we have then

$$\frac{m_2^3}{4} \left[1 + (b_{s1})_{Aplan} \right] = -\frac{L}{2d_1} ,$$

and

$$(b_{s1})_{Aplan} = -1 - \frac{2L}{d_1 m_2^3} , \tag{3.114}$$

whereby the second term gives the supplement relative to the parabolic form of the classical primary. It can be written in the form

$$(b_{s1})_{Aplan} = (b_{s1})_{cl} - \frac{2L}{d_1 m_2^3} \tag{3.115}$$

From the signs given in Tables 2.2 and 2.3, the supplement term is negative for the Cassegrain and positive for the Gregory, giving values near the parabola, but hyperbolic for the Cassegrain and elliptical for the Gregory (Cases 4 and 8 of Table 3.2). The difference from the parabolic form only becomes appreciable for small values of m_2.

In the limit, afocal case, Eq. (3.115) is singular. If we substitute $L = m_2(f_1' - d_1)$ from Eq. (2.75), the supplement is again zero in the afocal case, confirming that the classical and aplanatic forms are identical, i.e. parabolic.

We have seen that the aplanatic supplement for the primary is small for normal values of m_2. In manufacture, this is insignificant unless the primary is tested against a full-aperture flat in autocollimation, giving a null-test (RTO II, Chap. 1) in the classical case. The supplement for the secondary amounts to about 14% in the cases of Table 3.2 with $m_2 = \pm 4$. It increases rapidly for smaller values of m_2. It is important to note that, while the aplanatic forms of the secondaries are still hyperbolae or ellipses to a third order, these conic sections are no longer such that the foci are at the primary and secondary images, as is the case for the classical forms. This is a consequence of the fact that the primary image is not free from spherical aberration in the aplanatic case.

The amount of spherical aberration at the prime focus is given from Table 3.5 and Eq. (3.30):

$$(S_I)_{1\,Aplan} = \left(\frac{y_1}{f'}\right)^4 \left(-f'\zeta_{Aplan}\right) = -\left(\frac{y_1}{f'}\right)^4 f'\frac{m_2^3}{4}\left[1 + (b_{s1})_{Aplan}\right] \tag{3.116}$$

From Eq. (3.114), this reduces to

$$(S_I)_{1\,Aplan} = \left(\frac{y_1}{f'}\right)^4 \frac{f'L}{2d_1}, \tag{3.117}$$

a negative value for the Cassegrain (f' positive) and a positive value for the Gregory (f' negative). The differences between the values for surfaces 1 and 1* in Cases 4 and 8 of Table 3.3 for S_I correspond to Eq. (3.117). They are small (-0.581 in the RC case) compared with the total contribution of the primary as a spherical mirror ($+16$) but larger than that of a spherical primary of the same final focal length f' (compare Case 1 of Table 3.3). The aberration thus makes the prime focus unusable without a corrector. However, it is fortunate that the primary form in the (commonly used) RC case is hyperbolic, since this is favourable in the design of PF field correctors, correcting above all the field coma (see Chap. 4).

We now consider the astigmatism of the aplanatic 2-mirror telescope. The general relation for $\sum S_{III}$ is given by Eq. (3.61) and in Table 3.5. Comparison with Eqs. (3.22) shows that the term in s_{pr1}^2 contains the spherical aberration while that in s_{pr1} contains the coma, both of which are zero. So, even if $s_{pr1} \neq 0$, there is no change in astigmatism. If, therefore, for IR observation the stop is placed at the secondary, there is no effect on $\sum S_I$, $\sum S_{II}$, $\sum S_{III}$, $\sum S_{IV}$ in an aplanatic telescope, whereas in the classical telescope $\sum S_{III}$ is changed because $\sum S_{II} \neq 0$. The astigmatism of the aplanatic telescope is

$$\left(\sum S_{III}\right)_{Aplan} = \left(\frac{y_1}{f'}\right)^2 \left[f'\frac{(f'+d_1)}{L} + \frac{d_1^2}{L}\xi_{Aplan}\right]u_{pr1}^2 \tag{3.118}$$

Substituting from (3.108) gives

$$\left(\sum S_{III}\right)_{Aplan} = \left(\frac{y_1}{f'}\right)^2 \frac{f'}{L}\left(f' + \frac{d_1}{2}\right)u_{pr1}^2 \tag{3.119}$$

Comparing with the classical telescope, Eq. (3.96), we find a similar result except that d_1 is replaced by $d_1/2$ in the bracket. In the RC case, f' is positive, d_1 negative, so there is a modest increase of astigmatism. In the aplanatic Gregory, there is a modest reduction. Cases 4 and 8 of Table 3.3 reveal these effects in the normalized case.

Equation (3.119) is expressed, as in Eq. (3.96) for the classical telescope, in terms of the constructional parameters f', L and d_1. These relations, and all the others derived in this section, can, if preferred, be equally well expressed in terms of the parameters f_1', m_2 and R_A. Chap. 2 (§ 2.2.5.2) gives all the necessary conversion formulae. Here we make use of

$$f' = m_2 f_1' \tag{2.55}$$

$$R_A = \frac{L}{f'} \tag{2.58}$$

$$R_A = 1 - \frac{d_1}{f_1'} \tag{2.72}$$

Substitution in Eq. (3.119) gives the form

$$\left(\sum S_{III}\right)_{Aplan} = \left(\frac{y_1}{f_1'}\right)^2 \frac{f_1'}{m_2 R_A} \left[1 + \frac{(1 - R_A)}{2m_2}\right] u_{pr1}^2 \tag{3.120}$$

In the afocal case, since $m_2 \rightarrow \infty$ and the other quantities remain finite, the astigmatism is zero from (3.120), in agreement with Table 3.6 and the convergence of the classical and aplanatic solutions in this limit case.

The field curvature $\sum S_{IV}$ is independent of the aspheric form of the mirrors and is therefore identical in the aplanatic case to that of the classical telescope treated above.

The RC aplanatic form of the Cassegrain telescope, is, and will always remain, the ultimate optical design for a 2-mirror telescope, since it combines the compactness and telephoto effect of Cassegrain geometry with optimum aberration correction possibilities (2 conditions) according to the Schwarzschild theorem (§ 3.2.6.1). Its significance increases as $|m_2|$ is reduced and the angular field is increased. The development of arrays of electronic detectors, above all CCDs, and multi-object spectroscopy increase the advantages of the RC solution over the classical Cassegrain.

A typical angular field for a modern RC telescope is $\pm$ 15 arcmin, as with the ESO 3.5 m NTT. This has a final aperture ratio of f/11 and $m_2 = -5$. Figure 3.4 shows the spot-diagrams[2] for this field for the cases of a classical

---

[2] Spot-diagrams in this book set up to illustrate examples of systems considered to be fundamental in the evolution of the theory of the reflecting telescope are represented in formats given by the ZEMAX-EE optical design program. In the case of all reflecting (catoptric) systems, the "through-focus" matrix representation (see § 3.2.5.3) is used; for systems including refracting elements (catadioptric) the "single-focus field-wavelength" matrix is used. In all cases, the spot-diagram matrix is accompanied by the computer drawing of the system and the input calculation data. The latter has been taken over from computer optimization results *without* removal of redundant digits in the output data. The same also applies to system data given in tables in the text in this chapter. The removal of redundant digits from the output data is part of the design process of tolerancing and preparation of manufacturing specifications, a process which has not been undertaken in the design of the system examples given in Chapters 3 and 4 of this book.

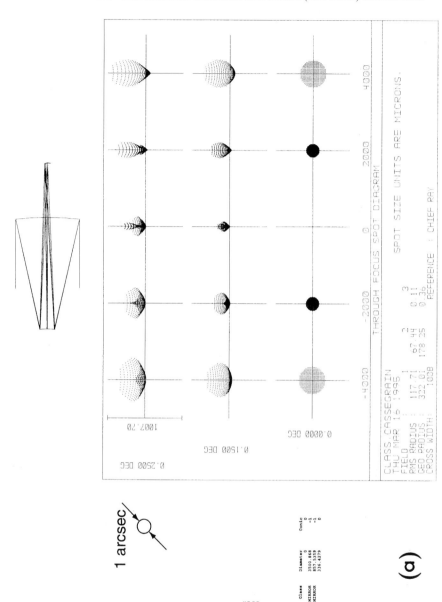

Fig. 3.4. (a) Spot-diagrams for a classical Cassegrain telescope with the geometry of the ESO 3.5 m NTT (f/11; $m_2 = -5$) for an optimum field curvature $r_c = -1955$ mm (concave to the incident light)

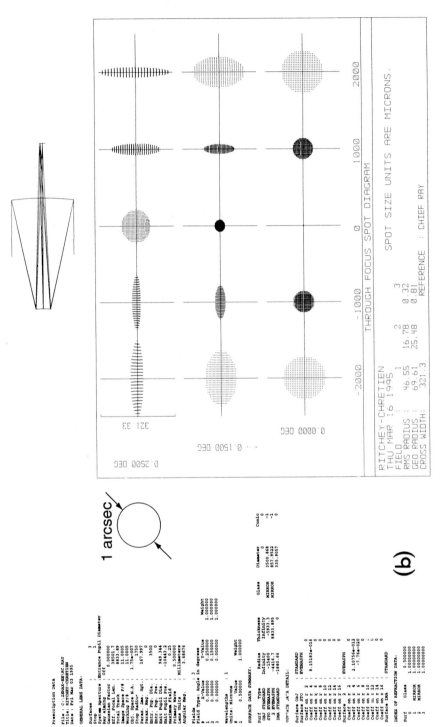

Fig. 3.4. (b) Spot-diagrams for an RC aplanatic telescope with the geometry of the ESO 3.5 m NTT (f/11; $m_2 = -5$) for an optimum field curvature $r_c = -1881$ mm

Cassegrain and an RC aplanatic telescope. The coma in the former case is the dominant field aberration. The form of the spot-diagrams for the basic aberrations is given in § 3.3. The point of the arrow-shaped coma figure points towards the axis (field centre) of the telescope. Such comatic images are common in star-field photographs with older telescopes. Bahner [3.5] gives an example from a PF photograph with an f/4 parabolic primary. By contrast, the spot-diagrams in the RC case are completely symmetrical. This is because the field curvature is chosen to give the best mean focus for the residual astigmatism, giving a round geometrical image (see § 3.3). If the focus is shifted slightly one way or the other, the astigmatic lines appear. The absence of symmetry about the horizontal axis makes the coma pattern the most objectionable of all aberrations.

The Gregory aplanatic form is only interesting if its intermediate image is useful (solar telescopes); or its extra length is unimportant so that the technical simplification of a concave secondary for testing is decisive, or its overcorrected field curvature is considered determinant.

A number of publications deal with the whole range of 2-mirror aplanatic forms. An excellent treatment was given in 1948 by Theissing and Zinke [3.15]. The properties of all aplanatic solutions are given in analytical and graphical form for the case in which the final image is formed in the plane of the primary. A more recent paper by Krautter [3.16] extends this general treatment, covering some systems with corrector plates and also the grazing incidence variants of the Mersenne forms for X-ray telescopes, due to Wolter [3.17]. Lens forms are also treated.

c) Telescopes with a spherical secondary: the Dall-Kirkham (DK) Cassegrain form. As is the case for many practical advances in reflecting telescope technology, the suggestion for a Cassegrain with a spherical secondary came out of the amateur telescope-making movement. It is clear that the Schwarzschild theorem given in § 3.2.6.1 can be applied to the general case of a 2-mirror telescope for which we require the fulfilment of only *one* condition, $\sum S_I = 0$, in the final focus. There are an infinite number of solutions, of which the classical telescope is a special case, for which $\sum S_I = 0$ for the prime focus as well as for the final focus. Two other cases are of special technical interest: telescopes with a spherical secondary and those with a spherical primary. As before, both Cassegrain and Gregory forms exist, but again the Cassegrain form is the more important in practice.

Although the first proposal was probably made by Dall, the first clear description was apparently given by Kirkham [3.18]: it is generally known as the Dall-Kirkham (DK) form of the Cassegrain. In Chap. 5 and RTO II, Chap. 1, we shall discuss the problems created by the manufacture and test of the classical convex hyperbolic secondary: for amateurs the task has always been formidable, so the technical advantages of a spherical secondary, tested by a concave negative, are enormous. However, as we shall see, this ease of manufacture is accompanied by a high price in field coma. Nevertheless, in certain circumstances, the DK form has legitimate application even for professional telescopes and a number of such instruments are in use: at the ESO La Silla Observatory, for example, the 1.4 m Coudé Auxiliary Telescope (CAT) and the Dutch 0.9 m national telescope.

The condition for spherical aberration requires from Eq. (3.59)

$$\left(\sum S_I \right)_{DK} = \left(\frac{y_1}{f'} \right)^4 \left(-f' \zeta_{DK} + L\xi_{DK} \right) = 0 \, , \tag{3.121}$$

in which the suffix DK should be understood to cover both Cassegrain and Gregory forms. By definition, for the DK form, we have

$$(b_{s2})_{DK} = 0 \, , \tag{3.122}$$

giving from Eq. (3.41)

$$\xi_{DK} = \frac{(m_2 + 1)^3}{4} \left(\frac{m_2 - 1}{m_2 + 1} \right)^2 = \frac{(m_2 + 1)}{4}(m_2 - 1)^2 \tag{3.123}$$

Also, from Eq. (3.30), we have

$$\zeta_{DK} = \frac{m_2^3}{4} \left[1 + (b_{s1})_{DK} \right] \tag{3.124}$$

Substitution in (3.121) gives

$$(b_{s1})_{DK} = -1 + \frac{L}{f'} \frac{(m_2 + 1)(m_2 - 1)^2}{m_2^3} \tag{3.125}$$

as the form required on the primary to correct the spherical aberration. For the normalized Cassegrain DK of Case 5 in Table 3.2 with $L=+0.225$, $f' = +1$ and $m_2 = -4$, Eq. (3.125) gives $(b_{s1})_{DK} = -0.73633$, an ellipse. The primary for the DK form is thus somewhat less aspheric than the parabola of the classical telescope. In the Gregory case, the primary would be a hyperbola but nearer the parabola, because $m_2 = +4$ and $f' = -1$.

The form of Eq. (3.125) becomes singular in the afocal case. If we substitute from (2.55) and (2.75) the relations

$$f' = m_2 f_1' \quad \text{and} \quad L = f' - m_2 d_1 \, ,$$

then we have the generally determinate form

$$(b_{s1})_{DK} = -1 + \left(1 - \frac{d_1}{f_1'} \right) \left[\frac{(m_2 + 1)(m_2 - 1)^2}{m_2^3} \right] \tag{3.126}$$

In the afocal limit case with $m_2 \to \infty$, the square bracket term becomes unity, giving

$$(b_{s1})_{DK}_{Afoc} = -1 + \left(1 - \frac{d_1}{f_1'} \right) = -\frac{d_1}{f_1'} = -1 + R_A \tag{3.127}$$

from Eq. (2.72). From Table 2.3 we see that R_A is positive in the Cassegrain (normal DK) case, again giving the elliptical form for the primary; while R_A is negative for the Gregory case, giving a hyperbola.

The function

$$f(m_2) = \frac{(m_2 + 1)(m_2 - 1)^2}{m_2^3} \; , \tag{3.128}$$

contained in the square bracket of (3.126) and representing the limit case of ξ in (3.41) when $b_{s2} = 0$, has very interesting properties. If we set $f(m_2) = +1$, then (3.128) gives

$$m_2^2 + m_2 - 1 = 0 \tag{3.129}$$

Apart from the solution $m_2 = \pm\infty$, given above from Eq. (3.128) for the limit case of a Mersenne-type, afocal telescope with a spherical secondary, Eq. (3.129) has two other roots $m_2 = -1.618\,034$ and $m_2 = +0.618\,034$, the first corresponding to the Cassegrain (normal DK) case, the second to the Gregory case. The first value is the famous magic number, τ, whose remarkable geometrical properties were first recognised by Pythagoras and led to the Fibonacci series in plant growth [3.19] [3.20]. The second root is, accordingly, simply the reciprocal of the first.

Another important limit case is given by $m_2 = \pm 1$, whereby $f(m_2) = 0$. From (3.126) the resulting form of the primary is parabolic: this must be the case since the spherical secondary has no optical power and has become a folding flat in the Cassegrain case (Fig. 2.15). The Gregory limit case with $m_2 = +1$ was discussed in §2.2.5 and corresponds to a secondary concentric with the primary image, a useless case in practice because of detector obstruction.

Figure 3.5 shows the complete function $f(m_2)$. The left-hand curve refers to the normal Cassegrain DK case. It has a very flat maximum at $m_2 = -3$, but the function is virtually constant between $+1.10$ and $+1.185$ for the entire range of practical m_2 values. From Eq. (3.126), this means the form of the primary is largely determined by the obstruction ratio R_A in all practical cases. Magnifications numerically less than unity give in both Cassegrain and Gregory cases a very steep rise to ∞ at $m_2 = 0$; but practical systems in this range are uncommon. The unit values at $m_2 = -\tau$ and $+1/\tau$ mean that Eq. (3.127) also applies in these cases: the asphericity of the primary depends strictly only on the obstruction ratio R_A. So the magic number also retains its powerful link with geometry in this optical case.

The arms of the function refer to the Cassegrain on the left and the Gregory on the right only in the normal case of a real final image. For virtual images they invert these roles, but such cases are rarely of practical significance.

We must now consider the *field coma* of the DK telescope. From Eq. (3.60) or Table 3.5, again neglecting the second term because of the condition $\sum S_I = 0$, we have

$$\left(\sum S_{II}\right)_{DK} = \left(\frac{y_1}{f'}\right)^3 \left[-d_1\xi_{DK} - \frac{f'}{2}\right] u_{pr1} \tag{3.130}$$

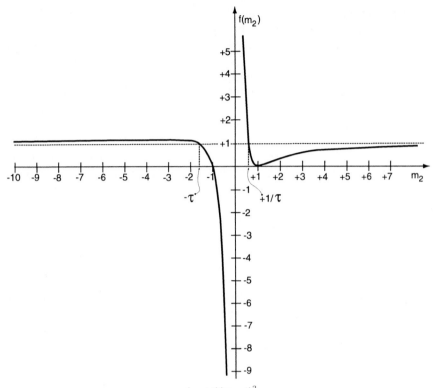

Fig. 3.5. The function $f(m_2) = \frac{(m_2+1)(m_2-1)^2}{m_2^3}$ for DK telescopes. The left-hand curve refers to Cassegrain solutions, the right-hand curve to Gregory solutions, if the image is real

Substituting from (3.123) gives

$$\left(\sum S_{II}\right)_{DK} = \left(\frac{y_1}{f'}\right)^3 \left[-\frac{d_1}{4}(m_2+1)(m_2-1)^2 - \frac{f'}{2}\right]u_{pr1} \qquad (3.131)$$

for the normal focal case. Similarly, Eq. (3.68a) or Table 3.6 give for the afocal case the particularly simple form in terms of f_1' and d_1:

$$\left(\sum S_{II}\right)_{DK \atop Afoc} = \left(\frac{y_1}{f_1'}\right)^3 \left(-\frac{d_1}{4}\right)u_{pr1} \qquad (3.132)$$

With the normalized geometry of Table 3.2, the DK case with $m_2 = -4$ gives from (3.131)

$$\left(\sum S_{II}\right)_{DK} = -3.63281 - 0.5 = -4.13281 \ ,$$

the value appearing in Table 3.3, Case 5. This must be compared with the value -0.5 for a classical Cassegrain and shows the high price that must be paid for the manufacturing convenience of the DK solution: the field coma is increased by a factor of over 8 compared with the classical Cassegrain with this typical geometry! Since field coma as a third order aberration varies linearly with field, it follows that, for a given acceptable coma limit, the field area is reduced by almost two orders of magnitude.

The *field astigmatism* of the DK telescope is given in the focal case by Eq. (3.61) or Table 3.5. The fourth term is again zero with $\sum S_I = 0$, but the third term is only zero if s_{pr1} is zero, i.e. the stop is at the primary. *Assuming* $s_{pr1} = 0$, we have

$$\left(\sum S_{III}\right)_{DK,0} = \left(\frac{y_1}{f'}\right)^2 \left[\frac{f'}{L}(f'+d_1) + \frac{d_1^2}{L}\xi_{DK}\right] u_{pr1}^2 , \qquad (3.133)$$

giving with (3.123)

$$\left(\sum S_{III}\right)_{DK,0} = \left(\frac{y_1}{f'}\right)^2 \left[\frac{f'}{L}(f'+d_1) + \frac{d_1^2}{4L}(m_2+1)(m_2-1)^2\right] u_{pr1}^2 \qquad (3.134)$$

With the geometry and normalization of Table 3.2, Eq. (3.134) gives with $m_2 = -4$

$$\left(\sum S_{III}\right)_{DK,0} = +3.58333 - 3.12826 = +0.45507 ,$$

the value in Table 3.3, Case 5. Thus, whereas the RC solution produced a modest increase (about 12%) of the astigmatism of the classical Cassegrain, the DK solution virtually compensates it, giving a value of only about one eighth with this geometry. However, the value of this virtue is limited in practice, since the image quality is totally dominated by the coma.

In the afocal case, Eq. (3.69a) or Table 3.6 give with $s_{pr1} = 0$

$$\left(\sum S_{III}\right)_{DK,0 \atop Afoc} = \left(\frac{y_1}{f_1'}\right)^2 \left[\frac{d_1^2}{4(f_1'-d_1)}\right] u_{pr1}^2 \qquad (3.135)$$

The field curvature is unaffected by mirror asphericities and is identical with that of the other solutions of the same geometry.

Figure 3.6 shows the spot-diagrams for the DK solution with the same geometry (NTT) as in Fig. 3.4. The circle again represents 1 arcsec but the maximum semi-field angle is only one tenth, i.e. ± 1.5 arcmin instead of ± 15 arcmin. These spot-diagrams are for a flat field, since the field curvature effect is negligible over this small field. Since the astigmatism is also negligible, these spot-diagrams effectively represent pure coma.

d) Telescopes with a spherical primary. As with the other solutions, this form also exists in both Cassegrain and Gregory form. The Cassegrain form with a spherical primary is termed the Pressmann-Camichel form by Rutten and van Venrooij [3.12(a)], but this nomenclature has not been so

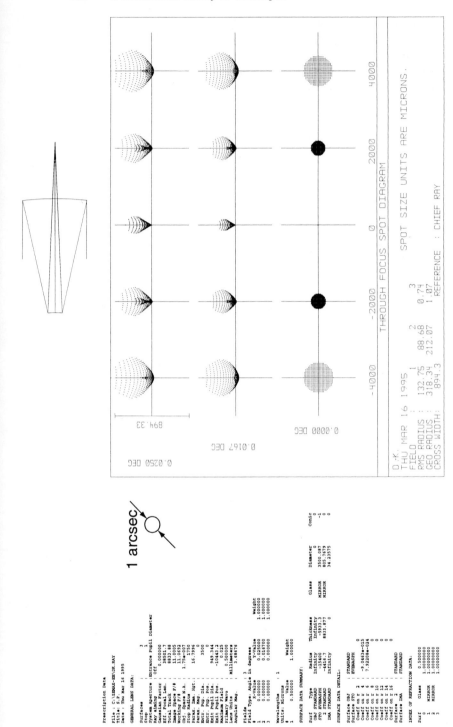

Fig. 3.6. Spot-diagrams for a DK Cassegrain telescope with the geometry of the ESO 3.5 m NTT (f/11; $m_2 = -5$) for a flat field. Compare with Fig. 3.4 where the field is 10 times larger

generally applied as that of the Dall-Kirkham (DK) above. We shall designate it as the SP (spherical primary) solution. The manufacturing advantage of a spherical primary is obvious. However, the ease of testing concave primaries compared with convex Cassegrain secondaries makes this solution, in general, less attractive for amateurs than the Dall-Kirkham, particularly as the field coma is far worse, as we prove below. Spherical primaries have more interest for the largest *professional* sizes, particularly in segmented construction (see RTO II, Chap. 3). The SP form has great generic interest in connection with more sophisticated telescope solutions using more mirrors, as we shall see in § 3.6.5.3. But 2-mirror SP solutions have also been proposed, for the giant (25 m) USSR telescope in the 1970's, or more recently for the French TEMOS project of more modest size, for which the enormous field coma was to have been corrected by refracting field correctors near the image, a very difficult undertaking as we shall see in Chap. 4, or by additional mirror correctors avoiding chromatic aberration.

The 2-mirror SP telescope is defined by $(b_{s1})_{SP} = 0$ in the equation (3.59)

$$\left(\sum S_I\right)_{SP} = \left(\frac{y_1}{f'}\right)^4 \left[-f'\zeta_{SP} + L\xi_{SP}\right] = 0$$

ζ and ξ are defined by (3.30) and (3.41), giving

$$\left(\sum S_I\right)_{SP} = \left(\frac{y_1}{f'}\right)^4 \left[-f'\frac{m_2^3}{4} + \frac{L(m_2+1)^3}{4}\left\{\left(\frac{m_2-1}{m_2+1}\right)^2 + (b_{s2})_{SP}\right\}\right]$$
$$= 0 \tag{3.136}$$

This leads to the necessary form of the secondary as

$$(b_{s2})_{SP} = \frac{f'}{L}\left(\frac{m_2}{m_2+1}\right)^3 - \left(\frac{m_2-1}{m_2+1}\right)^2 \tag{3.137}$$

From (3.93), the second term is the form for the secondary of the classical telescope so that we may write (3.137) in the form

$$(b_{s2})_{SP} = (b_{s2})_{cl} + \frac{f'}{L}\left(\frac{m_2}{m_2+1}\right)^3 \tag{3.138}$$

With the normalization and geometry of Table 3.2, Case 6, with $m_2 = -4$ for the Cassegrain form, $(b_{s2})_{cl} = -2.77778$ and the additional term is $+10.53498$ giving

$$(b_{s2})_{SP,Cass} = +7.75720 \,,$$

as given in Table 3.2. From Eqs. (3.12) and (3.13), this corresponds to an oblate spheroid of high asphericity. As for the DK case, Eq. (3.138) can be transformed into the non-singular, general form for the afocal case

$$(b_{s2})_{SP} = (b_{s2})_{cl} + \frac{1}{R_A}\left(\frac{m_2}{m_2+1}\right)^3 \,, \tag{3.139}$$

in which

$$\frac{1}{R_A} = \frac{f_1'}{f_1' - d_1}$$

from Eq. (2.72).

In the Gregory case, $(b_{s2})_{cl} = -0.36000$, while the second term in (3.138) gives with $f' = -1$ and $m_2 = +4$ in the normalized case of Table 3.2 the value -2.27556. The total gives

$$(b_{s2})_{SP,Greg} = -2.63556,$$

a modest hyperbola, a much easier form to test and make than the convex oblate spheroid of the equivalent Cassegrain form.

Since $\sum S_I = 0$, the *field coma* of the SP form is given from Eq. (3.60) or Table 3.5 by

$$\left(\sum S_{II}\right)_{SP} = \left(\frac{y_1}{f'}\right)^3 \left[-d_1 \xi_{SP} - \frac{f'}{2}\right] u_{pr1} \tag{3.140}$$

From (3.41) and (3.137), this reduces to

$$\left(\sum S_{II}\right)_{SP} = \left(\frac{y_1}{f'}\right)^3 \left[-\frac{f'}{2}\left(1 + \frac{d_1 m_2^3}{2L}\right)\right] u_{pr1}, \tag{3.141}$$

the second term in the round bracket being the supplement compared with the classical telescope. With the normalized geometry of Tables 3.2 and 3.3, this supplement is very large, giving in the Cassegrain case

$$\left(\sum S_{II}\right)_{SP,Cass} = -14.27778,$$

a value some three times higher than that of the equivalent DK and nearly thirty times higher than that of the classical telescope (Table 3.3, Cases 3 and 6).

In the afocal case, substitution of the relations (2.55) and (2.72) in (3.141) leads to the non-singular form

$$\left(\sum S_{II}\right)_{SP} \atop Afoc = \left(\frac{y_1}{f_1'}\right)^3 \left(\frac{-d_1}{4R_A}\right) u_{pr1}, \tag{3.142}$$

which can also be deduced at once from Table 3.6 and (3.137).

We now derive the field astigmatism of the SP form. We assume, as for the DK form, that the stop is at the primary: otherwise there will be a very large stop-shift term from the third term of (3.61) because of the huge value of $(\sum S_{II})_{SP}$. Then

$$\left(\sum S_{III}\right)_{SP,0} = \left(\frac{y_1}{f'}\right)^2 \left[\frac{f'}{L}(f' + d_1) + \frac{d_1^2}{L}\xi_{SP}\right] u_{pr1}^2, \tag{3.143}$$

giving from (3.41) and (3.137)

$$\left(\sum S_{III}\right)_{SP,0} = \left(\frac{y_1}{f'}\right)^2 \left[\frac{f'}{L}(f'+d_1) + \frac{f'}{4}\left(\frac{d_1}{L}\right)^2 m_2^3\right] u_{pr1}^2 \qquad (3.144)$$

With the geometry and normalization of Tables 3.2 and 3.3

$$\left(\sum S_{III}\right)_{SP,0 \atop Cass} = +3.58333 - 11.86420 = -8.28087\,,$$

the second term being the change compared with the classical telescope. Apart from the huge coma, we see there is also a serious increase in astigmatism.

In the afocal case, Eq. (3.144) reduces with the usual substitutions from (2.55) and (2.72) to the non-singular form

$$\left(\sum S_{III}\right)_{SP,0 \atop Afoc} = \left(\frac{y_1}{f_1'}\right)^2 \left[\frac{1}{4R_A}\frac{d_1^2}{(f_1'-d_1)}\right] u_{pr1}^2\,, \qquad (3.145)$$

a result which can also be derived at once from Table 3.6 and Eq. (3.137). Also, from (2.72),

$$R_A = (f_1' - d_1)/f_1'$$

We see that the SP form is afflicted not only with much worse coma than the DK form, but also with far more astigmatism. The field curvature is the same as for the other solutions. It is therefore rarely a useful practical form as such, but has great interest as a feeder telescope for more complex systems (see § 3.6.5.3).

Finally, Fig. 3.7 shows the spot-diagrams for the SP solution with the same geometry (NTT) as in Fig. 3.6. Again with a circle of 1 arcsec, the field is limited by the coma to only about ± 20 arcsec. As with the DK solution, the spot-diagrams show pure coma only, since both astigmatism and field curvature are negligible at this field.

3.2.6.4 General conclusions on 2-mirror telescopes from Tables 3.2 and 3.3.

The aberration theory above, combined with the geometrical advantages of the telephoto effect treated in Chap. 2, reveal why the Cassegrain geometry, above all in RC or classical form, has become the standard solution for modern telescopes. The advantage of the RC form increases with angular field and inversely with m_2. The DK solution is of interest for amateurs, and also for low-cost professional telescopes if a limited angular field is acceptable. The SP solution will rarely have practical interest as such, but the theory has great interest for its use as a feeder telescope for other solutions (see § 3.6.5.3).

The equivalent Gregory forms are longer and will only be interesting because of three properties: the real primary image for solar telescopes, the

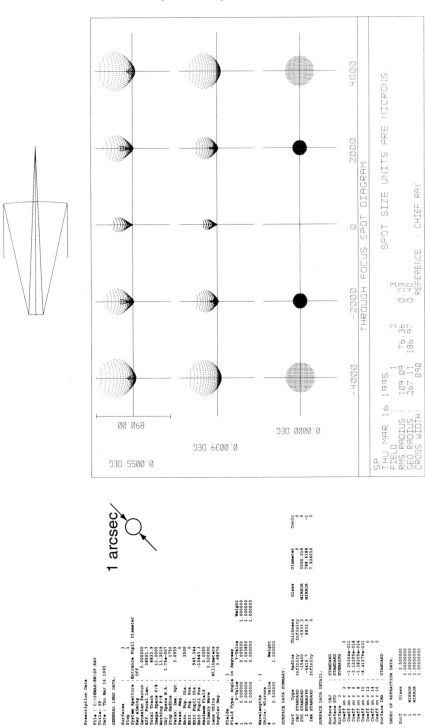

Fig. 3.7. Spot-diagrams for an SP Cassegrain telescope with the geometry of the ESO 3.5 m NTT (f/11; $m_2 = -5$), for a flat field. Compare with Fig. 3.6 with field $4\frac{1}{2}$ times larger and Fig. 3.4 with field 45 times larger

technical simplification in manufacture and test of the concave secondary compared with the convex secondary in the Cassegrain case, and the sign of the field curvature which is favourable for the compensation of that of auxiliary instruments.

It should be remembered that the technical price for the compactness of the Cassegrain telescope compared with the longer 1-mirror telescope is the associated problem of the severe *centering tolerances* (see § 3.7.2) unless the centering is actively controlled (see RTO II, Chap. 3). But the RC form of the Cassegrain offers a field correction which is impossible with a 1-mirror telescope.

3.2.7 Other forms of aplanatic 2-mirror telescopes (Schwarzschild, Couder)

3.2.7.1 The Schwarzschild original aplanatic telescopes. The fundamental contribution of Schwarzschild in 1905 [3.1] with the extension of aberration theory to field aberrations of reflecting telescopes was briefly discussed in § 3.2.6.1. Its significance has been further discussed from a historical viewpoint by the author [3.13]. Schwarzschild set up Eqs. (3.109) and (3.114) defining an aplanatic telescope in a somewhat different form and thereby recognised their validity for *any* chosen geometry. His initial aim, however, was more ambitious: he sought a 2-mirror solution permitting the correction of the first 4 Seidel conditions, $S_I = S_{II} = S_{III} = S_{IV} = 0$. In fact, his equations led him to such a solution, but he realised at once that it was not practicable for telescopes since the primary is convex, giving a secondary larger than the primary. Nevertheless, this solution has found application in somewhat modified form in spectrograph optics under the name "Bowen camera".

Having satisfied himself that no practical 2-mirror solution was available which satisfied more than the two conditions given by the "Schwarzschild theorem", Schwarzschild then sought an aplanatic solution ($S_I = S_{II} = 0$) giving a "fast" telescope (f/3.0) for the slow photographic emulsions of the time. Had he applied his equations to the Cassegrain geometry of the 60-inch reflector for Mt. Wilson, under construction at that time by Ritchey, he would certainly have identified the RC solution seventeen years before it was formally published by Chrétien [3.14]; although Chrétien states elsewhere [3.21] that studies were already proceeding at Mt. Wilson in 1910, confirming the review of the situation given in § 5.2. Since the 60-inch reflector had a primary of f/5 and $m_2 = -3$, the field coma as a classical telescope was by no means negligible, so the advantage of an aplanatic solution would already have been clear in 1905. But, for Schwarzschild, the resulting aperture ratio of f/15 was much too slow.

Before we consider Schwarzschild's aplanatic system, it is most instructive to repeat his argument proving the impossibility of a practical telescope fulfilling the four conditions $\sum S_I = \sum S_{II} = \sum S_{III} = \sum S_{IV} = 0$.

The aplanatic requirement $\sum S_I = \sum S_{II} = 0$ is embodied in the aplanatic solution of § 3.2.6.3 (b) above, by the Eqs. (3.109) for b_{s2} and (3.114) for b_{s1}. For $\sum S_{III} = 0$, we must satisfy from (3.119)

$$f' + \frac{d_1}{2} = 0 , \tag{3.146}$$

giving

$$d_1 = -2f' \tag{3.147}$$

For $\sum S_{IV} = 0$, we have the condition from Table 3.5

$$\frac{1}{f_1'} = \frac{m_2 + 1}{L} = \frac{m_2 + 1}{m_2 f_1' - m_2 d_1} ,$$

from (2.75) and (2.55), which reduces to

$$f_1' = -m_2 d_1 \tag{3.148}$$

or

$$(f_1')^2 = -f'd_1 \tag{3.149}$$

Substitution from (3.147) then gives

$$f_1' = \pm f' \sqrt{2} \tag{3.150}$$

Now from (2.72)

$$R_A = 1 - \frac{d_1}{f_1'} , \tag{3.151}$$

giving with (3.147)

$$R_A = 1 \pm \sqrt{2} \tag{3.152}$$

or

$$R_A = -0.4142 \text{ or } +2.4142 \tag{3.153}$$

Consider the first value $R_A = -0.4142$ from the negative root of (3.150) and (3.153). This is normal in that $|R_A| < 1$, implying that the secondary is smaller than the primary and that the primary is concave. From (3.150)

$$f_1' = -f' \sqrt{2} \tag{3.154}$$

Since f_1' is negative, f' must be positive and from (3.147)

$$d_1 = +f_1' \sqrt{2} \tag{3.155}$$

As must be the case, d_1 is negative like f_1' and $|d_1| > |f_1'|$. The telescope is then a Gregory form with a real primary image. However, the secondary must be convex with the same curvature as the primary to satisfy $\sum S_{IV} = 0$, giving a *virtual* final image and the positive final f' required by (3.154). The form is therefore useless for a 2-mirror telescope since it does not form a real image.

Consider now the positive value $R_A = +2.4142$ arising from the positive root of (3.150) and (3.152). $R_A > 1$ implies that the secondary is *larger* than the primary, which is now *convex* with f'_1 positive, i.e. no intermediate real image. From (3.150)

$$f'_1 = +f'\sqrt{2}, \qquad (3.156)$$

implying that f' is positive with f'_1 positive. From (3.147), it follows that

$$d_1 = -2f' = -f'_1\sqrt{2} \qquad (3.157)$$

and is the required negative quantity. Figure 3.8 shows the resulting system. To give $\sum S_{IV} = 0$, the secondary must be *concave* with the same curvature as the primary. The magnification m_2 of the secondary is given from (3.148) and (3.157) by

$$m_2 = +0.7071 \qquad (3.158)$$

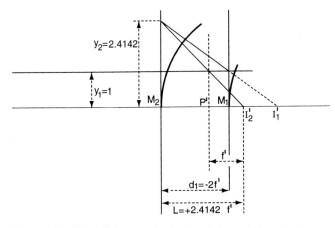

Fig. 3.8. Karl Schwarzschild's first impractical telescope solution fulfilling four Seidel conditions [3.1]

The "axial obstruction ratio" was given by (3.152) for the positive root as

$$R_A = 2.4142, \qquad (3.159)$$

meaning that, without field and with the stop at the primary, the hole in M_2 produces a linear obstruction of $1/2.4142$ on M_2 or $(R_A)_2 = 0.4142$. Since the final image is formed behind M_1, there is also a linear obstruction ratio $(R_A)_1$ from the hole in M_1 given by

$$(R_A)_1 = 2.4142 \left(\frac{L - 2f'}{L} \right), \qquad (3.160)$$

giving from (2.75)

$$(R_A)_1 = 0.4142 \,, \tag{3.161}$$

the same value as for $(R_A)_2$. Since Schwarzschild was interested in an angular field of several degrees, in order to profit from the excellent theoretical correction potential, the actual obstruction seemed prohibitive. Also, a secondary larger than the primary at once excludes the solution in practice for telescopes of significant size. Its physical length $L = 2.4142f'$ is also unfavourable. However, it does have the characteristic, essential for Schwarzschild at the time of being "fast"; a merit which explains its use in spectrograph cameras.

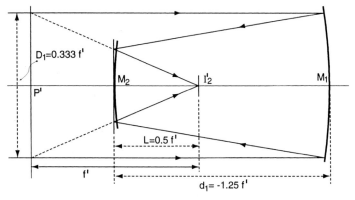

Fig. 3.9. Karl Schwarzschild's original aplanatic telescope (1905) [3.1][3.13]

Schwarzschild's conclusion that no practical 2-mirror telescope solution existed correcting four Seidel conditions was of fundamental importance in the development of reflecting telescope theory.

Having abandoned the aim of a 2-mirror telescope system correcting four Seidel aberrations, Schwarzschild decided to seek a solution based on the Schwarzschild theorem (§ 3.2.6.1) in which 2 aspheric mirrors could always correct the two conditions $\sum S_I$ and $\sum S_{II}$. He accepted the presence of finite astigmatism and field curvature, but imposed the condition that their optimum combination should give a flat field. He sought a system of a "speed" (f/3.0) which was quite revolutionary at the time, bearing in mind that a primary of f/5 was considered "fast" and that Cassegrain foci were rarely faster than f/15. He therefore considered a number of Cassegrain systems using a *concave* secondary [3] to achieve the desired high relative aperture of f/3.0, a speed becoming available in photographic objectives. This had two consequences: the final image position lay between the mirrors (Fig. 3.9) and the

---

[3] Because of the concave secondary, the Schwarzschild aplanatic telescope is sometimes referred to as a Gregory form. This is quite wrong as there is no real intermediate image: it is simply a Cassegrain telescope with $|m_2| < 1$.

magnification of the secondary m_2 was in the range $0 > m_2 > -1$. The relatively short focal length implied that existing plate sizes could cover a field of several degrees and thereby fully exploit optimum field correction. After a careful analysis of the obstruction and vignetting aspects, including those of the plateholder, Schwarzschild proposed the system of Fig. 3.9 which bears his name.

Complete data, as given by Schwarzschild and normalized to $f' = +1$, are given in Table 3.7. The aplanatic condition is defined from the given geometry by Eqs. (3.109) and (3.114) for $(b_{s2})_{Aplan}$ and $(b_{s1})_{Aplan}$ respectively. The secondary magnification m_2 is $+1/-2.5 = -0.4$ from (2.55), while $L = +0.5$ from (2.75) and $R_A = +0.5$ from (2.72). The power of the secondary is defined by (2.90) as $+0.83333$.

Table 3.7. Schwarzschild's data for the aplanatic telescope of Fig. 3.9

f'	$=$	$+1$
D_1	$=$	0.33333
Final f/ratio (ignoring obstruction)	$=$	f/3.0
Effective final f/no (with obstruction)	$=$	f/3.5
f_1'	$=$	-2.5 (f/7.5)
d_1	$=$	-1.25
L	$=$	$+0.5$
R_A	$=$	$+0.5$
f_2'	$=$	$+0.83333$
b_{s1}	$=$	-13.5
b_{s2}	$=$	$+1.96297$
Sagittal astigmatism (field $\pm 1°$)	$=$	-7 arcsec
Tangential astigmatism (field $\pm 1°$)	$=$	$+9$ arcsec

In its geometry, the Schwarzschild telescope suffers from the major drawbacks of a relatively high obstruction ratio (> 0.5 with the field effect), a system length of $1.25 f'$ and, above all, the exposed position of the final image which is difficult for baffling and awkward of access. For the telescope sizes of his time, however, these latter two drawbacks were less serious than they are today. The given geometry was determined by the effective "flat-field" condition as follows.

The astigmatism of an aplanatic telescope is given from Eq. (3.119) by

$$\left(\sum S_{III} \right)_{Aplan} = \left(\frac{y_1}{f'} \right)^2 \frac{f'}{L} \left(f' + \frac{d_1}{2} \right) u_{pr1}^2 , \qquad (3.162)$$

giving for the Schwarzschild values of f', d_1 and L and the usual normalization with $y_1 = u_{pr1} = 1$ the value

$$\left(\sum S_{III} \right)_{Schw.} = +0.75$$

Table 3.8. Comparison of the essential parameters in the evolution of the aplanatic telescope (from [3.13])

Form	Primary f/no	m_2	Final f/no	b_{s1}	b_{s2}
Schwarzschild (1905)	f/7.5	-0.4	f/3.0	-13.5	$+1.963$
Chrétien (1922)	f/2.6	-2.5	f/6.5	-1.147	-7.605
Modern RC	f/2.0	-4.0	f/8.0	-1.036	-3.160

From Table 3.5, the field curvature is given by

$$\sum S_{IV} \;=\; H^2 \left[\left(\frac{m_2}{f'} \right) - \left(\frac{m_2 + 1}{L} \right) \right] , \tag{3.163}$$

giving for the normalized system

$$\left(\sum S_{IV} \right)_{Schw.} = -1.6$$

From the last column of Table 3.3, the *effective field curvature* is given by

$$\text{EFC} \;=\; 2 \sum S_{III} + \sum S_{IV} , \tag{3.164}$$

giving

$$(\text{EFC})_{Schw.} = +1.5 - 1.6 = -0.1 ,$$

a value accepted by Schwarzschild as negligible. This small undercorrection of $\sum S_{IV}$ by $2 \sum S_{III}$ accounts for the small imbalance of the angular sagittal and tangential astigmatism values quoted in Table 3.7. The very large residuals (by today's standards) of -7 arcsec and $+9$ arcsec respectively reflect the coarse emulsions of the time. They also demonstrate how desirable was reduction of astigmatism in addition to aplanatism.

If m_2 is maintained unchanged (that is, the final f/no is unchanged) while L and R_A are reduced to improve the obstruction, then $|d_1|$ increases from (2.75) and $\sum S_{III}$ reduces from (3.162). But the negative value of $\sum S_{IV}$ increases rapidly, giving a rapid increase in the negative EFC residue from (3.164). Thus the effective flat-field condition forced Schwarzschild to accept the relatively high obstruction ratio. We shall see below that the Couder modification, which abandons the flat-field condition, opens up further possibilities.

The asphericities represented by b_{s1} and b_{s2} are relatively extreme forms. However, the hyperbolic form of the primary is not excessive in its deformation because the curvature is weak (f/7.5), as given by Eq. (3.11). The oblate spheroid of the secondary is, in practice, much more difficult because of test problems of this form.[4]

Table 3.8 gives a comparison of some fundamental parameters showing the evolution of the aplanatic telescope from Schwarzschild to modern telescopes, taken from ref. [3.13]. The modern RC telescope has a high value of m_2 and a primary form scarcely departing from the parabola. The hyperbola on the secondary is also modest compared with the original RC of Chrétien. The trend is determined by the increasing value of $|m_2|$, resulting from steeper primaries and a final f/no which would have been too slow for the emulsions available to Schwarzschild. With large $|m_2|$, the aplanatic solution is converging rapidly towards the common aplanatic/classical form of the afocal telescope.

Having derived his aplanatic, effective flat-field solution from third order theory, Schwarzschild applied the Abbe sine condition and his "Winkeleikonal" to the calculation of the total optical path of the rays forming the image in the field – see § 3.2.5.2. He thus calculated the higher order aberration effects, a beautiful and complete analysis of the imaging properties of the system in the pre-computer age. The constant optical path requirement was expressed as a differential equation in terms of polar coordinates of the secondary mirror. This led to an explicit analytical form for the polar equation of the secondary and then for the primary. Conversion to Cartesian coordinates and application of the sine condition of Eq. (3.84) led to explicit expressions for the forms of the two mirrors as a function of the aperture angle U'. These were then converted into infinite series giving the requirements for the correction of successive aperture orders of spherical aberration and coma, the terms up to y^4 agreeing with those deduced from third order theory. An excellent treatment of the general approach of such methods, all a consequence of the Fermat principle discussed briefly in § 2.2.2, is given by Schroeder [3.22].

Figure 3.10 shows the spot-diagrams for the original Schwarzschild telescope for an aperture of 1 m with f/3.0 and a field of $\pm 1°$, plotted on a flat field as he intended. The astigmatic limitation, growing with the square of the field according to Eq. (3.21), is very evident.

Schwarzschild's proposed form was the logical solution for his time, a fast system with a relatively slow primary. According to Dimitroff and Baker

---

[4] The reason why an oblate spheroid form is normally more difficult to test lies in the technology of *null-testing*. This fundamentally important test technology is treated in §§ 1.3.4 and 1.3.5 of RTO II. A *negative* Schwarzschild constant b_s (the normal case for telescope mirrors) requires a *positive* lens to compensate its aberration. A positive lens also forms a real image, which is essential in the test procedure. However, a *positive* b_s, corresponding to an oblate spheroid, requires a *negative* lens for compensation, producing a virtual image. A real image can then, in most cases, only be produced by a further (spherical) concave mirror, since positive lenses neutralize the desired aberration of the negative lens. The natural compensation for a negative b_s is illustrated by Eq. (1.77) in RTO II.

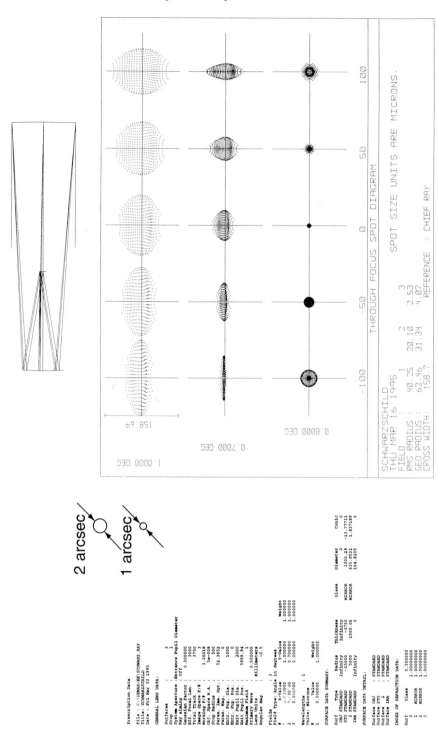

Fig. 3.10. Spot-diagrams for the Schwarzschild telescope 1905 [3.1] for an aperture of 1 m with f/3.0

[3.23(a)], two such telescopes of significant size with 24-inch and 12-inch apertures were made between the world wars in the United States. It seems unlikely this system will be manufactured again today, except perhaps in the Couder modification discussed below. But the precise form of the original Schwarzschild proposal is completely unimportant. His aims were later satisfied by Schmidt telescopes, and either by the primary foci of large telescopes with field correctors, or by Cassegrain foci with focal reducers. The fundamental importance of his work was the theoretical formulation in third order theory which opened the design path to all modern telescope solutions.

Before we leave the subject of the original layout of aplanatic telescopes, it is instructive to follow the elegant *graphical* demonstration of the application of the sine condition to their design given by Danjon and Couder [3.24(a)], from which Fig. 3.11 has been reproduced in our nomenclature. The illustration refers to the RC modification of a classical Cassegrain, but the construction would apply to any other aplanatic modification. In Fig. 3.11, M_1 and M_2 represent the poles of the mirrors of a classical Cassegrain telescope forming an image at I_2'. The principal plane in the Gaussian sense is $P_0'P_G'$, giving $f' = P_0'I_2'$. A ray at finite height Y_1 from infinity strikes the RC primary at A_{RC} and is reflected from the RC secondary at B_{RC} to I_2'. From the sine condition, obeyed by the aplanatic RC telescope, the projection of $I_2'B_{RC}$ back to the incident ray must cut that ray at P_{RC}', a point on the sphere $P_0'P_{RC}'$ centered on I_2', according to the relation

$$\frac{Y_1}{\sin U_{RC}'} = f' \tag{3.165}$$

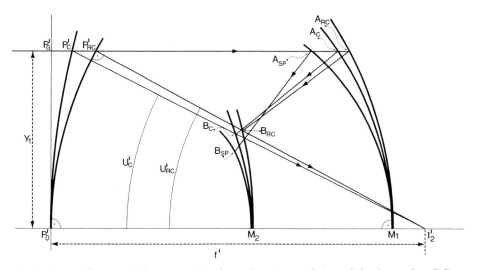

Fig. 3.11. Geometrical construction from the sine condition of the form of an RC telescope compared with a classical Cassegrain (from Danjon and Couder [3.24(a)])

from Eq. (3.85). Now, as stated in § 3.2.5.2 above, the elegant and simple form of the Abbe sine condition reveals that it takes account of *all* the aberration orders of coma in the *aperture*, although only the third order in the field. It is therefore a fundamentally different mathematical formulation of aperture aberrations from that of the Characteristic Function and third order Seidel theory. This is why the conversion from the sine condition to the corresponding third order coma, leading to the Staeble–Lihotzky condition, is by no means trivial. We require a similar transformation here to demonstrate the geometrical consequences of Fig. 3.11.

The treatment is much simplified if we apply an equivalent normalization to that of Table 3.2, but adapted to this special case, for which the second principal plane of the system $P_0'P_G'$ must be replaced by the reference sphere $P_0'P_{RC}'$ for aplanatism. For the aperture aberrations spherical aberration and coma, the only limitation to the aperture is when the reference sphere becomes a complete hemisphere. The maximum possible aperture angle U' is then 90°. We shall therefore start with the normalization $f' = Y_1 = \sin U' = 1$, where Y_1 and $\sin U'$ are real ray parameters normalized to 1 at the maximum aperture. The enormous reduction of Y_1 and $\sin U'$ in a real system will be applied later to link the conclusions with normal formulae for third order coma.

For the classical Cassegrain with its slightly steeper (parabolic) primary, the incident ray meets M_1 earlier at the point A_C. The classical secondary is also less eccentric than the RC form and therefore also lies to the left of the RC secondary. The result is, as shown, that the classical Cassegrain gives a slightly smaller U_C' angle than U_{RC}', the difference being $\delta U'$. This leads to the error $\delta f'$ in the sine condition of

$$\frac{Y_1}{\sin U_C'} = f' + \delta f' = 1 + \delta f' \tag{3.166}$$

Treating $\sin U_{RC}' = 1$ as a constant and differentiating Eq. (3.165) with respect to f' shows at once that

$$\delta Y_1 = \delta f'$$

This is the excess of the length of the chord of the projection of the reference sphere $P_0'P_C'$ to the vertical through I_2' beyond the radius $f' = 1$ of the reference sphere $P_0'P_{RC}'$. We now require the angular error $\delta U'$ corresponding to this offence against the sine condition $\delta f'$. Differentiating Eq. (3.165) with respect to U' with constant f' leads to an indeterminate result (infinity) for $\delta U'$ because at $U' = 90°$ the reference sphere is *orthogonal* to the Gaussian Y_1-axis $P_0'P_G'$! We must therefore replace the vertical axis by the reference sphere $P_0'P_{RC}'$ and measure Y_1 along this circle up to $U_{RC}' = 90°$. Setting then δY_1 as the increment at $U_{RC}' = 90°$, we have at once

$$\delta U' = \frac{\delta Y_1}{f'} \quad \text{and} \quad f' = 1$$

It follows that, for this extreme normalized case with $U' = 90°$ and U' *negative* (see Table 2.2 and Fig. 3.11: with U' negative, $\delta U'$ is positive, also $\sin U'$ is positive as shown in Eq. (3.165)),

$$\delta U' = \delta f' \tag{3.167}$$

Now the sine condition requires that this condition be met by *all* zonal apertures smaller than this maximum. However, as we have seen, the reference to a Gaussian principal plane represented by the Y_1-axis breaks down completely at the limit angle $U' = 90°$. In fact, the definitions relating to δY_1 are only strictly valid for the paraxial region. Beyond that, as Fig. 3.11 shows for a large angle U', the reference to the Y_1-axis no longer gives the same simple results for $\delta f'$ and $\delta U'$ in terms of δY_1, as revealed at once by Eqs. (3.165) and (3.166). This is because of higher order aberrations above the third order, which are automatically taken account of with the sine condition and the reference sphere $P_0' P_{RC}'$. For zonal rays, the terms emerging from Eqs. (3.165) and (3.166) involve trigonometrical functions which are inconvenient for conversion to normal third order aberration formulae. However, Eq. (3.167) provides a very convenient link. Since third order formulae are entirely based on *paraxial* parameters, the approximations involved are completely acceptable. We must also bear in mind for the construction of Fig. 3.11 that normal modern RC telescopes rarely have a final image $f/no\ N < 10$. Ritchey's first major RC telescope (see §5.2) was exceptionally fast ($f/6.8$) for photography at that time. With $N = 10$, $\sin U_{RC}' = 0.05$, $U_{RC}' = 3.18°$ and $\cos U_{RC}' = 0.9987$, a negligible difference from 1 and proving the high accuracy of third order theory for such telescopes when marginal ray data is used to establish the paraxial parameters used in third order aberration formulae.

We consider now Fig. 3.11 from the point of view of such real apertures. For the classical telescope, the projection of the emergent ray meets the incident ray at P_C' instead of P_{RC}' and the circle $P_0' P_C'$ has a longer radius than that required for aplanatism $P_0' I_2' = 1$. Since $\cos U' \simeq 1$, the extension $\delta f'$ is given by

$$\delta f' = P_{RC}' P_C'$$

To the first order, $P_{RC}' P_C'$ is given from Eq. (3.11) by

$$P_{RC}' P_C' = \frac{Y_1^2}{2}(1 - c_c) = \delta f', \tag{3.168}$$

where c_c is the curvature for the classical Cassegrain and 1 the curvature required for aplanatism.

If we now consider an oblique beam at a small field angle producing field coma in the classical Cassegrain, then the angular aberration of the equivalent oblique finite ray to the ray $P_{RC}' A_{RC}$ which defines the sagittal coma is simply $\delta U'$, the equivalent angular aberration to $\delta f'$ in Eq. (3.168).

We shall show in § 3.3.2 that, with the focal length normalization of Table 3.3 and of Eq. (3.166), the *radius* of the angular coma in the sagittal section is [5]

$$\delta U' = -\frac{\sum S_{II} Y_1^2}{2},$$ (3.169)

giving, from Eqs. (3.167) and (3.168)

$$\sum S_{II} = -(1 - c_c)$$ (3.170)

But, from Eq. (3.94), we have for the classical Cassegrain

$$\sum S_{II} = -\frac{1}{2} f' = -\frac{1}{2}$$ (3.171)

in our normalized case, giving finally from Eq. (3.170)

$$c_c = +\frac{1}{2}$$ (3.172)

It follows that, for the classical Cassegrain, the locus of points P'_C for different incident ray heights falls on the circle of radius $2f'$ instead of f' as required for aplanatism, i.e. that

$$P'_{RC} P'_C = \frac{1}{2}(P'_{RC} P'_G)$$ (3.173)

The above construction shows clearly the difference in geometrical function of the aplanatic RC telescope from the classical Cassegrain. If the same construction is performed for the spherical primary (SP) telescope, the incident ray is reflected from the primary at A_{SP} and meets the oblate spheroidal secondary at the point B_{SP} well below B_C. The corresponding sphere $P'_0 P'_{SP}$ then lies well to the left of the plane $P'_0 P'_G$.

3.2.7.2 The Couder 2-mirror (aplanatic) anastigmat In 1926, Couder [3.25] [3.24(b)] proposed a modification of the Schwarzschild telescope above in which $\sum S_{IV}$ was left uncorrected, but $\sum S_{III}$ was reduced to zero by applying the condition (3.146) also given by Schwarzschild

---

[5] Equation (3.169) is the normalized equivalent of Eq. (3.198) below for $(\delta u'_p)_{Coma_s}$, which, however, gives the *diameter* of the angular aberration, whereas (3.169) gives the *radius*. Eq. (3.198) also supposes $y_1 = y_m$, the maximum aperture height. If $y_1 \neq y_m$, then (3.198) assumes the general form

$$(\delta u'_p)_{Coma_s} = -\frac{S_{II}}{y_m} \left(\frac{y_1}{y_m}\right)^2$$

If, with our present normalization, we set $y_m = f' = 1$, then (apart from the factor 2 because we now express the *radius* of the angular aberration), the general form of (3.198) is identical with (3.169). As a result of the normalization, Y_1 is dimensionless, so that Eq. (3.169) is dimensionally correct.

$$d_1 = -2f' , \tag{3.174}$$

which results from Eq. (3.119) for $(\sum S_{III})_{Aplan}$. Couder argued that $\sum S_{IV}$ could be corrected by a positive field-flattening lens. We shall consider such correctors in detail in Chap. 4. This is optically a much superior solution to Schwarzschild's. However, a severe price is paid in the *length* of the system: $2f'$ instead of $1.25f'$ in the case of Schwarzschild. In practice, a further tube extension is required in both cases to prevent direct light reaching the detector. In the Couder case, a total tube length of about $2.5f'$ is needed.

The Couder telescope is defined by the zero-astigmatism condition (3.174) of the aplanatic telescope, together with the conditions (3.109) and (3.114) defining the mirror forms to give aplanatism. If (3.174) is substituted in these, together with (2.75), we can derive at once

$$(b_{s1})_{Coud.} = -1 + \left(\frac{1 + 2m_2}{m_2^3} \right) \tag{3.175}$$

$$(b_{s2})_{Coud.} = -\left(\frac{m_2 - 1}{m_2 + 1} \right)^2 + \frac{1}{(m_2 + 1)^3} , \tag{3.176}$$

relations which leave m_2 as a free parameter to be chosen. Couder gave this system the same final f/no as Schwarzschild, f/3.0, and for a normalized $f' = 1$ defined $f_1' = -3.25$, thereby determining

$$m_2 = -0.30769$$

from (2.55). The primary thus has an even weaker curvature (f/9.75) than the Schwarzschild telescope. The values of $(b_{s1})_{Coud.}$ and $(b_{s2})_{Coud.}$ are, from (3.175) and (3.176), -14.204 and -0.554 respectively. The first value is not excessive because of the weak curvature. From (2.75) and (3.174), we have

$$L_{Coud.} = f'(1 + 2m_2) , \tag{3.177}$$

giving $L_{Coud.} = R_A = +0.38462$ since $f' = 1$. Similarly, from (2.90) and (3.177)

$$(f_2')_{Coud.} = f' \left(\frac{2m_2 + 1}{m_2 + 1} \right) , \tag{3.178}$$

giving $f_2' = +0.55556$.

This leads to the Couder telescope shown in Fig. 3.12 and the constructional data of Table 3.9. Although the design is essentially similar to the Schwarzschild telescope, the Couder system is obviously markedly longer.

The field curvature of the Couder telescope is given from Table 3.5 with $H^2 = 1$ as

$$\left(\sum S_{IV} \right)_{Coud.} = \left(\frac{m_2}{f'} \right) - \left(\frac{m_2 + 1}{L} \right) = \left(\frac{m_2}{f'} \right) - \left(\frac{m_2 + 1}{f'(2m_2 + 1)} \right) , \tag{3.179}$$

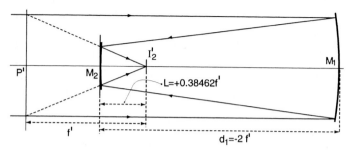

Fig. 3.12. The Couder (aplanatic) anastigmatic telescope (1926) [3.25]

from (3.177). The above data give

$$\left(\sum S_{IV}\right)_{Coud.} = -2.108 , \tag{3.180}$$

a higher value than for the Schwarzschild telescope because M_2 is steeper although the Couder primary is less steep. The Couder system has the important advantage that the obstruction R_A is significantly less. Even with the field supplement for a field of $\pm 1.5°$ (with the stop at the primary), the total linear obstruction is only 0.49.

In the Couder version with $D_1 = 0.8$ m, the above angular field gave a diameter of 126 mm for the field flattener. With a refractive index of 1.52, he gave a thickness supplement at the centre relative to the edge for a plano-convex lens of 5.07 mm. Of course, refractive correctors introduce chromatic aberrations, above all transverse aberration – see Chap. 4.

The Couder telescope, fulfilling the three conditions $\sum S_I = \sum S_{II} = \sum S_{III} = 0$, is the best 2-mirror solution in focal form from the point of view of imaging potential. But it has been hardly used because of its excessive length and poor baffling characteristics. It is highly significant that its optical length (without baffle extension) is $2f'$, *exactly the same as that of the Schmidt telescope* which also fulfils the above three aberration conditions. But the Schmidt has a spherical primary and field curvature less than half as great. Also the detector in the Schmidt is facing away from the incident light, a much more favourable baffling situation. On the other hand, the concave secondary of the Couder telescope, with its modest elliptical form, is far simpler to make than a large achromatic Schmidt plate.

Of course, the Couder telescope becomes more compact if its focal length is reduced by making it steeper, say f/1. But the solutions with more mirrors, discussed in § 3.6.5.3, are more compact and flexible in their designs. It seems, therefore, that the Couder telescope will remain an important theoretical limit case but will rarely be utilised in practice.

Figure 3.13 shows the spot-diagrams for the Couder telescope for an aperture of 1 m with f/3.0 and a field of $\pm 2.0°$, plotted for the optimum curved field of radius +1418 mm. The improvement in quality over the Schwarzschild

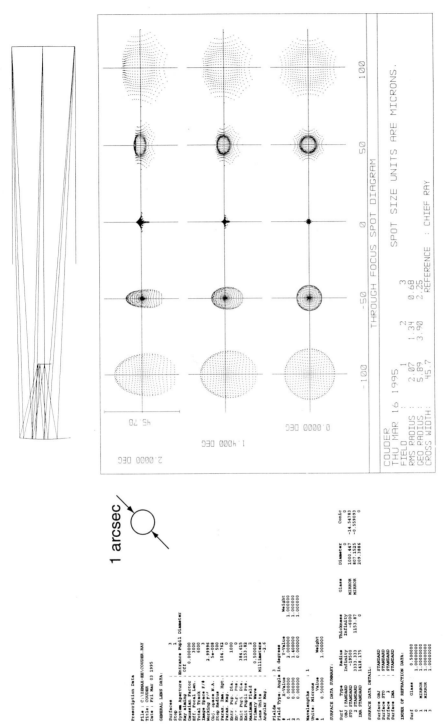

Fig. 3.13. Spot-diagrams for the Couder telescope (1926) [3.25] for an aperture of 1 m with f/3.0

Table 3.9. Constructional data of the Couder anastigmatic telescope (1926) [3.25]

f'	$=$	$+1$
D_1	$=$	0.33333 (f/3.0)
d_1	$=$	-2
f_1'	$=$	-3.25 (f/9.75)
m_2	$=$	-0.30769
L	$=$	$+0.38462$
R_A	$=$	$+0.38462$
D_2	$=$	0.1633 (for field $\pm 1.5°$)
f_2'	$=$	$+0.55556$
$(b_{s1})_{Coud.}$	$=$	-14.20358
$(b_{s2})_{Coud.}$	$=$	-0.55417

solution is striking, but the curved field must be borne in mind in making this comparison. Also, at this relative aperture, the obstruction supplement due to the field is so large that the image potential is not exploitable in practice without serious vignetting.

Finally, the question may legitimately be posed, since the equations for an aplanatic telescope are perfectly general for both Cassegrain and Gregory forms, whether a Couder anastigmatic form exists as a Gregory telescope with a real intermediate image. The answer is given by Eq. (3.174) defining the condition for freedom from astigmatism in an aplanatic telescope. Since f' must be *negative* for the Gregory form, it follows from (3.174) that d_1 must be *positive*. But this means that M_2 must lie *behind* the concave primary and can only form a *virtual* image of the real primary image. Indeed, the secondary mirror is itself virtual, since the light from the primary cannot reach it! The virtual secondary M_2 is then larger than the primary. This geometry is confirmed from Eqs. (3.177) and (3.178) giving negative values for L and f_2'.

It follows that a Gregory equivalent to the Couder anastigmatic telescope, giving a real final image, does not exist. This is the analogous situation to the virtual image Gregory form of Schwarzschild's impractical telescope form of Fig. 3.8.

3.2.8 Scaling laws from normalized systems to real apertures and focal lengths

The above formulation of third order aberration theory is quite general, but we have seen that it is useful for the comparison of different systems to normalize as in Tables 3.2 and 3.3 with

$$y_1 = u_{pr1} = +1$$

$$f' = \eta' = \pm 1$$

for the focal case. The general form of the wavefront aberration function in the third order approximation was given by Eq. (3.21) as

$$
\left.
\begin{aligned}
W_3'(y_1, \eta') = {} & \frac{1}{8}\left(\frac{y_1}{y_{m1}}\right)^4 \sum S_I + \frac{1}{2}\left(\frac{y_1}{y_{m1}}\right)^3\left(\frac{\eta'}{\eta_m'}\right)\sum S_{II}\cos\phi \\
& + \frac{1}{4}\left(\frac{y_1}{y_{m1}}\right)^2\left(\frac{\eta'}{\eta_m'}\right)^2\Big[(3\sum S_{III} + \sum S_{IV})\cos^2\phi \\
& \qquad\qquad\qquad\qquad + (\sum S_{III} + \sum S_{IV})\sin^2\phi\Big] \\
& + \frac{1}{2}\left(\frac{y_1}{y_{m1}}\right)\left(\frac{\eta'}{\eta_m'}\right)^3\sum S_V\cos\phi
\end{aligned}
\right\}
$$

$$(3.181)$$

The various powers of the ratios $\left(\frac{y_1}{y_{m1}}\right)$ and $\left(\frac{\eta'}{\eta_m'}\right)$ correspond simply to the terms $\rho^{(m+n)}$ and $\sigma^{(l+n)}$ of the Characteristic Function of Eq. (3.16), expressing the aperture and field dependence functions for the five Seidel aberrations. The numerical factors and terms $\sum S_I, \sum S_{II}$, etc. correspond to the constants $_{(l+n)}k_{(m+n)}, n$ in Eq. (3.16) and contain the information on the angular semi-aperture u', the focal length f' of the system and the semi-field angle u_{pr1} in radians, as expressed in the formulae of Tables 3.4 and 3.5. There, the terms in $\left(\frac{y_1}{f'}\right)$ represent the various powers of u', the power laws being, of course, identical to those of $\left(\frac{y_1}{y_{m1}}\right)$ in Eq. (3.181). Since all the other terms are dimensionless, while the wavefront aberrations all have the dimensions $(length)^1$, the aberration coefficients $\sum S_I$ etc. also have dimensions $(length)^1$ and are linearly proportional to the focal length of the system. Similarly, all the geometrical quantities are proportional to the focal length, which thus defines the scale of the system. This is illustrated by the relation for $\sum S_I$ in Table 3.5 for a 2-mirror system

$$\sum S_I = \left(\frac{y_1}{f'}\right)^4\left(-f'\zeta + L\xi\right),\tag{3.182}$$

in which ζ and ξ are dimensionless quantities. Substituting from (2.75) and (2.55) for L gives the form

$$\sum S_I = \left(\frac{y_1}{f'}\right)^4 f'\left[-\zeta + \left(1 - \frac{d_1}{f_1'}\right)\xi\right],\tag{3.183}$$

expressing the linear dependence on f' with otherwise only dimensionless quantities.

As an illustration of simple scaling, we will take Case 1 of Table 3.3, a single-mirror telescope with a spherical primary, giving for the normalized case

$$(S_I)_{norm} = +0.25, \quad (S_{II})_{norm} = -0.50$$

Suppose a practical system works at f/10 with a semi-field angle u_{pr1} of $1/100$ rad and has $f' = 1000$ mm. Then

$$S_I = +0.25 \left(\frac{1000}{20^4} \right) = +0.001\,562\,5 \text{ mm}$$

$$S_{II} = -0.50 \left(\frac{1000}{20^3 \cdot 100} \right) = -0.000\,625\,0 \text{ mm}$$

These results illustrate how the very large coefficients for the normalized systems of Table 3.3, corresponding to the enormous aperture and semi-field of f/0.5 and $u_{pr} = 1$ radian respectively, reduce to values comparable with the wavelength of visible light when scaled to the dimensions of real systems, even though the real focal lengths may be much greater in the units chosen. The actual wavefront aberrations are further reduced by the numerical factors in Eq. (3.181).

In the next section, the conversion formulae between wavefront aberration and the other common forms of aberration will be given.

3.3 Nature of third order aberrations and conversion formulae from wavefront aberration to other forms

The first four third order (Seidel) monochromatic aberrations are of fundamental importance in the optical layout and design of modern telescopes. We shall see in Reflecting Telescope Optics II, Chap. 3.5, that they are equally important in performance and maintenance aspects. It is therefore essential to have a clear idea of the physical meaning of these aberrations. Eq. (3.21) gave the general relation between the wavefront aberration and the third order aberration coefficients. Here we shall give the essential conversion formulae: a more complete treatment is given by Welford [3.6].

3.3.1 Spherical aberration (S_I)

The first term of Eq. (3.21) gives for the spherical aberration term at the Gaussian focus:

$$(W_1')_{GF} = \frac{1}{8} \left(\frac{y}{y_m} \right)^4 S_I \tag{3.184}$$

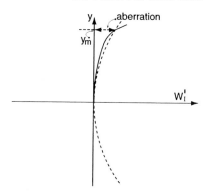

Fig. 3.14. Third order spherical aberration as wavefront aberration

W_I' varies, from Eq. (3.21), with y^4, where y is the height of the ray in the aperture. y_m will be the value of y taken for paraxial calculations and is the maximum value defining the aperture of the system. Figure 3.14 shows this wavefront aberration relative to the nominal reference sphere (Fig. 3.1). The choice of focus is free: it does not have to be the Gaussian focus represented by the y-axis as the reference sphere to that focus. A change of focus is represented by the parabola shown by the dashed curve, i.e. a function of y^2. If the parabola is such that it cuts the W_I' curve at y_m, the wavefront aberration is clearly reduced. It is easily shown that

$$(W_I')_{BF} = \frac{1}{4}(W_I')_{GF} , \qquad (3.185)$$

where $(W_I')_{BF}$ is the optimum (best focus) value corresponding to the case shown and $(W_I')_{GF}$ is the basic value referred to the Gaussian focus. The height in the pupil corresponding to the maximum of the residual zonal aberration $(W_I')_{BF}$ is $y_m/\sqrt{2} = 0.7071 y_m$ and the reference sphere shown cutting the wavefront at height y_m in Fig. 3.14 corresponds to a focus exactly *halfway* between the paraxial and marginal ray foci. It is also the focus of the zonal ray. This result is in fundamental disagreement with the best focus according to geometrical optics based on focusing rays, given below, and leading to a best focus (disc of least confusion) *three quarters* of the distance from the paraxial to the marginal focus. However, both the wavefront aberration and ray aberration treatments give a reduction of a factor of four of the paraxial focus aberration at their respective "best foci". An admirably clear demonstration of the wavefront relationships is given by Conrady [3.167]. The wavefront aberrations of Eqs. (3.184) and (3.185) have historically been applied to tolerancing on the basis of the Rayleigh $\lambda/4$ criterion. This matter is treated in detail in §3.10.5. The Rayleigh criterion is only a rough criterion according to physical optics, the physically correct treatment being the Strehl Intensity Ratio based on the variance of the wavefront – see Eq. (3.464). However, Table 3.26 shows that, in the cases of defocus, third order spherical aberration and their combination, there is close agreement between the Rayleigh approximation and the more rigorous Strehl approach. The fac-

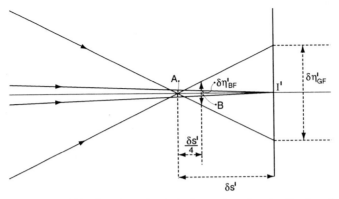

Fig. 3.15. Third order spherical aberration: longitudinal and lateral forms

tor 4 gain due to the optimum focus at the halfway focus is also given almost exactly by the Strehl treatment.

The fact that geometrical optics loses its validity as one approaches the diffraction limit should always be borne in mind in the geometrical treatment of aberrations using ray paths near the focus.

In elementary texts, spherical aberration is usually introduced as *longitudinal aberration*, a definition of value since there is no ambiguity of interpretation regarding focus compensation. Figure 3.15 shows the longitudinal and lateral forms of third order spherical aberration. The process of image formation from the pupil to the image is a Fourier transformation [3.26] following differential equations given by Nijboer [3.27] [3.3]. The paraxial rays focus at the Gaussian image position I', the marginal rays at A. The longitudinal aberration is

$$\delta s' = I'A$$

If all rays are considered, the smallest diameter of the "tube" containing the rays is at B where

$$AB = \delta s'/4$$

The diameter of the image (containing 100% of the geometrical energy) at the paraxial focus is $\delta\eta'_{GF}$ and the smallest diameter at B is $\delta\eta'_{BF}$. Then obviously

$$\delta\eta'_{BF} = \tfrac{1}{4}\delta\eta'_{GF} , \tag{3.186}$$

a result which corresponds to (3.185). $\delta\eta'_{BF}$ is called the *disk of least confusion*. From the Nijboer equations, it is easily shown that the *longitudinal aberration* is

$$\delta s' = -\frac{4}{n'u'^2}(W'_I)_{GF} , \tag{3.187}$$

where u' is the semi-aperture angle and n', the refractive index in the image space, also takes account of the sign inversions resulting from the change of light direction caused by an odd number of reflections. Clearly

$$\delta\eta'_{GF} = -2u'\delta s' \, ,$$

giving, with $u' = -y_1/f'$,

$$\delta\eta'_{GF} = -\frac{8f'}{n'y_1}(W'_I)_{GF} = -\frac{f'}{n'y_1}S_I \, , \qquad (3.188)$$

where y_1 is the semi-aperture of the telescope and is equal to y_m in Eq. (3.184). This is the *lateral spherical aberration (image diameter) at the Gaussian focus*. The disk of least confusion is therefore

$$\delta\eta'_{BF} = -\frac{2f'}{n'y_1}(W'_I)_{GF} = -\frac{f'}{4n'y_1}S_I \qquad (3.189)$$

The angular form is usually more convenient. For the Gaussian and best focus, the *angular aberration (image diameter)* is from (2.101) and (3.184):

$$\left.\begin{aligned}
(\delta u'_p)_{GF} &= -\frac{S_I}{n'y_1} \text{ rad} = -\frac{S_I}{n'y_1}(206\,265) \text{ arcsec} \\
(\delta u'_p)_{BF} &= -\frac{S_I}{4n'y_1} \text{ rad} = -\frac{S_I}{4n'y_1}(206\,265) \text{ arcsec}
\end{aligned}\right\}, \qquad (3.190)$$

where $(W'_I)_{GF} = \frac{1}{8}S_I$ if $y_1 = y_{m1}$ in (3.184).

3.3.2 Coma (S_{II})

The second term of Eq. (3.21) gives for the coma term at the Gaussian focus:

$$(W'_{II})_{GF} = \frac{1}{2}\left(\frac{y}{y_m}\right)^3 S_{II}\cos\phi \qquad (3.191)$$

The linear field dependence (η'/η'_m) in Eq. (3.21) can be ignored as we are only concerned with the aperture effects governing the size of the image for any given field position giving the value S_{II}. For the wavefront aberration, the aperture law for coma is, from (3.191), a cube law giving the S-shaped aberration shown in Fig. 3.16. The vertical y-axis shows the wavefront aberration relative to the Gaussian focus, which means in this case relative to its principal ray. If an oblique line is drawn through the origin, then this is equivalent to a tilt of the reference sphere, or simply that the aberration is referred to a height in the image plane different from that of the principal ray. If the tilted line is drawn at an angle such that it cuts the wavefront at y_m, then it can be shown that the resultant wavefront error is minimized. This is the "best focus" equivalent of the spherical aberration case, although here the "focus shift" is a lateral one. The remarks made above concerning wavefront and ray aberrations for spherical aberration also apply here. Reference

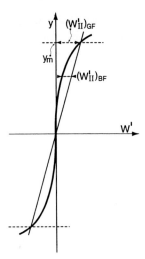

Fig. 3.16. Third order coma as wavefront aberration

to Table 3.26 reveals that the Rayleigh criterion for the wavefront aberration in the case of coma is only a crude approximation to the true physical effect, although the optimum tilt factor of 3 is exactly reproduced.

Since the tangential section of the optical imagery (the plane of the paper) has the aperture azimuth angles 0 and π, the full wavefront aberration as shown in Fig. 3.16 is, from (3.191)

$$(\overline{W}'_{II})_{GF} = 2(W'_{II})_{GF} = S_{II} \tag{3.192}$$

It can be shown that

$$(\overline{W}'_{II})_{BF} = 2(W'_{II})_{BF} = \tfrac{1}{3}S_{II} \tag{3.193}$$

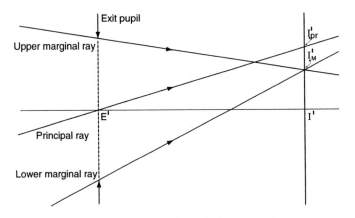

Fig. 3.17. Third order coma: lateral aberration form

The ray path for coma is shown in Fig. 3.17. The principal ray $E'I'_{pr}$ emerging from the centre of the exit pupil cuts the image plane at I'_{pr}. The marginal rays intersect at the point I'_M, either above or below I'_{pr}, depending on the sign of the coma. The upper ray has $\phi = 0$, the lower ray $\phi = \pi$. The Nijboer equations lead to the following ray coordinates in the image plane, referred to I'_{pr}, η' being in the tangential section, η'_s in the sagittal section:

$$
\left.
\begin{aligned}
\delta\eta' &= -\frac{S_{II}}{2n'u'}\left(\frac{y}{y_m}\right)^2 (\cos 2\phi + 2) \\
\delta\eta'_s &= -\frac{S_{II}}{2n'u'}\left(\frac{y}{y_m}\right)^2 \sin 2\phi
\end{aligned}
\right\}
\tag{3.194}
$$

If we take $y = y_m$, we have the coordinates of rays distributed round the edge of the pupil at azimuth angle ϕ. This leads to the classical transverse aberration of Fig. 3.18, known as the "coma patch". The principal ray intersects at the point I'_{pr}. This corresponds to the point E' at the centre of the pupil (right-hand diagram). Because of the double angle 2ϕ, both the marginal rays 1 in the t-section cut at the same point 1 in the coma patch. The ray coordinates lie on circles of radius

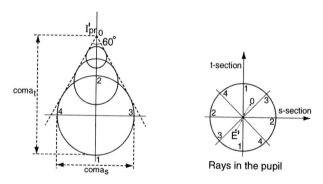

Fig. 3.18. Third order coma: the "coma patch"

$$
\frac{1}{2n'u'}S_{II}\left(\frac{y}{y_m}\right)^2
$$

whose centres are shifted from I'_{pr} by

$$
\frac{1}{n'u'}S_{II}\left(\frac{y}{y_m}\right)^2 ,
$$

i.e. by the diameter of the circle concerned. The circles nest within two lines tangent to all circles which meet at I'_{pr} with an angle of 60°.

The total extent of the coma patch in the t-section is called the *tangential coma* and is given for $y = y_m$ by

$$I'_{pr}I'_M = Coma_t = \frac{3}{2n'u'}S_{II} = \frac{3}{n'u'}(W'_{II})_{GF} , \qquad (3.195)$$

from (3.194). With $u' = -y_1/f'$:

$$Coma_t = -\frac{3f'}{2n'y_1}S_{II} = -\frac{3f'}{n'y_1}(W'_{II})_{GF} \qquad (3.196)$$

The *sagittal coma*, $Coma_s$, is the *diameter* of the circle corresponding to y_m, giving

$$Coma_s = -\frac{f'}{n'y_1}S_{II} = -\frac{2f'}{n'y_1}(W'_{II})_{GF} \qquad (3.197)$$

As angular aberration, by the same conversion as for spherical aberration, we have

$$\left. \begin{array}{l} (\delta u'_p)_{Coma_t} = -\dfrac{3}{2}\dfrac{S_{II}}{n'y_1}(206\,265) = -3\dfrac{(W'_{II})_{GF}}{n'y_1}(206\,265) \text{ arcsec} \\[3mm] (\delta u'_p)_{Coma_s} = -\dfrac{S_{II}}{n'y_1}(206\,265) \quad = -2\dfrac{(W'_{II})_{GF}}{n'y_1}(206\,265) \text{ arcsec} \end{array} \right\} ,$$
$$(3.198)$$

if the parameter u_{pr1}, in the expression for S_{II} (Tables 3.4 and 3.5), defining the field, is measured in radians.

It should be noted that $(\overline{W}'_{II})_{GF}$, the full peak-to-peak wavefront aberration, is *twice* the wavefront aberration *coefficient* $(W'_{II})_{GF}$ of Eq. (3.191). This factor of 2 exists for all aberrations dependent on $\cos^n \phi (\equiv \cos n\phi)$, i.e. all aberrations except the axisymmetric ones with $n = 0$ in Table 3.1. For most technical purposes, the wavefront coefficient $(W'_{II})_{GF}$ is the more basic quantity, so it is used in the above relations.

The coma patch, as defined above, contains 100% of the geometrical energy. The 60° triangle enclosing the circles gives the characteristic asymmetric "flare" (Greek "coma" = hair) which makes coma the most damaging of the monochromatic aberrations. The best way to identify the direction (sign) is to remember that a Newton or Cassegrain telescope gives S_{II} *negative* from Table 3.3 and has "inward" coma, i.e. the point of the coma patch corresponding to the principal ray is pointing towards the field centre, the flare away from it.

Let us consider, as an example, the coma coefficients for a 1-mirror telescope or a classical 2-mirror telescope from Tables 3.4 and 3.5 respectively, for which the normalized coefficient with $s_{pr1} = 0$ is simply $-\frac{f'}{2}$. Then we have from (3.198)

$$(\delta u'_p)_{Coma_t} = \frac{3}{4}\left(\frac{y_1}{f'}\right)^2 u_{pr1} \text{ arcsec} = \frac{3}{16}\left(\frac{1}{N^2}\right) u_{pr1} \text{ arcsec} , \qquad (3.199)$$

where u_{pr1} is expressed in arcsec and N is the f/no.

3.3.3 Astigmatism (S_{III}) and field curvature (S_{IV})

The third term of (3.21) gives the combined effect of astigmatism and field curvature at the Gaussian focus:

$$(W'_{III} + W'_{IV})_{GF} = \frac{1}{4}\left(\frac{y}{y_m}\right)^2\left(\frac{\eta'}{\eta'_m}\right)^2\left[(3S_{III} + S_{IV})\cos^2\phi \atop + (S_{III} + S_{IV})\sin^2\phi\right] \tag{3.200}$$

The factor $(y/y_m)^2$ implies that astigmatism is essentially a defocus effect in the aperture, but the effect is dependent on the section in the pupil because of the $\cos^2\phi$ and $\sin^2\phi$ terms. As before, we can ignore the field dependent factor $(\eta'/\eta'_m)^2$ since we are only concerned with aperture effects for a given field position. Eq. (3.200) transforms to

$$(W'_{III} + W'_{IV})_{GF} = \frac{1}{4}\left(\frac{y}{y_m}\right)^2\left[(S_{III} + S_{IV}) + 2S_{III}\cos^2\phi\right],$$

or, with $\cos^2\phi = \frac{1}{2}(\cos 2\phi + 1)$:

$$(W'_{III} + W'_{IV})_{GF} = \frac{1}{4}\left(\frac{y}{y_m}\right)^2 S_{III}\cos 2\phi + \frac{1}{4}\left(\frac{y}{y_m}\right)^2(2S_{III} + S_{IV}) \tag{3.201}$$

The first term here is the astigmatism effect, the second term the *effective* field curvature at the mean astigmatic image:

$$(W'_{III})_{GF} = \frac{1}{4}\left(\frac{y}{y_m}\right)^2 S_{III}\cos 2\phi \tag{3.202}$$

$$(W'^{*}_{IV})_{GF} = \frac{1}{4}\left(\frac{y}{y_m}\right)^2(2S_{III} + S_{IV}) \tag{3.203}$$

This is why the quantity $(2S_{III} + S_{IV})$ was calculated in Table 3.3.

Consider first the astigmatism term. As with coma, the total wavefront aberration is

$$(\overline{W}'_{III})_{GF} = \frac{1}{2}\left(\frac{y}{y_m}\right)^2 S_{III} \tag{3.204}$$

because $\cos 0 = +1$ and $\cos\pi = -1$. The *positive* focus change at azimuth 0 is accompanied by a *negative* focus change at azimuth $\pi/2$ because of the term $\cos 2\phi$. Figure 3.19 shows the wavefront aberration $(W'_{III})_{GF}$ for the sections $\phi = 0$ and $\phi = \pi/2$.

The practical consequence is shown in Fig. 3.20. The principal ray $E'I'_{pr}$ leaves the axis at the exit pupil at E' and cuts the image plane at I'_{pr}. Rays in the *tangential section* (plane of the paper) focus at the point I'_t on the *tangential astigmatic surface*, whereas rays in the sagittal section (at right angles to the plane of the paper) focus at I'_s on the *sagittal astigmatic surface*.

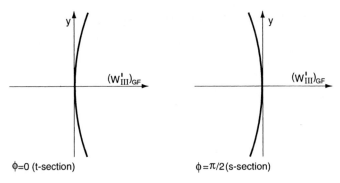

ϕ=0 (t-section) ϕ=π/2 (s-section)

Fig. 3.19. Third order astigmatism: wavefront aberration reversal in the t- and s-sections due to the $\cos 2\phi$ term

Fig. 3.20 shows marginal rays, but the effect is also present for rays paraxial to the principal ray since defocus is a Gaussian error (first order), as was shown in Table 3.1. The *longitudinal astigmatism* $I'_t I'_s$ is therefore independent of aperture. The *lateral astigmatism*, by contrast, depends on the aperture: it reveals itself as the *astigmatic lines*.

At the tangential focus I'_t, the rays in the s-section are out of focus and form a line perpendicular to the plane of the paper; at I'_s the rays in the t-section are out of focus and form a line in the plane of the paper along the s-surface. Figure 3.21 shows the effect in a symmetrical field and explains the origin of the terms "tangential" and "sagittal". If we start by focusing well inside the tangential focus, the image will be a round defocus patch made slightly elliptical by the astigmatism with the long axis tangential. At the t-focus, it becomes a pure tangential line. Beyond this it becomes elliptical

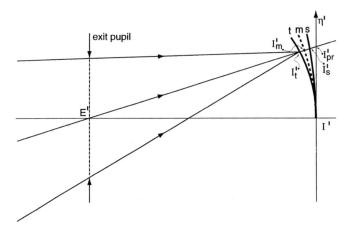

Fig. 3.20. Third order astigmatism: astigmatic surfaces and lines

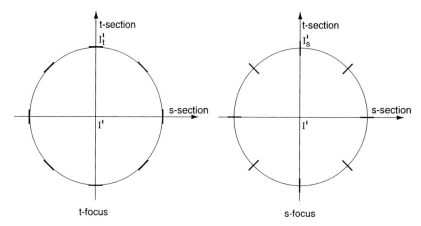

Fig. 3.21. Tangential and radial astigmatic lines at the *t*-focus and *s*-focus respectively

again with growing minor axis: at I'_m the *geometrical* image is *exactly round* (if diffraction is considered, this is not true for small values of astigmatism – see §3.10). After I'_m the image becomes elliptical again with long axis radial, passing through the sagittal astigmatic line at I'_s. The behaviour of the astigmatic patch is most important in the practical analysis of telescopes since the changes through focus permit a good qualitative assessment of the astigmatism in the image.

The distances $I'_m I'_t$ and $I'_{pr} I'_m$ are given by

$$
\left.
\begin{aligned}
I'_m I'_t &= -\left(\frac{y}{y_m}\right)^2 \frac{1}{2n'u'^2} S_{III} \\
I'_{pr} I'_m &= -\left(\frac{y}{y_m}\right)^2 \frac{1}{2n'u'^2} (2S_{III} + S_{IV})
\end{aligned}
\right\}
\tag{3.205}
$$

Since $1/u'^2 = (f'/y_1)^2$ and the coefficients S_{III} and S_{IV} always contain the factor $(y_1/f')^2$ (see Tables 3.4 and 3.5), we see that these longitudinal measures are independent of aperture. If we now set $y = y_m$ for the edge of the pupil, then:

$$
\left.
\begin{aligned}
I'_m I'_t &= -\frac{1}{2n'} \left(\frac{f'}{y_1}\right)^2 S_{III} \\
I'_{pr} I'_m &= -\frac{1}{2n'} \left(\frac{f'}{y_1}\right)^2 (2S_{III} + S_{IV})
\end{aligned}
\right\}
\tag{3.206}
$$

At the *mean astigmatic focus* I'_m, the *image diameter* is given by

$$
\delta\eta'_{ast,m} = -(I'_m I'_t) 2u'
$$

or

$$\delta\eta'_{ast,m} = -\frac{f'}{n'y_1}S_{III} \tag{3.207}$$

The *astigmatic lines* have the length $2\delta\eta'_{ast,m}$.

The *angular aberration* (diameter) of the round image at best astigmatic focus is

$$(\delta u'_p)_{ast,m} = -\frac{S_{III}}{n'y_1} \text{ rad} = -\frac{S_{III}}{n'y_1}(206\,265) \text{ arcsec}, \tag{3.208}$$

again on the assumption, as for the coma case, that the field angle u_{pr1} contained (as u_{pr1}^2) in the expressions for S_{III} is given in radians.

Similarly, the *image diameter* produced by the field curvature effect $I'_{pr}I'_m$ is

$$\delta\eta'_{FC,m} = -\frac{f'}{n'y_1}(2S_{III} + S_{IV}), \tag{3.209}$$

or in angular measure

$$(\delta u'_p)_{FC,m} = -\frac{(2S_{III} + S_{IV})}{n'y_1}(206\,265) \text{ arcsec} \tag{3.210}$$

3.3.4 Distortion (S_V)

The distortion coefficient S_V was not calculated in Table 3.3 or considered in the general formulations because it has little significance for the one- or two-mirror telescopes we have dealt with. Distortion can become more important if elements are introduced near the image, as we shall see in Chap. 4. In any event, the coefficient S_V can be readily calculated from the formulae in Eqs. (3.20) and (3.21).

The wavefront form of distortion is very useful as the starting formula for calculating its value but is not, in itself, a very practical measure. Distortion is not an error of quality of an image point: it is simply in the wrong place compared with the linear scale of the system as defined by its focal length f' according to Eq. (2.102). So the wavefront aberration corresponds simply to a wavefront tilt whose value depends on the cube of the field size (see Eq. (3.21).

Depending on its sign, distortion will be either "barrel" or "pincushion", as shown by the distortion of a square field in Fig. 3.22.

For *astrometric* work, fixed distortion in a system can normally be calibrated out. If large distortion is present, a more serious consequence is *photometric*. The pincushion distortion of prime focus correctors, for example, leads to a light intensity fall-off with increasing field radius, an effect which can be a nuisance with non-linear detectors like photographic plates.

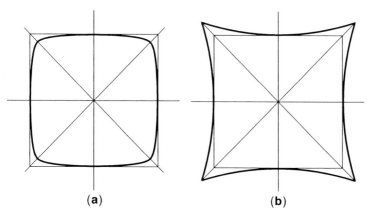

Fig. 3.22. Distortion: (**a**) barrel, (**b**) pincushion

3.3.5 Examples of conversions

It is instructive to consider examples of the above conversion formulae for some of the cases of Table 3.3.

The *angular aberrations*, which are normally the most practical form for astronomical telescopes because of the direct comparison with the seeing, are independent of the size (linear scale) of the telescope and depend only on the f/no, field angle and aberration coefficients. Since u_{pr1} is expressed in radians, the dependence on u_{pr1}^2 in the case of astigmatism brings, for practical field angles, a large reduction factor compared with the linear factor u_{pr1} in the case of coma.

The angular aberrations are given in Table 3.10 for a 1-mirror telescope with spherical primary (Case 1 of Table 3.3), a classical Cassegrain (Case 3) and an RC telescope (Case 4). In each case, the telescope has a final image beam of f/10 and the semi-field angle u_{pr1} is 30 arcmin or $(1800/206\,265)$ rad. For the Cassegrain cases 3 and 4, $m_2 = -4$ as in Table 3.3.

The signs are the opposite of those of the coefficients S_I, S_{II}, S_{III} if $n' = +1$, i.e. for the two 2-mirror cases. This simply reflects the fact that a positive wavefront aberration corresponds to a negative ray height change $\delta\eta'$ in the image plane. The sign is most important in the case of the asymmetric aberration coma, where a positive value in Table 3.10 with $n' = +1$, and a negative coefficient S_{II} in Table 3.3 imply that the point of the coma patch is directed towards the field centre as in Fig. 3.4 (a).

3.3.6 Conversions for Gaussian aberrations

Although the two first order (Gaussian) aberrations are not usually treated in books on geometrical optics as "aberrations", because they can be removed by small positional adjustments, they are, nevertheless, in practical telescopes

Table 3.10. Angular spherical aberration, coma and astigmatism for three telescope cases of Table 3.3, with an f/10 image beam and a semi-field angle u_{pr1} of 30 arcmin

Case	$(\delta u_p')_{BF}$ (arcsec)	$(\delta u_p')_{Coma_t}$ (arcsec)	$(\delta u_p')_{Ast,m}$ (arcsec)
1-mirror telescope spherical primary (Case 1)	+ 1.611	−3.375	+ 0.785
Classical Cassegrain $m_2 = -4$ (Case 3)	0	+ 3.375	−2.814
RC $m_2 = -4$ (Case 4)	0	0	−3.152

two of the most serious sources of image error, namely defocus error and tracking error.

It is easily seen from Eq. (3.23) that the conversions are as follows:

$$\delta W_L' = -\frac{1}{2}n'\left(\frac{y_1}{f'}\right)^2 \delta z \tag{3.211}$$

$$\delta W_T' = -n'\left(\frac{y_1}{f'}\right)\delta\eta'\cos\phi \tag{3.212}$$

These are the wavefront aberration changes resulting from a longitudinal focus shift δz or a transverse focal (image height) shift $\delta\eta'$, respectively.

3.4 The theory of aspheric plates

The theory of aspheric plates has great elegance and simplicity. It has fundamental importance in understanding the function of wide-field telescope solutions, as we shall see in § 3.6. In addition, it is a powerful tool for dealing with supplementary correctors for correcting field aberrations (Chap. 4) or aberrations to correct manufacturing errors. A classic example of the latter is the analysis of possible correctors for the spherical aberration found, after launch, to afflict the Hubble Space Telescope (HST). This application will be discussed below and further in Chap. 4 and RTO II, Chap. 3.

Although the term "plate theory" or "plate diagram" was coined later by Burch [3.28], the basic theory was laid down by Schwarzschild in his classic paper of 1905 [3.1]. Burch's approach is well expounded by Linfoot [3.29]. The theory given below follows more closely the original approach of Schwarzschild which is essentially embodied in the aspheric terms of Eq. (3.20).

The theory of aspheric plates is derived directly from the concept of "stop shift". Suppose for any optical system the "stop" is at a certain location, for example at the prime mirror in a telescope system. If the effective stop is shifted by introducing a diaphragm or element sufficiently small that it takes over the role of the stop (see Chap. 2), then the aberrations of the system are modified according to the following formulae – see the works of Hopkins [3.3] or Welford [3.6] for the derivation (a more general form was given in Eq. (3.22)):

$$\left.\begin{aligned}
\partial S_I &= 0 \\
\partial S_{II} &= (H\partial E)S_I \\
\partial S_{III} &= 2(H\partial E)S_{II} + (H\partial E)^2 S_I \\
\partial S_{IV} &= 0 \\
\partial S_V &= (H\partial E)(S_{IV} + 3S_{III}) + 3(H\partial E)^2 S_{II} + (H\partial E)^3 S_I
\end{aligned}\right\} \qquad (3.213)$$

S_I, S_{II}, S_{III}, S_{IV}, S_V are the aberrations present with the original stop position. We have already encountered the factor $(H\partial E)$ in Eqs. (3.20) and (3.22). ∂E is a normalized factor expressing the shift in stop position, H the Lagrange Invariant (Chap. 2) which is a measure of the throughput of the system. If $S_I \ldots S_{IV}$ are all zero, a stop shift has no effect on any of these third order aberrations. If – as in the case of the error in HST which is on the primary mirror – the system has a finite S_I but the other terms are zero, then a change in coma, astigmatism and distortion occurs through the right-hand terms containing S_I in (3.213). The other terms will be zero in such a case.

The above "stop-shift" effect is closely linked to the effect of inserting an aspheric plate into the stop of a system and then shifting the plate away from the stop by a normalized distance ∂E. This is shown in Fig. 3.23. If the

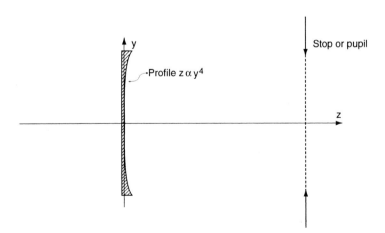

Fig. 3.23. Stop-shift effect for a single third order aspheric plate shifted from the pupil

aspheric plate is inserted at the stop (or its image, the pupil), it acts, to third order accuracy, as a "pure" element in its monochromatic function: it only affects S_I (wavefront aberration is also a fourth power function of aperture), all field effects are zero, so that we have:

$$\underline{\partial E = 0} : \left.\begin{aligned} S_I &= \delta S_I \\ S_{II} &= 0 \\ S_{III} &= 0 \\ S_{IV} &= 0 \\ S_V &= 0 \end{aligned}\right\} \tag{3.214}$$

If now the plate is shifted from the stop by ∂E, then substituting (3.214) in (3.213) gives:

$$\underline{\partial E \neq 0} : \left.\begin{aligned} \delta S_I &= S_I \\ \delta S_{II} &= (H\partial E)S_I \\ \delta S_{III} &= (H\partial E)^2 S_I \\ \delta S_{IV} &= 0 \\ \delta S_V &= (H\partial E)^3 S_I \end{aligned}\right\} \tag{3.215}$$

This result shows the origin of the right-hand, aspheric, terms in Eqs. (3.20). The term "aspheric plate" has a much more general significance than the normal refracting plate shown in Fig. 3.23. It applies equally to *any fourth power figuring modification* to a refracting lens surface, plane mirror or mirror with optical power (curvature).

The influence of a given corrector plate, producing a change S_I in spherical aberration, on the field aberrations depends then only on S_I and the parameter $(H\partial E)$. This has a very simple geometric optical significance. It can be shown ([3.3] [3.6]) that

$$H\partial E = \frac{y_{pr}}{y} \, , \tag{3.216}$$

where y_{pr} and y are simply the heights of the paraxial principal ray and normal (aperture) paraxial ray at the plane of the system where the plate is inserted. Fig. 3.24 shows schematically the way the heights of these rays vary as they pass through a typical Cassegrain telescope. The aperture ray is shown by a continuous line, the principal ray by the dashed line, the stop being placed at the primary. Clearly, the ratio y_{pr}/y is small in the object space above the primary for normal small semi-fields u_{pr} and grows linearly with the distance above the primary. After reflection at the primary, y_{pr} at the secondary is the same as in the object space, but y suffers a dramatic reduction as the aperture ray converges towards the prime focus. The smaller the secondary (i.e. the higher the secondary magnification m_2), the bigger the increase of y_{pr}/y at the secondary reflection. In such circumstances, δS_{II} in (3.215) can be quite large. As will be shown in RTO II, Chap. 3, this is the

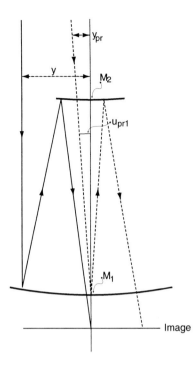

Fig. 3.24. Heights of the paraxial aperture and paraxial principal rays as they pass through a Cassegrain telescope

reason why the theoretical possibility of correcting the spherical aberration of the HST primary by modifying the secondary must be ruled out, even if it were possible in practice, because the field coma δS_{II} in (3.215) is far too large at a field radius of 10 arcmin to be acceptable. However, y_{pr}/y is in this case still $\ll 1$. This means the field astigmatism δS_{III} is much smaller than δS_{II}.

After reflection at the secondary, the principal ray diverges rapidly because of the telephoto effect of the Cassegrain form, whereas the aperture ray converges to the final image. At some point, depending on the field chosen, the rays intersect giving $y_{pr}/y = 1$. Further down, $y_{pr}/y > 1$ and the effect on the astigmatism increases rapidly.

From (3.214), if correction of spherical aberration can be achieved at the pupil, monochromatically and to a third order the correction is perfect. But if a single corrector plate is inserted elsewhere with $H\partial E = y_{pr}/y \neq 0$, then field aberrations appear according to (3.215). However, if two plates *with significant separation* are introduced, two conditions can be fulfilled; with three separated plates, three conditions. In other words, with three separated plates in any geometry, monochromatic residues of S_I, S_{II} and S_{III} can theoretically be corrected in any telescope system. Whether the solution is practical will depend on the aberration values and the separations available in practice. This general theorem is of great importance in understanding the possibilities of wide-field telescopes ($\S 3.6$).

Let us consider again the practical example of the spherical aberration of the HST. One possible means of correction is a corrector in the baffle ("stovepipe") above the image. A single aspheric plate generates unacceptable field aberrations because of the large y_{pr}/y values. But three separated plates can fulfil the three third order correction conditions

$$\left.\begin{array}{rclcl}\sum S_I & = & (S_I)_1 + (S_I)_2 + (S_I)_3 & = & \delta S_I \\ \sum S_{II} & = & k_1(S_I)_1 + k_2(S_I)_2 + k_3(S_I)_3 & = & 0 \\ \sum S_{III} & = & k_1^2(S_I)_1 + k_2^2(S_I)_2 + k_3^2(S_I)_3 & = & 0 \end{array}\right\} , \qquad (3.217)$$

in which k_1, k_2, k_3 represent $H\partial E = y_{pr}/y$ for the positions of the three plates. Inversion of the matrix (3.217) gives the required aspheric powers of the plates. As a plausible example, k_1, k_2 and k_3 were given values 1, 1.5 and 2 respectively. Solution of (3.217) gives

$$\left.\begin{array}{rcl}(S_I)_1 & = & +6(\delta S_I) \\ (S_I)_2 & = & -8(\delta S_I) \\ (S_I)_3 & = & +3(\delta S_I) \end{array}\right\} , \qquad (3.218)$$

giving $\sum S_I = +\delta S_I$ as required. However, the price paid for the field correction with plates in these positions is an increase of individual aspheric powers up to eight times the required correction. This is simply the equivalent phenomenon to the fact that the powers of the individual lenses of an achromatic doublet or triplet camera objective are much higher than the resulting total power of the system. Of course, in practical optical design, compensations with quadratic and even sixth power terms must also be taken into account rather than the simple plate shown in Fig. 3.23 with a pure fourth power term. Aspheric plates have the merit of correcting aberrations without affecting the Gaussian optical terms associated with the optical power of the system: specifically they do not change the f/no of the transmitted or reflected beam.

Following Bahner [3.5], we can use the general formulation of § 3.2.4 for the third order aberrations of 1-mirror and 2-mirror telescopes to derive relations for such telescopes when an aspheric plate is added. Tables 3.4 and 3.5 gave the general formulae, whereby the terms in the square brackets correspond to the normalization of Tables 3.2 and 3.3 with

$$y_1 = u_{pr1} = +1 \quad \text{and} \quad f' = \pm 1$$

The formulation of (3.215) and (3.216) above assumes that the pupil is *initially* at the primary (i.e. $s_{pr1} = 0$), the plate shift from it being expressed by $H\partial E = y_{pr}/y$. (We show below that, *finally*, the pupil position has no effect on the first three third order aberrations, provided that the first two are corrected to zero). With the normalization, the quantity $y_{pr}/y = s_{pl}/f'$, where s_{pl} is the plate shift relative to the primary. If the plate introduces third order spherical aberration δS_I^*, then the equations of Table 3.4 for a 1-mirror telescope with an aspheric plate become from (3.215), with $s_{pr1} = 0$:

$$\sum S_I = \left(\frac{y_1}{f_1'}\right)^4 \left[-\frac{f_1'}{4}(1+b_{s1}) + \delta S_I^*\right]$$

$$\sum S_{II} = \left(\frac{y_1}{f_1'}\right)^3 \left[-\frac{f_1'}{2} + \frac{s_{pl}}{f_1'}\delta S_I^*\right] u_{pr1} \qquad (3.219)$$

$$\sum S_{III} = \left(\frac{y_1}{f_1'}\right)^2 \left[-f_1' + \left(\frac{s_{pl}}{f_1'}\right)^2 \delta S_I^*\right] u_{pr1}^2$$

We shall see in §3.6 that these equations lead at once to the Schmidt telescope.

Similarly, if the same addition of an aspheric plate is made to a 2-mirror telescope, the equations of Table 3.5 give, by the same reasoning:

$$\sum S_I = \left(\frac{y_1}{f'}\right)^4 \left[-f'\zeta + L\xi + \delta S_I^*\right]$$

$$\sum S_{II} = \left(\frac{y_1}{f'}\right)^3 \left[-d_1\xi - \frac{f'}{2} + \frac{s_{pl}}{f'}\delta S_I^*\right] u_{pr1} \qquad (3.220)$$

$$\sum S_{III} = \left(\frac{y_1}{f'}\right)^2 \left[\frac{f'}{L}(f'+d_1) + \frac{d_1^2}{L}\xi + \left(\frac{s_{pl}}{f'}\right)^2 \delta S_I^*\right] u_{pr1}^2$$

If these three equations are set to zero, then such a system of 2 aspheric mirrors and an aspheric plate can always give a solution with $\sum S_I = \sum S_{II} = \sum S_{III} = 0$ for any geometry (generalized Schwarzschild theorem [3.13]). Now we can apply a very important consequence of the stop-shift equations (3.22). If the first three Seidel aberrations have been made zero in Eqs. (3.220), then all three are independent of a stop shift in the total system. This means that *the entrance pupil can be shifted to the aspheric plate (or elsewhere) without changing the third order correction of $\sum S_I$, $\sum S_{II}$, $\sum S_{III}$.* This is a very important general principle in the theory of wide-field telescopes. For the 1-mirror telescope with an aspheric plate of Eqs. (3.219), *in the general case,* only two conditions can be fulfilled with two aspherics, giving $\sum S_I = \sum S_{II} = 0$. But a stop shift will still not affect $\sum S_I$, $\sum S_{II}$ and $\sum S_{III}$, even if $\sum S_{III} \neq 0$. Of course, higher order aberrations are not independent of stop shifts and may have a determinant influence.

If the aspheric plate is in the object space, as in the Schmidt telescope, then s_{pl} can be applied directly as a negative quantity, i.e. with the same sign as f_1' in (3.219). For a plate inside the system, i.e. to the right of the primary for the positive light direction from left to right, a virtual positive value of s_{pl} can be calculated from the Gaussian image of the plate backwards into the object space [3.28] [3.29]. The necessary expressions are given in Chap. 4 (see Fig. 4.2) in connection with field correctors consisting of aspheric plates, applied to 1-mirror or 2-mirror telescope foci.

Since third order spherical aberration varies, as wavefront aberration, with y^4, the plate profile must obey the same law. From Eq. (3.21), its coefficient is given for a plate of refractive index n' placed in a medium of index n by

$$\delta S_I^* = 8(n' - n)ay^4 \, , \tag{3.221}$$

where a is a constant and $n = 1$ for the normal case of a glass plate in air. If a virtual plate in the object space is calculated as described above, the profile constant a in (3.221) must be multiplied by m_{pl}^4, where m_{pl} is the magnification of the plate imagery back into object space.

In §3.6, we shall see the direct application of Eqs. (3.219) and (3.220) to wide-field telescopes, and in Chap. 4 to field correctors.

3.5 The role of refracting elements in modern telescopes: chromatic variations of first order and third order aberrations

Although the refracting telescope and oculars are not considered as belonging to the domain of modern telescope optics and are not treated in this book (older books such as König-Köhler [3.30] or Bahner [3.5] deal excellently with this material), this does not mean that chromatic aberrations no longer play an important role in modern telescope systems. This role is generally that of correcting aberrations without contributing significantly to the optical power of the system. The optical power is essentially provided by the mirror system. This does not mean that lenses with significant power will not be used: it means that powered lenses will normally be used in a corrective combination which has a small total power. Since, for thin lenses, primary chromatic aberration (the variation of longitudinal focus with wavelength for a simple lens) is dependent on the total power, the main curse of lenses is thus avoided. Of course, different materials may also be used to produce achromatism, but the effects of *secondary spectrum*, longitudinal aberration residues due to the different dispersions of the materials, remain. Such effects can only be reduced by the use of "special glasses" or crystal materials which may have other design limitations or limitations of availability (diameter).

Refractive corrector elements can produce not only simple longitudinal chromatic aberration but also other chromatic effects. The general theory is given by Hopkins [3.3] and Welford [3.6]. By analogy with Eq. (3.21) for the monochromatic aberrations, the wavefront form for the chromatic aberrations of the first and third orders can be expressed by

$$
\delta W'_{(\lambda_2 - \lambda_1)} = \left[\frac{1}{2} C_1 \left(\frac{y}{y_m} \right)^2 + C_2 \left(\frac{y}{y_m} \right) \left(\frac{\eta'}{\eta'_m} \right) \cos \phi \right]
$$
$$
+ \left[\frac{1}{8} (\delta S_I)_C \left(\frac{y}{y_m} \right)^4 + \frac{1}{2} (\delta S_{II})_C \left(\frac{y}{y_m} \right)^3 \left(\frac{\eta'}{\eta'_m} \right) \cos \phi \right.
$$
$$
+ \frac{1}{2} (\delta S_{III})_C \left(\frac{y}{y_m} \right)^2 \left(\frac{\eta'}{\eta'_m} \right)^2 \cos^2 \phi \right] \, , \tag{3.222}
$$

in which the chromatic coefficients are, by analogy with Eq. (3.20):

Longitudinal chromatic
variation of focus: $C_1 \quad = \sum_\nu Ay\Delta\left(\frac{\delta n}{n}\right)$
(longitudinal colour)

Chromatic variation
of wavefront tilt: $C_2 \quad = \sum_\nu \overline{A}y\Delta\left(\frac{\delta n}{n}\right)\ldots\ldots\left[+H\partial E C_1\right]$
(lateral colour)

Chromatic variation
of spherical aberration: $(\delta S_I)_C \; = \sum_\nu A^2 y\Delta\left(\frac{u}{n}\frac{\delta n}{n}\right) + \sum_\nu(\delta S_I^*)_C$
(spherochromatism,
Gauss-error)

Chromatic variation
of coma (colour coma): $(\delta S_{II})_C \; = \sum_\nu A\overline{A}y\Delta\left(\frac{u}{n}\frac{\delta n}{n}\right) + \sum_\nu HE(\delta S_I^*)_C$

Chromatic variation
of astigmatism: $(\delta S_{III})_C = \sum_\nu \overline{A}^2 y\Delta\left(\frac{u}{n}\frac{\delta n}{n}\right) + \sum_\nu(HE)^2(\delta S_I^*)_C$
(colour astigmatism)

$$(3.223)$$

The second group of terms in (3.222) are identical wavefront functions to the monochromatic aberrations S_I, S_{II}, S_{III}. Note that we have omitted the chromatic variation of S_{IV}, although in the general use of refracting optics, for example in photographic objectives, this term may not be negligible. It is omitted here because, as stated above, the total optical power of refracting elements in modern telescopes is normally small. The Petzval sum is correspondingly small and its chromatic variation a negligible factor. By contrast, we shall see in Chap. 4 that the chromatic variations of spherical aberration and, above all, of coma and astigmatism, are often the limiting factors of corrector systems.

The first group of terms in (3.222) are the chromatic variations of the first order (Gaussian) terms in Table 3.1. While these *monochromatic* terms can be focused out by movement of the secondary mirror in a Cassegrain telescope or by a change of pointing, their *chromatic variations* given in (3.222) cannot be eliminated. They are therefore serious factors limiting the performance if a wide spectral band is employed. For narrow spectral bands they can be "focused" or "pointed" out. This is not possible for the chromatic aberrations of the second group.

The second term, the lateral colour, produces – for a centered optical system with symmetry axis – a radial spectrum of a point image whose length depends on the coefficient C_2 and the field radius η'. Atmospheric dispersion is a similar aberration phenomenon produced by the atmosphere except that the spectrum is radially symmetrical to the zenith point in the sky. This will be dealt with in Chap. 4.

The chromatic aberrations $(\delta S_I)_C$, $(\delta S_{II})_C$, $(\delta S_{III})_C$ in (3.223) each consist of two terms, as for the monochromatic aberrations of (3.20). The first terms result from the optical power of lenses, as do the first order aberrations C_1 and C_2. The second terms $(\delta S_I^*)_C$, $(\delta S_{II}^*)_C$ and $(\delta S_{III}^*)_C$ are due to aspheric surfaces. For aspheric plates, these terms are the only chromatic effects. From (3.221) we have

$$\delta S_I^* = 8(n' - n)ay^4 = 8\Delta(n)ay^4 \ ,$$

the symbol Δ in all these formulae meaning the change $(n' - n)$ causing refraction. The chromatic variation is then

$$(\delta S_I^*)_C = 8\Delta(\delta n)ay^4 \ , \tag{3.224}$$

in which $\delta n = (n_2' - n_1') - (n_2 - n_1)$ for the wavelengths λ_2 and λ_1 in the two media before and after refraction. In the case of an aspheric glass plate in air, for which the index is taken to be 1 with negligible dispersion, the quantity $\Delta(\delta n)$ is simply the dispersion $(n_2' - n_1')$ of the glass. An analogue interpretation applies to the quantities $\Delta\left(\frac{\delta n}{n}\right)$ and $\Delta\left(\frac{u}{n} \cdot \frac{\delta n}{n}\right)$ in the terms for spherical surfaces with optical power.

Since the stop-shift effect due to ∂E on C_2 was not given in Eqs. (3.22), it has been added to the equation for C_2 in (3.223). The equivalent stop-shift terms for $(\delta S_{II})_C$ and $(\delta S_{III})_C$ have been omitted, as they are the same as those given in (3.22).

3.6 Wide-field telescopes

3.6.1 The symmetrical stop position: the Bouwers telescope

The problem of making a satisfactory primary had so dominated the development of the reflecting telescope that it was axiomatic for about 270 years that it would also form the stop (pupil) of the system, since this used its surface with maximum efficiency. Recognition that field aberrations (coma, astigmatism and distortion) followed the formulation of third order aberration theory by Seidel in 1856 [3.4], and many of his successors involved above all in the development of photographic objectives, led to deep understanding of the significance of pupil position relative to the constructional elements of the system. It therefore seems almost amazing that nobody before B. Schmidt in 1931 [3.31] considered the possibilities of stop shift into the object space in front of the primary. Schwarzschild and Chrétien, both concerned to correct field aberrations to produce wider fields and in possession of the necessary aberration theory, failed to realize this simple possibility.

Figure 3.25 shows the fundamental form of a wide-field telescope without correction of spherical aberration. It consists of a stop at the centre of curvature of a spherical mirror. The principal rays are normals to the sphere. The system has no axis except that of the stop. Its field with constant image

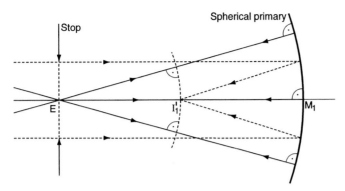

Fig. 3.25. Fundamental form of a wide-field telescope without correction of spherical aberration

quality is 180°, only limited by vignetting of the stop. If the stop were rotated about its centre to remain normal to a principal ray of varying field angle, the theoretical field is 360°, though the complete spherical mirror would, in practice, block the entry of any light. The system has only two aberrations, field curvature and a spherical aberration, which is identical for all field directions apart from stop vignetting. This reduces the beam width in the t-section giving an asymmetrical effect on the spherical aberration looking like astigmatism; but the system has no astigmatism in the real sense. It is also free from coma. It has the field curvature of the concave primary equal to its focal length, giving an image surface on a sphere also centered on the stop. These properties are obvious from the geometrical symmetry of the arrangement. Schmidt's great achievement was to recognize this: the addition of a device to correct the spherical aberration was simply a corollary, although technically very difficult to achieve.

The equations of Table 3.4 must lead to the same conclusions. Setting $b_{s1} = 0$ for the spherical primary, we have

$$
\left.
\begin{aligned}
S_I &= -\left(\frac{y_1}{f_1'}\right)^4 \left[\frac{f_1'}{4}\right] \\
S_{II} &= -\left(\frac{y_1}{f_1'}\right)^3 \left[\frac{f_1'}{2} - \frac{s_{pr1}}{4}\right] u_{pr1} \\
S_{III} &= -\left(\frac{y_1}{f_1'}\right)^2 \left[f_1' - s_{pr1} + \left(\frac{s_{pr1}}{2f_1'}\right)^2 f_1'\right] u_{pr1}^2
\end{aligned}
\right\}
\tag{3.225}
$$

Now $s_{pr1} = 2f_1'$, both being negative. This gives at once:

$$
\left.
\begin{aligned}
S_I &= -\left(\frac{y_1}{f_1'}\right)^4 \frac{f_1'}{4} \\
S_{II} &= 0 \\
S_{III} &= 0
\end{aligned}
\right\}
\tag{3.226}
$$

This result is also evident from the basic equations (3.20) since $\overline{A} = 0$ for principal rays normal to the mirror at all field angles.

It is often useful to deal with all 1-mirror and 2-mirror solutions with one single set of formulae. The 1-mirror case can be treated with the 2-mirror formulae of Table 3.5 by treating it as the limit case where the secondary becomes a folding flat with $m_2 = -1$ (Fig. 2.15). Then $\xi = 0$ and, with $b_{s1} = 0$, $\zeta = -\frac{1}{4}$. The equations become:

$$
\left.
\begin{aligned}
S_I &= \left(\frac{y_1}{f'}\right)^4 \left[\frac{f'}{4}\right] \\[2mm]
S_{II} &= \left(\frac{y_1}{f'}\right)^3 \left[-\frac{f'}{2} - \frac{s_{pr1}}{f'}\left(\frac{f'}{4}\right)\right] u_{pr1} \\[2mm]
S_{III} &= \left(\frac{y_1}{f'}\right)^2 \left[\frac{f'}{L}(f' + d_1) + s_{pr1} + \left(\frac{s_{pr1}}{f'}\right)^2 \left(\frac{f'}{4}\right)\right] u_{pr1}^2
\end{aligned}
\right\}
\qquad (3.227)
$$

Inserting $s_{pr1} = -2f' = 2f_1'$, since $f' = -f_1'$ because of the folding flat, and $L = f' - m_2 d_1 = f' + d_1$ with $m_2 = -1$, Eq. (3.227) reduces to the same result as (3.226).

The above two results (3.226) and (3.227) show that Schwarzschild's own formulation implicitly gave the Schmidt solution for freedom from field aberrations.

Of course, some solution must be found to correct the spherical aberration. In fact, the most general way of doing this, following the symmetry concept, was not the Schmidt method with an aspheric corrector but the *concentric meniscus* proposed by Bouwers in 1941 [3.32]. The Bouwers telescope is shown in Fig. 3.26. As with the basic, stop-shifted form without spherical aberration correction of Fig. 3.25, it is completely symmetrical and without axis, again giving a field of 180° of identical performance apart from vignetting effects. However, the system has a fundamental problem: the concentric meniscus corrects the third order spherical aberration but introduces primary longitudinal chromatic aberration defined by C_1 of (3.223). In its pure form, therefore, the system is only useful for narrow spectral bandwidths. It also exhibits spherochromatism, $(\delta S_I)_C$ in (3.223). Furthermore, the relative aperture is limited to about f/4 by fifth order spherical aberration (zonal error). For these reasons, the Bouwers form has rarely found application. But as a generic type leading to other modified "meniscus-type" solutions, it is of great significance. The most important of these is the Maksutov telescope. The basic theory of the Bouwers form leading to the Maksutov modification is given in § 3.6.3. An important modification, proposed by Bouwers to produce chromatic correction while preserving concentricity, is dealt with in § 3.6.4.2.

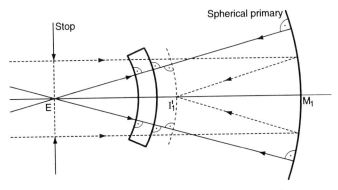

Fig. 3.26. The Bouwers concentric telescope (1941)

3.6.2 The Schmidt telescope

3.6.2.1 The basic principles of the Schmidt telescope. Bernhard
Schmidt's publication in 1931 [3.31] represented the second revolutionary advance in the theory of telescope optics of this century. Schwarzschild had made
the first such advance in 1905 [3.1] with his analysis of the possibilities of field
correction by varying the aspheric forms of two mirrors with optical power as
a function of the stop position. The bulk of the theory of § 3.2 derives from
Schwarzschild's work. Schmidt's advance consisted of the systematic shift of
the stop to the centre of curvature of a spherical primary, as discussed in
§ 3.6.1, combined with the insertion of an aspheric refracting corrector plate
in the plane of the stop, as shown in Fig. 3.27. The Schmidt concept simply
transfers the aspheric correction term from the Newton paraboloidal primary
to the refracting corrector at its centre of curvature. Monochromatically, as
we have seen in § 3.4, such a corrector plate with a fourth power figuring term

$$z = ay^4$$

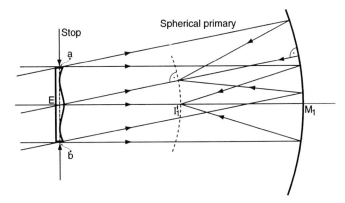

Fig. 3.27. The Schmidt telescope (1931)

will correct the third order spherical aberration S_I. Because of its symmetry at the pupil ($H\partial E = 0$), it is free from third order field coma and astigmatism S_{II} and S_{III} (Eq. (3.215)). Thus, monochromatically, it achieves the same correction (proven formally, as for the Bouwers telescope, in Eq. (3.227))

$$\left.\begin{array}{rcl} S_I & = & 0 \\ S_{II} & = & 0 \\ S_{III} & = & 0 \end{array}\right\} \qquad (3.228)$$

as the Couder telescope (§ 3.2.7.2). It is a remarkable demonstration of the law of nature that you never get something for nothing, that both the Schmidt and Couder telescopes achieve this monochromatic correction at the cost of a physical tube length $= 2f'$, about three times the length of a Cassegrain telescope using a primary with the same geometry. However, the Schmidt telescope has a more convenient image position and does not suffer from baffling problems which, in practice, increase the length of the Couder telescope further. From the point of view of stray light from the sky, the Schmidt telescope is as good as the classical refractor.

The Schmidt telescope has a major monochromatic advantage over the Bouwers telescope (§ 3.6.1): the corrector plate does not introduce its own zonal error (fifth order spherical aberration), whereas the concentric Bouwers meniscus in combination with its spherical primary does. Furthermore, since the aspheric form is theoretically free, it can compensate fifth and even seventh order spherical aberration of very steep primaries, so very steep primaries of the order of f/1 can be compensated, *if the correctors can be made to sufficient accuracy.*

Unlike the Bouwers telescope, the Schmidt telescope does possess an *axis*, albeit only weakly defined by the axis of symmetry of the aspheric form on the plate. This leads to asymmetries in the angles of incidence of oblique beams on the aspheric surface of the plate at points a and b in Fig. 3.27. These asymmetries lead to higher order field aberrations. This limits the theoretical field, the limitation being more severe the higher the relative aperture of the system, but the chromatic limitations are usually more serious in practice.

The corrector plate introduces uncorrected chromatic aberrations, the most important of which is $(\delta S_I^*)_C$ of Eq. (3.223): spherochromatism. There are also chromatic differences of the monochromatic higher order field effects. Since, from Eq. (3.214), the plate produces no third order S_{II} or S_{III}, there are also no chromatic differences: spherochromatism is the only third order chromatic aberration.

3.6.2.2 Corrector plate profile and spherochromatism. As we saw from Eqs. (3.219) or (3.220), the required correction is achieved by

$$\delta S_I^* = -S_I = +f_1'/4 \qquad (3.229)$$

in our normalized system with $f_1' = -y_1 = -1$, where S_I is the third order spherical aberration of the primary and δS_I^* the compensating aberration of the plate. From (3.184), the spherical aberration of the spherical mirror, expressed as wavefront aberration at the Gaussian focus, is given by:

$$(W_I')_{GF} = \frac{1}{8} S_I \left(\frac{y}{y_m} \right)^4 \tag{3.230}$$

From Table 3.4 we have, setting $y_m = y$ for the edge of the aperture,

$$(W_I')_{GF} = -\frac{1}{32} \frac{y^4}{f_1'^3} , \tag{3.231}$$

referred to the Gaussian focus. In § 3.3.6 we gave in Eq. (3.211) the conversion formula for longitudinal focus shift δz to equivalent wavefront aberration as

$$\delta W_L' = -\frac{1}{2} n' \left(\frac{y}{f_1'} \right)^2 \delta z \tag{3.232}$$

Since $n' = -1$ after reflection at the mirror, the total wavefront aberration to be corrected by the plate is

$$W' = -\frac{y^4}{32 f_1'^3} + \frac{y^2}{2 f_1'^2} \Delta z , \tag{3.233}$$

whereby the second term introduces a wavefront compensation by a focus shift produced by the plate. If $W_{y_m}' = 0$ for the edge of the pupil y_m, then (3.233) gives, for example,

$$\Delta z_0 = +\frac{y_m^2}{16 f_1'} , \tag{3.234}$$

a negative distance since f_1' is negative. From (3.187), the longitudinal spherical aberration is (with sign reversal because $n' = -1$)

$$\delta s' = \frac{4}{u'^2} (W_I')_{GF} = +4 \left(\frac{f_1'}{y_m} \right)^2 \left(-\frac{1}{32} \frac{y_m^4}{f_1'^3} \right) = -\frac{1}{8} \frac{y_m^2}{f_1'} , \tag{3.235}$$

a positive distance since f_1' is negative. The focus shift Δz_0 from (3.234) is therefore half the longitudinal aberration $\delta s'$ and of opposite sign.

For the figured plate of thickness d_{pl}, we can generalise (3.221) to give the general plate form for compensation

$$\delta W_I'^* = \frac{1}{8} \delta S_I^* = (n' - n) d_{pl} , \tag{3.236}$$

in which d_{pl} is a function of y. This gives from (3.233) if $n = 1$ for a glass plate in air:

$$d_{pl} = \left(\frac{1}{n' - 1} \right) \left[-\frac{y^4}{32 f_1'^3} + \frac{y^2}{2 f_1'^2} \Delta z \right] + (d_{pl})_0 , \tag{3.237}$$

where $(d_{pl})_0$ is the axial thickness of the plate. Following Bahner [3.5], we now define a dimensionless profile constant k_{pl} in units of the Δz_0 focus shift defined by (3.234) as

$$k_{pl} = +\Delta z \frac{16 f_1'}{y_m^2} \tag{3.238}$$

together with the normalized height ρ_{pl} as

$$\rho_{pl} = y/y_m \tag{3.239}$$

Eq. (3.237) can now be written

$$d_{pl} = \left(\frac{1}{n'-1}\right)\left(-\frac{y_m^4}{32 f_1'^3}\right)\left[\rho_{pl}^4 - k_{pl}\rho_{pl}^2\right] + (d_{pl})_0 \tag{3.240}$$

With $D = 2y_m$ and the f/no N given by

$$N = -f_1'/D , \tag{3.241}$$

Eq. (3.240) can then be expressed in the convenient form

$$d_{pl} = \frac{1}{512}\left(\frac{1}{n'-1}\right)\frac{D}{N^3}\left[\rho_{pl}^4 - k_{pl}\rho_{pl}^2\right] + (d_{pl})_0 \tag{3.242}$$

The thickness constant $(d_{pl})_0$ for the axial thickness has no optical effect since a plane-parallel plate in the parallel incident beam produces no angular aberration on axis or in the field. The profile of the function $(\rho_{pl}^4 - k_{pl}\rho_{pl}^2)$ is shown in Fig. 3.28. The abscissa is ρ_{pl}, the ordinate the function for various values of the parameter k_{pl}.

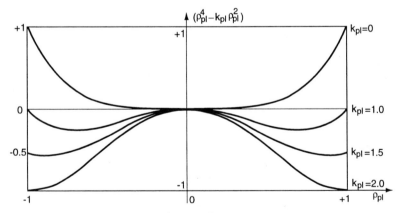

Fig. 3.28. Profile function $(\rho_{pl}^4 - k_{pl}\rho_{pl}^2)$ for Schmidt corrector plates with various values of the form profile parameter k_{pl}. The glass plate is formed by considering the area under the curves to be filled with glass down to an abscissa tangential to the curve in question. To the resulting axial thickness, the constant thickness $(d_{pl})_0$ is added to give the necessary minimum plate thickness. (After Bahner [3.5])

The choice of the optimum form is determined by the best balance for the *spherochromatism*. The assumption is made in Eq. (3.242) that the ray heights at the mirror are identical with those at the plate. In practice, all such small discrepancies are taken care of by the standard ray tracing and optimization procedures of optical design programs.

From (3.236) the plate profile (3.242) gives the required wavefront aberration

$$
\left.
\begin{aligned}
(\delta W_I'^*)_0 &= \frac{1}{512} \frac{D}{N^3} (\rho_{pl}^4 - k_{pl}\rho_{pl}^2) \\[2mm]
\text{or} & \\[2mm]
(\delta W_I'^*)_0 &= -\frac{1}{512} \frac{f_1'}{N^4} (\rho_{pl}^4 - k_{pl}\rho_{pl}^2)
\end{aligned}
\right\}
\tag{3.243}
$$

for the refractive index n_0' corresponding to the chosen correction wavelength λ_0. For a wavelength λ_1 and index n_1', the wavefront aberration produced is

$$
(\delta W_I'^*)_1 = \left(\frac{n_1' - 1}{n_0' - 1} \right) (\delta W_I^*)_0 \;,
$$

giving a chromatic differential wavefront error between λ_0 and λ_1 of

$$
\delta(\delta W_I'^*) = (\delta W_I'^*)_1 - (\delta W_I'^*)_0 = \left(\frac{n_1' - n_0'}{n_0' - 1} \right) (\delta W_I'^*)_0 = \frac{1}{\nu_A}(\delta W_I'^*)_0 \;,
\tag{3.244}
$$

in which $(n_1' - n_0')$ is the dispersion and ν_A the Abbe number for the glass (see § 3.6.2.6 below), assuming λ_0 and λ_1 are chosen for the standard definition of ν_A. If $\lambda_1 > \lambda_0$ (i.e. λ_1 more into the red), there will be undercorrection of the mirror aberration, overcorrection into the blue. The transverse aberration generated by this spherochromatism is determined by the slope of the wavefront of (3.243):

$$
\left.
\begin{aligned}
\frac{d(\delta W_I'^*)_0}{dy} &= \frac{d}{dy}\left[\frac{D}{512N^3}(\rho_{pl}^4 - k_{pl}\rho_{pl}^2) \right] \\[2mm]
&= \frac{d}{dy}\left[\frac{y_m}{256N^3}\left(\frac{y^4}{y_m^4} - k_{pl}\frac{y^2}{y_m^2} \right) \right] \\[2mm]
\text{or} & \\[2mm]
\frac{d(\delta W_I'^*)_0}{dy} &= \frac{1}{256N^3}(4\rho_{pl}^3 - 2k_{pl}\rho_{pl})
\end{aligned}
\right\}
\tag{3.245}
$$

The slope function $(4\rho_{pl}^3 - 2k_{pl}\rho_{pl})$ has a minimum at

$$
12\rho_{pl}^2 - 2k_{pl} = 0 \;,
$$

giving

$$
(\rho_{pl}^2)_{min} = \frac{k_{pl}}{6}
\tag{3.246}
$$

The minimum transverse aberration will result for that plate profile in Fig. 3.28 giving the same numerical value of the slope at the edge of the pupil with $\rho_{pl} = \pm 1$ as for the minimum given by (3.246). This is the case for $k_{pl} = 1.5$, as is easily confirmed as follows.

Inserting $k_{pl} = 1.5$ in (3.246) gives $(\rho_{pl})_{min} = \pm 0.5$. The slope function at the point $(\rho_{pl})_{min} = +0.5$ is then

$$4(\rho_{pl})^3_{min} - 2k_{pl}(\rho_{pl})_{min} = -1$$

At the point $\rho_{pl} = +1$, the slope is $+1$. This shows that the form given in Fig. 3.28 with $k_{pl} = 1.5$ is the most favourable for the focus balance of spherochromatism if this is measured as the transverse (or angular) aberration. The effect of this optimum focus term can be understood in the following way. The central wavelength λ_0 is corrected for a certain axial focus position. For any other wavelength λ_1, the focus shift is such that the disk of least confusion of its spherical aberration arising from the spherochromatism is at the same focus position. This brings the same factor-of-4 advantage compared with a pure y^4-plate with $k_{pl} = 0$ that we had in Eqs. (3.186) and (3.190).

The *diameter* $\delta\eta'_{BF}$ of the transverse aberration at this optimum focus can be derived from

$$\delta\eta'_{BF} = 2\frac{f'_1}{\nu_A}\frac{\mathrm{d}(\delta W'^*_I)_0}{\mathrm{d}\rho_{pl}}, \qquad (3.247)$$

the factor 2 in this equation arising from the fact that the maximum slope of the wavefront slope function is present at $\rho_{pl} = \pm 0.5$ and ± 1.0 with opposite signs. From (3.245) this gives

$$\delta\eta'_{BF} = 2\frac{f'_1}{\nu_A}\frac{1}{256N^3}(4\rho^3_{pl} - 2k_{pl}\rho_{pl}) \qquad (3.248)$$

Since, with the optimum focus $k_{pl} = 1.5$, the maximum slope is also at the edge of the pupil (plate), we can put $\rho_{pl} = 1$ in (3.248), giving

$$\delta\eta'_{BF} = \frac{f'_1}{\nu_A}\frac{1}{128N^3} \qquad (3.249)$$

for the diameter of the transverse aberration.

This important result can be confirmed from the general formula (3.189) for the transverse diameter of the disk of least confusion

$$\delta\eta'_{BF} = \frac{2f'_1}{y_1}(\delta W'^*_I)_{GF},$$

where GF refers to the Gaussian focus and BF to the best focus. At the Gaussian focus $k_{pl} = 0$, giving with $y_1 = D/2$ from (3.243) and (3.244)

$$\delta\eta'_{BF} = \frac{2f'_1}{(D/2)}\frac{1}{\nu_A}\frac{D}{512N^3}\rho^4_{pl}$$

or

$$\delta\eta'_{BF} = \frac{f'_1}{\nu_A}\frac{1}{128N^3} \, , \tag{3.250}$$

since $\rho_{pl} = 1$ for the edge of the pupil, thus confirming (3.249).

In the usual measure of angular diameter, the diameter of the spherochromatic aberration is

$$(\delta u'_p)_{BF} = \frac{\delta\eta'_{BF}}{f'_1} = \frac{1}{\nu_A 128N^3} \text{ rad} = \frac{1611}{\nu_A N^3} \text{ arcsec} \tag{3.251}$$

In using this formula, it should be remembered that ν_A is only the classical Abbe number if the same dispersion range is used. The general form is given from (3.244) as

$$\frac{1}{\nu_A} = \frac{n'_1 - n'_0}{n'_0 - 1} \, , \tag{3.252}$$

in which n'_0 refers to the correction wavelength λ_0 in laying down the plate profile and n'_1 to any given subsidiary wavelength λ_1 for which the spherochromatism is to be calculated. Bahner [3.5] gives the example of a Schmidt plate made of the Schott glass UBK7, for which the value of ν_A in (3.251) is 100 between $\lambda_0 = 430$ nm and $\lambda_1 = 386$ nm in the blue or 490 nm in the red. With a typical value of $N = 3$, Eq. (3.251) gives an aberration diameter (100% geometrical energy) of 0.60 arcsec. Bearing in mind that the spectral bandwidth of 104 nm is not very large, *it is clear that spherochromatism is by far the most significant residual error in the normal Schmidt telescope with a singlet corrector plate.*

The largest Schmidt telescopes have apertures of 1 m or more. The two largest are at Palomar Observatory with dimensions for plate diameter (aperture)/mirror diameter/focal length of 1.22/1.83/3.07 m and a maximum field of 6.5° × 6.5° [3.33], and the Universal Telescope of the Karl Schwarzschild Observatory at Tautenburg in Germany with 1.34/2.00/4.00 m and a maximum field of 3.4° × 3.4° [3.34]. This latter instrument can also be used as a quasi-Cassegrain.

It is instructive to see the extent of the aspheric deformation required for a Schmidt plate of large size. If the values $k_{pl} = 1.5$, $N = 3$, $n' = 1.5$ and $D = 1000$ mm are inserted in Eq. (3.242) defining the plate profile, the maximum departure from the plane-parallel plate occurs at the "neutral zone" where the plate thickness is a minimum. This radius is given from Eq. (3.245) for the slope function by

$$4\rho^3_{pl} - 2k_{pl}\rho_{pl} = 0 \, ,$$

with $k_{pl} = 1.5$. This gives the neutral zone at $\rho_{pl} = 0.866$. Inserting this value into (3.242), with the other parameters defined as above, gives the maximum asphericity of the plate as only $(d_{pl}) - (d_{pl})_0 = 0.081$ mm.

3.6.2.3 Aberrations and compensation possibilities of a plane-parallel plate (normally filter) in the convergent beam.

Plane-parallel glass plates in the form of colour filters are an essential adjunct of astronomical observations. Such a plate in the incident (parallel) beam has no effect, as mentioned above in connection with the finite thickness of Schmidt corrector plates. However, a plane-parallel plate placed in a converging (or diverging) beam has important aberration effects, *all of which depend linearly on its thickness d.* These are illustrated here with reference to the Schmidt telescope but are quite general for any telescope system. Since the commonest use is for filters, we will characterize the aberrations of such a plate by the suffix Fi.

A plate of thickness d and refractive index n' produces a focus shift

$$(\Delta s')_{Fi} = \left(\frac{n'-1}{n'}\right)d \tag{3.253}$$

It also produces the following third order aberrations [3.3] [3.6]:

$$\left.\begin{aligned}
(S_I)_{Fi} &= -\left(\frac{n'^2-1}{n'^3}\right)u'^4 d\,, & (S_{II})_{Fi} &= \left(\frac{u'_{pr}}{u'}\right)(S_I)_{Fi}\,, \\
(S_{III})_{Fi} &= \left(\frac{u'_{pr}}{u'}\right)^2 (S_I)_{Fi}\,, & S_{IV} &= 0\,, \\
S_V &= \left(\frac{u'_{pr}}{u'}\right)^3 (S_I)_{Fi}
\end{aligned}\right\} \tag{3.254}$$

Apart from the specific form of $(S_I)_{Fi}$, the influence on $(S_{II})_{Fi} \ldots (S_V)_{Fi}$ is determined by the same factor $(\overline{A}/A)$ which we encountered in (3.20) and in the general third order theory of telescopes: for a plate in a centered system, $i'_{pr} = u'_{pr}$ and $i' = u'$, where u' is the semi-angle of the exit beam. In most practical cases with small angular fields, the factor $(u'_{pr}/u') \ll 1$ so that the effect on the spherical aberration is dominant. The sign of $(S_I)_{Fi}$ implies overcorrection i.e. the marginal ray focuses too long compared with the paraxial ray. The wavefront aberration is, from (3.21):

$$(W'_I)_{Fi} = -\frac{1}{8}\left(\frac{n'^2-1}{n'^3}\right)u'^4 d \tag{3.255}$$

If this is compared with the aberration of (3.233), the aberration corrected by the Schmidt plate, then by similar transformations we can express the third order spherical aberration of the filter in the same form as in Eq. (3.243), giving

$$(W'_I)_{Fi} = \left(\frac{n'^2-1}{n'^3}\right)d\left[\frac{\rho_{pl}^4 - k_{pl}\rho_{pl}^2}{128N^4}\right]\,, \tag{3.256}$$

assuming a similar defocus term. This spherical aberration can effectively compensate for the spherochromatism at a secondary wavelength λ_1, as defined by $\delta(\delta W'^*_I)$ in Eq. (3.244). Setting

$$(W_I')_{Fi} = -\delta(\delta W_I'^*)$$

gives from Eqs. (3.243), (3.244) and (3.256) the compensation condition

$$d = \left(\frac{n'^3}{n'^2 - 1}\right) \frac{f_1'}{4\nu_A} , \qquad (3.257)$$

or

$$\nu_A \simeq \frac{2}{3} \frac{f_1'}{d} , \qquad (3.258)$$

if $n' \simeq 1.5$. The sign of this compensation effect is important. The Schmidt primary has *undercorrection* of spherical aberration. To correct this, the Schmidt corrector plate introduces compensating *overcorrection* which is stronger in the blue, weaker in the red. The filter plate adds further *overcorrection*, thereby pushing the corrected wavelength of the total system further into the *red*. Since $d > 0$ in all real cases, addition of filters will therefore always worsen the spherical aberration correction for wavelengths on the UV-side (shorter) of the central correction wavelength λ_0. In deciding on the value of λ_0 for a new Schmidt telescope, the mean filter thickness should be taken into account. If red filters are thicker than blue ones, this will improve the total correction.

3.6.2.4 Field aberrations of a Schmidt telescope. As stated in § 3.6.2.1, the Schmidt telescope is corrected monochromatically for S_I, S_{II} and S_{III} but still has, from its symmetrical nature, a field curvature equal to the focal length of the primary (Fig. 3.27). As in the Couder telescope, this can be corrected by a *field-flattening lens*. This device was first suggested by the Scottish astronomer Piazzi Smyth in 1874 [3.35]. In Eq. (3.20), the factor y for the aperture beam height appears in S_I, S_{II} and S_{III}: at the image therefore, a powered surface has no effect. However, it does affect the Petzval sum S_{IV}. In the case of the Schmidt telescope with a field curvature in the same sense as the mirror and equal to $1/f_1'$, the error can be corrected by a convex lens placed as near to the image as is technically acceptable. The general form of the refraction (or reflection) expression given in Eq. (2.33) can be written for a thin lens in the form

$$\frac{n'}{s'} - \frac{n}{s} = (n' - n)\left(\frac{1}{r_1} - \frac{1}{r_2}\right) = K$$

For a plano-convex lens in air, the power K is

$$K = (n' - 1)\left(\frac{1}{r}\right)$$

The Petzval sum of such a field-flattening lens is, from the definitions preceding (3.20),

$$(P_C)_{FFL} = \frac{K_{FFL}}{n'} = \frac{n' - 1}{n'}\left(\frac{1}{r_{FFL}}\right) \qquad (3.259)$$

For compensation of the field curvature of a Schmidt telescope of focal length f_1', we require

$$(P_C)_{FFL} = 1/f_1' ,\qquad (3.260)$$

which gives

$$r_{FFL} = \left(\frac{n'-1}{n'}\right) f_1' \qquad (3.261)$$

Such field-flattening lenses (singlets) are not free from other optical aberrations, in particular transverse chromatic aberration (C_2 in Eq. (3.223)) and distortion. In practice, because the lens has a certain finite distance from the image surface, there is also some coma. If the whole system is designed to include the field flattener, compensations are possible such that the only effective disadvantage is a slight increase in chromatic aberrations. A detailed treatment is given by Linfoot [3.29].

For larger Schmidt telescopes with $f' > $ ca. 0.5 m, it is possible to bend the photographic plates to the required spherical form. Alternatively, film can be used.

Figure 3.27 shows that the plateholder inevitably obstructs the incident beam. This limits the free field according to the obstruction ratio D_{PH}/D, where D is the telescope diameter and D_{PH} that of the plateholder:

$$\frac{D_{PH}}{D} = 2(u_{pr})_{max} N ,\qquad (3.262)$$

where $(u_{pr})_{max}$ is the semi-field in radians and N the f/no. Eq. (3.262) shows that large fields are only possible with reasonable obstructions if N is small, i.e. with high relative apertures. In practice, plateholders take up a larger angle than the free field. The diameter D_1 of the primary required to avoid vignetting of the field is

$$D_1 = D\left(1 + 2\frac{D_{PH}}{D}\right) \qquad (3.263)$$

In most practical cases, $D_1/D \sim 1.5$. This relation assumes, of course, that the pupil is at the corrector plate. Although this is the normal case, it should be remembered that the correction of the terms S_I, S_{II} and S_{III} makes the Schmidt telescope very insensitive to *stop shift* (see §3.4). Nevertheless, there are good reasons for retaining the basic geometry. Any shift of the stop from the corrector will increase the necessary diameter of this most difficult optical element. Also field beams pass asymmetrically through the plate giving chromatic variations over the field.

There are two obvious physical sources of higher order field aberrations. Firstly, an oblique beam traverses the plate with an increase of optical path length of

$$\delta W' = n'\left(\frac{1}{\cos u_{pr}} - 1\right) = n'\left(\frac{1 - \cos u_{pr}}{\cos u_{pr}}\right) \simeq n'\frac{u_{pr}^2}{2} ,\qquad (3.264)$$

compared with unity for the path length of equivalent axial rays. Secondly, there is a projection effect of $y/\cos u_{pr}$ in the effective height of the plate for oblique beams because the corrector does not rotate to be perpendicular to the principal rays as it is for the axis. Bahner [3.5] gives the formula

$$\frac{\Delta\eta'}{f_1'} = \frac{u_{pr}^2}{16n'N^3} \text{ rad} \qquad (3.265)$$

for the angular aberration diameter produced by these two effects together, where u_{pr} is the semi-field angle in radians, n' the refractive index of the corrector and N the f/no. With $u_{pr} = 3°$, $N = 3$, $n' = 1.5$, Eq. (3.265) gives a diameter of 0.87 arcsec. Much work was done in the pre-computer era on the analytical theory of higher order aberrations of the Schmidt telescope, for example by Baker [3.36] and Linfoot [3.29]. Although this work retains great interest for specialists, the universality of computers and sophisticated optical design programs has reduced its practical importance: higher order aberration effects are revealed exactly by ray tracing and *spot-diagram representations* [3.7] [3.37] (see also § 3.2.5.3). Third order theory, by contrast, will always retain its validity and importance because it reveals the fundamental possibilities and limitations of a given system.

Figure 3.29 shows spot-diagrams of the theoretical image quality of the ESO 1 m, f/3.0 Schmidt telescope with its original singlet corrector plate, plotted over the field for 24 cm × 24 cm plates bent to the optimum field curvature (±3.20° field).

3.6.2.5 Tolerances. Tolerances and alignment procedures will be dealt with in detail in RTO II, Chap. 2. Suffice it to say here that the Schmidt telescope is a favourable design from the point of view of the tolerances on position of the corrector plate except for lateral decentering. It will be shown in RTO II, Chap. 2 that the tolerances in this respect are a function only of the primary mirror and the distance of any secondary element (mirror or plate) from it. Sag or bending of the corrector plate is completely uncritical, since the optical path in the glass remains unchanged. Tilt of the corrector shifts the axis of the corrected image away from the centre of the photographic plate, leading to field asymmetry, but this is also not critical. The tolerance on the axial position of the plate can be deduced directly from Eqs. (3.225) in which s_{pr1} is set to $2f_1' + \delta z$ instead of $2f_1'$. It is obvious that the coma introduced is $\delta z/2f_1'$ of that of the uncorrected primary, i.e. $\delta z/4$.

3.6.2.6 The achromatic Schmidt telescope. The price paid for the great gain in field coverage of the Schmidt telescope is its physical length and the spherochromatism introduced by transferring the asphericity on the primary (reflecting) to the singlet corrector plate (refracting). The latter can be virtually entirely removed if the singlet corrector is replaced by a doublet. This is the same principle as an achromatic doublet objective with the important

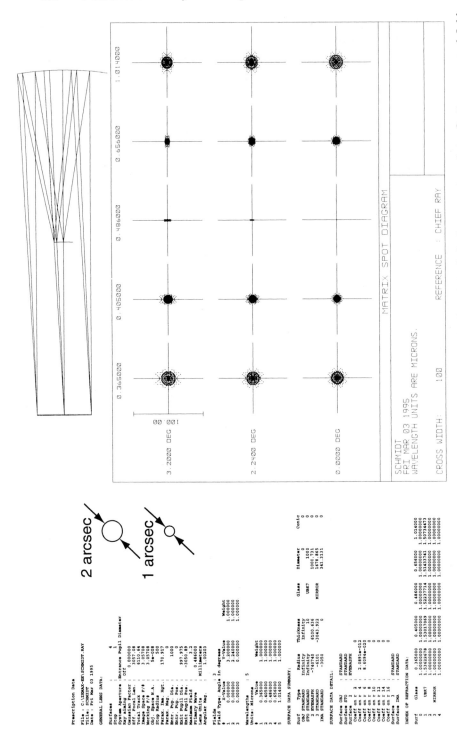

Fig. 3.29. Spot-diagrams for the ESO 1 m, f/3.0 Schmidt telescope with the original singlet corrector plate. Optimum curved field of radius 3050 mm and ±3.20° field for 24 cm × 24 cm plates

difference that the doublet objective is concerned to correct the first order longitudinal chromatic aberration C_1 of (3.223), whereas the achromatic
Schmidt corrector is concerned to correct $(\delta S_I)_C$, the third order chromatic
effect on S_I. We saw, from (3.244) that the spherochromatism of the singlet
plate is given by

$$\delta(\delta W_I^{'*}) = \left(\frac{n_1' - n_0'}{n_0' - 1}\right)(\delta W_I^{'*})_0 \ ,$$

in which n_0' is the refractive index at the central (correction) wavelength λ_0
and n_1' that for any other wavelength λ_1. The dispersion function

$$\frac{1}{\nu_A} = \frac{n_1' - n_0'}{n_0' - 1} \tag{3.266}$$

of the glass of the corrector is therefore the function determining the spherochromatism for a constant $(\delta W_I^{'*})_0$ resulting from the primary mirror of the
Schmidt telescope. For a typical glass, e.g. Schott UBK7 chosen for favourable
UV transmission, this dispersion curve is shown in Fig. 3.30. We shall profit
from the excellent example of the achromatic Schmidt telescope to illustrate
the dispersion properties of optical glasses, leading to the chromatic effects
of "primary and secondary" spectrum.

The slope of the mean straight line of this function gives the *primary
spectrum* effect. So-called "normal" glasses have a similar departure from
the straight line called a "normal" *secondary spectrum* effect. The significance
in the correction of achromatic doublets is dealt with by Bahner [3.5] and
Welford [3.6]. The same physical situation as in Fig. 3.30 applies except that
the ordinate is replaced by the back focal distance s' of the image as a function of wavelength. In both cases, the process of achromatisation combines
a dispersion function with positive slope with another one of equal negative slope to give a resulting function of minimum slope over the wavelength

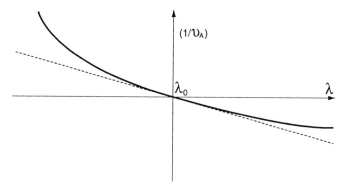

Fig. 3.30. The dispersion function for a typical optical glass
$\left(\frac{1}{\nu_a} = \frac{n_1' - n_0'}{n_0' - 1}\right)$

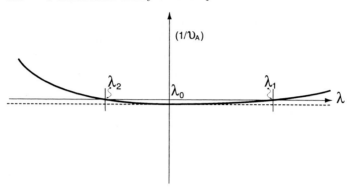

Fig. 3.31. The effect of achromatisation: the dispersion function is rotated to minimize its slope, giving desired zero points λ_1 and λ_2

range used, as shown in Fig. 3.31. If two "normal" glasses are combined, the resulting curvature of the dispersion function will be little changed, leaving a "normal" uncorrected secondary spectrum effect. This achromatisation process is only possible if the dispersion functions $(\nu_A)_\alpha$ and $(\nu_A)_\beta$ for the two elements α and β (glasses) are different: otherwise, the residual optical power in the case of a lens doublet, or the residual correction of δS_I^* in the Schmidt plate case, would be zero. Fortunately, a wide variety of glasses with different dispersions and refractive indices n_0' are available (Fig. 3.32). The larger the difference in $1/\nu_A$, the easier the combination, in that the individual powers (achromatic doublet) or asphericities (achromatic Schmidt corrector) will be less extreme.

For simplicity, we will now omit the plate stop shift δ of δS_I, $\delta W_I'$ etc., corresponding to Eq. (3.215), since the spherical aberration is independent of stop shift and $\delta S_I = S_I$, $\delta W_I' = W_I'$. Then from (3.244), the conditions for achromatisation between λ_0 and λ_1 for the Schmidt corrector are

$$\left.\begin{array}{rcl} \dfrac{(W_I'^*)_{0\alpha}}{\nu_{A\alpha}} & = & -\dfrac{(W_I'^*)_{0\beta}}{\nu_{A\beta}} \\[2mm] \text{with } (W_I'^*)_0 & = & (W_I'^*)_{0\alpha} + (W_I'^*)_{0\beta} \end{array}\right\} \tag{3.267}$$

for the required total correction, giving

$$\left.\begin{array}{rcl} (W_I'^*)_{0\alpha} & = & (W_I'^*)_0 \left(\dfrac{\nu_{A\alpha}}{\nu_{A\alpha} - \nu_{A\beta}}\right) \\[4mm] (W_I'^*)_{0\beta} & = & -(W_I'^*)_0 \left(\dfrac{\nu_{A\beta}}{\nu_{A\alpha} - \nu_{A\beta}}\right) \end{array}\right\} \tag{3.268}$$

The asphericities are both proportional to $1/(\nu_{A\alpha} - \nu_{A\beta})$. Note that the wavelengths λ_0 and λ_1 used in (3.266) will not normally coincide with the conventional definition of ν_A used in Fig. 3.32. The wavelengths and corresponding definitions of $\nu_{A\alpha}$ and $\nu_{A\beta}$ must be adapted to the astronomical requirements of the Schmidt telescope. For large sizes of the order of 1 m, the choice of glasses is limited by availability and considerations of absorption in the UV and blue. For the ESO 1m Schmidt telescope, the original singlet corrector was replaced by an achromat made from the glasses UBK7 and LLF6, the latter giving an adequate difference $(\nu_{A\alpha} - \nu_{A\beta})$ together with acceptable transmission. Glasses further to the right in Fig. 3.32 than LLF6 contain increasing amounts of lead with unacceptable absorption in the blue.

The manufacture of such an achromatic plate is a very difficult technical undertaking. Apart from making the individual aspheric plates, the manufacturer must normally combine them in a total precision unit with optical cement between the elements. The cement should ideally have a refractive index intermediate between those of the plates in order to reduce light loss and, above all, ghost images at two extra reflecting surfaces. According to the law of Fresnel, the intensity of reflected light at the boundary between media of refractive indices n_1' and n_2' is

$$I = \left(\frac{n_2' - n_1'}{n_2' + n_1'} \right)^2 \tag{3.269}$$

If $n_2' = 1.5$ and $n_1' = 1$ for air, (3.269) gives the normal figure for (untreated) glass surfaces in air of $I \simeq 4\%$. If the cement reduces $(n_2' - n_1')$ by a factor of ten to 0.05, the reflections are reduced by a factor of 100.

Figure 3.33 shows the spot-diagrams of the (theoretical) ESO 1 m *achromatic* Schmidt in the same format as Fig. 3.29, but with a scale five times as large. The residual spherochromatism is small compared with the grain of the photographic plates. In practice, this gain must be offset by manufacturing errors of the corrector plates which produce a uniform degradation over the field. This is a matter to be defined by contractual tolerances.

3.6.3 The Maksutov telescope

We return now to the Bouwers telescope dealt with in § 3.6.1, which, in its pure concentric form, suffers from first order chromatic aberration apart from fifth order spherical aberration and spherochromatism. The first order chromatic aberration is fatal unless a minimal wavelength bandpass is used. Fortunately, this aberration can be corrected, at the cost of perfect symmetry, by the *Maksutov form* [3.38]. Meniscus solutions have two great advantages over the Schmidt telescope. Firstly, the constructional length is about one third less – unless the stop is maintained strictly at the centre of curvature, an important point discussed below. Secondly, above all for smaller telescopes, a meniscus with spherical surfaces is far easier and cheaper to produce than a Schmidt corrector plate.

In its pure form, the Maksutov telescope only departs from the symmetrical concept of Bouwers (Fig. 3.26) by changing the radii of the meniscus to give a slightly convex form, as shown schematically in Fig. 3.34. This slight

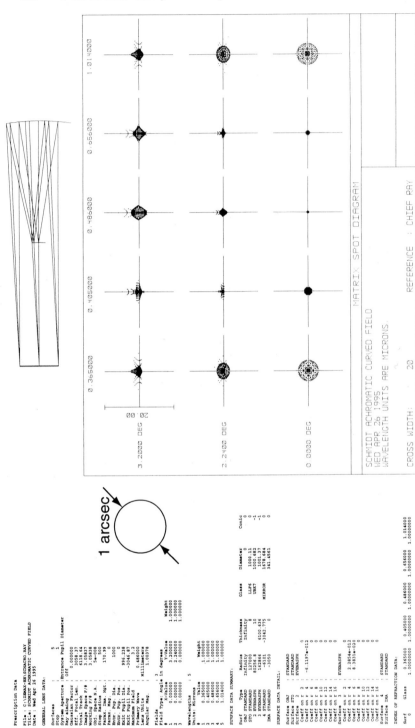

Fig. 3.33. Spot-diagrams for the ESO 1 m, f/3.0 Schmidt telescope with the achromatic (doublet) corrector plate (glasses UBK7 and LLF6). Optimum curved field of radius 3050.5 mm and ±3.20° field for 24 cm × 24 cm plates (format identical with Fig. 3.29, but scale five times larger)

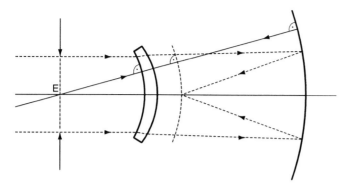

Fig. 3.34. The basic form of the Maksutov telescope (1944)

convex form is sufficient to compensate for the negative first order chromatic aberration generated by the pure Bouwers meniscus, which has slight negative power.

Excellent treatments of the basic theory are given by Bahner [3.5] and Schroeder [3.22]. Essentially, the following formulation is similar to that of Schroeder following Bouwers [3.32] and Maksutov [3.38]. A general analytical solution for the third order spherical aberration introduced by the thick *concentric meniscus* of the Bouwers telescope is possible, but complex. Bouwers himself simplified the formulation by assuming the approximation of a thin lens for an object at infinity. He starts from the formula [3.32], valid for the third order aberration of low power ($1/f'_{men}$) and strongly "bent" lenses, for the angular aberration[6]. In our notation, the angular diameter of the aberration is

$$(\delta u'_p)_{GF,men} = \left. -\frac{y_1^3}{f'^3_{men}} \left[\left(\frac{n'+2}{n'} \right) \left(\frac{f'_{men}}{r_1} \right)^2 \right. \right. \\ \left. \left. - \left(\frac{(2n'+1)}{n'-1} \right) \left(\frac{f'_{men}}{r_1} \right) + \left(\frac{n'}{n'-1} \right)^2 \right] \right\} , \qquad (3.270)$$

---

[6] This formula for the angular third order spherical aberration due to a thin lens for an object at infinity was taken by Bouwers from the classical work "The Principles and Methods of Geometrical Optics" by J. P. C. Southall (Macmillan 1910), page 388. The same formula is given by Czapski-Eppenstein [3.11(b)], who give its origins as far back as Huygens' "Dioptrica" (Ed. 1703), Prop. 27. The general form for a thin lens with an object at any finite distance contains 6 terms: it is derived for longitudinal spherical aberration by A. E. Conrady in "Applied Optics and Optical Design", Part 1 (Oxford 1929 and 1943), page 95. This is the well-known "G-sum" formula. Hopkins [3.3] also derives it, in the wavefront aberration form.

where n' is the index of the meniscus and r_1 the radius of its first surface. From (3.190), bearing in mind that the index in the image space is 1 and (3.21), this gives

$$(W'_I)_{GF,men} = \frac{y_1^4}{8f'^3_{men}} \left\{ \left[\left(\frac{f'_{men}}{r_1} \right)^2 \left(\frac{n'+2}{n'} \right) \right. \right.$$
$$\left. \left. - \left(\frac{f'_{men}}{r_1} \right) \left(\frac{2n'+1}{n'-1} \right) + \left(\frac{n'}{n'-1} \right)^2 \right] \right\} \tag{3.271}$$

As a further legitimate approximation in view of the thin lens assumption and the fact that, in practice $f_{men} \gg r_1$, he reduces this to

$$(W'_I)_{GF,men} \simeq \frac{y_1^4}{8f'_{men}r_1^2} \left(\frac{n'+2}{n'} \right) \tag{3.272}$$

Eq. (2.51) gave the expression for the total power of a system of two thin lenses in air, separated by a distance d. If the lenses are refracting surfaces separated by d in a medium of index n', we have a "thick lens", the case of our concave meniscus in the Bouwers telescope. The focal length is given by

$$K = K_1 + K_2 - \frac{d}{n'} K_1 K_2 , \tag{3.273}$$

K referring to the power $1/f'$. For the concentric meniscus

$$d = r_1 - r_2 , \tag{3.274}$$

in which $d > 0$ and radii r_1 and r_2 are negative. Now

$$K_1 = \frac{(n'-1)}{r_1} , \quad K_2 = \frac{(1-n')}{r_2} \tag{3.275}$$

Substituting (3.274) and (3.275) in (3.273), we derive for the focal length of the concentric meniscus

$$\frac{1}{f'_{men}} = -\frac{d}{r_1 r_2} \left(\frac{n'-1}{n'} \right) \tag{3.276}$$

Now f'_{men} is, in Eq. (3.271), the determinant parameter for the spherical aberration of a thin lens, for given values of y_1, n' and shape determined by r_1. If we give the real, thick meniscus the same focal length, we have, within the approximation defined, the condition for setting its aberration to compensate that of the primary mirror. From Table 3.4 and Eq. (3.21), the third order spherical aberration of the spherical mirror M_1 is given by

$$(W'_I)_{GF,M1} = \frac{(S_I)_{M1}}{8} = - \left(\frac{y_1}{f'_1} \right)^4 \frac{f'_1}{32} \tag{3.277}$$

Combining (3.276) with (3.272), we have the condition for correction of spherical aberration in the total system as

$$(W'_I)_{Bou} = (W'_I)_{men} + (W'_I)_{M1} ,$$

giving

$$(W_I')_{GF,Bou} \simeq -\frac{(n'-1)(n'+2)}{n'^2}\left(\frac{dy_1^4}{8r_1^3 r_2}\right) - \frac{y_1^4}{32 f_1'^3} , \tag{3.278}$$

in which f_1', r_1 and r_2 are negative, while d is positive, giving compensation between the first and second terms. This gives the Bouwers condition for the correction of third order spherical aberration as

$$d_{Bou} \simeq -r_1 \left/ \left[\frac{(n'-1)(n'+2)}{n'^2}4\left(\frac{f_1'}{r_1}\right)^3 - 1\right]\right. , \tag{3.279}$$

in which r_1 will be given some arbitrary value, usually as large as possible, taking account of the detector, to reduce higher order aberration.

In the Maksutov modification, the spherical aberration compensation must be maintained while the concentric meniscus form must be slightly modified to compensate its primary chromatic aberration. This arises from the dependence of its focal length f_{men}' on the index n'. Differentiating (3.276) gives

$$\frac{df_{men}'}{f_{men}'} = \frac{-dn'}{n'(n'-1)} \tag{3.280}$$

Since the mirror images the virtual image of the object formed by the meniscus, shifts of its focus df_{men}' will be reproduced as longitudinal shifts of the final focus with wavelength. The Maksutov solution therefore requires a thick lens fulfilling the condition

$$\frac{df_{men}'}{dn'} = 0 \tag{3.281}$$

If (3.275) is inserted in (3.273), the "thick lens" formula converts to

$$f_{men}' = \left(\frac{n'}{n'-1}\right)\left[\frac{r_1 r_2}{n'(r_2-r_1)+(n'-1)d}\right] \tag{3.282}$$

Differentiation and (3.281) give then as the Maksutov condition for achromatism of the primary aberration C_1 in (3.222):

$$d = (r_1 - r_2)\left(\frac{n'^2}{n'^2-1}\right) \tag{3.283}$$

This gives a positive d for negative radii with the usual sign convention. The separation of the centres of curvature of the two surfaces of the Maksutov meniscus is

$$\Delta z = (r_1 - r_2) - d = -\frac{d}{n'^2} , \tag{3.284}$$

the minus sign indicating that the centre of curvature of surface 2 is closer to the mirror.

The final design of a Maksutov telescope requires an interactive process typical of optical design and implicit in modern optimization programs. Essentially, for a given thickness d, the three parameters power, "bending" (i.e. shape for a given power) and axial position must be optimized to give optimum correction of the three aberrations C_1 (longitudinal chromatic aberration), S_I (spherical aberration) and S_{II} (coma). It should be borne in mind that, once S_I and S_{II} are corrected, the stop-shift formulae (3.22) and (3.213) show that, to the third order approximation, S_I, S_{II} and S_{III} are *independent of the stop position*. The stop-shift term for C_2, given in brackets in Eqs. (3.223), shows that this is also true of C_2 because C_1 is corrected.

It follows that we can draw a very important conclusion for all such Maksutov-type systems: with C_1, S_I and S_{II} corrected, *the front stop position of Fig. 3.34 near the centre of curvature of the mirror ("long" Maksutov) can be abandoned in favour of a stop position at the meniscus* without changing C_1, C_2, S_I, S_{II} and S_{III}. Only S_V is influenced, from Eqs. (3.22), since $S_{IV} \neq 0$ (and by a small S_{III} residual), but this is of little practical consequence. This leads to the *"short"* Maksutov, a form which, with slight optimization, gives virtually the same image quality except for a minor increase in astigmatism. According to Maksutov [3.38], the optimum axial distance of the meniscus from the mirror is about $1.3f'$ to $1.4f'$. Not only is the reduction in length of the "short" form of great significance, but also that of the diameters of the meniscus and mirror.

Table 3.11. Data for the "short" Maksutov telescope giving the results of Fig. 3.35 with $D = 400$ mm, f/3.0

r	d	n' ($\lambda = 486$ nm)	Medium
(stop)	1.0	1	air
-522.2100	50.0	1.522 387	BK7
-551.5188	1579.784	1	air
-2496.391	-1281.084	-1	air
-1233.465 (image)			

The data of such a "short" Maksutov for a pupil diameter of $D = 400$ mm and $N = 3.0$ are given in Table 3.11 and the corresponding spot-diagrams in Fig. 3.35. The quality, for a broad spectral band, is in no way inferior to the "long" system because it is limited by the lateral chromatism C_2 in both cases. This is shown in Fig. 3.36, which represents the change to the principal ray path due to the stop shift. With the "concentric" stop (a), the principal ray cuts the first surface normally, and also the second surface 2_C if the meniscus is concentric. C_2 is then zero. With the stop shifted to (b)

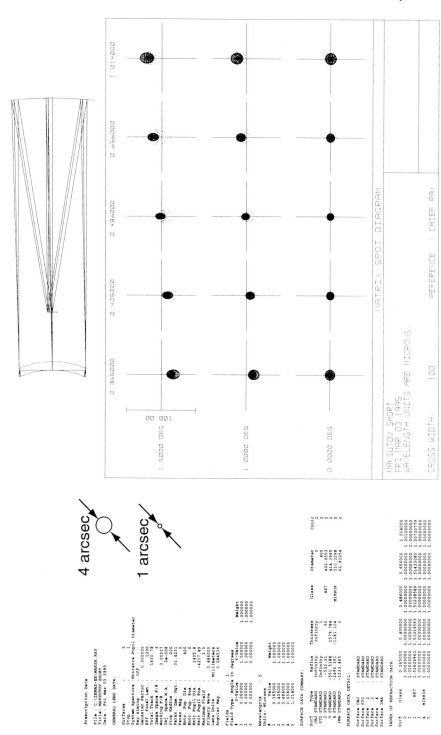

Fig. 3.35. Spot-diagrams for the "short" Maksutov telescope of Table 3.11 with aperture 400mm and f/3.0

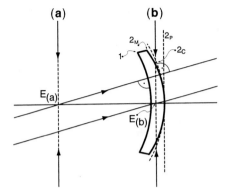

Fig. 3.36. Effect of stop shift on transverse chromatic aberration C_2 in a Maksutov meniscus: the "short" version (**b**) causes refraction and dispersion of the principal ray at the first surface, but the effect is largely (though not entirely) compensated at the second surface

at the first meniscus surface, refraction and dispersion of the principal ray take place at this surface. C_2 would only be zero if the second surface were plane, 2_P, equivalent to a plane-parallel plate for the principal ray. For a concentric meniscus, C_2 would correspond solely to the stop-shift term in Eqs. (3.223) for the value of C_1 given by the concentric (Bouwers) meniscus. If the second surface is steepened to 2_M, the Maksutov form, correcting C_1, there is no stop-shift term for C_2; but the finite curvature of 2_M, combined with the finite thickness, produce a finite C_2 residual. With the stop at (a), the incidence angle of the principal ray at 2_M produces the same net effect. This gives a numerically almost identical C_2 aberration residual in both cases, which is determinant for the image quality with typical fields of $\pm 1.5°$, as we see in Fig. 3.35.

Further, more complex variants of the simple meniscus corrector are discussed below. An extremely useful extension is the addition of an *achromatic field flattener*. Table 3.12 gives the data of the "short" Maksutov above, op-

Table 3.12. Data for the "short" Maksutov system ($D = 400$ mm and f/3.0) optimized with an additional achromatic field flattener, giving the results of Fig. 3.37

r	d	n' ($\lambda = 486$ nm)	Medium
(stop)	1.0	1	air
−523.5586	60.0	1.522 387	BK7
−557.2162	1512.377	1	air
−2497.425	−1271.319	−1	air
678.388	−8.0	−1.522 387	BK7
132.567	−5.0	−1.632 103	F2
241.8428	−3.9831	−1	air
∞ (image)			

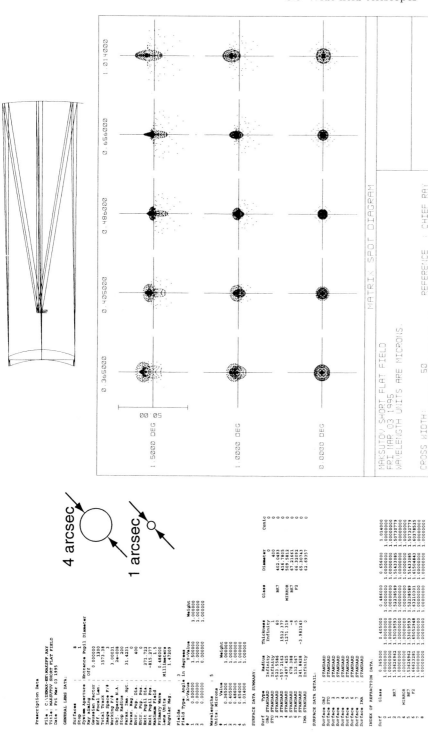

Fig. 3.37. Spot-diagrams for a "short" Maksutov telescope with aperture 400mm and f/3.0 optimized with an achromatic field flattener (Table 3.12)

timized with such a field flattener. The extra degrees of freedom enable C_2 to be reduced to about one quarter and the residual S_{III} to about one third. Fig. 3.37 shows the spot-diagrams to twice the scale of Fig. 3.35. If the relative aperture is reduced from f/3.0 to f/3.5, a further significant improvement is achieved by reduction of fifth order aberrations.

The residual C_2 shown in the spot-diagrams of Fig. 3.37 is largely due to the secondary spectrum of the "normal" glasses BK7 and F2 composing the field flattener. This leads to a curvature of the locus of the centres of gravity of the spot-diagrams in the field. The curvature is convex to the axis abcissa, corresponding to overcorrection of C_2 in the blue, undercorrection in the red, compared with the uncorrected C_2 of the meniscus shown in Fig. 3.35.

3.6.4 More complex variants of telescopes derived from the principles of the Schmidt, Bouwers and Maksutov systems

The telescope forms dealt with so far have consisted of one element or two separated elements. As we have seen, the analytical theory of the 2-element telescope is by no means trivial and was only initiated and completed in the first half of this century through the pioneer work of Schwarzschild, Chrétien, Schmidt, Bouwers and Maksutov.

The analytical theory of sophisticated forms containing more than three elements inevitably grows rapidly in complexity. Third order theory is still of utmost value in understanding the potential and limitations of any given system type, but the final layout and optimized design will be performed with one of the many programs available for a wide range of computers, including PCs. However, for the systems dealt with in §3.6.4.1 below consisting of an aspheric plate and one or two mirrors (aspheric in the general case), the analytical equations derived in the general theory above provide a simple and powerful basis for complete understanding of their optical properties.

A complete treatment of the systems proposed and investigated in the past 50 years would itself require a whole book. The purpose of the account given in this section is to describe the characteristics of the basic types and refer to the literature for further detail. The books referred to are all excellent in their way and give a rich diversity of treatment reflecting the particular interests or style of the author.

3.6.4.1 Further developments of Schmidt-type systems. *Solid and semi-solid Schmidt cameras*: This form is really only of interest for spectrograph cameras rather than telescopes but I include it here for completeness and because it is instructive optically. I follow the brief but excellent account by Schroeder [3.22]. Fig. 3.38(a) shows the basic optical property of the solid Schmidt and is to be compared with Fig. 3.36(b) for the Maksutov form. The solid Schmidt consists of a glass block with the aspheric form (front surface) at the centre of curvature of the spherical concave back. In glass with refractive index n', therefore, the Schmidt geometry is fully maintained. The essential feature is the single refraction of the principal ray with the semi-field angle u_{pr}, giving a refracted angle of $u'_{pr} = u_{pr}/n'$. Clearly, the image height at the spherical focal surface is reduced by the factor $1/n'$. In other words, the effective focal length $f'_{sol} = f'/n'$, where f' is the equivalent focal length of a normal Schmidt in air of the same physical length. The "speed" of the system as a camera is therefore $(n')^2$ higher, an important advantage in spectrographs. Alternatively, one can interpret the effect as an increase in the field coverage in object space.

The image height is given by

$$\eta' = \frac{u_{pr}}{n'} f'$$

from Fig. 3.38(a). Differentiating with respect to n' gives

$$\mathrm{d}\eta' = -u_{pr}f' \frac{\mathrm{d}n'}{(n')^2} \tag{3.285}$$

Schroeder quotes the application of this equation to the case of a solid Schmidt made of SiO_2 for the wavelength range 400-700 nm and $f' = 500$ mm at an angle $u_{pr} = 1°$, giving $\mathrm{d}\eta' = 61$ μm. Since the scale is $f'/206\,265\,n' \simeq 1.66$ μm/arcsec, this transverse colour aberration amounts to about 37 arcsec, a value which excludes use for direct imagery unless very narrow band filters are used. Transverse colour aberration is of no consequence in a spectrograph camera because it only produces a slight distortion in the spectrum.

A conventional Schmidt plate, even if it is thick, produces no transverse colour because a plane parallel plate produces no change of angle of an incident beam, irrespective of wavelength. For a Maksutov telescope with its stop at the meniscus (Fig. 3.36(b)), the refraction at its first surface is the same as in the solid Schmidt; but this huge error is almost entirely removed by the refraction at the second surface.

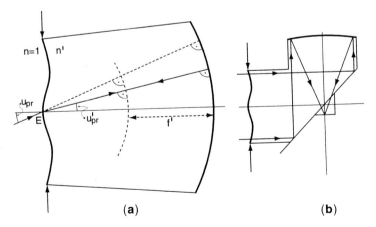

Fig. 3.38. The solid Schmidt camera in the direct form (**a**) and folded form (**b**), with effective focal length f'/n'

Schroeder shows that the solid Schmidt reduces the field aberrations by a factor $(n')^2$ compared with the normal Schmidt in air.

The focal surface can be made accessible by drilling a cylindrical hole into the block towards the primary mirror from the corrector face to the image surface, but the access remains inconvenient. Figure 3.38(b) shows how this problem can be solved by including a plane mirror in the block, a device first proposed (according to Riekher [3.39(a)]) by Hendrix in 1939.

A semi-solid Schmidt is shown in Fig. 3.39. Its advantage is that the focal surface is accessible without an additional reflection but is still rigidly attached to the mirror by a single glass block. The focal length is f'/n' as with the solid Schmidt; the aberration characteristics are also similar.

Wright-Väisälä system: This was the first important modification [3.40] [3.41] of the basic invention by Schmidt. Accounts are also given by Bahner [3.5], Baker [3.23(b)] and König-Köhler [3.30]. Essentially, the aim was to overcome the major weakness of the classical Schmidt, its excessive length. Wright and Väisälä independently discovered the important property that the conditions for the correction of spherical aberration (S_I) and field coma (S_{II}) can be fulfilled for *any* position of the corrector plate along the axis, if the mirror is given a suitable aspheric form. In fact, this is simply a further application of the important law discussed above as a consequence of Schwarzschild's theory [3.1] that, in any telescope consisting of two elements, the two conditions $S_I = S_{II} = 0$ can be fulfilled for *any* (significant) axial separation by appropriate aspherics on the elements (see Eqs. (3.219)). The third condition fulfilled by the classical Schmidt ($S_{III} = 0$) *cannot* be satisfied, the price to be paid for the more compact form. The Wright-Väisälä solution places the correcting plate at the focus, halving the length of the classical Schmidt (Fig. 3.40).

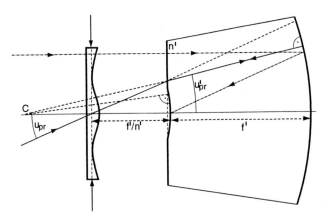

Fig. 3.39. The semi-solid Schmidt camera with effective focal length f'/n'

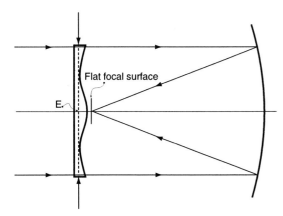

Fig. 3.40. The Wright-Väisälä telescope (1935)

The third order aberrations of such systems can be derived at once from the general equations (3.219) for an aspheric mirror and an aspheric plate. Since, by definition of the Wright-Väisälä telescope, the plate is effectively at the focus, we have $s_{pl} = +f_1'$ with f_1' negative. Eqs. (3.219) can then be written:

$$
\left.
\begin{aligned}
\sum S_I &= \left(\frac{y_1}{f_1'}\right)^4 \left[-\frac{f_1'}{4}(1 + b_{s1}) + \delta S_I^*\right] \\[2mm]
\sum S_{II} &= \left(\frac{y_1}{f_1'}\right)^3 \left[-\frac{f_1'}{2} + \delta S_I^*\right] u_{pr1} \\[2mm]
\sum S_{III} &= \left(\frac{y_1}{f_1'}\right)^2 \left[-f_1' + \delta S_I^*\right] u_{pr1}^2
\end{aligned}
\right\} \tag{3.286}
$$

Setting $\sum S_I = \sum S_{II} = 0$ in the first two equations gives

$$
\left.
\begin{aligned}
\delta S_I^* &= +\frac{f_1'}{2} \\[2mm]
b_{s1} &= +1
\end{aligned}
\right\}, \tag{3.287}
$$

thereby defining the third order optical design. The third equation of (3.286) then gives for the astigmatism

$$
\sum S_{III} = -\left(\frac{y_1}{f_1'}\right)^2 \frac{f_1'}{2} u_{pr1}^2 \tag{3.288}
$$

If we now take the *normalized* data of Table 3.3 with $y_1 = -f_1' = u_{pr1} = 1$, then (3.288) reduces to

$$
\sum S_{III} = +\frac{1}{2} \tag{3.289}
$$

With the same normalization, the field curvature is, from Table 3.4

$$\sum S_{IV} = -1 \tag{3.290}$$

The *effective field curvature* is given from Eq. (3.203) and corresponding to Table 3.3 by

$$2\sum S_{III} + \sum S_{IV} = 0 , \tag{3.291}$$

from (3.289) and (3.290). Thus the optimum field is flat in the Wright-Väisälä telescope, as in the original Schwarzschild telescope of Fig. 3.9 but with a far more favourable constructional length.

From (3.289) and Table 3.4, the astigmatism of the Wright-Väisälä telescope is only half that of a spherical or parabolic primary with the pupil at its pole and only about 1/8 of that of an RC telescope of the same focal length (see Table 3.3), as well as giving the further advantage of a flat optimum field. Its field can therefore be more than twice as large as that of an RC telescope, comfortably covering 1° or more. But, because of its astigmatism, it cannot compete with the field of the classical Schmidt.

The classical Schmidt gives with Eqs. (3.219), setting $s_{pl} = +2f_1'$ and $b_{s1} = 0$, the plate deformation $\delta S_I^* = +\frac{f_1'}{4}$. Eqs. (3.287) show therefore that the Wright-Väisälä telescope requires *twice* the plate asphericity of the classical Schmidt and that the mirror has an asphericity of the same amount as a parabola *but of opposite sign*. The required form is therefore an *oblate spheroid* – see Eq. (3.13). Dimitroff and Baker [3.23(a)] list three cameras of the Wright type built in the United States with apertures up to 20 cm at f/4. Väisälä built one in Turku, Finland [3.42]. In general, the steep aspherics have been a disadvantage compared with Maksutov solutions, in spite of the favourable length.

Figure 3.41 shows spot-diagrams for a Wright-Väisälä telescope of aperture 400 mm and f/4.0. The image quality is solely limited by the uncorrected astigmatism, growing as the square of the field.

Schmidt-Cassegrain systems: In the general case, since Schmidt-Cassegrain systems provide three separated aspheric surfaces, the three aberrations $\sum S_I$, $\sum S_{II}$ and $\sum S_{III}$ can be corrected according to Eqs. (3.220) *for any geometry* of the optical system. Of course, if the separations or power parameters of the mirrors are unfavourable, very high asphericities can result. Conversely, favourable choices for the geometrical parameters can produce excellent solutions with zero or low asphericity on the mirrors. Such possibilities were first systematically analysed in a classic paper by Baker [3.43]. Shorter, but excellent accounts are also given by Bahner [3.5], Dimitroff and Baker [3.23(c)], König-Köhler [3.30(a)], Slevogt [3.44], Schroeder [3.22] and Riekher [3.39(b)].

Baker proposed four different systems, types A, B, C, D (Fig. 3.42), which largely covered the range of interest of the 3-element Schmidt-Cassegrain. The systems in Fig. 3.42 are shown on the same scale of the equivalent focal length, the fifth case being the classical Schmidt for comparison. In all four Baker systems, the mirrors have equal radius of opposite signs, giving zero

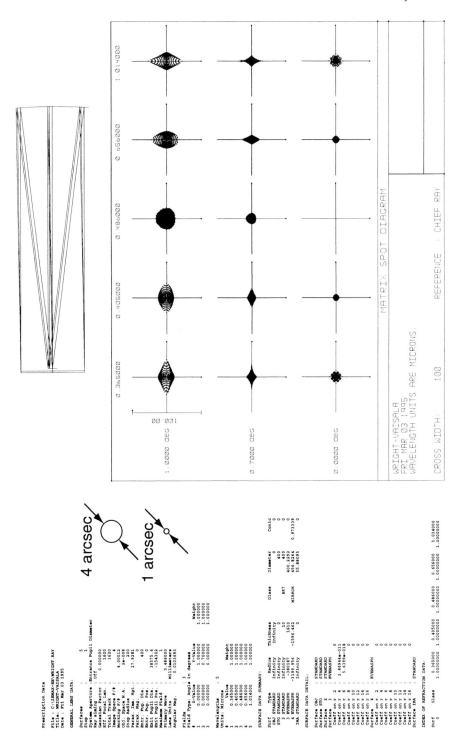

Fig. 3.41. Spot-diagrams for a Wright-Väisälä telescope of aperture 400 mm and f/4.0

Petzval sum ($\sum S_{IV} = 0$). All systems fulfil the conditions $\sum S_I = \sum S_{II} = \sum S_{III} = 0$, giving therefore a corrected flat field in front of the primary mirror. The different types differ only in the position of the corrector plate. Type A is the shortest – the length from plate to primary can be reduced to only $\frac{2}{3}f'$ or less – but requires appreciably aspheric mirrors and a very aspheric plate. Baker also proposes a front baffle which loses some of the advantage of the very short plate length. (It is important to remember that, as a result of the stop-shift formulae (3.22), the stop position is – to third order accuracy – of no consequence in these systems because $\sum S_I \ldots \sum S_{IV}$ are all zero). Types B and C are dimensionally almost identical: type B has a spherical secondary and lightly aspheric primary, type C a spherical primary and lightly aspheric secondary. Type D corrects the further condition of distortion ($\sum S_V = 0$) but requires significant asphericities on the mirrors, as with type A. As a consequence of the fulfilment of the condition $\sum S_{IV} = 0$ requiring $|f_2| = |f_1|$, *all* the Baker systems require an image position *in front of* the primary to achieve an acceptable obstruction ratio R_A. For a Cassegrain telescope with the following quantities defined as positive from Tables 2.2 and 2.3, we have from (2.57) and (2.58)

$$T = \frac{f'}{L} = \frac{1}{R_A},$$

giving $R_A = L/f' = L$ with our normalization $f' = 1$. Baker's systems have $L = 0.350$ with the geometry he proposed. This is a rather high value, bearing in mind the field supplement required which brings the real physical obstruction up to about 0.5. But it can still be acceptable in view of the other advantages. If the image position were in the more conventional position *behind* the primary, the obstruction ratio would normally be unacceptable.

Baker considered *Type B* the most interesting because it reduced the length to some two thirds of that of the classical Schmidt while allowing a spherical secondary and only slightly aspheric primary. The spherical secondary is a great advantage in manufacture: Baker pointed out that the primary can be used as an interference proof plate, since its central area hardly departs from the sphere. We will now take the Type B system as an example for third order analysis. The data of the system [3.5] [3.23(c)] [3.43]

Table 3.13. Optical data of Baker Schmidt-Cassegrain Type B with f/3.0 and $f' = 1$

r	d	D'	b_s
(Plate)	+ 1.394	0.333	(δS_I^*)
−1.300	−0.4225	0.333	+ 0.016
−1.300	+ 0.350	0.167	0
	(to image = L)		

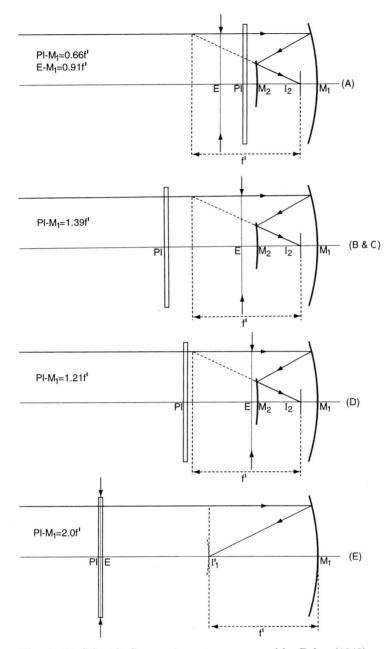

Fig. 3.42. Schmidt-Cassegrain systems proposed by Baker (1940)

are given in Table 3.13, normalized for $f' = 1$. We can apply this data using Eqs. (3.220) to deduce the necessary asphericity δS_I^* for the aspheric plate and confirm the correction

$$\sum S_I = \sum S_{II} = \sum S_{III} = 0,$$

giving

$$
\left.
\begin{array}{l}
-f'\zeta + L\xi + \delta S_I^* = 0 \\[2mm]
-d_1\xi - \dfrac{f'}{2} + \dfrac{s_{pl}}{f'}\delta S_I^* = 0 \\[2mm]
\dfrac{f'}{L}(f' + d_1) + \dfrac{d_1^2}{L}\xi + \left(\dfrac{s_{pl}}{f'}\right)^2 \delta S_I^* = 0
\end{array}
\right\}
\tag{3.292}
$$

Now from Table 3.13, we have $s_{pl} = -1.394 f'$ with $f' = 1$. Also $f_2' = -0.65$ and from Eq. (2.90)

$$f_2' = \frac{L}{m_2 + 1},$$

giving with $L = 0.350$ the value $m_2 = -1.538$. The mirror separation $d_1 = -0.4225$. Now the secondary is defined as spherical in this system, giving from (3.41) with $b_{s2} = 0$

$$\xi = \frac{(m_2 + 1)^3}{4}\left(\frac{m_2 - 1}{m_2 + 1}\right)^2 = -0.866$$

The first two equations of (3.292) then give

$$\delta S_I^* = -0.621$$
$$\zeta = -0.924$$

and substitution of the values in the third equation shows that the conditions are fulfilled. Eq. (3.30) gives

$$b_{s1} = +0.016 ,$$

a slight oblate spheroid but only 1.6% of a parabolic deformation of opposite sign.

In general, such a system can be set up with any geometry by solving Eqs. (3.292).

For the Baker system B above, the aspheric plate requires $\delta S_I^* = -0.621$, about 2.5 times the value of -0.25 given by Eq. (3.229) with $f_1' = -1$ for the Schmidt of the same focal length. This is the price to be paid for reducing the length of the system. Since the chromatic errors increase linearly with the asphericity on the plate, Baker proposed an achromatic plate of the type discussed in § 3.6.2.6. Without this, it is clear from the stop-shift equations (3.22) that the spherical aberration induced by the spherochromatism will lead to coma and astigmatism if the stop is shifted from the plate. Such chromatic effects are considered in detail by Schroeder [3.22(a)]. Because of these, it is no longer considered normal good practice to shift the stop from the plate as proposed by Baker (Fig. 3.42). The same remark applies to any system with a strong singlet aspheric plate producing appreciable spherochromatism.

A Baker-type telescope with plate diameter 84 cm was built in 1949 for the Boyden station (South Africa) of the Harvard College Observatory in association with the Armagh and Dublin Observatories (ADH – telescope) [3.39(c)]. The flat field has a diameter of 4.8° (26 cm) with relative aperture f/3.7. A similar system following a design by Linfoot [3.29] with an aperture of 50 cm was built for St. Andrews University in 1953. Linfoot [3.45] proposed a design with 2 spherical mirrors, but with appreciable astigmatism and chromatic aberration. Slevogt's modification [3.44] [3.46] of Baker's Type B system to make both mirrors spherical and accept a slight amount of astigmatism seems a more practical solution.

Further details of the performance of the Baker flat-field cameras are given by Schroeder [3.22].

Linfoot [3.45] [3.47] proposed a further interesting basic form, the *concentric* (monocentric) Schmidt-Cassegrain (Fig. 3.43). From its basic geometry, this is the most fundamental, indeed obvious, extension of the basic Schmidt to the Schmidt-Cassegrain. However, the monocentric form puts a special constraint on the radius of the secondary relative to that of the primary and rules out immediately the possibility of the flat field given by the Baker constraint of $r_2 = r_1$. In the general case, without restriction on the final image position, we have from the concentricity:

$$2f_2' = 2f_1' - d_1 \tag{3.293}$$

Also, from Eq. (2.54), we have for the equivalent focal length of a Cassegrain system in air

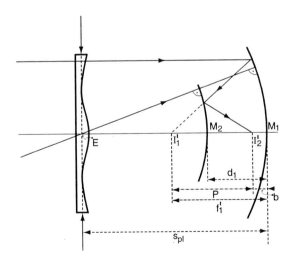

Fig. 3.43. Monocentric (concentric) Schmidt-Cassegrain proposed by Linfoot (1944)

$$\frac{1}{f'} = -\frac{1}{f_1'} + \frac{1}{f_2'} - \frac{d_1}{f_1'f_2'} \tag{3.294}$$

Eliminating d_1 from these two equations gives

$$\frac{1}{f'} = \frac{1}{f_1'} - \frac{1}{f_2'} , \tag{3.295}$$

showing that the monocentric Cassegrain system has the same equivalent focal length for all finite separations d_1 as the "thin-lens" combination with $d_1 = 0$. Now from Table 3.5 the right-hand side of (3.295) also represents the Petzval sum P_c of the system, so that

$$P_c = 1/f' \tag{3.296}$$

and the field curvature S_{IV} is the reciprocal of the equivalent focal length of the system, exactly as for a classical single-mirror Schmidt, normalizing $H^2 = 1$ in Table 3.5. This is a price to be paid for the concentric form, but the system must be anastigmatic from its spherical symmetry about the plate centre (stop) and has the advantage of having purely spherical mirrors.

We will now derive the asphericity required on the plate. For simplicity, we shall assume that the final image is at the pole of the primary, i.e. $b = 0$ in Fig. 3.43. Then from Eqs. (2.67) and (2.72)

$$L = -f_1'(1 - R_A) = -d_1 = f'R_A \tag{3.297}$$

with $b = 0$. From Table 2.2, the quantities f_1', f_2' and d_1 were defined as negative in a Cassegrain telescope. From the monocentricity and Eq. (3.293):

$$f_2' = \frac{2f_1' - d_1}{2} , \tag{3.298}$$

giving from (3.297)

$$f_2' = \frac{f_1'(1 + R_A)}{2} \tag{3.299}$$

If f_2' is now substituted in Eq. (3.295), we obtain

$$\frac{1}{f'} = -\frac{1}{f_1'}\left(\frac{1 - R_A}{1 + R_A}\right) \tag{3.300}$$

But from (3.297) we have

$$f_1' = -f'\left(\frac{R_A}{1 - R_A}\right) \tag{3.301}$$

Eliminating f_1' from (3.300) and (3.301) gives

$$R_A = 1/3 ,$$

a value slightly more favourable than the value 0.35 for the Baker systems with an image position in front of the primary. The obstruction ratio must be more favourable, since the secondary is more strongly curved than in the Baker flat-field solutions.

If $b = 0$ in Fig. 3.43, we have $P = -f_1'$ and Eq. (2.84) becomes

$$f_2' = -f_1' \left[\frac{m_2}{(m_2 + 1)(m_2 - 1)} \right] \qquad (3.302)$$

Combining this with (3.299) gives with $R_A = 1/3$

$$\frac{m_2}{(m_2 + 1)(m_2 - 1)} = \frac{1 + R_A}{2} = \frac{2}{3} \qquad (3.303)$$

This leads to the equation

$$2m_2^2 + 3m_2 - 2 = 0 \qquad (3.304)$$

for the value of the secondary magnification m_2, whose solution is, taking the negative root for the Cassegrain case,

$$m_2 = -2.0 \qquad (3.305)$$

Since $f_1' = f'/m_2$ and $s_{pl} = +2f_1'$, we now have the following parameters for calculating the third order aberrations for the normalization $f' = 1.0$:

$$f' = 1.0 , \qquad f_1' = -0.5 , \qquad f_2' = -0.33333$$
$$d_1 = -L = -R_A = -1/3 = -0.33333$$
$$s_{pl} = -1.0$$

Spherical mirrors give from Eqs. (3.30) and (3.41) with $b_{s1} = b_{s2} = 0$

$$\left. \begin{array}{l} \zeta = \dfrac{m_2^3}{4} = -2.0 \\[2mm] \xi = \dfrac{(m_2 + 1)^3}{4} \left(\dfrac{m_2 - 1}{m_2 + 1} \right)^2 = -2.25 \end{array} \right\}$$

The first equation of (3.220) requires

$$-f'\zeta + L\xi + \delta S_I^* = 0$$

for correction of $\sum S_I$ by the plate, giving

$$\delta S_I^* = -2.0 + 0.75 = -1.25 \qquad (3.306)$$

This is an important practical result to be compared with the value -0.25 for the classical Schmidt from Eq. (3.229) and -0.621 for the Baker Schmidt-Cassegrain Type B of Table 3.13. The monocentric Schmidt-Cassegrain thus requires a plate asphericity *five times* that of a classical Schmidt of the same equivalent focal length, and about twice that of the Baker Type B system. It is instructive to see from (3.306) why this is so: the first term ζ dominates and depends on $(m_2)^3$. Now m_2 is -1 for the classical Schmidt (effectively a plane mirror as secondary), -1.538 for the Baker Type B and -2.0 for the monocentric Schmidt-Cassegrain. Here we see that the advantages of increase in $|m_2|$ (reduction in effective length and reduction in obstruction ratio from the theoretical value of 0.5 for the classical Schmidt with a fictitious plane mirror to 0.333 for the monocentric Schmidt-Cassegrain) have to be paid for by a marked increase in asphericity on the plate. Of course, these effects are completely analogous to the higher aberration of a Cassegrain primary compared with a Newtonian of the same focal length because of the telephoto effect and the steeper primary (see Table 3.3). The very strong plate is the weakness of the monocentric system. It is highly desirable to achromatise it; but this produces much larger individual plate strengths unless highly absorbing flint glass is chosen to give a big difference in Abbe number. A further weakness is the length of the system compared with the mirror separation.

It is left to the reader to confirm from the second and third equations of (3.220) that

$$\sum S_{II} = \sum S_{III} = 0$$

The analysis above of the required asphericity on the corrector plate reveals the main reason why Schmidt-Cassegrain solutions have only been realized in sizes up to barely about 1 m effective aperture: the chromatic effects of a simple plate become unacceptably large. The classical Schmidt may be long but the plate manufacture is far easier and a single plate may be adequate, although large Schmidts are now often equipped with an achromatic plate. An achromatic plate of 1 m diameter for a monocentric Schmidt-Cassegrain is a major technical undertaking. Spot-diagrams showing the correction of such a system with a singlet (non-achromatic) plate are given in Fig. 3.44 and should be compared with those of the achromatic concentric meniscus system disussed in § 3.6.4.2. Apart from the field curvature, the spot-diagrams show that the spherochromatism is the only significant error giving absolutely uniform quality over the ±1° field.

If some astigmatism and a smaller field of the order of 1° are acceptable, then a wide variety of *aplanatic* solutions are possible. Such solutions are widely offered by professional suppliers for the amateur market for modest sizes and are also realized by amateurs. The advantages of compactness usually drive the plate position parameter s_{pl} down towards $+f'_1$, giving lengths about half that of the classical Schmidt. A primary of f/2 and Cassegrain focus of f/10 ($m_2 = -5$) is a common layout, typical apertures being 200 to 400 mm. Slevogt types [3.44] with both mirrors spherical are available from about f/4 to f/12. Excellent accounts of these possibilities are given by Rutten and van Venrooij [3.12(b)] and by Schroeder [3.22]. Usually, at least one mirror is spherical, most frequently the primary. Eqs. (3.220) provide the basis for rapidly analysing any such system to third order accuracy. The starting geometry for amateur use will be strongly dependent on whether visual observation is intended, where low central obstruction ratio R_A is important (see Sections 3.10.4 and 3.10.7 on physical optical aspects), or only photographic (flat field and anastigmatism for larger fields more important). Such aspects are discussed in detail in [3.12].

It is instructive to set up a typical modern aplanatic Schmidt-Cassegrain for amateur use to illustrate the ease and rapidity with which the paraxial and aberration formulae can be applied.

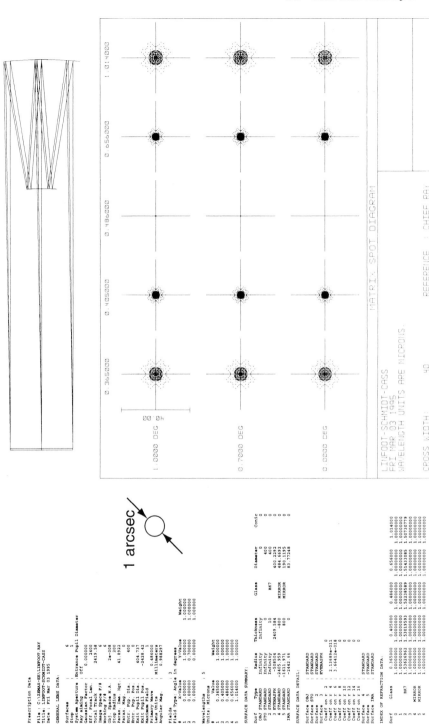

Fig. 3.44. Spot-diagrams for a Linfoot monocentric Schmidt-Cassegrain with spherical mirrors and a singlet (non-achromatic) corrector plate (400 mm, f/3.0 – f/6.0)

We first define the following basic parameters:

- Aperture ratios f/2 - f/10 giving $m_2 = -5$
- $f' = 1$, therefore $f_1' = -0.2$
- Corrector plate at prime focus giving $s_{pl} = +f_1' = -0.2$
- Primary defined as spherical, therefore $b_{s1} = 0$
- Cassegrain image $f'/10$ behind pole of primary, therefore $b = 0.1$

The system is shown in Fig. 3.45. We derive the remaining parameters as follows:

$$P = -f_1' + b = 0.3$$

From (2.84):

$$f_2' = P\left[\frac{m_2}{(m_2 + 1)(m_2 - 1)}\right] = -0.0625$$

From (2.82):

$$s_2 = P/(m_2 - 1) = -0.05$$

From (2.71):

$$d_1 = f_1' - s_2 = -0.15$$

From (2.89):

$$L = P\left(\frac{m_2}{m_2 - 1}\right) = 0.25$$

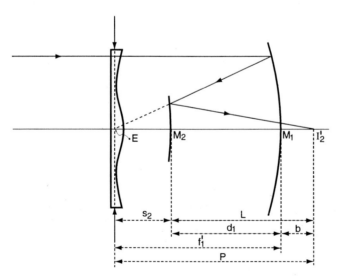

Fig. 3.45. Typical modern aplanatic Schmidt-Cassegrain for advanced amateur use with aperture 400 mm, f/2 - f/10

(*Check*: $f'_2 = L/(m_2 + 1) = -0.0625$ from (2.90))
From (2.58):

$$R_A = \frac{s_2}{f'_1} = 0.25$$

(*Check*: $R_A = L/f' = 0.25$)
Further checks are, from (2.64) and (Eq. (2.54):

$$f'_2 = \frac{R_A L f'_1}{L + R_A f'_1} = -0.0625$$

$$\frac{1}{f'} = -\frac{1}{f'_1} + \frac{1}{f'_2} - \frac{d_1}{f'_1 f'_2} = +1$$

From (3.30) with $b_{s1} = 0$:

$$\zeta = \frac{m_2^3}{4}(b_{s1} + 1) = -31.25$$

We can now apply the first two equations of (3.220) and solve for δS_I^* and ξ, the necessary eccentricities on the corrector plate and secondary to fulfil the aplanatic condition

$$\sum S_I = \sum S_{II} = 0 ,$$

giving

$$-f'\zeta + L\xi + \delta S_I^* = 0$$
$$-\left[d_1\xi + \tfrac{1}{2}f'\right] + \frac{s_{pl}}{f'}\delta S_I^* = 0$$

This leads to:

$$\delta S_I^* = -24.0625$$
$$\xi = -28.75$$

From (3.41), we can now derive b_{s2} for the asphericity on the secondary:

$$\xi = \frac{(m_2 + 1)^3}{4}\left[\left(\frac{m_2 - 1}{(m_2 + 1)}\right)^2 + b_{s2}\right] ,$$

giving

$$b_{s2} = -0.4531 ,$$

corresponding to an ellipse with less than half the deformation of a parabola.

We can deduce the astigmatism from the third equation of (3.220)

$$\sum S_{III} = \left(\frac{y_1}{f'}\right)^2 \left[\frac{f'}{L}(f' + d_1) + \frac{d_1^2}{L}\xi + \left(\frac{s_{pl}}{f'}\right)^2 \delta S_I^*\right] u_{pr1}^2 ,$$

giving, with $f' = 1$, $y_1 = 1/20$ for f/10, and $u_{pr1} = 0.01$ rad for a field diameter of 1.15°,

$$\sum S_{III} = \frac{1}{400}[-0.150]\left(\frac{1}{100}\right)^2 = -3.75 \cdot 10^{-8} ,$$

a modest value. This can be converted into angular aberration at best focus using Eq.(3.208):

$$(\delta u_p')_{ast,m} = -\frac{\sum S_{III}}{n' y_1}(206\,265) \text{ arcsec}$$

Then the angular astigmatism at best focus is, setting $n' = 1$ for the image space,

$$(\delta u_p')_{ast,m} = +0.155 \text{ arcsec} ,$$

an excellent value. With slight defocus of the detector, astigmatic lines of length 0.31 arcsec would appear. We must remember, from Table 3.1, that this grows with the square of the field u_{pr1}.

Because of the high secondary magnification m_2 and telephoto effect T, the only serious aberration is the field curvature. This is given from Table 3.5, with $H^2 = 1$ for a normalized system, by

$$S_{IV} = -\sum P_c = \frac{1}{f_1'} - \frac{1}{f_2'} = +11$$

The effective field curvature is (see Table 3.3 and Eq. (3.203)), for the normalized system

$$2S_{III} + S_{IV} = 2(-0.150) + 11 = +10.70 ,$$

giving a radius of curvature of the optimum image surface

$$r_{c(opt)} = -f'/10.70$$

With an aperture of 400 mm, $f' = 4000$ mm and $r_{c(opt)} = -373.8$ mm. For visual use, this is of little consequence, but for photography a field flattener would be used. Since the image surface is concave towards the light, the field flattener must be a *negative* lens (see §3.2.7.2 concerning field flatteners).

We saw above that the asphericity required on the correcting plate is

$$\delta S_I^* = -24.0625 ,$$

a very high value compared with -0.25 (Eq. 3.229) for a basic Schmidt telescope. This is the price for the steep (f/2) primary and the short constructional length. The determination of the plate form from the value δS_I^* was given in detail in §3.6.2.2. A singlet plate is limited by chromatic effects,

above all spherochromatism. However, it is still better than a refractor of
similar size using classical glasses. An achromatic plate gives better performance, but is a major manufacturing complication. A singlet plate is already
a difficult element to manufacture, particularly as fifth order correction will
also be necessary with an f/2 primary, but such a design gives a very powerful
and compact design for amateur use.

Figure 3.46 shows the spot-diagrams of the above design, somewhat modified by optimization, with an achromatic corrector plate. The quality is excellent: the field of $\pm 1°$ could be increased.

It is instructive to consider the evolution of such an aplanatic Schmidt-
Cassegrain design as the plate is moved further from the primary. The plate
asphericity δS_I^* weakens and b_{s2} becomes more positive, passing through
the Slevogt solution (see below) in which both primary and secondary are
spherical. With a doubled plate distance ($s_{pl} = 2f_1'$), the secondary is
already an oblate spheroid with $b_{s2} = +0.75$ and $\delta S_I^* = -19.25$. If the
plate distance is increased by a factor of five ($s_{pl} = -f'$), then $b_{s2} = +2.555$
and $\delta S_I^* = -12.03$. In the limit case with $s_{pl} = -\infty$, its lever arm is infinite
and $\delta S_I^* \to 0$, i.e. the plate is non-existent and we have from Eqs. (3.220)
the same result as the 2-mirror SP telescope without coma correction. From
Eq. (3.137), the secondary then has the extreme form $b_{s2} = +5.56$.

Reference has been made here and above to the Slevogt system [3.44]
[3.46], in which *both* primary and secondary of a Schmidt-Cassegrain have
a *spherical* form. In this respect, it resembles the monocentric Schmidt-
Cassegrain of Linfoot (Fig. 3.43). However, in the Slevogt design they are not
monocentric: the curvatures are nearly identical, giving only a small Petzval sum residual. The design is therefore nearer to the Baker-types shown in
Fig. 3.42, also from the point of view of the constructional length relative to
the equivalent focal length which is about $1.38f'$. However, it is formally an
aplanatic Schmidt-Cassegrain, not an anastigmat, because the astigmatism
is not fully corrected. This small residual is balanced against the residual
Petzval curvature so that $\sum S_{IV} = -2 \sum S_{III}$ to give an effectively flat field.
Because of the weak curvature of the secondary, the final image is placed just
in front of the primary, as in the Baker systems, to give reasonable obstruc-
tion. The value of the secondary magnification $|m_2|$ is bound to be low, giving
typically f/2 - f/3.25. With a singlet corrector plate, as with the Baker designs, the image quality is entirely limited by the large spherochromatism, but
becomes excellent with an achromatic plate. The quality is then fully comparable with the aplanatic Schmidt-Cassegrain of Fig. 3.46, also equipped with
an achromatic plate. Figure 3.47 shows the spot-diagrams for such a Slevogt
system of aperture 400 mm, f/2.0 - f/3.25.

The Slevogt system has the advantages of the spherical secondary and a
flat field and is fast because of its low $|m_2|$; but its increased length, high ob-
struction and final image position are much less favourable than the geometry
of Fig. 3.45.

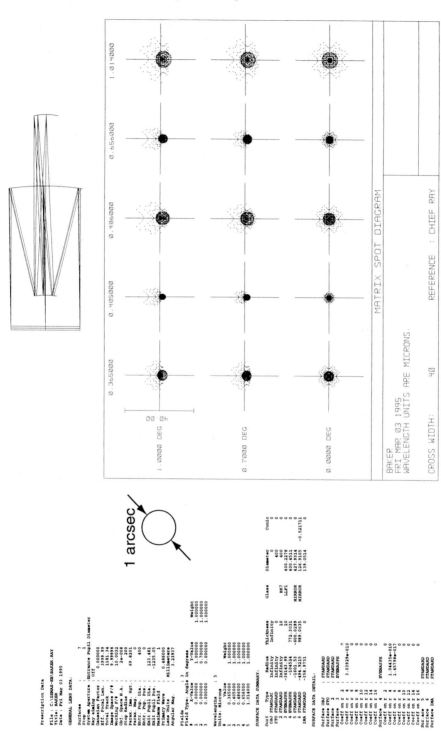

Fig. 3.46. Spot-diagrams for the aplanatic Schmidt-Cassegrain system of Fig. 3.45 with 400 mm, f/2 - f/10, and an achromatic corrector plate

3.6.4.2 Further developments of meniscus-type systems. The first obvious extension of the Bouwers-Maksutov concepts discussed in §§ 3.6.1 and 3.6.3 is the *Bouwers-Cassegrain* or the *Maksutov-Cassegrain*.

The basic principles of the layout of such a system are essentially the same as those of the Schmidt-Cassegrain, the role of the corrector plate being replaced by some meniscus-type system. However, the theory of aspheric plates is particularly elegant and simple, as was shown in § 3.4. Eqs. (3.270) and (3.282) showed that the effect of a meniscus is more complex, both monochromatically and chromatically. Nevertheless, the formulae given in § 3.6.3 enable a Maksutov-Cassegrain to be laid out in a way similar to the Schmidt-Cassegrain. The asphericity term δS_I^* of the plate is replaced by the meniscus contribution to $\sum S_I$, respecting of course, the chromatic condition (3.283) for the Maksutov-type meniscus.

Quasi-concentric (monocentric) Bouwers-Cassegrain: Bouwers [3.32] proposed a two-glass concentric form of meniscus to get round the problem of longitudinal chromatic aberration. This can be used either in a prime focus Bouwers telescope or in a Bouwers-Cassegrain, as shown in Fig. 3.48. Instead of departing from concentricity according to the Maksutov formula (3.283) to achieve longitudinal achromatism, Bouwers proposed an achromatic concentric meniscus with two glasses separated by a plane surface. Since the effect of the meniscus is that of a negative lens, the positive half of the meniscus will have the higher dispersion glass. Monochromatically, the system remains strictly monocentric (more so than the monocentric Schmidt-Cassegrain because of asymmetries of oblique pencils at the aspheric plate) if the refractive index for the central wavelength is the same for both glasses, i.e. on a horizontal line of the glass diagram of Fig. 3.32. The convex, right-hand half of the meniscus must have the same dispersive power with its (weaker) concentric surface as the non-concentric right-hand surface of the Maksutov meniscus provides. The requirement of a central wavelength index which is the same for both glasses (and preferably good blue transmission for the high dispersion glass) pushes the designer towards expensive optical glasses. This is no doubt the reason why this form has been little used by both amateurs and professionals. However, its correction possibilities are extremely interesting, since the complete symmetry together with the basic correction of S_I by the Bouwers concentric meniscus give for a theoretical field of 180° apart from vignetting by the stop:

$$S_I = S_{II} = S_{III} = 0$$
$$S_{IV} = 1/f',$$

for the normalization $H^2 = 1$. This is the same as for the monocentric Schmidt-Cassegrain, but is valid for a much larger field. Of course, the concentric meniscus will still limit the f/no because of fifth order spherical aberration.

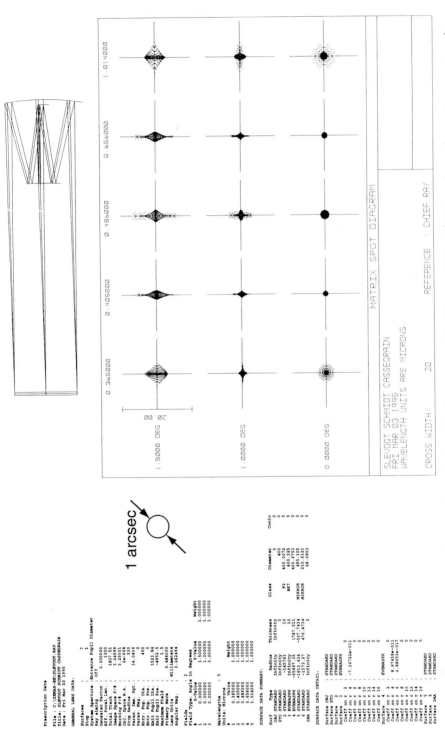

Fig. 3.47. Spot-diagrams for a Slevogt aplanatic Schmidt-Cassegrain with 400 mm, f/2.0 - f/3.25, and an achromatic corrector plate. The field is flat

It can easily be shown that the change of dispersion required for the second glass is very modest, so that this proposal by Bouwers can be a practical proposition if the glass costs are not a dominant matter. Eq. (3.284) gave the concentricity departure of the second surface of a Maksutov meniscus as

$$\Delta z = -d/n'^2$$

This is, then, the change dr_2 required to correct the longitudinal chromatic aberration C_1. Consider this as a change in power of the right-hand half of the concentric meniscus shown in Fig. 3.48, initially having the same glass as the left-hand half. With the approximation of thin lenses, the power K_{m+} of the right-hand positive half is given by

$$\frac{1}{f'_{m+}} = K_{m+} = \frac{(n'-1)}{r_2} \ ,$$

giving

$$dK_{m+} = -\frac{(n'-1)}{r_2^2}dr_2 \tag{3.307}$$

It can be shown [3.3] [3.6] that such a thin lens gives the longitudinal chromatic aberration

$$C_1 = \frac{y^2}{\nu_A}K_{m+} \ , \tag{3.308}$$

where y is the semi-aperture and ν_A the Abbe number

$$\nu_A = \frac{n'-1}{n'_F - n'_C} \tag{3.309}$$

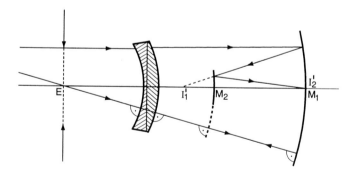

Fig. 3.48. Two-glass concentric (monocentric) Bouwers-Cassegrain telescope

measuring the dispersion between wavelengths F and C. Then the change in chromatism due to the Maksutov radius change is

$$(dC_1)_M = \frac{y^2}{\nu_A} dK_{m+} = -\frac{y^2}{\nu_A} \frac{(n'-1)}{r_2} \frac{dr_2}{r_2} \tag{3.310}$$

In the Bouwers proposal, the same change must be generated by a change of ν_A, giving from (3.308)

$$(dC_1)_B = -\frac{y^2 K_{m+}}{\nu_A} \frac{d\nu_A}{\nu_A} = -\frac{y^2}{\nu_A} \frac{(n'-1)}{r_2} \frac{d\nu_A}{\nu_A} \tag{3.311}$$

Setting $(dC_1)_B = (dC_1)_M$ gives with $dr_2 = \Delta z$

$$d\nu_A = -\nu_A \frac{d}{n'^2 r_2} \tag{3.312}$$

Since thin-lens theory is used, (3.312) is only an approximation. If ν_A is taken as 60 for typical crown glasses and d/r_2 is taken as $\simeq -1/10$ for a typical Bouwers meniscus with $n' \simeq 1.5$, we have

$$d\nu_A \simeq 2.7 \ ,$$

a very small change of dispersion which limits the glass choice.

Equations (3.223) show that the correction of C_1 will not normally correct the spherochromatism; but it reduces it compared with the simple meniscus.

Figure 3.49 shows spot-diagrams for a Bouwers achromatic, monocentric meniscus-Cassegrain, to be compared with the Linfoot monocentric Schmidt-Cassegrain of Figs. 3.43 and 3.44 using a singlet, non-achromatic plate. The mirror geometry was analysed in connection with the Linfoot monocentric system above. The glasses chosen (BK10 and BK3) reflect the limited choice and have $d\nu_A = 1.9$, less than that required from (3.312) above. The front stop is abandoned, for the same reasons given for the Maksutov. The finite thickness of the meniscus, together with the crude approximation for $d\nu_A$, lead to lateral chromatism C_2. For this reason, a singlet (BK7) field flattener has been introduced, allowing excellent correction with a flat field. The geometry is lightly modified to 400 mm, f/3.11 - f/6.0. The optical quality is virtually constant over the $\pm 1°$ field, which could be extended apart from the obstruction limitation of the concentricity.

Rutten and van Venrooij [3.12(c)] also give data for another achromatic variant of the concentric system, *a classical Bouwers telescope with an additional weak lens in the concentric stop* (Fig. 3.50).

The additional lens must be weakly positive. It has the advantage over the two-glass solution of Fig. 3.48 that a single glass type is used in both meniscus and additional lens. It has the same basic chromatic characteristics as the Maksutov but gives better performance because the concentric symmetry is preserved apart from small asymmetries of oblique pencils at the stop lens. Its performance is comparable at f/3 with a classical Schmidt with a singlet plate, as is shown by the spot-diagrams for such a prime focus system with aperture 400 mm in Fig. 3.51. This may be compared with the spot-diagrams

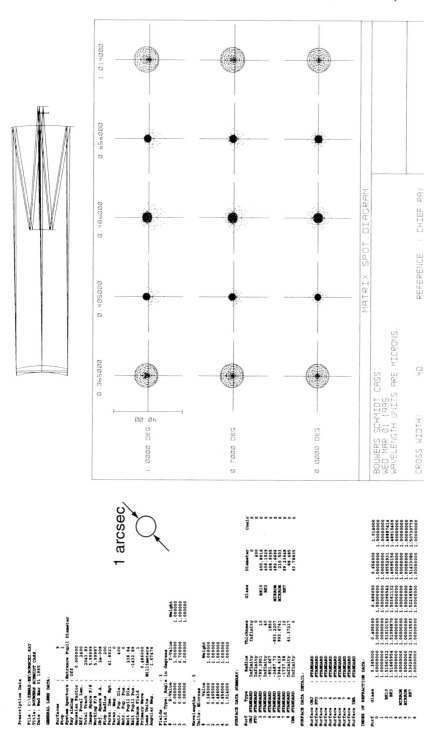

Fig. 3.49. Spot-diagrams for an achromatic, monocentric, Bouwers meniscus-Cassegrain telescope as in Fig. 3.48, but with a singlet field flattener added. The geometry is lightly modified to 400 mm, f/3.11 - f/6.0, and the stop is shifted to the meniscus

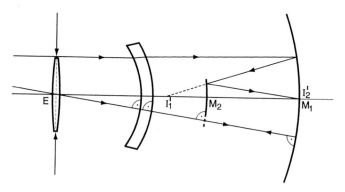

Fig. 3.50. Classical Bouwers telescope with additional weak lens at the stop

of the f/3 Schmidt with a singlet plate as Fig. 3.29. For the prime focus, we take the large angular field of $\pm 3°$. Fifth order spherical aberration causes an axial image core of about 1 arcsec. If the relative aperture is reduced to f/3.5, this effect declines rapidly to give about 0.6 arcsec.

Since the results of Fig. 3.51 are achieved with only spherical surfaces, this modified Bouwers system must rate as a very interesting solution. If further sophistication is desired, the fifth order spherical aberration can be corrected by aspherising one face of the weak lens, permitting higher relative apertures. This then becomes a mixed plate-meniscus solution, a type discussed in § 3.6.4.3.

Figure 3.51 can also be directly compared with the spot-diagrams of prime-focus Maksutov designs of the "short" type given in Figs. 3.35 and 3.37. Because of the weak lens at the pupil in the centre of curvature, the modified Bouwers form is appreciably longer than the "short" Maksutov.

As a concentric Cassegrain, the extended Bouwers telescope of Fig. 3.50 is also an attractive form. As with the Linfoot concentric Schmidt-Cassegrain, $m_2 = -2$. Figure 3.52 shows spot-diagrams for a system with aperture 400 mm and f/3.0 - f/6.0. The field correction is so good that the angular field of $\pm 1°$ is limited solely by the obstruction.

The Maksutov-Cassegrain: Maksutov solutions, in general, accept a departure from concentric symmetry. The most complete accounts of the possibilities are given in two books mainly intended for amateurs, since the amateur market with modest apertures has made most extensive use of such systems. The two references concerned are Rutten and van Venrooij [3.12(d)], frequently referred to above, and Mackintosh [3.48]. The latter is a collection of articles from the "Maksutov Circulars" for advanced amateurs.

Here we shall demonstrate the power of this solution by giving the results in detail for a "short" Maksutov-Cassegrain of the non-flat field type with an aperture of 400 mm and a primary working at f/3.5. This is slightly "slower" than the examples given in [3.12(d)] and relaxes the "zonal error" (fifth order spherical aberration).

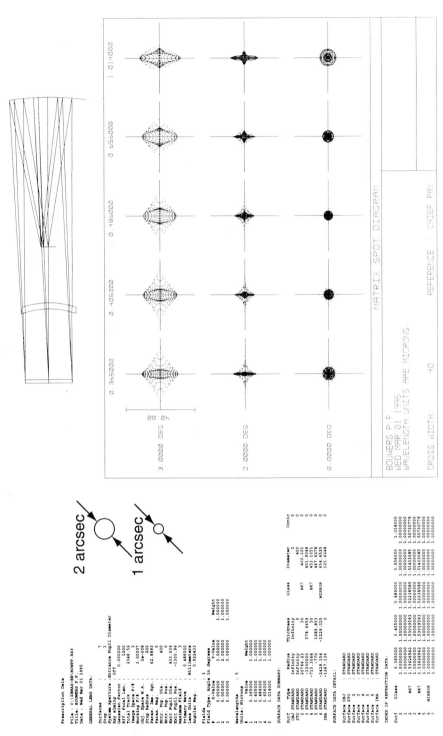

Fig. 3.51. Spot-diagrams of a classical Bouwers telescope (prime focus) with weak achromatising positive lens in the pupil (400 mm, f/3). The angular field is large ($\pm 3°$)

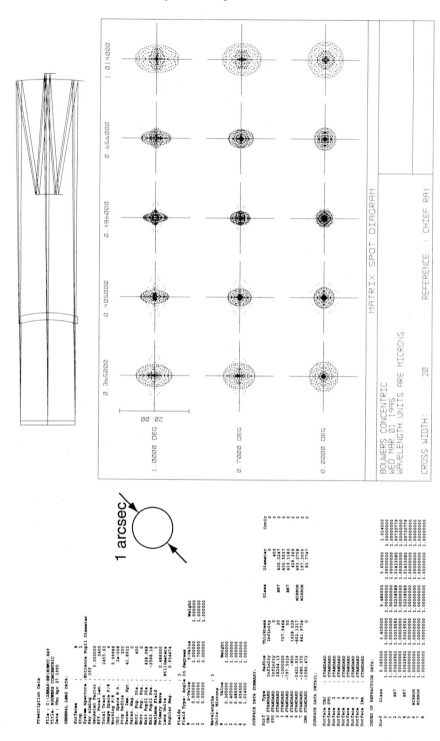

Fig. 3.52. Spot-diagrams of a classical Bouwers-Cassegrain with weak achromatising lens (400 mm, f/3.0 - f/6.0), as shown in Fig. 3.50

Figure 3.53 shows the preferred variant in two versions, "long" and "short". The only difference is the position of the stop, the "long" version having the conventional stop position quasi-concentric to the meniscus, the "short" version the stop at the meniscus. Since the system is aplanatic it is uncritical for the monochromatic third order aberrations where the stop is placed; and, as we saw in Fig. 3.36, the residual transverse chromatic aberration (C_2 in Eqs. (3.223)) is unaffected by the stop shift because C_1 is corrected.

Table 3.14. Design data for the Maksutov-Cassegrain of Fig. 3.53. Aperture 400 mm, f/3.5 - f/10.71

	Radius	Separation	Medium	n
Stop		10.0 (short)	Air	
		650.0 (long)		
	-635.758	52.153	Schott BK7	1.518 522
	-666.034	1089.798	Air	
	-2800.000	-1047.000	Refl.	
	-1122.990	1047.000	Refl.	
(Ghost surface)	∞	150.000	Air	
(Field surface)	-1000.000	0	Air	

The design data (in mm) are given in Table 3.14. The system was laid out for a miniature camera format with image diameter to the field corners 44 mm. The basic parameters are (mm):

$2y_1$	$=$	400
f'	$=$	4285.29
f'_1	$=$	-1400 (f/3.5)
b	$=$	150
N_{Cass} (f/no)	$=$	10.71
m_2	$=$	-3.06
Length	$\simeq$	1302 (stop (short) to Cass image)
Length	$\simeq$	1941 (stop (long) to Cass image)
Semi-field angle	$=$	$u_{pr1} = 0.30°$
Gaussian semi-image height	$=$	22.435
Central wavelength	$=$	486 nm
Subsidiary wavelengths	$=$	365, 405, 656, 1014 nm
		(for chromatic aberrations)
Effective field radius	$=$	-1000 mm
Scale	$=$	20.8 μm/arcsec

The total length of the "short" version is 30.4% of f'; of the "long" version 45.3%. Even the long version is a very compact telescope.

Figure 3.54 gives the spot-diagrams for the "short" version.

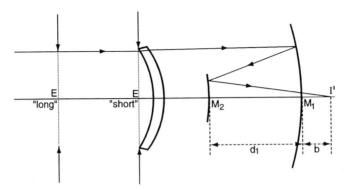

Fig. 3.53. Maksutov-Cassegrain in "long" and "short" versions with secondary separated from the meniscus. Example with aperture 400 mm, f/3.5 - f/10.71

The spot-diagrams of Fig. 3.54 are, *monochromatically*, almost entirely limited by fifth order spherical aberration. The aplanatic condition in this design is achieved by adjusting m_2: a value $|m_2| < 3.06$ produces "inward" coma, a value $|m_2| > 3.06$ "outward" coma. This has an important consequence for the second variant (secondary mirror coated on the back surface of the meniscus) analysed below. The constancy of the monochromatic spot-diagrams over the field is shown by the d_{80} values of the energy concentrations (80% encircled energy diameters) for the "short" system. These are given in Table 3.15 for three field heights and three wavelengths.

Table 3.15. d_{80} values (80% encircled energy diameters in μm) for the "short" system of Fig. 3.53 and Table 3.14

Wavelength	400 nm	550 nm	700 nm
Semi-field height			
0°	13.5	7.3	12.3
0.21°	12.3	8.4	12.6
0.30°	11.4	8.1	12.2

Clearly, the monochromatic field aberrations are effectively negligible *over the curved field* (this is *not* true for a flat field). Spherochromatism increases the central wavelength value of about 7 μm at the subsidiary wavelengths, showing that the f/no of f/3.5 for the primary is in good balance with the spherochromatism. The transverse colour error is the most serious error. The origin is solely the finite thickness of the meniscus, combined with its finite rear curvature (see Fig. 3.36). Since the finite thickness is essential to correct the spherical aberration and to give the meniscus stability, there is no means of correcting C_2 without a further optical element, such as a field flattener.

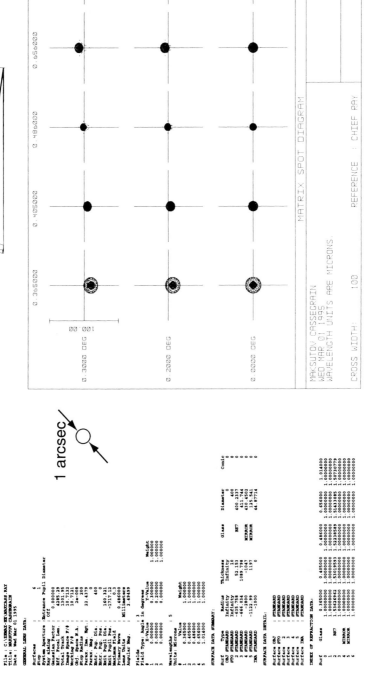

Fig. 3.54. Spot-diagrams for the "short" version of the Maksutov-Cassegrain of Fig. 3.53 and Table 3.14 for an optimum curved field

The "long" system not only increases the length by almost 50%; it also slightly increases the diameters of the elements unless vignetting is accepted. The slight improvement in astigmatism does not justify the major increase in length, because the colour error C_2 is determinant.

An attractive solution for manufacture is to combine the secondary with the back surface of the meniscus by evaporating aluminium on to its central area. The system is shown in Fig. 3.55, the data in Table 3.16 and the spot-diagrams in Fig. 3.56.

The radius of the secondary is no longer a free parameter except within the limits over which the meniscus can be optically "bent". The system given has $m_2 = -4.34$ and has a substantial amount of "outward" coma as a result, a total spread of about 2.0 arcsec at the corner of a miniature camera format (semi-field $= 0.21°$), with field curvature. The aperture and radius of the primary are the same as those of the previous system.

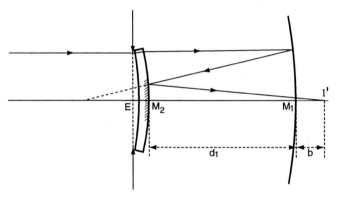

Fig. 3.55. A Maksutov-Cassegrain with secondary combined with the meniscus. Aperture 400 mm, f/3.5 - f/15.20

Table 3.16. Design data for the Maksutov-Cassegrain of Fig. 3.55

	Radius	Separation	Medium	n
Stop		10	Air	
	−689.537	90.000	Schott BK7	1.518 522
	−741.9692	1152.691	Air	
	−2800.000	−1152.691	Refl	
	−741.9692	1152.691	Refl	
(Ghost surface)	∞	150.000	Air	
Image surface	−1000	0	Air	

The lack of aplanatism is, of course, a normal situation in Newton and Cassegrain telescopes and can be quite acceptable for visual use. Because of

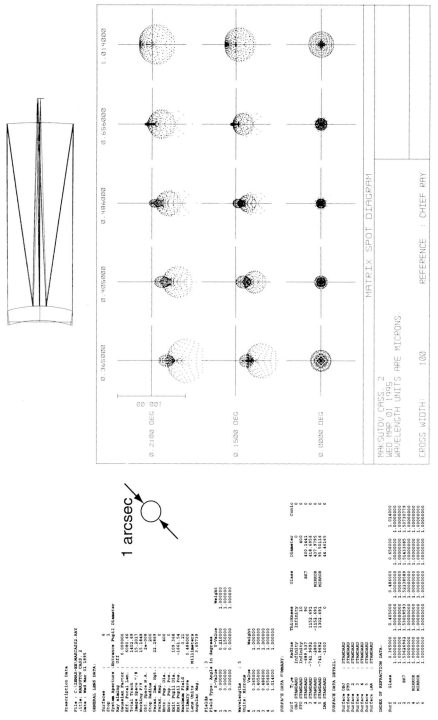

Fig. 3.56. Spot-diagrams of the Maksutov-Cassegrain of Fig. 3.55

the field coma, a "long" version of this variant makes no sense at all. The advantage of such a system for amateurs is the economy of one convex spherical surface. Since only the central area of the convex meniscus face is used for the secondary, a high quality of its figure is largely assured. The combined element also has advantages in maintaining centering. Nevertheless, the aplanatic solution of Fig. 3.53 is optically much superior and normally worth the extra effort of producing a separate secondary if photography (direct imagery) is intended.

Some comment is necessary on the *thickness* of the meniscus corrector in Maksutov telescopes. The situation is not the same as that of strictly concentric (*Bouwers*) designs. In the latter, the contribution of the meniscus thickness to the compensating spherical aberration is a direct function of the thickness: the first surface produces a massive overcorrection, the second a massive undercorrection. The undercorrection is smaller because the second surface has a radius longer by Δr with

$$d = r_1 - r_2 = \Delta r \;,$$

since the spherical aberration contribution of a surface depends for a parallel beam roughly on the third power of $(1/r)$ – see the first equation of (3.20) in which A and $\Delta \left(\frac{u}{n}\right)$ are proportional to $(1/r)$ for a parallel beam. The difference between the two contributions must compensate the spherical aberration of the mirror system. The shallower the meniscus, the thicker it must be.

In *Maksutov* designs, concentric symmetry has been abandoned to achieve the chromatic condition $C_1 = 0$, giving from (3.283)

$$d = \Delta r \left(\frac{n'^2}{n'^2 - 1} \right) \tag{3.313}$$

as the paraxial condition. For the glass BK7, this gives

$$d = (1.766)\Delta r$$

Maksutov [3.38] [3.12(e)] recommended

$$d = (1.70)\Delta r \;,$$

which compensates for the focus shift due to the spherochromatism. The data of Tables 3.14 and 3.16 above follow this formula quite closely, the factors being 1.723 and 1.716 respectively.

Since concentricity has been abandoned, a degree of freedom is available in Maksutov designs which is not available in the Bouwers telescope: the freedom to "bend" the meniscus optically irrespective of its position while maintaining the chromatic condition (3.313). The "bending" changes the "shape factor" involved as $(1/r_1)$ in Eq. (3.270) [3.3] [3.6] for the "thin-lens" approximation, giving a variation in the spherical aberration produced. To a third order, then, the thickness of a Maksutov meniscus becomes arbitrary. Maksutov [3.38] recommended a thickness ratio of $D/10$ which is often followed. Wright [3.49] and Rutten and van Venrooij [3.12(e)] show that the fifth order spherical aberration (zonal error) is reduced by increasing the thickness d, but the transverse chromatic aberration C_2 is made worse. So increase of d improves the axial quality at the cost of field correction. The improvement in zonal error comes essentially from the increase of ray height y at the second surface compared with the first.

The system of Table 3.14 has a fairly conventional thickness $d = D/7.67$. In Table 3.16, a thickness $d = D/4.44$ was taken, giving a massive support for the secondary on the convex surface. The d_{80} energy concentration of the axial spot-diagram of Fig. 3.56 is 0.33 arcsec; that of Fig. 3.54 with the thinner meniscus is 0.35 arcsec. This difference is too small to justify the additional cost and weight of a 90 mm meniscus in the one case compared with the 52 mm meniscus in the other case. However, the 52 mm meniscus has a somewhat lighter task, as the larger secondary mirror compensates somewhat more of the aberration generated by the primary.

Table 3.17. Surface contributions for the aplanatic Maksutov-Cassegrain of Table 3.14

Surface	$SA3 \, (= S_I)$	$SA5$
First meniscus	1.400 134	0.080 553
(difference)	(0.100 093)	(0.007 402)
Second meniscus	−1.300 041	−0.073 151
Primary	−0.159 686	−0.001 554
Secondary	0.051 701	0.000 841
Sum	−0.007 892	0.006 689

Finally, it is instructive to note the third and fifth order surface contributions $(SA3 = S_I, SA5)$ in such cases. The aplanatic system of Table 3.14 gives from calculations with the ACCOS V program the values of Table 3.17. The negative residual sum of $SA3$ balances the positive residual sum of $SA5$. The net correction of $SA3$ produced by the meniscus is 0.100 093, whereas its individual surface contributions are some $13\frac{1}{2}$ times higher. This is a warning that the meniscus is an element with critical radius tolerances. Although the surfaces are spherical, because they are steep it is not an easy element to manufacture.

The system of Table 3.14 has a curved field. In a normal Maksutov-Cassegrain, a flat-field requirement is incompatible with aplanatism. The field can be flattened by an additional field flattener.

For further details of such systems, the reader is referred to Chapters 10 and 11 of Rutten and van Venrooij [3.12] and Chap. 4 of Mackintosh [3.48].

3.6.4.3 Mixed plate-meniscus systems. The zonal (fifth order) spherical aberration of the Bouwers and Maksutov telescopes soon led to proposals with

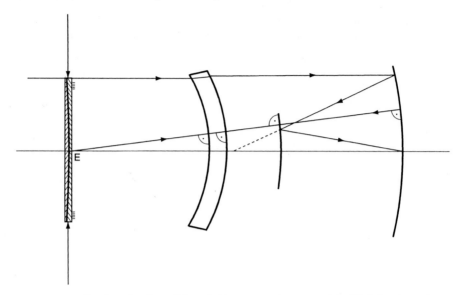

Fig. 3.57. Hawkins-Linfoot Schmidt-Bouwers telescope with f/1.2 in the prime focus

aspheric plate-meniscus combinations. Classic work was done by Hawkins and Linfoot [3.50] in 1945. The starting point was a concentric Bouwers telescope[7]. The chromatic aberration C_1 was compensated by a weak achromat in the stop, corresponding to the weak positive lens proposed by Bouwers. In order to achieve a high relative aperture (f/1.2), the zonal aberration was corrected by aspherising the plane external surface of the achromatic corrector (Fig. 3.57), which shows the system both in the original prime focus form and also in the concentric (monocentric) Cassegrain variant. The weak achromat corrector has similar properties to the two-glass Bouwers meniscus of Fig. 3.48, the two glasses having the same refractive index for the central wavelength but different dispersions, the convex part again being of higher dispersion glass. In contrast to the weak positive lens corrector of Fig. 3.50, this corrector has no optical power at the central wavelength: its function is purely limited to chromatic correction (C_1) and removal of fifth order spherical aberration. It should therefore be superior to the monocentric Schmidt-Cassegrain of Fig. 3.43, since the asphericity on the plate is far lower,

---

[7] Hawkins and Linfoot point out that D. Gabor [3.51] had already patented a meniscus corrector system of the Maksutov type with annular aperture in 1941 (see also Maxwell [3.52] for details). König-Köhler [3.30(b)] and Riekher [3.39(d)] point out that such a system was also patented by Penning [3.53] in 1941. According to Riekher [3.39(d)], Bouwers [3.32] sought a patent in the Netherlands for his concentric meniscus in 1941 and 1942 and in 1945 patented a system using a concentric meniscus with an aspheric correcting plate at its centre of curvature, i.e. a Hawkins-Linfoot system.

giving smaller asymmetries for oblique pencils. Spot-diagrams are shown in
Fig. 3.58, which should be compared with Figs. 3.44 and 3.49. Clearly, the
optical quality is superb, with spot-diagrams within 0.5 arcsec over the whole
field of $\pm 1.0°$.

Another possibility is to maintain a purely concentric meniscus with an
aspherised singlet positive lens, as shown in Fig. 3.50 without the aspherisation. This monocentric form should be nearly as good as the Hawkins-Linfoot
monocentric design.

A further modification is discussed by Rutten and van Venrooij [3.12(f)]
under the name "Companar". The corrector system is an aspherised weak
positive lens together with a concentric meniscus as above, but is otherwise
laid out as a flat-field design with mirrors of roughly equal curvature. The
system is designed for a very high aperture ratio in the Cassegrain focus of
f/2.5 giving a high obstruction ratio of about 60% of the diameter. At f/3, the
aspherisation on the corrector lens is considered unnecessary, so the system
is like Fig. 3.50, but in a flat-field version.

The Hawkins-Linfoot (prime focus) design is often referred to as a "Super-
Schmidt system". According to comparisons by König-Köhler [3.30(c)], such
a system still gives a monochromatic extension of a spot-diagram in the tangential section within 0.01 for a system normalized to $f' = 100$ at a field of
$\pm 30°$ and for a relative aperture of f/1.5. A classical Schmidt reaches this
limit at about $\pm 1.5°$. A concentric system is almost as good, monochromatically, with an extension of 0.02.

Another Super-Schmidt system of a similar, but even more sophisticated
form, was patented by Baker [3.54] [3.55] in 1945. The Hawkins-Linfoot aspheric achromatic corrector is at the centre of two concentric meniscus shells.
This system is discussed by Bahner [3.5], König-Köhler [3.30] and Maxwell
[3.52]. The Baker Super-Schmidt is shown in Fig. 3.59. The nominal relative
aperture is 0.67 and the effective relative aperture 0.82. Further developments
are described by Bradford [3.56] and Davis [3.57]. The image quality is remarkable, bearing in mind the extreme speed of the camera: even at the edge
of the field of $\pm 26°$, the d_{80} energy concentration diameter is within 50 μm
for the entire photographic wavelength band. The system length is $3.7f'$.

Another form of Super-Schmidt is the Baker-Nunn camera, designed for
satellite tracking. This is described in detail by Henize [3.58]. Briefer accounts
are given by Bahner [3.5] and Riekher [3.39(e)]. This system (Fig. 3.60) is
essentially a modified Schmidt, since it dispenses with the menisci of the
meteor camera and uses a symmetrical close triplet group of lenses in the
pupil instead of a Schmidt plate. The inner 4 of the 6 lens surfaces are all
aspherised. The focal length is 510 mm at f/1. The outer (convex) lenses are
of Schott short flint glass KzFS2, the inner (concave) lens of dense crown
SK14. KzFS2 is a "special glass" with abnormal dispersion, allowing correction of secondary spectrum (see Fig. 3.31 and Chap. 4). The two glasses have
roughly the same refractive index for the central wavelength. If the corrector

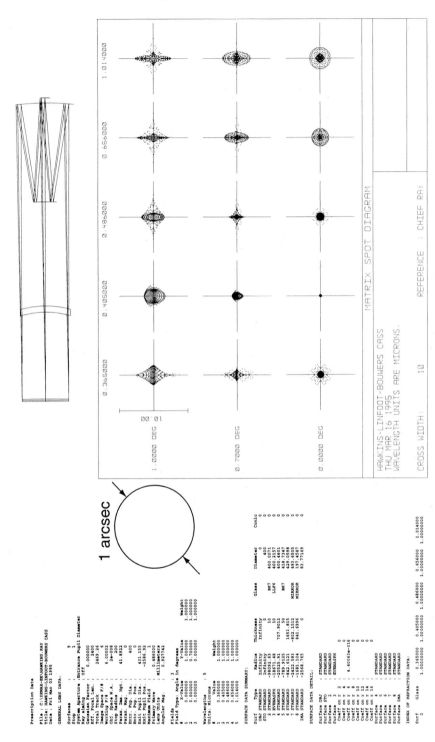

Fig. 3.58. Spot-diagrams for the Hawkins-Linfoot monocentric Cassegrain of Fig. 3.57, aperture 400 mm, f/3.0 - f/6.0

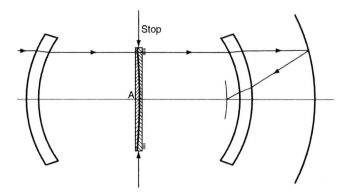

Fig. 3.59. Baker Super-Schmidt, with $f' = 200$ mm, effective aperture ratio f/0.82 and field $\pm 26°$

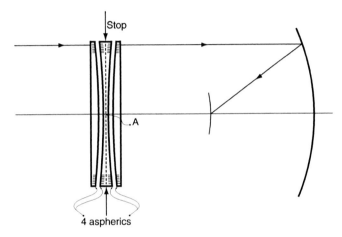

Fig. 3.60. Baker-Nunn camera, designed for satellite tracking, aperture 508 mm, f/1.0

is considered as two identical halves disposed about the central symmetry plane, each half can be seen as a refinement of the Hawkins-Linfoot corrector in which the powered surface is separated into two aspheric surfaces by an airspace. Each half is therefore like an achromatic Schmidt plate with power added. The symmetrical doubling of the system halves the asphericities and produces complete symmetry to the stop, thereby eliminating asymmetries of the oblique pencils. The d_{80} energy concentration diameter is within 20μm for a wavelength band from the near UV to the far IR even at the edge of the field strip of $5° \times 30°$.

The Baker-Nunn camera has been included here as one of the "Super-Schmidt" systems to be compared with the Baker Meteor Camera above (Fig. 3.59). However, generically, the Baker-Nunn system is also related to the Houghton-type lens correctors dealt with below in this section.

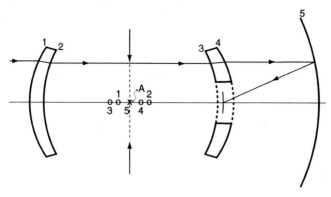

Fig. 3.61. Double-meniscus system due to Wynne [3.59] (schematic)

Double or multiple meniscus systems without corrector plate: The paper of Maksutov in 1944 [3.38] was remarkably complete with regard to the possibilities with a single meniscus. Shortly after, in 1947, Wynne [3.59] [3.60] extended the analysis in an important paper to two menisci more or less concentrically disposed round the stop. These systems are also discussed by Maxwell [3.52], who refers also to such a stop-symmetrical two-meniscus system designed earlier by Bennett [3.61]. Bennett stated that his system was self-achromatic. Wynne optimized the balance of the centres of curvature of the surfaces of the system with regard to the stop (Fig. 3.61). He showed that not only the longitudinal chromatic aberration C_1 could be much improved in this way, but also that the zonal (fifth order) spherical aberration could be considerably reduced for a given total meniscus thickness. Furthermore, the symmetry balance of the centres of curvature of the surfaces shown in Fig. 3.61 improves the field correction. The system can therefore be used at higher relative apertures than a single meniscus system, or the field may be extended since transverse chromatic aberration C_2 and astigmatism can also be influenced.

Wynne also investigated in detail the effect of an asymmetry of the thicknesses of the menisci, giving the system shown in Fig. 3.62. This asymmetry permits both the longitudinal and transverse chromatic aberrations C_1 and C_2 to be fully corrected. A further advantage is that fifth order spherical aberration can be largely eliminated, giving excellent axial performance at high apertures (f/1); but the off-axis correction is made worse. For wide fields, therefore, the more symmetrical design of Fig. 3.61 is to be preferred.

Maxwell [3.52] also mentions the principal advantages of an aspheric extension of the Wynne symmetrical double-meniscus system of Fig. 3.61, in which the two inner meniscus surfaces are aspherics. This gives an extremely powerful control of both fifth order spherical aberration and spherochroma-

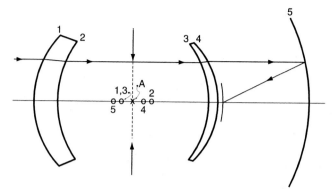

Fig. 3.62. Double-meniscus system due to Wynne with strongly asymmetric meniscus thicknesses [3.59] (schematic)

tism, as emerges from the analysis of Wynne [3.60]. This system rivals the Super-Schmidt systems discussed above.

Wynne [3.59] also analysed a system in which the front (thick) meniscus of Fig. 3.62 is split into two, giving a three element corrector with spherical surfaces. The basic purpose was to maintain a large total thickness of the menisci, which is favourable for the spherical aberration and field correction, but reduces the mass and thickness of the single front element. The nearly concentric airspace produced has some design advantages but these are offset by the increase in complexity and number of surfaces of the system.

Richter-Slevogt or Houghton-type lens correctors at the pupil: We have seen that the recognition by Schmidt in 1931 of the advantages of a corrector element in a pupil shifted in front of the primary mirror of a telescope unleashed a remarkable harvest of optical design solutions using aspheric plates and concentric or quasi-concentric menisci. In this period, a third generic variant was also discovered: the replacement of the Schmidt aspheric plate by a lens system with spherical surfaces. According to Köhler [3.30(d)] [3.46] and Riekher [3.39(f)], the first such system was probably published by Sonnefeld in 1936 [3.62]. This was an extremely fast system using a doublet corrector and a Mangin mirror (see § 3.6.4.4) and is not closely linked to the basic Schmidt geometry. However, in 1941 Richter and Slevogt [3.63] applied for a patent for a Schmidt-type solution replacing the aspheric plate by an afocal doublet with spherical surfaces, an important design form. In 1944, the same design principle was patented by Houghton [3.64]. In the English literature, such systems are usually called Houghton-type systems, but equal credit should be given to Richter and Slevogt.

For a full account of the many interesting possibilities of this generic type, the reader is referred to Rutten and van Venrooij [3.12(g)].

Houghton's patent gave a lens corrector for a spherical primary. Above all for smaller telescopes this was a very significant advance. His aim was to show

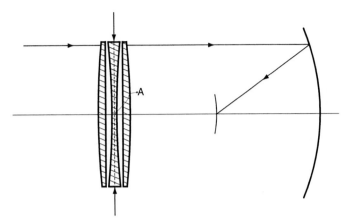

Fig. 3.63. Buchroeder design of a Houghton-type corrector in Schmidt geometry (200 mm, f/3) given by Rutten and van Venrooij [3.12(g)]

that a two- or three-lens corrector with spherical surfaces could achieve good correction instead of the technically difficult Schmidt plate. In principle, then, any design considered above with an aspheric plate combined with one or two mirrors can be considered as a candidate for a Houghton-type system where the plate is replaced by a lens corrector. If aspheric surfaces are introduced, such as in the Baker-Nunn system of Fig. 3.60, the system becomes a hybrid of Houghton and aspheric plate or plates.

A basic Houghton type described in [3.12(g)] is the Buchroeder design with the geometry of the basic Schmidt, i.e. the corrector at the centre of curvature of a spherical mirror (Fig. 3.63). This type of corrector is particularly attractive for amateurs since it not only has spherical surfaces but also only one glass type. The corrector is afocal, thereby eliminating longitudinal chromatic aberration C_1. Furthermore, four surfaces all have the same radius, either positive or negative, the other two being plane. This greatly simplifies the manufacture. Of course, as a single glass corrector generating spherical aberration, it is bound to suffer from spherochromatism proportional to the dispersion of the glass. Also, since the system only has spherical surfaces, it will – like the Maksutov – be limited in relative aperture by fifth order spherical aberration. Rutten and van Venrooij give spot-diagrams for their standard aperture of 200 mm at f/3. Over the curved field, the spot-diagrams remain well within 25 μm for the whole photographic spectral range (C – h) out to a field of ± 30 mm (nearly ± 3°). Although this is not as good as a classical Schmidt of this size, the quality is excellent compared with photographic grain for small sizes of telescope.

Another very interesting Houghton-type analysed by Rutten and van Venrooij is ascribed to Lurie [3.158, 3.159]. In this form, a doublet corrector is placed at a pupil somewhat *inside* the focus, even shorter than the Wright-

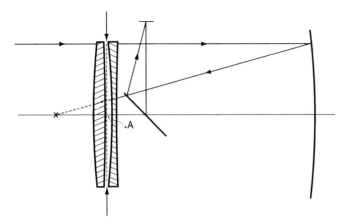

Fig. 3.64. Lurie design of a Houghton-type corrector in Wright-Väisälä camera geometry (200 mm, f/4) given by Rutten and van Venrooij [3.12(g)]

Väisälä camera (Fig. 3.40) which it resembles in a Newton form (Fig. 3.64). However, the Lurie design has an extra degree of freedom with its doublet corrector compared with a single plate in the Wright-Väisälä camera. This enabled Lurie to produce an aplanatic design with a *spherical* mirror. The field is limited by astigmatism, but this is less than that of the Wright-Väisälä camera. Such a system, designed for f/4, is virtually diffraction limited over the visual spectral range (C – F) and a field of 1.5° diameter. The improved axial performance compared with the f/3 Buchroeder version arises essentially from the layout at f/4 for the aperture of 200 mm, which reduces the fifth order spherical aberration and spherochromatism. In the Newton form the optical length of the f/4 Lurie system is $< f'$, i.e. about 650 mm, compared with an optical length $> 2f'$, i.e. about 1250 mm, for the f/3 Buchroeder system. The price of this compactness is the astigmatism limitation in the field for the Lurie system, although this to some extent flattens the field according to the effective field curvature $(2S_{III} + S_{IV})$ – see Table 3.3. In view of its compactness and use of only spherical surfaces, Rutten and van Venrooij are certainly right in drawing the attention of amateurs to the merits of this form. They also show an equally compact Maksutov-Newton telescope with the same geometry. Unlike the Lurie-Houghton design, the Maksutov-Newton they show is not aplanatic and therefore seems less interesting.

More sophisticated designs use two different glasses, enabling correction of spherochromatism.

Figure 3.65 shows spot-diagrams for a Lurie-type design, somewhat modified from that shown in Fig. 3.64 to allow a more direct comparison with the Wright-Väisälä system of Fig. 3.40. The Lurie system now has the afocal corrector just in front of the focus, as with the aspheric plate in Fig. 3.40. This increases the length somewhat, but also relaxes the design. In both systems

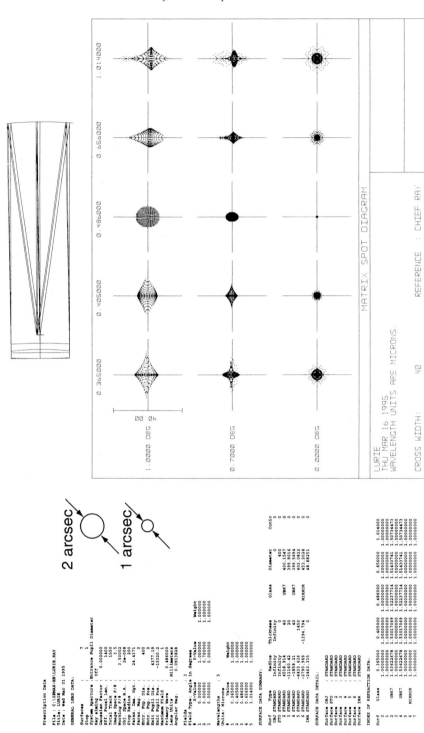

Fig. 3.65. Spot-diagrams for a modified Lurie-Houghton design with aperture 400 mm at f/3.5 and geometry like the Wright-Väisälä system of Fig. 3.40

the field is limited by astigmatism. However, even at f/3.5, the Lurie system (with all spherical surfaces) has only about half the astigmatism of the Wright-Väisälä system at f/4.0. At a field of $\pm 1°$ this Lurie gives an image spread on the optimally curved field of about 2 arcsec, compared with about 4 arcsec for the Wright-Väisälä (Fig. 3.41) on its optimum (flat) field.

It is possible to improve the astigmatism of the Lurie even further by bending the two corrector lenses to give a double-meniscus corrector, with its elements either concave or convex to each other. However, this worsens the correction of spherical aberration and coma.

Cassegrain solutions equivalent to the prime focus solutions of Figs. 3.63 and 3.64 also give excellent performance [3.12(g)].

In general, as stated by Rutten and van Venrooij, such Houghton-type systems are of great interest for amateur size telescopes. For professional instruments, the maximum size will be limited, as in the Maksutov case, by the limitations of high quality optical glass of appreciable thickness. All these solutions can find application in spectrograph optics.

3.6.4.4 Mangin types and refracting combinations. Although the use of a meniscus, as a separate element more or less concentric with the pupil for correcting the aberration of a concave spherical primary, was only invented in the 1940s [3.32] [3.38] [3.51] [3.53], its invention as a double-pass element attached to the primary is much older. It was first proposed by Mangin [3.65] in 1876 for systems projecting light beams (searchlights). Since such projection systems are simply small-field telescopes in reverse, many telescope forms involving the Mangin principle have been proposed.

The original form (Fig. 3.66) of the Mangin system for projection (searchlights) is analysed in detail in the older literature, e.g. Czapski-Eppenstein [3.11(c)]. Both the front and back (mirror) surfaces are spherical, the double-pass meniscus correcting the third order spherical aberration. The back surface reflection had the great advantage that the chemically silvered reflecting surface available at that time could be protected from behind by paint and

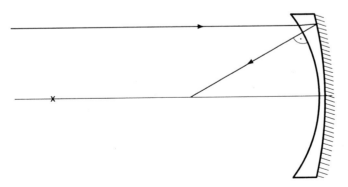

Fig. 3.66. Original Mangin system for searchlight projection

was far more robust than a front-silvered paraboloid. The system was designed to have the front surface concentric with the focus. Various improvements subsequently aspherised either or both the surfaces.

For telescopes, the basic Mangin form is very limited in application, since – with the pupil in the normal position at the primary – the system has the same order of non-aplanatism as the single paraboloid (Newton) telescope. Furthermore, the axial imagery suffers not only from spherochromatism but also from longitudinal chromatic aberration C_1. This arises because there are only two free parameters, the radius difference and the thickness of the meniscus, whereas a Maksutov meniscus has the further free parameters of its position and its optical "bending". In the Mangin form, the radius difference for a given thickness must be fixed to correct S_I, the only condition correctable. With spherical surfaces, the system also suffers from fifth order spherical aberration which is reduced, as with the Maksutov, by increasing the thickness of the meniscus.

Details of more modern developments of telescopes using the Mangin principle are given by Maxwell [3.52] and Rutten and van Venrooij [3.12(h)]. The first obvious improvement is to achromatise the meniscus in a way analogous to the two-glass achromatic Bouwers meniscus of Fig. 3.48. Maxwell [3.52] shows such an achromatic version due to Rosin and Amon [3.66]. It uses the Bouwers principle of two glasses with the same refractive index for the central wavelength (1.6134), but different dispersion ($\nu_A = 57.3$ and 43.9), the convex part being of higher dispersion glass as in the Bouwers system. Unless the stop is shifted from the Mangin mirror, even the achromatic form can only be used for limited fields at high relative apertures. If the stop is shifted to some compromise point of the centres of curvature of the two surfaces, a major improvement in field performance at the cost of doubling the length of the system is possible. Maxwell [3.52] shows a number of more sophisticated designs using Mangin elements. An interesting one by Silvertooth [3.67] uses both a Mangin primary and secondary, the secondary being on the *front* face of a full-aperture meniscus corrector. This system also has a 4-element corrector in the beam converging on the focus.

A practical design for an achromatic Mangin prime focus telescope is given by Rutten and van Venrooij [3.12(h)]. Here the two glasses do *not* have the same index at the central wavelength, so that an additional small refraction takes place at the glass interface. For their standard aperture of 200 mm, they select a relative aperture of f/5. Since the glasses have "normal" dispersions, the secondary spectrum limits the axial performance, according to their calculations, although the power of the Mangin element is weak compared with a normal refracting objective and therefore gives much less secondary spectrum. This defect could be much alleviated by using "special" glasses with better secondary spectrum characteristics, but such glasses are much more expensive. The field is limited by coma which is about half that of a Newton telescope. It can be improved by the glass choice and, they claim, removed by an airspace in the Mangin meniscus (compare secondary spectrum compensation in the Schupmann Medial telescope below). Rutten and van Venrooij state that a "normal glass" version would need to have its relative aperture reduced to f/10 to give fully acceptable secondary spectrum characteristics, too long to have much interest. This confirms the viewpoint of Maxwell [3.52] that the basic achromatic Mangin form is of limited interest. This is a pity, since the back-reflection remains a favourable feature for protecting an aluminium coating or, even more, silver with its superior reflectivity in the visible. Also, the basic Mangin consists of a single block and avoids centering problems between separated elements. The Mangin possibility should be borne in mind for other, more complex designs such as those cited by Maxwell [3.52] or the Sonnefeld system [3.62] referred to in § 3.6.4.3 above.

Lens telescopes with catadioptric compensation–Schupmann Brachymedial and Medial: Although we are entering here somewhat into the domain of refracting telescopes, these designs should be considered as an interesting consequence of the basic Mangin technique of combining lenses with mirrors. An excellent account of the historical development is given by Riekher [3.39(g)]. He describes the Dialyte telescope of Plössl (1850) in which a singlet objective is compensated for its aberrations by a smaller, nearly afocal doublet of two (normal) glasses, giving an achromatic combination (Fig. 3.67).

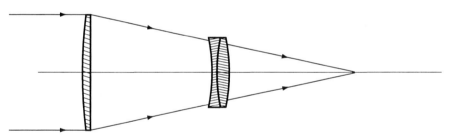

Fig. 3.67. Dialyte telescope due to Plössl (1850)

The longitudinal aberration $\sum C_1$ of a set of separated thin lenses is given, following Eq. (3.308) by

$$\sum C_1 = \frac{y_1^2 K_1}{(\nu_A)_1} + \frac{y_2^2 K_2}{(\nu_A)_2} + \frac{y_3^2 K_3}{(\nu_A)_3} \cdots \tag{3.314}$$

In the Dialyte corrector, we assume the ray height y to be the same for both elements, giving

$$\sum C_1 = \frac{y_1^2 K_1}{(\nu_A)_1} + y_2^2 K_2 \left(-\frac{1}{(\nu_A)_2} + \frac{1}{(\nu_A)_3} \right) = 0 \tag{3.315}$$

for chromatic correction. It is easily shown that the secondary spectrum is also reduced compared with a classical full-size achromat because the individual powers of such an achromat are higher than that of the Dialyte singlet. This is true of "normal" glasses: if special glasses are used, the secondary spectrum can be further reduced.

The medial systems of Schupmann took up these principles but used cata-dioptric compensation elements. The first such system was already proposed by Hamilton [3.68] in 1814 and consisted of a singlet objective compensated by a Mangin-type concave mirror (Fig. 3.68). Schupmann [3.69] analyzed such possibilities exhaustively and derived two classic solutions, the Schupmann Brachymedial and Medial forms. The Brachymedial (Fig. 3.69) is, in principle, close to the Dialyte and Hamilton Brachymedial in that the compensating element is nearly half the size of the singlet objective. An important differ-ence from the Hamilton version is the broken contact (air space) between the compensating lens and the mirror. The use of the concave mirror permits virtually complete compensation of secondary spectrum (as well as the com-pensation of the primary spectrum C_1) because of an important consequence of the theory of secondary spectrum. The secondary spectrum of a "normal" glass was shown in Fig. 3.30 and is simply the departure from linearity of the dispersion function with wavelength. First order chromatic correction (C_1) rotates the function so that two wavelengths λ_1 and λ_2 are corrected as shown in Fig. 3.31. The correction of secondary spectrum requires that a third wavelength should also be corrected. It was shown by König [3.11(d)]

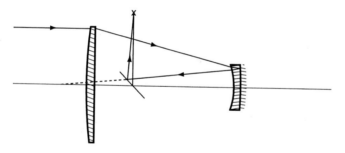

Fig. 3.68. Brachymedial due to Hamilton (1814)

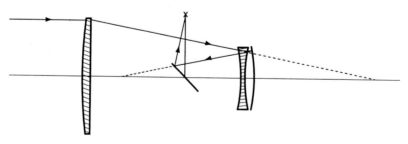

Fig. 3.69. Brachymedial due to Schupmann (1899)

that *correction of the secondary spectrum with "normal" glasses* obeying the
linear relationship between index n and Abbe number ν_A for the dispersion

$$n = A + B\nu_A ,$$ (3.316)

which is roughly true for the main sequence of classical glasses, *is only possible
for a train of separated "thin" lenses if either object or image are virtual.* For
telescopic systems, the object is always real, even at infinity. The above law
implies, then, in practice, *that correction of secondary spectrum in such a
refracting system is only possible if the image is virtual.* The extension of
this theorem to normal glasses not obeying Eq. (3.316) has been given by
Kerber [3.70]. The Schupmann systems accept this situation and convert the
virtual image of the refracting system into a real image by reflection at a
spherical concave mirror, as shown in Fig. 3.69. As in the Dialyte of Fig. 3.67,
the lens-mirror corrector is roughly afocal in its final effect except that the
magnification is ~ -1 instead of $\sim +1$. The third order spherical aberration
can be corrected by adjusting the "bending" of the concave corrector lens.

A more sophisticated form due to Schupmann is the Medial telescope
(Fig. 3.70). Here the singlet objective is compensated by a double-meniscus
compensator with spherical concave mirror placed *after* the focus. This re-
images the primary image without magnification with a slight tilt to make
it accessible to one side. Such imagery at a concave sphere without magnification is aplanatic due to symmetry and the astigmatism is small for small
tilts. The 90° deviating prism also forms a convex lens. The aperture of the
compensating system is about 1/4 of that of the objective. Schupmann investigated four possible compensation systems. The one shown in Fig. 3.70
has a separated meniscus in front of a Mangin mirror, a form used for the
335 mm aperture system built for the Urania Observatory in Berlin in 1902
[3.39(g)]. The secondary spectrum was only 1/20 of that of a normal equivalent refractor. If a further airspace is introduced in front of the mirror, even
finer correction is possible.

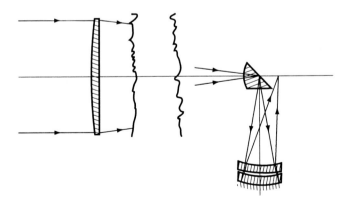

Fig. 3.70. The Medial telescope due to Schupmann

The Schupmann designs are essentially special forms of the refracting telescope which has not been included in our definition of modern telescope optics. They have been discussed here because they are an elegant example of correction by double-pass Mangin-type compensators. For amateur sizes they certainly retain considerable interest, as indicated by Rutten and van Venrooij [3.12(h)]. They quote a modern work by Daley [3.71] dealing with the amateur construction of such systems.

Systems with Mangin secondaries and Medial geometry: Many designs of great variety exist. Fig. 3.71 shows a very compact design with all spherical surfaces due to Delabre [3.72]. Here, the primary is f/2 working in a Cassegrain type arrangement with a plane secondary as the back of a negative Mangin lens. Although the secondary is plane, the obstruction is quite low because of an internal image which is transferred out in the Brachymedial geometry by a 5-element refracting system. The length between the reflecting surfaces of primary and secondary is only 636 mm for an aperture of 400 mm and an equivalent focal length of 3000 mm (f/7.5). The field is about 0.8° diameter for miniature camera format. The quality is limited by coma and transverse chromatic effects, 80% of the energy being within 1 arcsec (14.5 μm) at a field of ±0.3°. The purpose of such designs is to replace full-aperture menisci by smaller elements: but corrector elements near the pupil are always the most effective.

Further interesting developments of the basic optical concept of the Brachymedial designs of Figs. 3.68 and 3.69 have recently been published by Busack [3.160]. In a normal uniaxial form, an additional convex lens, near the image, is added to the basic design of Fig. 3.68. This enables the removal of the fundamental optical weakness of the Brachymedial, the lateral colour or chromatic difference of magnification (see Eq. (3.223)), using only spherical surfaces. Busack claims that this gives a performance equivalent to the

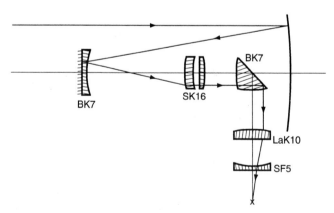

Fig. 3.71. Compact system using Mangin secondary and Brachymedial geometry due to Delabre

Schmidt telescope, with apertures up to $f/2$ and fields up to $5°$, but with the advantage of a flat field and shorter overall length. The disadvantage of such solutions is that they require high quality optical glass not only for the front lens, but also for the Mangin mirror, whose optical thickness is doubled by the back reflection. This limits the size of practical application. However, the back reflection permits better protection of the reflecting coat. A number of modifications of the basic system are given in the patent.

In a later patent [3.161], Busack extends the above prime focus system to a Cassegrain version. He refers to a patent by Gallert of a similar system, whereby Gallert uses a positive front lens of convex-concave form, the concave back surface then serving in its reflecting centre as a *concave* secondary mirror. This system gives good correction at $f/3$ over a $4°$ field. Busack's system uses a similar front lens, but has a separated conventional Cassegrain *convex* secondary, close to the front lens. This is essentially a normal Cassegrain extension of his prime focus system and gives similar excellent correction at $f/4$ over a $4°$ field. The advantages claimed over the Gallert system are a shorter overall length, a longer back focal distance, a smaller central obstruction and correction of distortion. If some field curvature is permitted, the central obstruction can be so far reduced that the system is considered eminently suitable for visual observation. All surfaces are again spherical and only one glass type is used.

3.6.5 Three- or multi-mirror telescopes (centered)

3.6.5.1 Various three-mirror solutions. Most designs using 3 mirrors are unattractive in practice because of obstruction problems. However, there are two solutions due to *Korsch* [3.73] of considerable practical interest. In general, the theory of 3-mirror systems is of fundamental importance for the extension to 4 mirrors.

By analogy with the theory of 2-mirror-plate telescopes above and from Eqs. (3.220), it is clear that any axial distribution of 3 powered, aspheric mirrors can give correction of S_I, S_{II} and S_{III} (see §§ 3.6.2.1 and 3.4, and also [3.13]). The general analytical formulation is given by Schroeder [3.22(b)] and inevitably has a complexity exceeding that of Eqs. (3.220) because the third mirror has power as well as asphericity in the general case. A closed-form solution for centered 3-mirror telescopes giving correction of all 4 third order conditions

$$S_I = S_{II} = S_{III} = S_{IV} = 0$$

has been given by Korsch [3.73]. His formulation, lightly modified to bring it into line with our notation and sign convention, is reproduced here. Korsch's formulae assume the entrance pupil is at the primary and the object distance is infinite, i.e. $m_1 = 0$. His free design parameters are then

$$m_2, m_3 \text{ and } p_2$$

where m_ν is the object to image magnification of surface ν, and p_ν is the pupil object to image magnification of surface ν. m_ν is determined by the paraxial ray, p_ν by the paraxial principal ray. For the formulation of a 2-mirror system in Table 3.5 we used the strict sign convention of Tables 2.2 and 2.3. The Korsch formulae, adapted to this system, are:

$$
\begin{aligned}
p_2 &= -\frac{m_2 R_A}{1 + m_2 - R_A} \\[2mm]
p_3 &= \frac{m_3}{1 - p_2 m_2 m_3 - m_2 m_3 + p_2 m_3} \\[2mm]
b_{s3} &= -\Bigg[2 m_2 m_3 \left(\frac{m_2}{m_2 + 1} - \frac{2 p_2}{p_2 + 1} \right) \\
&\quad + \frac{(m_2 + 1)^2}{m_2 - p_2}(p_2 - 1) m_3 \left(\frac{m_2 - 1}{m_2 + 1} - \frac{p_2 - 1}{p_2 + 1} \right) \\
&\quad + \frac{(m_3 + 1)^2}{m_3 - p_3}(p_3 - 1) \left(\frac{m_3 - 1}{m_2 + 1} \frac{1}{m_3} - \frac{p_3 - 1}{p_2 + 1} \frac{1}{p_3} \right) \Bigg] \\
&\quad \div \left[\frac{(m_3 + 1)^2}{m_3 - p_3}(p_3 + 1) \left(\frac{m_3 + 1}{m_2 + 1} \frac{1}{m_3} - \frac{p_3 + 1}{p_2 + 1} \frac{1}{p_3} \right) \right] \\[2mm]
b_{s2} &= -\Bigg[\frac{2 m_2^2 m_3}{m_2 + 1} + \frac{(m_2 + 1)^2}{m_2 - p_2}(p_2 - 1)\frac{m_2 - 1}{m_2 + 1} m_3 \\
&\quad + \frac{(m_3 + 1)^3}{m_3 - p_3} \frac{p_3 + 1}{m_2 + 1} \frac{1}{m_3} \left(b_{s3} + \frac{(m_3 - 1)(p_3 - 1)}{(m_3 + 1)(p_3 + 1)} \right) \Bigg] \\
&\quad \div \left[\frac{(m_2 + 1)^2}{m_2 - p_2}(p_2 + 1) m_3 \right] \\[2mm]
b_{s1} &= -\Bigg[(m_2 m_3)^3 + \frac{(m_2 + 1)^4}{m_2 - p_2} p_2 m_3^3 \left(b_{s2} + \left(\frac{m_2 - 1}{m_2 + 1} \right)^2 \right) \\
&\quad + \frac{(m_3 + 1)^4}{m_3 - p_3} p_2 p_3 \left(b_{s3} + \left(\frac{m_3 - 1}{m_3 + 1} \right)^2 \right) \Bigg] \div (m_2 m_3)^3
\end{aligned}
\qquad (3.317)
$$

$$
r_1 = \frac{2 f'}{m_2 m_3} \qquad (3.318)
$$

$$
r_2 = -\frac{2 f' p_2}{m_3 (m_2 - p_2)} \qquad (3.319)
$$

$$
r_3 = \frac{2 f' p_2 p_3}{m_3 - p_3} \qquad (3.320)
$$

$$d_1 = \frac{f'(1 + p_2)}{m_3(m_2 - p_2)} \tag{3.321}$$

$$d_2 = -f'\left(\frac{m_2 + 1}{m_2 - p_2} + p_3\frac{m_3 + 1}{m_3 - p_3}\right)\frac{p_2}{m_3} \tag{3.322}$$

$$L = \frac{f'p_2p_3(m_3 + 1)}{m_3 - p_3} \tag{3.323}$$

Korsch gives two examples of the application of his formulae above to two three-mirror geometries, shown in Fig. 3.72 (a) and (b). The application of the Korsch formulae to his system (a) is shown in Case 9 of Tables 3.2 and 3.3 for the usual normalization with $y_1 = 1$ and $f' = -1$. The set-up data for the Korsch proposal with f/4.5 are:

Primary f/no $=$ f/3
Final f/no $=$ f/4.5 $(y_1 = 1/9.0 = 0.111\,111)$
f' $= -1$
m_2 $= -10$ (negative for Cassegrain)
f_1' $= -0.666\,667$
r_1 $= -1.333\,333$
R_A $= +0.25$ (positive for Cassegrain)

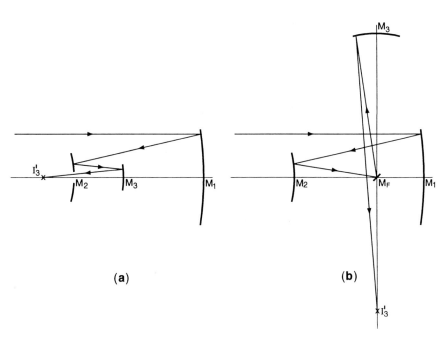

Fig. 3.72. Two 3-mirror anastigmatic, flat-field solutions proposed by Korsch (1972): (**a**) single-axis system, (**b**) 2-axis system

Then, from (3.318),

$$m_3 = -0.150 ,$$

an exact value differing from the value -0.149 given by Korsch. This small rounding error has a significant effect on the remaining values deduced from the formulae, which therefore differ somewhat from those given by Korsch.

Table 3.18. Data of the system of Fig. 3.72(a), adapted from Korsch [3.73]

N_1	3.0
N_{system}	4.5
f'	-1
f_1'	$-0.666\,667$
r_1	$-1.333\,333$
m_2	-10
R_A	$+0.25$
m_3	-0.150
p_2	$-0.270\,270$
p_3	$+2.775\,003$
r_2	$-0.370\,370$
r_3	$-0.512\,821$
d_1	$-0.500\,000$
d_2	$+0.213\,675$
L	$-0.217\,949$
b_{s3}	$-1.401\,477$
b_{s2}	$-2.843\,216$
b_{s1}	$-1.262\,939$

Inserting the above parameters in Eqs. (3.317) to (3.323) gives the result of Table 3.18 for the complete data of the system. The Schwarzschild constants $b_{s\nu}$ for the three mirrors show that all three are hyperbolic with eccentricities not excessive compared with normal 2-mirror solutions. The above data give exact correction for Case 9 in Table 3.3 of all four third order aberrations S_I, S_{II}, S_{III}, S_{IV}.

The complexity of the Eqs. (3.317) is instructive. We saw in §3.2.4 that the third order theory of a 1-mirror telescope is trivially simple; that of a 2-mirror telescope considerably more complex. The 3-mirror formulation

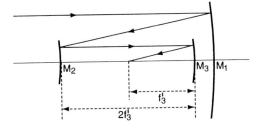

M_2

M_3 M_1

f_3'

$2f_3'$

Fig. 3.73. 3-mirror system due to Paul (1935)

above is instructive, but too unwieldy to be of wide application in practice, even though it is simplified by the assumption of the entrance pupil at the primary. (Of course, the stop position is uncritical, to the third order, for an anastigmatic telescope). However, the forms of Fig. 3.72 proposed by Korsch are the two most interesting 3-mirror systems: (a) because of the final image position behind the secondary, giving stray light baffling by M_2; and (b) because of its 2-axis nature and the general potential of 2-axis systems – see § 3.6.5.3.

The system of Fig. 3.72 (a) given by Korsch is, in fact, a more generalised form of a system already proposed in 1935 in a classic paper by *Paul* [3.74], shown in Fig. 3.73. The essential properties of this system and further developments are excellently treated by Schroeder [3.22(b)]. The original Paul form starts off with a Mersenne afocal anastigmatic beam compressor defined by Eq. (3.97) and Table 3.6 as a Cassegrain system of two confocal paraboloids (Fig. 1.3). Paul added a spherical tertiary mirror to this beam compressor as shown in Fig. 3.73, *placed so that its centre of curvature is at the vertex of the secondary.* This tertiary, of course, introduces spherical aberration, but otherwise functions as a Schmidt primary receiving an anastigmatic beam from the exit pupil of the beam compressor, if the exit pupil is placed at the secondary. Paul then corrected the spherical aberration by the following elegant and simple concept: he defined $r_3 = r_2$ and, instead of correcting $\sum S_I$ by making the tertiary parabolic, he achieved the same correction by *removing* the parabolic form of the secondary, thereby making it also spherical.

We can now apply the "aspheric plate theory" of § 3.4 to show that the field aberrations coma and astigmatism are also corrected. Let the exit pupil of the beam compressor be placed at the vertex of the secondary. (This assumption has no effect on the anastigmatism of the beam compressor, since from Eqs. (3.213) a stop shift has no effect on the third order terms apart from distortion). Then the removal of the parabolic form from the secondary to correct $\sum S_I$ is the equivalent of adding an aspheric plate to the unchanged secondary to achieve the same effect. Since the pupil is at the secondary and the tertiary is concentric with it, the addition of such an equivalent "Schmidt system" has no effect on $\sum S_{II}$ and $\sum S_{III}$ which are zero from the original anastigmatic beam compressor, as follows from Eqs. (3.214). If the exit pupil of the beam compressor is now shifted to its normal position behind the secondary, corresponding to the usual stop position at the primary, Eqs. (3.213) show there is no effect on $\sum S_I$, $\sum S_{II}$ and $\sum S_{III}$ due to this stop shift. The Paul system is therefore anastigmatic with $\sum S_I = \sum S_{II} = \sum S_{III} = 0$, but the field is not flat, and has a curvature of $P_C = 1/f_1'$.

Baker [3.23(d)] [3.75] modified the Paul system into what is known as the *Paul-Baker system* by setting the condition $|r_3| > |r_2|$ such that the Petzval sum is zero and a flat field is achieved. The third order theory of this Paul-Baker form is simple and elegant.

From the definitions leading to Eq. (3.20) we have for the field curvature

$$\sum P_C = \sum c_\nu \left(\frac{1}{n_\nu'} - \frac{1}{n_\nu} \right)$$

with $c_\nu = 1/r_\nu$. Then

$$\sum P_C = -2 \left(\frac{1}{r_1} - \frac{1}{r_2} + \frac{1}{r_3} \right) \tag{3.324}$$

For the afocal, 2-mirror beam compressor

$$R_A = \frac{y_2}{y_1} = \frac{r_2}{r_1}$$

and

$$\sum P_C = -\frac{2}{r_1} \left[1 - \frac{r_1}{r_2} \left(1 - \frac{r_2}{r_3} \right) \right] = -\frac{2}{r_1} \left[1 - \frac{1}{R_A} \left(1 - \frac{r_2}{r_3} \right) \right] \tag{3.325}$$

The condition for $\sum P_C = 0$ in the Paul-Baker system is therefore

$$\frac{r_2}{r_3} = 1 - R_A \tag{3.326}$$

Since $|r_3| > |r_2|$, the spherical aberration introduced by the tertiary is reduced compared with the Paul system. Baker compensates this by modifying the aspheric on the secondary. Since it receives an afocal beam, $(S_I)_3$ can be calculated from Eqs. (3.25) for the simple case of a primary with paraxial ray height $y_3 = y_2$. Then, in the first equation of (3.20) we have

$$y_3 \quad = y_2$$

$$A_3 \quad = +\frac{y_2}{r_3}$$

$$\Delta \left(\frac{u}{n} \right)_3 = +2 \frac{y_2}{r_3} ,$$

giving

$$(S_I)_3 = -y_2^4 \frac{2}{r_3^3} = -y_1^4 \left(\frac{y_2}{y_1} \right)^4 \frac{2}{r_3^3}$$

With $y_2/y_1 = r_2/r_1$, this reduces to

$$(S_I)_3 = -2 \frac{y_1^4}{r_1^3} \left(\frac{r_2}{r_1} \right) \left(\frac{r_2}{r_3} \right)^3 \tag{3.327}$$

Similarly, for the secondary

$$A_2 = -\frac{y_2}{r_2} \text{ and } \Delta\left(\frac{u}{n}\right)_2 = -\frac{2y_2}{r_2} ,$$

giving

$$(S_I)_2 = +\frac{2y_1^4}{r_1^3}\left(\frac{r_2}{r_1}\right) \tag{3.328}$$

for its contribution as a spherical mirror. For the basic afocal beam compressor, corrected for aberration, $(S_I)_2$ is compensated by $(S_I)_2^*$ by making it parabolic. In (3.20) we have for an aspheric form defined by b_{s2}

$$(S_I)_2^* = \tau_2 = \frac{1}{r_2^3}(n'-n)_2 b_{s2} y_2^4$$

With $(n'-n)_2 = +2$, this reduces to

$$(S_I)_2^* = +2\frac{y_1^4}{r_1^3}\left(\frac{r_2}{r_1}\right)b_{s2} , \tag{3.329}$$

giving, with $b_{s2} = -1$, the required compensation of $(S_I)_2$ for the normal Mersenne beam compressor with the parabolic form. For the Paul-Baker system we require for the correction of $\sum S_I = 0$, the condition from (3.327), (3.328) and (3.329) with a parabolic primary form

$$(S_I)_2 + (S_I)_2^* + (S_I)_3 = 0 ,$$

since the tertiary is defined as spherical. This gives

$$2\frac{y_1^4}{r_1^3}\left(\frac{r_2}{r_1}\right)\left[1+b_{s2}-\left(\frac{r_2}{r_3}\right)^3\right] = 0 , \tag{3.330}$$

leading to the required form of the secondary as

$$b_{s2} = -1 + \left(\frac{r_2}{r_3}\right)^3 = -1 + (1-R_A)^3 , \tag{3.331}$$

from (3.326) and (3.330). If $R_A = 0.25$, this gives $b_{s2} = -0.578$, a moderate elliptical form. For normal values of R_A, the value of r_3 from (3.326) will require M_3 to be placed *behind* the primary in order to provide a reasonable final image position. This implies a *perforated* primary, as proposed by Baker.

The "aspheric plate theory" of §3.4 can be applied in exactly the same way as for the Paul system above to prove that field coma and astigmatism are zero, so that the Paul-Baker system gives

$$\sum S_I = \sum S_{II} = \sum S_{III} = \sum S_{IV} = 0$$

Schroeder [3.22(b)] comments on the first large practical realisation of such a Paul-Baker telescope as a transit instrument with a 1.8 m primary working at f/2.2 and an axial obstruction ratio $R_A \sim 0.32$. He states near-diffraction-limited imagery is given over a 1° field. As with all such three-

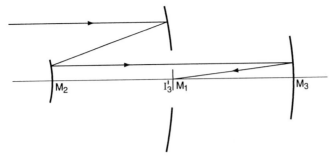

Fig. 3.74. The Willstrop Mersenne-Schmidt telescope with f/1.6 and a 4° diameter field (1984)

mirror solutions, careful baffling is required to prevent light reaching the detector directly from M_1 or M_3, giving appreciable central obscuration of about 22% of the light. The system is described by McGraw et al. [3.76].

An interesting extension of the Paul or Paul-Baker concept is a proposal by *Willstrop* [3.77]. He also perforates the primary, but places the third (spherical) mirror sufficiently far behind it that the final image is formed in the plane of M_1 (Fig. 3.74). The optical design theory is not changed in principle, but allows a compact telescope with moderate field curvature and excellent correction over a 4° field diameter. Willstrop points out that Mersenne's original proposal in 1636 (Fig. 1.3) showed a perforated primary. He therefore calls his system a Mersenne-Schmidt telescope. The system is admirable since the Mersenne telescope does not suffer from chromatic aberrations. Its one weak feature is the inevitable large central obstruction. This is a more severe restraint than in the Paul or Paul-Baker concepts because of the large separation M_2M_3 which defines $|r_3|$ with the Schmidt geometry centered on the secondary, as with Paul. If, following Paul, both M_2 and M_3 are spherical to the third order, then $r_2 = r_3$ and is relatively long. Since $R_A = r_2/r_1$ in the afocal Mersenne telescope, the obstruction is high. In the Willstrop proposal, $R_A = 0.5$ because $d_1 = f_3' = f_2' = f_1'/2$. This axial obstruction is inevitably further increased at the primary perforation by the field supplement. Willstrop analyses the total obstruction in detail for a field of 2° or 4° in diameter. Depending on the stop position and baffle arrangement, he deduces total obstructions (by *area*) between 41 and 45% for the 4° diameter field. He makes the entirely valid comment that the effective surface area loss at the primary of typical large Schmidt telescopes due to the pupil (plate) shift and field is even larger, about 53%.

Epps and Takeda [3.78] had already proposed a similar system with a perforated primary and third mirror a short distance behind it. We shall see in § 3.6.5.3 that these 3-mirror concepts of Epps and Takeda, Willstrop and Korsch can be extended to give excellent designs with 4 powered mirrors, particularly with 2-axis geometry.

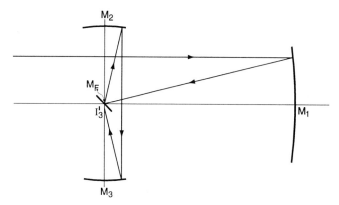

Fig. 3.75. Baker 3-mirror, 2-axis anastigmatic telescope (1945)

Baker [3.23(d)] also gave design details of a 2-axis (brachy-type), anastigmatic predecessor of the concept of Fig. 3.72 (b) suggested by Korsch. A primary concave paraboloid has a 45° flat mirror at its focus which deflects the diverging beam to a concave collimating hyperboloid. The collimated beam is then focused by a concave sphere as in the Paul-Baker telescope. Apart from the brachy-form in which the beam is turned at right angles, this Baker form (Fig. 3.75) is generically the Gregorian equivalent of the Paul-Baker telescope above. The hyperbolic form of the concave secondary has the same function as the elliptical convex secondary in the Paul-Baker telescope. The properties of this system, as a fundamental generic type of 2-axis telescope, are further discussed in § 3.6.5.3. As a practical form for modern 3-mirror telescopes, the Korsch design of Fig. 3.72 (b) is much superior.

A modification of the Paul-Baker 3-mirror telescope which is of interest, above all, in smaller sizes for amateurs is the *Loveday* telescope [3.79]. Details are also given by Rutten and van Venrooij [3.12(i)]. Instead of an independent tertiary, the Loveday design uses the primary with a second reflection, as shown in Fig. 3.76. The final image is at the prime focus since the first two mirrors again form a beam-compressor of the classical Mersenne afocal, anastigmatic pair of confocal paraboloids. The primary can either be used directly as a relatively fast Newton telescope or as a slow Newton telescope with 3 reflections. If the f/no of the primary is f_1'/N_1, then that of the complete system is $f_1'/(N_1/R_A)$. Typical values for an amateur telescope given by Rutten and van Venrooij are $N_1 = 6$, $N_{fin} = 24$, with $R_A = 0.25$.

The similarity with the Paul-Baker telescope ceases with the 3-mirror geometry, since the tertiary (i.e. M_1) is parabolic in form, not spherical. It therefore suffers from the field aberrations of a parabolic primary with its entrance pupil at the exit pupil of the beam compressor. However, the relative aperture is so weak that the performance is still excellent over the 40 mm field given by Rutten and van Venrooij for a Loveday telescope of

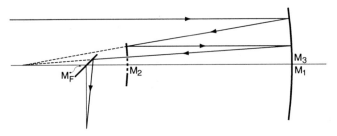

Fig. 3.76. Dual-purpose Newton telescope due to Loveday (1981)

200 mm aperture working at f/24. As a tertiary in this Loveday geometry, the primary has a radius nearly three times that of a Paul-Baker telescope.

Recently (2000), a complete and masterly analysis of all potential 3-mirror telescopes was initiated by Rumsey and carried out by Rakich [3.162] [3.163]. The basis of his approach is the Burch "plate diagram" or "see-saw diagram" (see § 3.4). He also uses a theorem enunciated in 1900 by Aldis [3.164], stating that four spherical surfaces are sufficient to produce an anastigmatic image, and extended by Burch in 1942 [3.28] to show that 2 spherical surfaces and one aspheric surface or 2 aspheric surfaces alone are sufficient for anastigmatism. (This general theorem is not applicable to concentric systems such as the Schmidt telescope or more complex forms of concentric mirrors). Thus Rakich deals with the general case of anastigmatic solutions containing 3 powered mirrors of which 2 are spherical, the object conjugate being infinite. The theorems of Aldis and Burch, as used by Rakich, can also be derived as limit cases of the *generalised Schwarzschild theorem* given in § 3.6.5.3.

The extension of the Burch method is developed into an automatic program which, for the predefined input parameters t_1 (separation of primary and secondary mirrors – d_1 in our notation) and c_2 (curvature of the secondary mirror), generates and solves a cubic equation. The solutions of this cubic equation define anastigmats. Different cubics are generated depending on which mirror, primary, secondary or tertiary, is aspherized. This enables a survey of the plane in parameter space defined by t_1 and c_2. The use in three programs of the three separate solutions gives rise to 3 families of solutions for each type of cubic equation. Each family is divided into sub-families, separate regions in the (t_1, c_2) plane between which no real anastigmats exist.

In this way, a total of 15 sub-families are generated. A further division defines sub-regions containing solutions with positive or negative Petzval sum, flat field solutions lying on curves where these regions touch.

"Filters" are built in as conditions which reject impractical systems because of too extreme parameters. As a result, many solutions are rejected and 5 potentially useful families remain. Rakich presents, as normal uniaxial or folded systems, 9 novel anastigmats, some of which are closely related to known systems and others completely new.

Finally, some very interesting "Schiefspiegler" systems are presented which are discussed in § 3.7.

By covering the whole effective parameter space of his 3-mirror solutions, using Rumsey's basic ideas, Rakich has presented a most elegant analysis which may be seen as the definitive triumph of classical third order theory as a tool for setting up telescope solutions.

Rakich [3.165] has subsequently attempted to extend his approach to 4-mirror (all spherical) telescopes. He points out that the dimensionality of the solution space goes from one dimension for 2 aspheric mirrors, through two dimensions for 3 mirrors with 1 aspheric to three dimensions for 4 spherical mirrors. Most solutions thrown up are blocked by his filters and *none* were attractive because of high central obstruction, etc. The solutions discussed in § 3.6.5.3 do indeed suggest that, at this level of parametric complication, intuition and experience may well be more fruitful than a general analytical solution. However, according to latest information from Rakich, he is still hoping to set up useful Schiefspiegler solutions with 4 spherical mirrors. The spherical mirror form makes the manufacture of such excentric systems much easier than that of excentric aspherics.

3.6.5.2 Recursion formulae for calculating aberration coefficients for the general case of $\tilde{n}$ mirror surfaces.

In § 3.2.4.3 analytical expressions were developed for the general case of 2-mirror telescopes. As was pointed out in connection with Tables 3.2 and 3.3, calculations based on the direct tracing of the paraxial aperture and paraxial principal rays are highly desirable as a check on systems set up from analytical formulae. Ray-trace calculations must follow the systematic sign convention given in § 2.2.3.

The analytical formulae given by Korsch in § 3.6.5.1 for the general solution of a 3-mirror system satisfying all four aberration conditions ($S_I = S_{II} = S_{III} = S_{IV} = 0$) show the complexity of a general analytical solution even for 3 mirrors. Furthermore, these formulae apply the simplifying conditions that $s_1 = \infty$ and $s_{pr1} = 0$, i.e. object at infinity and entrance pupil at the primary.

Rather than extend such general analytical solutions to even more complex forms for $\tilde{n} > 3$, we will now give *recursion formulae* which enable the basic parameters of Eqs. (3.20) to be determined rapidly and simply without tracing paraxial rays. These formulae are derived, in fact, by tracing a paraxial ray in the general case and expressing the result only in basic constructional parameters or natural parameters of the system. These parameters are similar to those used by Korsch in Eqs. (3.317)–(3.323). Such recursion formulae represent a special case for reflecting surfaces of a general formulation known as *Gaussian brackets* [3.7(a)] [3.80] [3.81]. According to Herzberger [3.7(a)], Gauss devised this algorithm for the solution of a linear diophantine equation [3.81]. In spite of its apparently evident applicability to the linear paraxial equations that Gauss himself had formalised, he failed, according to Zimmer [3.80(a)], to recognize the utility of his algorithm in this case.

The general starting parameters (see Eqs. (2.36)) for a paraxial ray (either aperture or principal) incident on a telescope system are s_1 (or s_{pr1}) and u_1 (or u_{pr1}). Otherwise, all quantities must be expressed in terms of

$$r_1, r_2, \ldots, r_{\tilde{n}} \text{ and } d_1, d_2, \ldots, d_{\tilde{n}},$$

the constructional parameters. The refractive indices $n_1, n_2 \ldots, n_{\tilde{n}}$ are not included because our purpose is to establish formulae for a system of $\tilde{n}$-mirrors in air, giving only sign inversions with $n = \pm 1$. This brings a considerable simplification, but for $\tilde{n} \geq 3$, the formulae still become complex in terms of these parameters. They can be simplified to elegant, recursive forms by introducing, similarly to Korsch, the parameters

$$\mu_1, \mu_2, \ldots, \mu_{\tilde{n}} \quad \text{and} \quad \mu_{pr1}, \mu_{pr2}, \ldots, \mu_{pr\tilde{n}},$$

in which μ_ν expresses the *inverse* magnification at the ν-surface of the aperture ray and $\mu_{pr\nu}$ the *inverse* magnification at the ν-surface of the principal ray (inverse pupil magnification). Then

$$\mu_\nu = \frac{u'_\nu}{u_\nu} \quad \text{and} \quad \mu_{pr\nu} = \frac{u'_{pr\nu}}{u_{pr\nu}} \tag{3.332}$$

These are therefore the reciprocals of the quantities m_ν and p_ν in Eqs. (3.317). Our sign convention also strictly follows the ray-trace rules: because $n'/n = -1$ at the reflections, d_ν is negative if ν is odd, positive if even. Radii are negative if concave to the incident light direction (from left to right). The signs of μ_ν and $\mu_{pr\nu}$ are given by the ratios defined in (3.332).

The *general recursion formulae* for the parameters of Eqs. (3.20) are, then, for mirror surface ν with $\nu \geq 1$:

$$y_\nu = \left[-s_1 + \mu_0 d_0 + (\mu_0\mu_1)d_1 + (\mu_0\mu_1\mu_2)d_2 + \ldots + (\mu_0\ldots\mu_\nu)d_\nu \right] u_1$$

$$A_\nu = \left[(-1)^\nu \tfrac{1}{2} \left(\mu_0\mu_1\mu_2 \ldots \mu_{(\nu-1)} \right) (\mu_\nu - 1) \right] u_1$$

$$\Delta\left(\tfrac{u}{n}\right)_\nu = \left[(-1)^\nu (\mu_0\mu_1\mu_2 \ldots \mu_{(\nu-1)})(\mu_\nu + 1) \right] u_1$$

$$\mu_\nu = \frac{1}{r_\nu} \left[\left(\frac{2}{(\mu_0\mu_1\mu_2 \ldots \mu_{(\nu-1)})} \right) \left(s_1 - (\mu_0)d_0 - (\mu_0\mu_1)d_1 \right. \right.$$
$$\left. \left. - (\mu_0\mu_1\mu_2)d_2 - \ldots - (\mu_0\mu_1\mu_2 \ldots \mu_{(\nu-1)})d_{(\nu-1)} \right) - r_\nu \right]$$

$$\text{with } d = d_0, d_1, d_2, \ldots d_{(\nu-1)}$$
$$\text{and } \mu_0 = 1, \; d_0 = 0$$

$$\tag{3.333}$$

These equations apply to both the paraxial aperture and principal rays. Starting with s_{pr1} and u_{pr1}, they then give the field parameters $y_{pr\nu}$ and $\overline{A}_\nu$ in

Eqs. (3.20). If the object is at a finite distance s_1 and the aperture ray has the angle u_1 to the axis, then the formulae give the aperture parameters y_ν, A_ν and $\Delta\left(\frac{u}{n}\right)_\nu$.

For a 4-mirror telescope, (3.333) gives:

$$
\begin{aligned}
y_1 &= -s_1 u_1 \\
y_2 &= (-s_1 + \mu_1 d_1)u_1 \\
y_3 &= (-s_1 + \mu_1 d_1 + \mu_1\mu_2 d_2)u_1 \\
y_4 &= (-s_1 + \mu_1 d_1 + \mu_1\mu_2 d_2 + \mu_1\mu_2\mu_3 d_3)u_1 \\[6pt]
A_1 &= -\tfrac{1}{2}(\mu_1 - 1)u_1 \\
A_2 &= +\tfrac{1}{2}\mu_1(\mu_2 - 1)u_1 \\
A_3 &= -\tfrac{1}{2}\mu_1\mu_2(\mu_3 - 1)u_1 \\
A_4 &= +\tfrac{1}{2}\mu_1\mu_2\mu_3(\mu_4 - 1)u_1 \\[6pt]
\Delta\left(\tfrac{u}{n}\right)_1 &= -(\mu_1 + 1)u_1 \\
\Delta\left(\tfrac{u}{n}\right)_2 &= +\mu_1(\mu_2 + 1)u_1 \\
\Delta\left(\tfrac{u}{n}\right)_3 &= -\mu_1\mu_2(\mu_3 + 1)u_1 \\
\Delta\left(\tfrac{u}{n}\right)_4 &= +\mu_1\mu_2\mu_3(\mu_4 + 1)u_1 \\[6pt]
\mu_0 &= +1 \quad \text{(by definition)} \\
\mu_1 &= \frac{1}{r_1}\,(2s_1 - r_1) \\
\mu_2 &= \frac{1}{r_2}\left(\frac{2s_1}{\mu_1} - 2d_1 - r_2\right) \\
\mu_3 &= \frac{1}{r_3}\left(\frac{2s_1}{\mu_1\mu_2} - \frac{2d_1}{\mu_2} - 2d_2 - r_3\right) \\
\mu_4 &= \frac{1}{r_4}\left(\frac{2s_1}{\mu_1\mu_2\mu_3} - \frac{2d_1}{\mu_2\mu_3} - \frac{2d_2}{\mu_3} - 2d_3 - r_4\right)
\end{aligned}
\tag{3.334}
$$

Putting $s_1 = 0$ and applying these formulae to the paraxial principal ray, the quantities y_{pr} and $\overline{A}_\nu$ can be checked for any of the systems given in Tables 3.2 and 3.3. With the normalization used there, we set up $u_{pr1} = +1$.

The above general form is, in practice, of more direct use for the principal ray for which s_{pr1} is normally finite, often zero. For astronomical telescopes, of course, the normal case is $s_1 = \infty$, and $s_1 \neq \infty$ has little practical significance. Because of the singularity introduced with $s_1 = \infty$, the general recursion form of (3.333) requires modification. The recursion formulae resulting break down for $\nu = 1$, but are generally valid for $\nu > 1$:

Recursion formulae for $s_1 = \infty$

$$\left.\begin{aligned}
y_\nu &= y_1\left[1 - \frac{2}{r_1}\Big((\mu_0)d_0 + (\mu_0\mu_1)d_1 + (\mu_0\mu_1\mu_2)d_2 \ldots\right.\\
&\qquad\left.+(\mu_0\mu_1\mu_2\ldots\mu_{(\nu-1)})d_{(\nu-1)}\Big)\right]\\[6pt]
A_\nu &= (-1)^{(\nu-1)}\left[(\mu_1\mu_2\ldots\mu_{(\nu-1)})(\mu_\nu - 1)\left(\frac{y_1}{r_1}\right)\right]\\
&\quad\text{with the condition } \nu > 1\\[6pt]
\Delta\left(\tfrac{u}{n}\right)_\nu &= (-1)^{(\nu-1)}\left[(\mu_1\mu_2\ldots\mu_{(\nu-1)})(\mu_\nu + 1)\left(\frac{2y_1}{r_1}\right)\right]\\
&\quad\text{with the condition } \nu > 1\\[6pt]
\mu_\nu &= \frac{1}{r_\nu}\left[\left(\frac{2}{\mu_1\mu_2\ldots\mu_{(\nu-1)}}\right)\left(\frac{r_1}{2} - (\mu_1)d_1 - (\mu_1\mu_2)d_2\ldots\right.\right.\\
&\qquad\left.\left.-(\mu_1\mu_2\ldots\mu_{(\nu-1)})d_{(\nu-1)}\right) - r_\nu\right]\\
&\quad\text{with the condition } \nu > 1
\end{aligned}\right\} \quad (3.335)$$

In these formulae the following quantities are pre-defined:

$$\mu_0 = \mu_1 = +1, \quad d_0 = 0$$

Because of the singularity, apart from the case of y_1, the formulae break down for $\nu = 1$. The recursive nature requires $\mu_1 = +1$, but the specific case of $\nu = 1$ gives $\mu_1 = 0$, which also gives the correct result from the terms $(\mu_\nu - 1)$ and $(\mu_\nu + 1)$ for A_1 and $\Delta\left(\frac{u}{n}\right)_1$.

For a 4-mirror telescope working in the normal way with $s_1 = \infty$, the formulae (3.335) then give the following results, in which $\mu_1 = +1$ by definition:

$$y_1 \quad = y_1$$

$$y_2 \quad = y_1 \left[1 - \frac{2}{r_1} (\mu_1 d_1) \right]$$

$$y_3 \quad = y_1 \left[1 - \frac{2}{r_1} \left(\mu_1 d_1 + (\mu_1 \mu_2) d_2 \right) \right]$$

$$y_4 \quad = y_1 \left[1 - \frac{2}{r_1} \left(\mu_1 d_1 + (\mu_1 \mu_2) d_2 + (\mu_1 \mu_2 \mu_3) d_3 \right) \right]$$

$$A_1 \quad = +\frac{y_1}{r_1}$$

$$A_2 \quad = -\mu_1 (\mu_2 - 1) \left(\frac{y_1}{r_1} \right)$$

$$A_3 \quad = +\mu_1 \mu_2 (\mu_3 - 1) \left(\frac{y_1}{r_1} \right)$$

$$A_4 \quad = -\mu_1 \mu_2 \mu_3 (\mu_4 - 1) \left(\frac{y_1}{r_1} \right)$$

$$\Delta \left(\tfrac{u}{n} \right)_1 \quad = +\frac{2y_1}{r_1}$$

$$\Delta \left(\tfrac{u}{n} \right)_2 \quad = -\mu_1 (\mu_2 + 1) \left(\frac{2y_1}{r_1} \right)$$

$$\Delta \left(\tfrac{u}{n} \right)_3 \quad = +\mu_1 \mu_2 (\mu_3 + 1) \left(\frac{2y_1}{r_1} \right)$$

$$\Delta \left(\tfrac{u}{n} \right)_4 \quad = -\mu_1 \mu_2 \mu_3 (\mu_4 + 1) \left(\frac{2y_1}{r_1} \right)$$

$$\mu_1 \quad = +1 \text{ by definition}$$

$$\mu_2 \quad = \frac{1}{r_2} \left[\left(\frac{2}{\mu_1} \right) \left(\frac{r_1}{2} - (\mu_1) d_1 \right) - r_2 \right]$$

$$\mu_3 \quad = \frac{1}{r_3} \left[\left(\frac{2}{\mu_1 \mu_2} \right) \left(\frac{r_1}{2} - (\mu_1) d_1 - (\mu_1 \mu_2) d_2 \right) - r_3 \right]$$

$$\mu_4 \quad = \frac{1}{r_4} \left[\left(\frac{2}{\mu_1 \mu_2 \mu_3} \right) \left(\frac{r_1}{2} - (\mu_1) d_1 - (\mu_1 \mu_2) d_2 - (\mu_1 \mu_2 \mu_3) d_3 \right) - r_4 \right]$$

$$(3.336)$$

For the normal case of $s_1 = \infty$, Eqs. (3.335) or (3.336) give the aperture parameters y_ν, A_ν and $\Delta \left(\frac{u}{n} \right)_\nu$ in Eqs. (3.20). From (3.333) or (3.334) we can calculate $y_{pr\nu}$ and $\overline{A}_\nu$. The parameter $(HE)_\nu$ in Eqs. (3.20) is given by

$$(HE)_\nu = (y_{pr}/y)_\nu \,,$$

as given in the definitions for (3.20). The remaining parameter is the aspheric parameter τ_ν given in the definitions for (3.20) as

$$T_\nu = +c_\nu^3(n' - n)b_{s\nu}y_\nu^4 = \frac{1}{r_\nu^3}(n' - n)b_{s\nu}y_\nu^4$$

For our purely catoptric system with reflexions only, this becomes

$$T_\nu = (-1)^\nu 2b_{s\nu}\left(\frac{y_\nu^4}{r_\nu^3}\right) \qquad (3.337)$$

Finally, we have for the field curvature

$$(P_C)_\nu = \frac{1}{r_\nu}\left(\frac{1}{n_\nu'} - \frac{1}{n_\nu}\right)$$

or

$$(P_C)_\nu = (-1)^\nu\left(\frac{1}{r_\nu}\right) \qquad (3.338)$$

for a purely catoptric system.

This completes the formulae enabling the calculation of third order aberrations for any system consisting only of mirrors in a quite general way, with or without normalization of aperture and field, without the need of numerical ray tracing. They provide a powerful tool for the analysis of any of the purely catoptric systems treated above and may be preferred by the informed reader, because of their generality, to the specific formulations for 2-mirror systems given earlier in this chapter.

An interesting analysis of the theory of 3-mirror telescopes is given by Robb [3.82]. Robb is concerned specifically with 3-mirror solutions like Korsch [3.73], and notes the complexity of the Korsch equations (3.317) et seq. He uses first order combinations of the constructional parameters to simplify the formulation of third order aberrations for the three mirror case. The same procedure is possible with the recursion formulae above if the parameters are inserted in Eqs. (3.20). But the simple recursion properties are then lost and the complexity increases rapidly for $\tilde{n}$ surfaces where $\tilde{n} > 2$.

3.6.5.3 Other 3-mirror and 4-mirror solutions. Robb [3.82] considers a number of 3-mirror designs and gives an optimized form for the geometry shown in Fig. 3.77, for an angular field of 2.30°. The geometry is similar to that of the Willstrop telescope of Fig. 3.74, except that the strict afocal feeder is replaced by a slightly convergent beam and an optimization performed by the technique referred to at the end of § 3.6.5.2 above. This optimization was performed balancing third, fifth and seventh orders, the primary having a relative aperture f/1.94 and the final value being f/5.0. The mirror forms are hyperbolic to the third order, as in the Korsch system of Table 3.18 and all such flat-field, 3-mirror solutions.

Robb also draws attention to 3-mirror designs by Shack and Meinel [3.83] and Rumsey [3.84] [3.85]. The latter design is similar to the Loveday telescope, treated above, in that the primary and tertiary mirrors are combined; but with the important difference from Loveday that the central part for the tertiary is figured differently according to its specific requirements. The

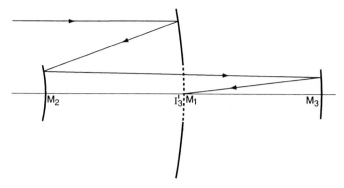

Fig. 3.77. 3-mirror system proposed by Robb (1978)

Rumsey solution is therefore nearer to a Paul-Baker telescope, but with the position of the tertiary linked to that of the primary.

What, then, is the optimum geometry of a 3-mirror telescope? This will depend upon several parameters, particularly the angular field, the acceptable obstruction ratio and the baffling solution. An attractive form has been optimized by Laux [3.8(b)], combining aspects of the Paul-Baker system (Fig. 3.73) with those of the Korsch single-axis system (Fig. 3.72 (a)). It is shown in Fig. 3.78. The final image is behind the secondary, as in the Korsch system, giving a decisive advantage for baffling. However, M_3 is virtually in the plane of the primary, following Baker or Rumsey. This is probably the most convenient constructional position and is more compact than the Willstrop or Robb geometries. However, a price must be paid with a relatively high *obstruction ratio* if a fast system (f/4) and large field of about 2.5° diameter are aimed for. The image quality over a flat field is excellent, within 0.2 arcsec for a field diameter of 2°; but the linear obstruction ratio is about 0.5. This system has been proposed for a 2.5m wide-field survey telescope, the LITE project [3.86], using a large CCD-array detector.

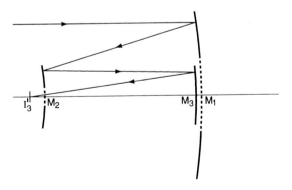

Fig. 3.78. 3-mirror system given by Laux (1993) for a fast, flat-field 2.5m wide-field survey telescope with f/2.18 primary and f/4.0 final image, with a field diameter of 2.0° to 2.5°

Korsch [3.87] has recently proposed a 4-reflection solution using 3-mirrors, the secondary in double pass, in connection with possible forms for a Next Generation Large Space Telescope. Korsch proposes two possible solutions with the same 4-reflection geometry shown in Fig. 3.79. The initial Cassegrain telescope forms an intermediate image between the secondary and tertiary mirrors, the tertiary being placed in the same plane as the primary as in the Rumsey form, but forms a separate mirror in the primary central hole. Korsch has pointed out that a 4-reflection solution is more favourable for the final image position and baffling than a normal 3-mirror solution [3.88]. In the best solution using an f/1.25 primary all three mirrors are aspheric [3.87]. The primary is then ellipsoidal and the system gives a maximum residual aberration of 10 nrad (effectively diffraction limited for 16 m aperture at $\lambda = 150$ nm) over a field of 0.5° diameter. If the primary is made spherical, this quality is limited to a field of 1 arcmin diameter.

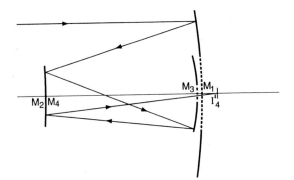

Fig. 3.79. 3-mirror, 4-reflection telescope proposed by Korsch (1991) for a future large space telescope

A further theoretical possibility on these lines would be a straight double pass Cassegrain, an extension of the Loveday telescope to 4 reflections with two mirrors. To produce a real image, the first reflection at the secondary must produce a slightly convergent beam. As the telephoto effect of the single-pass Cassegrain is greatly increased, the emergent beam will have a very slow f/no, since slight convergence implies a high telephoto effect. Furthermore, the third and fourth reflections can give no practical contribution to aberration correction, as is also the case in the Loveday telescope with the third reflection at the primary. So a double-pass Cassegrain normally has no practical interest.

The statement by Korsch [3.88] that the geometry of a telescope with 4 reflections is fundamentally more favourable (above all because of the final image position) than with 3 reflections is a profound truism. An *even* number of reflections places the image at the end of the system from the point of view of the incident light. Apart from the telephoto advantage, this was a major advantage of the Cassegrain over the Newton or prime focus forms. One-reflection or 3-reflection solutions will only normally be interesting for direct imagery, where the detector is compact and not too heavy; for spectroscopy or general purpose instruments they will rarely be suitable.

In § 3.2.6.1, we referred to the *generalised Schwarzschild theorem* [3.13]. This states that, for any geometry with reasonable separations between the elements, n Seidel monochromatic conditions can be fulfilled by n powered mirrors or plates, in the general case with aspheric forms. We have seen (for example, the Mersenne afocal 2-mirror telescope or the Schmidt telescope) that a favourable natural geometry can satisfy n conditions with fewer than n aspheric elements. The generalised Schwarzschild theorem tells us that a system of 4 powered mirrors with 4 aspherics can anyway satisfy the four conditions $S_I = S_{II} = S_{III} = S_{IV} = 0$. However, the Petzval condition S_{IV} requires only the sum of the powers to be zero: if the geometry is predetermined to fulfil this, three aspherics are sufficient to correct the other three conditions, as in the systems of Paul-Baker (Fig. 3.42), Korsch (Fig. 3.72 (b)) or Laux (Fig. 3.78). It follows that, if the geometry is fixed to give a zero or acceptable field curvature, one of the mirrors in a telescope with 4 powered mirrors can be spherical, the other three correcting the first three Seidel conditions. Such design principles are well known and an increasing number of designs with 4 powered mirrors (or 4 reflections) have been published over the last 20 years or so. However, hardly any have been realised, above all in larger diameters. The reason has been partly the *obstruction problem*, which limits the reasonable geometries available for 4 mirrors on a single axis, but, above all, the *reflectivity problem*. The most backward aspect of reflecting telescope technology today is the inadequate reflectivity of the simple unprotected aluminium coat, still the standard solution for *large* mirrors (see RTO II, Chap. 6). The notable advance of evaporating aluminium was perfected by Strong in 1933 [3.89], which means there has been no advance for large optics in over 60 years! In view of the immense technological advances in all other areas, this is a surprising and unbalanced situation. However, there is clear evidence that technical solutions giving enhanced reflectivity R are known [3.90]. Investment in their practical application is still required, but it seems clear that various forms of protected (or enhanced) silver coats seem very promising. It seems a reasonable expectation that durable coats, maintained by modern cleaning routines (RTO II, Chap. 6), will become available within the next decade with $R \geq 0.95$ instead of $R \simeq 0.80$ for simple aluminium, even for very large mirrors. The new "Optikzentrum" in Bochum, Germany, is giving this technology considerable weight in its research and development program [3.91]. If $R = 0.95$ can be achieved, the loss from 4 reflections will only be 19%, much less than in a currently normal Cassegrain telescope with $R = 0.80$. In view of their excellent optical design potential, telescopes with 4 powered mirrors could then be extremely attractive, provided the geometry is otherwise favourable.

The geometry of 4-mirror telescopes is best approached using the elegant properties of some of the 3-mirror solutions due to Paul, Baker, Willstrop and Korsch, discussed in §3.6.5.1 above. The most fundamental are those of Paul (Fig. 3.73) and Willstrop (Fig. 3.74), exploiting the properties of the Mersenne afocal telescope and the Schmidt principle. If we consider extending the Willstrop concept to 4 mirrors, with an additional powered mirror, then, from the generalised Schwarzschild theorem, we can make one mirror spherical. Logically, this will be the *primary*, the largest and most expensive mirror in the system. Although the aspheric (parabolic) primary of the 10m Keck telescope is a notable success in the manufacture and active control of aspheric segments, it is generally agreed that a *spherical* primary of large size and steep modern form for a compact construction has significant technical and, above all, cost advantages, whether it be segmented or monolithic [3.13] [3.88] [3.92] [3.93] [3.94] [3.95] [3.96] [3.97]. Now, the Willstrop telescope used a classical Mersenne feeder telescope with confocal paraboloids, anastigmatic according to the equations of Table 3.6. These equations show that $\sum S_{II}$ and $\sum S_{III}$ remain zero, even if $\sum S_I \neq 0$, provided that (initially) the entrance pupil is at the primary, i.e. $s_{pr1} = 0$. This is simply a formal proof of the fact that the aspheric form of an optical element has no effect on S_{II} and S_{III} if it is placed at the pupil. The Mersenne afocal feeder, equipped with a spherical primary and a secondary of parabolic form, then delivers a beam with field curvature and enormous spherical aberration from a steep primary, but no third order coma or astigmatism. The large spherical aberration S_I must be corrected by the remaining two mirrors of the system, without spoiling the field correction. This is clearly most effectively achieved if the strongly aspheric mirror producing this correction of S_I is *at the transferred pupil*. If we follow the Paul-Willstrop concept, the mirror M_3 will be a Schmidt-type sphere centered on the exit pupil (behind M_2) of the feeder telescope. This re-images the pupil back on itself, so that the transferred pupil is not accessible for a fourth mirror M_4. This is an important conclusion: *it is impossible for a single-axis, 4-mirror system to respect the Schmidt geometry of the Paul-Willstrop concept.* Of course, other solutions with the three aspherics M_2, M_3 and M_4 following the Schwarzschild theorem are possible (see below), also with a pupil transfer to M_4, but not with Paul-Willstrop geometry. However, this geometry can be fully maintained, together with perfect pupil transfer, if a *2-axis* solution is used [3.13] [3.94] [3.95] with a small flat mirror at the intermediate image I_3' as shown in Fig. 3.80. The radius of curvature of M_3, concentric to the exit pupil P_2, is chosen such that the image I_3', with $P_2 I_3' = I_3' M_3 = I_3' P_3$, is placed at a convenient axial position for the altitude axis of an alt-az telescope and that M_4 and P_3 are conveniently placed outside the incident beam and with respect to the bearing of the alt-axis. The function of M_F is simply to image the pupil to an accessible position, so that the high (hyperbolic) asphericity of M_4 has no effect on the field aberrations. M_4 then images I_3' with a magnification of about 2 to the normal "Nasmyth-type" focus I_4' on the other side. The optical power of M_4 will introduce modest field aberrations because its imagery is free from neither

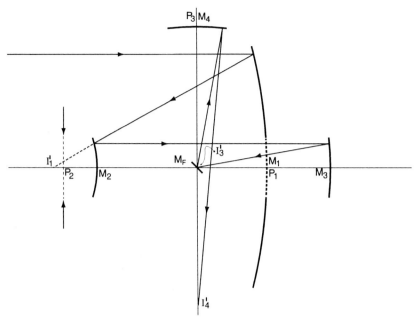

Fig. 3.80. *First solution* of a 2-axis system with 4 powered mirrors (spherical primary and secondary) and a folding flat (f/1.5 and f/7.29), proposed by Wilson and Delabre (1993, 1995)

coma nor astigmatism. But these relatively small aberrations can easily be corrected by lightly modifying the forms of M_2, M_3, M_4 and the separations.

This extension of the Willstrop telescope, with a spherical primary and the other three mirrors M_2, M_3, M_4 to a third order roughly parabolic, roughly spherical and hyperbolic respectively, gives such excellent imagery at about f/7 with an f/1.5 primary and the necessary higher order correction terms on account of the steep primary, that the field correction cannot be exploited for a large telescope with foreseeable detectors. It can be seen as a generic type of 4-mirror telescope with natural, relaxed geometry, giving a particularly well-conditioned solution matrix. In fact, the solution is so relaxed that the secondary M_2 *can also be made spherical*, still giving excellent imagery with aspherics on M_3 and M_4 only. In the basic solution with an f/1.5 spherical primary (*first, 2-axis solution*), the Schwarzschild constants are $b_{s3} = -0.951$ and $b_{s4} = -11.116$. The data of this system are given in Table 3.19 and the spot-diagrams in Fig. 3.81. We see that the spot-diagrams have $d_{80} \leq 0.1$ arcsec for a field of $\pm$ 9 arcmin and $d_{80} \leq 0.5$ arcsec for $\pm$ 18 arcmin.

A further advantage of the 2-axis geometry is that the obstruction ratio is much more favourable than with a single axis. Table 3.19 is laid out for a linear obstruction ratio of about 0.25, both from M_2 and M_F, giving some vignetting by M_F beyond $\pm$ 9 arcmin. But even for $\pm$ 18 arcmin, the ob-

Table 3.19. Optical design data of the first, 2-axis solution with 4 powered mirrors and flat of Fig. 3.80 with primary f/1.5 and final image f/7.29

Mirror	Radius (mm)	Separation (mm)	Free diameter (mm) (paraxial) (± 9 arcmin field)
1	− 48 000	− 18 000	16 000
2	− 12 000	+ 21 000	3 928
3	− 25 000*	− 11 000	3 786
Flat	∞	+ 12 214.20	723 × 512
4	− 15 000*	− 25 000.52	3 454
		(to image)	
Image	− 19 200		611

EFL = 116 670 mm (f/7.292)
* Aspherics

Mirror	$1/r = c$	b_s	A_s
3	− 0.000 040 00	0	+ 0.7290E–14
4	− 0.000 066 67	0	+ 0.4093E–12

Mirror	B_s	C_s	D_s
3	+ 0.2770E–21	− 0.1810E–28	+ 0.1710E–34
4	+ 0.3400E–20	− 0.4380E–29	+ 0.4360E–35

$$z = \frac{cy^2}{1 + \left[1 - (1 + b_s)c^2 y^2\right]^{1/2}} + A_s y^4 + B_s y^6 + C_s y^8 + D_s y^{10}$$

struction is only about 0.32 without vignetting. Even for a very large (16m) telescope, the size of M_F remains small enough that a multi-dielectric coating, giving very high reflectivity, should be possible. Part of its dimensional requirements arises from the high spherical aberration, but this has the advantage that it is much less sensitive to dust.

The fact that M_3 has become almost parabolic in form ($b_{s3} = -0.951$) shows that the parabola has effectively been shifted from M_2 to M_3 by making M_2 spherical. Its spherical form compensates about a quarter of the spherical aberration produced by the primary. The loss of compensation of coma and astigmatism is easily rectified by the aspheric forms of M_3 and M_4 with appropriate small axial shifts. The coma is above all compensated by M_4 while M_3 produces a major compensation of astigmatism because it is far from the pupil.

For the initial set-up of the geometry, the entrance pupil was placed at the primary. However, since the final system has effectively $S_I = S_{II} = S_{III} = 0$, the stop position of the complete system is uncritical and can be placed as desired. The field curvature S_{IV} is small, but not zero. M_2 is too weak to

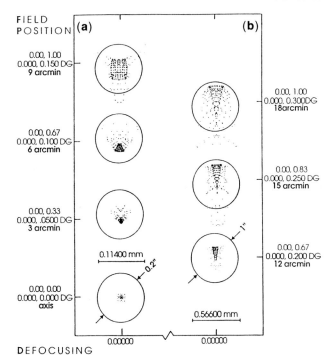

FIELD POSITION

(a)

0.00, 1.00
0.000, 0.150 DG
9 arcmin

0.00, 0.67
0.000, 0.100 DG
6 arcmin

0.00, 0.33
0.000, .0500 DG
3 arcmin

0.00, 0.00
0.000, 0.000 DG
axis

0.11400 mm

0.2"

0.00000

(b)

0.00, 1.00
0.000, 0.300DG
18arcmin

0.00, 0.83
0.000, 0.250 DG
15 arcmin

0.00, 0.67
0.000, 0.200 DG
12 arcmin

1"

0.56600 mm

0.00000

DEFOCUSING

Fig. 3.81. Spot-diagrams of the first, 2-axis solution of Table 3.19 and Fig. 3.80: (**a**) axis to ± 9 arcmin with circle 0.20 arcsec; (**b**) ± 12 arcmin to ± 18 arcmin with circle 1.00 arcsec

compensate the three concave mirrors, so there is a slight undercorrection in the Schmidt sense. This is (to some extent) favourable for matching the field curvature of instruments. It could be compensated by CCD arrays. In principle, a flat field is possible by weakening M_3 and M_4: but M_3 moves further behind M_1 and M_4 is further from the axis, giving less compact geometry.

Figure 3.82 shows a symmetrical version with two identical mirrors M_4 and M_4' fed by appropriate rotation of M_F, giving two "Nasmyth foci". Each mirror M_4 and M_4' baffles with its central hole the final image produced by the other. This 2-axis arrangement is perfectly adapted to active optics, because M_4 and M_4' are at the pupil and are in a fixed (vertical) plane, unaffected by telescope movements in the alt-az mount. Thus, a push-pull system, independent of cosine effects of the zenith angle, can be applied for the active correction.

The centering tolerances are no more severe than for conventional telescopes with a similar M_1 and M_2. Active centering can either be done at M_2, as with the NTT [3.98] (see RTO II, Chap. 3), or by a rotation of M_F, which produces coma by shearing the pupil at M_4. The pointing shift can

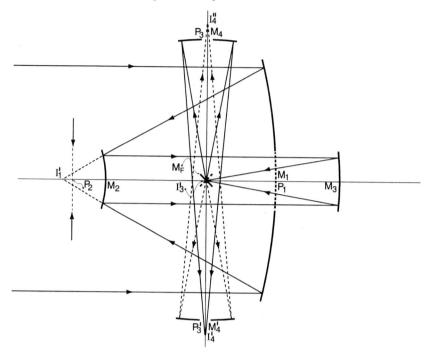

Fig. 3.82. First solution, 2-axis system as in Fig. 3.80, but with two identical "Nasmyth-type" foci

be corrected by rotating M_4 about its pole. Korsch [3.87] has pointed out the advantage of active centering by actuating 2 mirrors in his 4-reflection system of Fig. 3.79.

Analogue to the above *first, 2-axis solution*, there is a *second, 2-axis solution* which also uses 4 powered mirrors and a 45° flat. This is an extension of the Korsch 3-mirror solution of Fig. 3.72 (b) in the same way that the first solution is an extension of the Willstrop telescope of Fig. 3.74. Instead of an afocal feeder, a real image is formed after M_2 at the alt-axis (Fig. 3.83). M_3, again concentric to the pupil P_2, is now on the second axis, thereby producing a collimated beam and a real, accessible pupil image P_3 at which M_4 is placed. The final image is formed behind M_3, which acts as a baffle as in Fig. 3.82. The geometrical properties of pupil transfer are thus very similar to those of the first solution.

The optical design data are given in ref. [3.94]. The optical quality is similar to that of the first solution and is shown in Fig. 3.84.

Since the correction potential is similar, the advantages and disadvantages of the first and second solutions depend on the geometry. The *second solution* has the advantage that the alt-axis can be located at will without regard to the projection of M_3 beyond M_1 inevitable in the first solution. However, if M_1 is very steep, the beam aperture angle from M_2 blows up the diameter

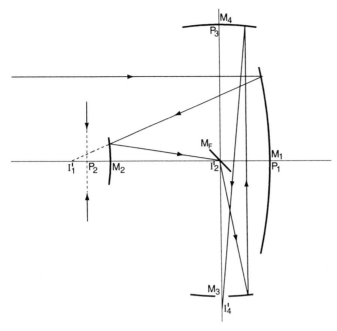

Fig. 3.83. *Second,* 2-axis solution with 4 powered mirrors (spherical primary and secondary) and a folding flat (f/1.5 and f/6.01), proposed by Wilson and Delabre (1993, 1995)

of M_3 and M_4 unless M_2 is made small with a larger magnification m_2. But this increases the obstruction due to the field at M_F. By contrast, the *first solution* fixes the diameters of M_3 and M_4 to be effectively the same as that of M_2. These aspects are treated in detail in ref. [3.94]. The broad conclusion is that the second solution is probably better if the primary is not too steep, say f/2.0 to f/1.5. For steeper primaries of about f/1.3 to f/1.0, the geometry of the first solution becomes more and more favourable and compact. Both solutions can still give excellent field performance with f/1.2 primaries, probably even f/1.0. The second solution has the disadvantage that the "double Nasmyth" focus of Fig. 3.82 is not possible.

It is now clear that there are essentially 3 ways of establishing a second axis:

a) The Baker 3-mirror telescope of Fig. 3.75, placing M_F after *one* reflection;
b) The Korsch 3-mirror telescope of Fig. 3.72 (b) or the 4-mirror telescope of the second solution above (Fig. 3.83), placing M_F after *two* reflections;
c) The first solution above (Fig. 3.80) for a 4-mirror telescope, placing M_F after *three* reflections.

There are no other practical possibilities for mirror telescopes, although the idea of a second axis can be realised in many other forms and goes back to

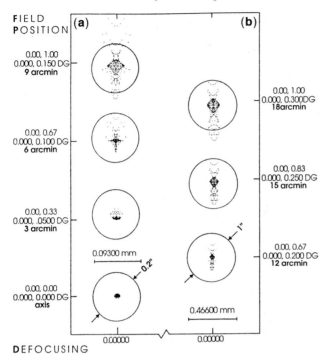

Fig. 3.84. Spot-diagrams of the second, 2-axis solution of Fig. 3.83: (**a**) axis to ±9 arcmin with circle 0.2 arcsec (**b**) ± 12 arcmin to ± 18 arcmin with circle 1.00 arcsec

the brachymedial and medial forms of Schupmann [3.69]. The Baker form of Fig. 3.75 is largely of historic interest because steep, modern primaries would lead to mirrors M_2 and M_3 which are too large relative to the primary to be practicable. However, the basic Baker concept of the concave mirror pair as a corrector system can be used in single-axis systems in the Korsch geometry, as shown in Fig. 3.85, a design by Delabre [3.94]. This design is seen as a fast, wide-field telescope for extending Schmidt-type operations up to sizes of 4 m or more. The image quality has $d_{80} \leq 0.5$ arcsec over the whole field of ±0.75°, as shown in Fig. 3.86. The limitations of this system are the diameter of M_3 and the strong field curvature (undercorrected). The transferred pupil is near M_4 and the generalised Schwarzschild theorem enables the wide-field correction with the f/1.2 spherical primary.

The system of Fig. 3.85 has been widely used in designs given in the literature. Already in 1979, Robb [3.99] clearly recognized the advantages of such a system with a spherical primary and a transferred pupil at the fourth mirror to correct the spherical aberration. His design, working at f/10 with an f/2.0 primary, covered a field of 1.0° diameter with diffraction limited quality for a 4 m aperture. Notable are the more recent TEMOS proposals

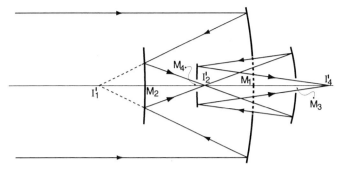

Fig. 3.85. Single-axis, 4-mirror system with f/1.2 - f/2.657 giving a field diameter of 1.50°. The primary is spherical

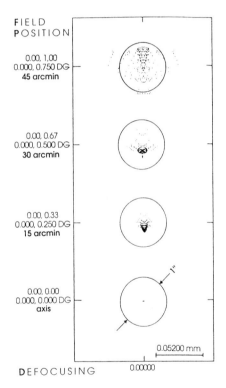

Fig. 3.86. Spot-diagrams for the fast, wide-field, 4-mirror design of Fig. 3.85. The circle diameter is 1 arcsec

by Baranne and Lemaître [3.96] [3.97] and the proposals by Ardeberg et al. [3.92] [3.93] for a 25 m telescope with a very steep spherical primary (f/0.8),the latter at present (early 1994) with a field limited to about 3 arcmin diameter[8]. A segmented primary and segmented M_4 are envisaged.

---

[8] This field has subsequently been increased, mainly by increasing the relative size of the corrector mirrors.

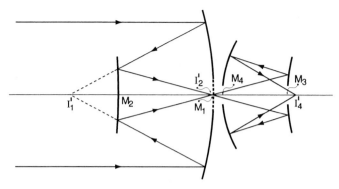

Fig. 3.87. Single-axis, 4-mirror concept for a fast, wide-field telescope with improved field curvature

A 4-mirror system which might eliminate or significantly reduce the field curvature [3.94] is shown in Fig. 3.87. The Baker concave mirror pair is replaced by a Schwarzschild-Bowen pair, as used in spectrographs. Similar proposals have been made by Baranne and Lemaître [3.96] [3.97], but the potential of this system does not appear to have been analysed in detail. The pupil transfer situation is certainly less favourable than Baker-type form of Fig. 3.85 because the convex M_3 forms a virtual image of the pupil far from M_4.

Similarly, a single-axis solution is possible by analogy with the first, 2-axis solution above, using an afocal feeder telescope (Fig. 3.88). As in the system of Fig. 3.85, it is possible to image the pupil more or less on to M_4. But the final f/no is always a tied function of the diameter of M_2: the smaller M_2 becomes, the larger the final f/no. This is the same law as that governing the 2-axis solutions of Figs. 3.80 and 3.83. The steeper the spherical primary, the more important correct pupil transfer to M_4 will be if this mirror is correcting the bulk of the spherical aberration.

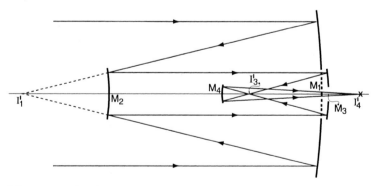

Fig. 3.88. Single-axis, 4-mirror system using an afocal feeder and a spherical primary

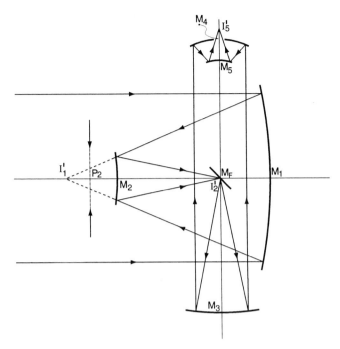

Fig. 3.89. A 2-axis system with 5 powered mirrors capable of a fast output beam (faster than f/3.0) and a flat field. The primary and secondary mirrors are spherical as in Fig. 3.83

In spite of the analogy of Fig. 3.88 to the first, 2-axis solution of Fig. 3.80, and of Fig. 3.85 to the second, 2-axis solution of Fig. 3.83, it is important to realise that the 2-axis solutions are fundamentally more relaxed and natural because of the Schmidt characteristics of M_3. These characteristics are *impossible* in the single-axis solutions, although the pupil transfer to M_4 can still be achieved. In the single-axis cases, therefore, *three* aspheric mirrors will be necessary following the generalised Schwarzschild theorem, whereas, as we saw above, the two-axis solutions can give good field correction with only 2 aspherics. This important difference will always mean that the aspheric forms in single-axis solutions are more complex and more extreme.

The two basic 2-axis solutions given above will lead, with normal (i.e. not excessively small) obstruction by M_2, to final f/ratios of the order of f/6-f/7, as we have seen. Their geometry precludes fast f/nos. Suppose, however, M_4 in Fig. 3.83 is withdrawn somewhat from the primary beam, then it could be made steeper and the image removed sideways with a Newton flat. But this final image position is no longer on the alt-axis and has an inconvenient position. Furthermore, a steeper concave M_4 worsens the field curvature. If, therefore, a further mirror is to be introduced, it seems better to give it a more constructive role than that of a Newton flat. Figure 3.89 shows a 2-axis system with five powered mirrors where $M_4 M_5$ form a Cassegrain telescope. If M_4 is made very steep ($\sim$f/1.0), then high relative apertures (equivalent to Schmidts) are possible with modest obstruction; furthermore the field curvature can be corrected. A lightly modified version of this system has been set up [3.94], giving at f/2.805 spot-diagrams with $d_{80} \leq 0.2$ arcsec over 30 arcmin field diameter and $d_{80} \leq 0.7$ arcsec over 60 arcmin diameter. The field curvature has a small overcorrection residual from the convex M_5, but it would probably be possible to eliminate this. The primary and secondary mirrors are spherical, as in Fig. 3.83, and the pupil is transferred to a plane between M_4 and M_5.

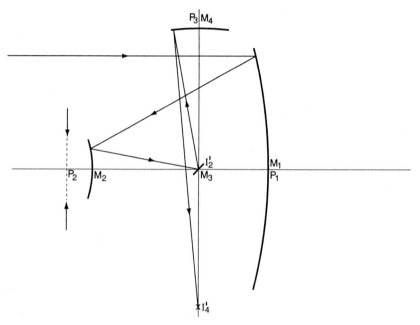

Fig. 3.90. A 2-axis solution with 4 powered mirrors proposed by Sasian (1990). Either M_1 or M_2 is spherical, M_3 is toroidal

A very interesting 2-axis solution (Fig. 3.90) with 4 powered mirrors and correct pupil transfer to the final mirror has been proposed by Sasian [3.100]. He recognizes the importance of active correction by an element at the pupil, in his case the last mirror of the system as in the 2-axis concept of Fig. 3.80. Sasian's system resembles the Korsch geometry of Fig. 3.72 (b), but the plane mirror M_F in that system is replaced by a powered, concave mirror M_3. Because it is inclined at 45° to form the second axis (Sasian terms this a "bilateral" configuration), M_3 has to have a toroidal form. It images the pupil P_2 on to the final mirror M_4. Because of the pupil transfer, Sasian obtains interesting solutions with only 2 aspherics: the toroidal M_3 is not aspheric and either M_1 or M_2 is spherical. The d_{80}-values of the spot-diagrams remain well within 1 arcsec for fields of 6 arcmin diameter, the case with the spherical secondary being somewhat better. This solution is to be compared with the 2-axis system of Fig. 3.83, which gives markedly better imagery with both primary and secondary spherical because the toroidal mirror for transferring the pupil is replaced by a flat. But Sasian's system has one mirror less, albeit at the cost of manufacturing an accurate toroidal surface. He saw his system as of particular interest with a spherical primary, because of the simplification of manufacture.

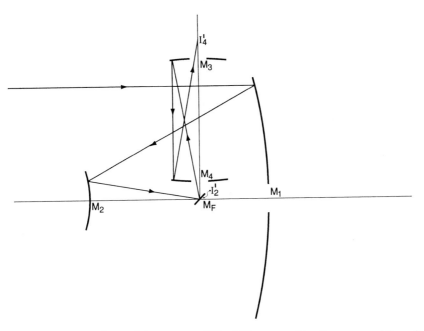

Fig. 3.91. 2-axis form of the system of Fig. 3.85 proposed by Baranne and Lemaître (1986), the mirror pair $M_3 M_4$ forming a corrector and focal transfer system with a magnification of -1 in the TEMOS concept, giving f/2.0 - f/4.5 - f/4.5

The pupil transfer to the fourth mirror is fundamental to successful field correction if a steep primary of spherical form is used. An interesting 2-axis variant of the single-axis system of Fig. 3.85, also permitting (in principle – it is not clear whether it is the case here) such pupil transfer to M_4, is given by Baranne and Lemaître as one of the corrector proposals for the TEMOS telescope [3.96] [3.97]. This is shown in Fig. 3.91. In this concept, the form of M_2 can be varied actively according to the requirements of the various corrector systems. For the corrector shown, M_1 is spherical and M_2, M_3, M_4 aspheric. The spot-diagrams are within 0.4 arcsec over a field of 20 arcmin diameter. M_4 is placed within the central obstruction and this geometry cannot permit a Schmidt-type pupil transfer by M_3, as given by the geometry of Fig. 3.83. The latter permits a better image quality with only two aspherics, the secondary being spherical. Since M_4 is symmetrically placed on the opposite side of the axis from M_3, it follows that the final focal length will be doubled in comparison with Fig. 3.91.

If imagery of the pupil on M_4 is abandoned, many other variants of 4-mirror telescopes are possible. If the primary is not spherical, but parabolic or some similar form, the pupil transfer is not important. A system with favourable geometry is the *double Cassegrain* (Fig. 3.92) proposed by Korsch [3.88]. This has 4 independent mirrors, in distinction to the (normally impracticable) *double-pass Cassegrain* referred to above, which has 4 reflections from 2 mirrors. Since pupil transfer to M_4 cannot be achieved, the field performance is vastly better with a primary of parabolic form. With a spherical primary of f/1.7, superb performance (50 nrad) for the rms blur diameter is possible over only 1 arcmin field diameter, whereas a larger field with this quality is possible with an f/0.6 primary of parabolic form.

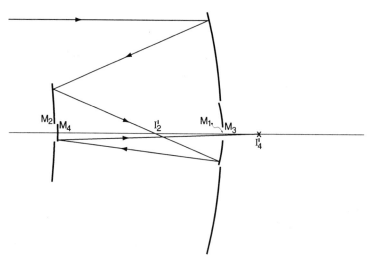

Fig. 3.92. Double-Cassegrain 4-mirror telescope with intermediate image after M_2, proposed by Korsch (1986)

Other designs using fast spherical primaries, above all in an Arecibo (fixed primary) concept, have been proposed by Schafer [3.101] and Shectman [3.102].

In summary, it appears clear that designs with 4 powered mirrors, either as 2-axis or single-axis solutions, have great promise in the further development of the astronomical telescope. They permit excellent field correction for both conventional and fast f/nos and are, unlike designs involving aspheric plates in the pupil, no more restricted in aperture than classical Cassegrain or

RC telescopes. Furthermore, the image position and baffling characteristics are just as favourable, the application of active optics even more so. Of particular interest are designs with a spherical primary, a spherical secondary as well being a further bonus for the basic 2-axis designs given above. The more general application of such designs requires improvement in the reflectivity of the mirrors: protected silver seems to offer the greatest promise for large optics, multi-coating for smaller mirrors.

Single-axis solutions of the types shown in Figs. 3.85, 3.87, 3.88, 3.91 and 3.92 will all have increasing problems of higher order aberrations as the corrector system formed by M_3 and M_4 is made smaller or more compact (steeper curvatures on these mirrors).

3.7 Off-axis (Schiefspiegler) and decentered telescopes

3.7.1 Two- and three-mirror Schiefspiegler

Up till now we have been concerned solely with *centered* optical systems in which the elements with optical power all have a common axis or, in the case of spherical surfaces, are normal to the common axis. Of course, centered optical systems do not exclude deflections of the beam by *plane* mirrors which have no optical power and hence no optical axis. The restriction to a unique axis of symmetry leads to the Hamilton Characteristic Function dealt with in §3.2.1 and the aberration types of Table 3.1.

If powered optical elements are tilted in one plane, which we will define as the tangential plane normally given by the plane of the paper on which the system section is represented, then symmetry to a line in space, the common optical axis, is abandoned. There is, however, still symmetry about the tangential plane. If the elements are tilted in two dimensions, then this symmetry is also lost. All those aberration terms which were eliminated by symmetry about the optical axis in Hamilton theory are present in the general case, so the theory becomes much more complex.

The first use of a tilted element in telescope design was, in fact, the Herschel type "front-view" single mirror reflector of Fig. 1.1. This was afflicted by the field aberrations of the paraboloid as the stop of the system. For the Herschel mirrors working at f/10 or longer, the limiting field coma was small and offset by the gain in light gathering power through avoidance of a second reflection at a speculum mirror with reflectivity of about 60% at best.

According to Riekher [3.39(g)], the first Cassegrain-type telescope with tilted components was the Brachy telescope of Forster and Fritsch in 1876. Both mirrors were spherical, requiring low relative apertures to give adequate correction of spherical aberration and a correspondingly long construction.

The first systematic investigation of the possibilities of a telescope with two tilted mirrors was published in 1953 by Kutter [3.103], followed by a second book in 1964 [3.104]. He introduced the German term "Schiefspiegler" (oblique mirror) which has since become the generic name for this type of telescope. Kutter addressed himself particularly to amateurs, his specific purpose being the removal of the central obstruction in the Cassegrain telescope. It will be shown in § 3.10, dealing with physical optical aspects, that the central obstruction significantly reduces the contrast of imagery of surface detail in amateur size telescopes. Above all for observation of the moon or major planets this is a disadvantage for amateurs.

A detailed description of the Kutter Schiefspiegler and derivatives is given by Rutten and van Venrooij [3.12(j)]. Kutter also started with spherical mirrors like Forster and Fritsch. He set up the three solutions shown in Fig. 3.93. In solution I, both coma and astigmatism are present in the "central" ("axial") image point; in solution II, astigmatism is corrected but not coma; in solution III, coma is corrected but not astigmatism. As with Forster and Fritsch, the use of spherical mirrors imposed low relative aperture mirrors. This, in turn, imposed a low secondary magnification m_2 of about 1.7; but even then the final f/no was f/20-f/30, values normally only adapted to visual use with small telescopes. Kutter proposed a secondary with the same radius as the primary, giving simplified manufacture and testing as well as a flat field.

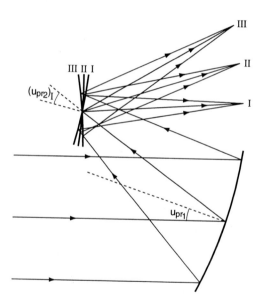

Fig. 3.93. The Kutter Schiefspiegler [3.103] [3.104] showing 3 solutions (after Rutten and van Venrooij [3.12(j)])

Although the general theory for the aberrations of a finite field is highly complex, this is not so for the imagery of the "axial" image point shown at I, II, III in Fig. 3.93. For this point, the field is effectively zero and both reflections can be considered as if the stop were at the mirror concerned. A simple and instructive example is the classical Czerny-Turner arrangement

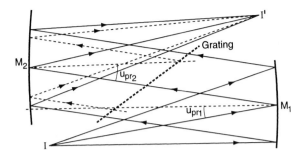

Fig. 3.94. The basis of coma compensation in a Czerny-Turner mono-chromator

for monochromators or spectrographs [3.105], the theory of which is also discussed by Schroeder [3.22(c)]. The basis of the Czerny-Turner system is a pair of concave spherical mirrors arranged in so-called Z-form and tilted in such a way that the coma is zero for the "axial" image point I' (Fig. 3.94). Normally a grating is placed in the parallel beam which, because of the asymmetry of the incident and diffracted angles of the beams, produces a compression or expansion of the beam incident on M_2. In Fig. 3.94 a beam compression is shown. We can apply the Eqs. (3.334) and (3.336) to derive the coma S_{II} as though we were dealing with two primaries with a parallel incident beam. From (3.20)

$$S_{II} = -y A \overline{A} \Delta \left(\frac{u}{n} \right) ,$$

giving from (3.334) and (3.336)

$$\sum S_{II} = -2 \frac{y_1^3 u_{pr1}}{r_1^2} + 2 \frac{y_2^3 u_{pr2}}{r_2^2} = 0 , \qquad (3.339)$$

for coma compensation. The Czerny-Turner condition is therefore

$$\frac{u_{pr2}}{u_{pr1}} = \left(\frac{r_2}{r_1} \right)^2 \bigg/ \left(\frac{y_2}{y_1} \right)^3 , \qquad (3.340)$$

in which (y_2/y_1) is the beam compression ratio and u_{pr1} and u_{pr2} are the tilt angles in Z-form.

An important point to note with all such tilted systems is that the above aberration calculations for the "axial" image point are performed as though the stop were at *each surface* of the system: stop shift terms only have meaning for the field of the Schiefspiegler. It follows that the coma and astigmatism of the "axial" image point are *independent of the form of the mirrors*, whether spherical or aspheric: they are solely functions of the mirror radii, the apertures and the geometry of the arrangement. The Czerny-Turner arrangement is simpler than the Schiefspiegler because there is a collimated beam between the mirrors, so that both can be treated as "primaries". The "Schiefspiegler" then provides an elegant application of the general recursion aberration formulae (3.334) and (3.336). We shall consider two cases, the coma-free Schiefspiegler and the anastigmatic Schiefspiegler.

Coma-free Schiefspiegler: Using, from (3.20),

$$(S_{II})_\nu = -y_\nu A_\nu \overline{A}_\nu \Delta \left(\frac{u}{n}\right)_\nu \tag{3.341}$$

with the parameters derived from (3.336), we have

$$(S_{II})_1 = -2\frac{y_1^3}{r_1^2}u_{pr1} \tag{3.342}$$

for the contribution of M_1. For M_2, Eqs. (3.336) give

$$
\left.
\begin{aligned}
\mu_2 &= \frac{1}{r_2}(r_1 - 2d_1 - r_2) \\[2mm]
y_2 &= y_1\left(1 - \frac{2d_1}{r_1}\right) \\[2mm]
A_2 &= \frac{-y_1}{r_1}\left[\left(\frac{r_1 - 2d_1}{r_2}\right) - 2\right] \\[2mm]
\Delta\left(\frac{u}{n}\right)_2 &= \frac{-2y_1}{r_1}\left(\frac{r_1 - 2d_1}{r_2}\right)
\end{aligned}
\right\} \tag{3.343}
$$

Since, for the "axial" image in the Schiefspiegler, the pupil can be considered as being at the mirror for both M_1 and M_2, we calculate $\overline{A}_2$ in this case as though M_2 were a primary, using (3.334). Then

$$\overline{A}_2 = \overline{A}_1^* = -\frac{1}{2}(\mu_{pr1}^* - 1)u_{pr2} , \tag{3.344}$$

in which $\overline{A}_1^*$ and μ_{pr1}^* refer to a paraxial principal ray calculation for M_2 as though it were a primary. Also

$$\mu_{pr1}^* = \frac{1}{r_2}(2s_{pr1}^* - r_2) ,$$

in which s_{pr1}^* is the pupil shift from M_2. But $s_{pr1}^* = 0$, so that

$$\mu_{pr1}^* = -1$$

From (3.344), this gives

$$\overline{A}_2 = +u_{pr2} \tag{3.345}$$

Substituting from (3.343) and (3.345) in (3.341) gives

$$(S_{II})_2 = -2\frac{y_1^3}{r_1^2}\frac{(r_1 - 2d_1)^2(r_1 - 2d_1 - 2r_2)}{r_1 r_2^2}u_{pr2} \tag{3.346}$$

For the coma-free Schiefspiegler in Z-form, where u_{pr1} and u_{pr2} have the same sign because $n_1' = -1$:

$$\sum S_{II} = (S_{II})_1 + (S_{II})_2 = 0 ,$$

giving from (3.342) and (3.346) the condition for freedom from coma in the "axial" image

$$\frac{u_{pr2}}{u_{pr1}} = -\frac{r_1 r_2^2}{(r_1 - 2d_1)^2(r_1 - 2d_1 - 2r_2)} \tag{3.347}$$

Anastigmatic Schiefspiegler: For astigmatism, we have from (3.20)

$$(S_{III})_\nu = -y_\nu \overline{A}_\nu^2 \Delta \left(\frac{u}{n}\right)_\nu \tag{3.348}$$

Then

$$(S_{III})_1 = -2\frac{y_1^2}{r_1} u_{pr1}^2 \tag{3.349}$$

and substitution from (3.343) and (3.345) in (3.348) gives

$$(S_{III})_2 = +2\frac{y_1^2}{r_1}\frac{(r_1 - 2d_1)^2}{r_1 r_2} u_{pr2}^2 \tag{3.350}$$

For the anastigmatic Schiefspiegler

$$\sum S_{III} = (S_{III})_1 + (S_{III})_2 = 0 \ ,$$

giving the condition

$$\frac{u_{pr2}}{u_{pr1}} = +\frac{\sqrt{r_1 r_2}}{r_1 - 2d_1} \tag{3.351}$$

for anastigmatism of the "axial" image.

Rutten and van Venrooij [3.12(j)] give designs and spot-diagrams for three sorts of Schiefspiegler, the first being an anastigmatic system of the Kutter type with an aperture of 110 mm working at f/26. The radii r_1 and r_2 are identical with -3240 mm, $d_1 = -965$ mm and the angles of deviation are given as 5.234° and 13.666° for the primary and secondary respectively. The ratio is 2.611 and should be same as the ratio (u_{pr2}/u_{pr1}) from (3.351). With the above numerical values, (3.351) gives 2.473. The coma-free condition (3.347) gives 3.834 for the ratio. The Schiefspiegler of Rutten and van Venrooij thus lies close to the geometry given by (3.351), slightly shifted towards the coma-free condition. This is a compromise recommended by Kutter. The "axial" spot-diagram of Rutten and van Venrooij shows a residual aberration, mainly coma, comparable in size with the Airy disk, a good correction.

Equations (3.347) and (3.351) prove one of Kutter's propositions concerning the basic Schiefspiegler: that it is impossible to set the secondary at an angle u_{pr2} which corrects both coma and astigmatism for the "axial" image. This proposition is only strictly true for the normal telescopic case with the object at infinity. For a finite object distance, the magnification of the primary provides a further degree of freedom so that geometries are possible giving correction of both coma and astigmatism. Such a system was designed by Wilson and Opitz [3.106] as a very high aperture 2-mirror condensor for use in chemical analysis equipment, in which absence of central obstruction in the pupil and achromatic performance were important. Unfortunately, this solution is not applicable to telescopes since the primary is used at a magnification $|m_1| > 1$.

The coma-free Schiefspiegler defined by (3.347) can be made anastigmatic by adding cylindrical power to the secondary mirror, giving a toroidal surface. Since such non-rotationally symmetrical surfaces are very difficult to manufacture, this solution is rarely of interest for amateurs. Kutter proposed a more practical solution by choosing a compromise between the coma-free and anastigmatic solutions of Eqs. (3.347) and (3.351) and adding a weak, tilted planoconvex lens in the exit beam about half way from the secondary to the image. This is to correct both the residual coma and astigmatism. The chromatic aberration is small and its effect can be reduced by adding a slight wedge to the lens. Rutten and van Venrooij [3.12(j)] give design data and spot-diagrams for such a Kutter catadioptric Schiefspiegler for an aperture of 200 mm working at f/20.

A further version developed by Kutter and also analysed by Rutten and van Venrooij is the Tri-Schiefspiegler using a weakly concave tertiary which folds the beam back at right-angles to the original incident beam. Their design for aperture 200 mm at f/14.7 does not seem fully optimized. In principle, with 3 powered mirrors having free tilt values, a correction for the "axial" image for both coma and astigmatism must be possible. The secondary and tertiary are spherical, while the primary has an elliptical aspheric form to correct the spherical aberration.

Further designs for amateurs are given by Leonard in Mackintosh [3.48] under the names Yolo and Solano reflectors. The Yolo is a Tri-Schiefspiegler consisting of 3 concave spherical mirrors arranged for freedom from coma. The astigmatism is additive for concave mirrors and Leonard corrects this by making the primary or secondary, or both, toroidal. If necessary, spherical aberration is corrected by aspherising the primary in addition. The Solano reflector is also a Tri-Schiefspiegler, but with a convex tertiary. By using long radii, toroidal surfaces are avoided in this design. Further references to such systems are given by Leonard in Mackintosh [3.48] or in Rutten and van Venrooij [3.12(j)]. Of particular interest are articles by Buchroeder [3.107] and Kutter [3.108].

Finally, it should be mentioned that a perfect Schiefspiegler can be made as an excentric (off-axis) section of a centered, 2-mirror telescope, as shown in Fig. 3.95. If the centered telescope form is an RC form, the Schiefspiegler will have the same field correction, but improved by the reduction to the small excentric aperture. Although this form is theoretically simple and gives optimum imagery, it is of little interest, in practice, for amateur telescopes because the off-axis aspherics are asymmetrical and extremely difficult to make. If the mirror powers are not too high and the off-axis angles are kept to a minimum, as Leonard proposes in his Schiefspiegler, the off-axis sections can be approximated by toroidal surfaces.

The pure form of Fig. 3.95 is sometimes used in spectrograph design at the professional level to overcome central obstruction problems.

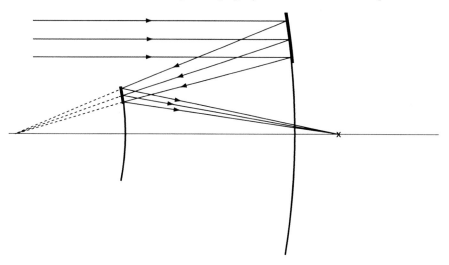

Fig. 3.95. Schiefspiegler achieved by off-axis sections of a centered, 2-mirror telescope

3.7.2 The significance of Schiefspiegler theory in the centering of normal telescopes: formulae for the effects of decentering of 2-mirror telescopes

3.7.2.1 Lateral decenter. The treatment above for the theory of Schiefspiegler can be taken over directly to analyse one of the most important errors in telescopes: field-uniform coma induced by decentering of the secondary relative to the primary.

Consider the special case of the Schiefspiegler of Fig. 3.93 in which the aspheric axes of the primary and secondary are parallel, but laterally translated by the amount δ (Fig. 3.96). A principal ray is drawn at incident angle u_{pr1}, such that the reflected ray strikes the axis of the secondary. As above for the Schiefspiegler, we apply our general recursion formulae (3.336) in the

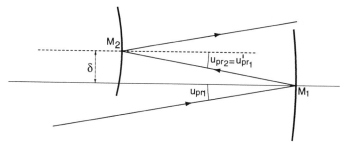

Fig. 3.96. Schiefspiegler interpretation of lateral decentering in a Cassegrain telescope

normal way for the paraxial aperture ray parameters y_ν, A_ν and $\Delta\left(\frac{u}{n}\right)_\nu$, and the formulae (3.334) for the paraxial principal ray to derive $\overline{A}_\nu$ and $(y_{pr})_\nu$ as *though the stop were at the mirror for both primary and secondary.* This gives immediately $y_{pr1} = y_{pr2} = 0$ and $(HE)_1 = (HE)_2 = 0$; or from (3.20) the consequence that the coma associated with the field angle u_{pr1} in Fig. 3.96 is independent of the aspheric figure on both mirrors. This gives, with the suffix δ, S implying decenter treated as a Schiefspiegler,

$$\left(\sum S_{II}\right)_{\delta,S} = (S_{II})_{\delta,S,1} + (S_{II})_{\delta,S,2}$$

Now, applying the conditions (3.342) and (3.343) with $\overline{A}_1 = \overline{A}_2 = u_{pr1}$ (since $u_{pr2} = -u'_{pr1}$ with $n'_1 = -1^9$), we derive at once

$$\left(\sum S_{II}\right)_{\delta,S} = -\frac{2y_1^3}{r_1^2}\left[1 + \left(1 - \frac{2d_1}{r_1}\right)(\mu_2^2 - 1)\right]u_{pr1} , \qquad (3.352)$$

where $\mu_2 = \frac{1}{r_2}(r_1 - 2d_1 - r_2)$ from (3.343) and also $\mu_2 = \frac{1}{m_2}$ by definition from (3.332). For the small angles u_{pr1} involved in decentering

$$\delta = -d_1 u_{pr1}$$

and, from (3.196), the lateral tangential coma is given by

$$(Coma_t)_{\delta,S} = -3\frac{f'}{d_1}\left(\frac{y_1}{r_1}\right)^2\left[1 + \left(1 - \frac{2d_1}{r_1}\right)(\mu_2^2 - 1)\right]\delta , \qquad (3.353)$$

---

[9] This sign definition for u_{pr2}, giving $u_{pr2} = u_{pr1}$, is in apparent conflict with the definition in Fig. 3.96 which corresponds to $u_{pr2} = -u_{pr1}$. The reason for this sign reversal is the following. Figure 3.96 gives the signs in our normal Cartesian ray tracing system, as applied to centered systems. But the Schiefspiegler is not a normal system, in so far as the application of our third order equations is concerned, with regard to the field parameter $\overline{A}_1 = u_{pr1}$ and $\overline{A}_2 = u_{pr2}$. In our Schiefspiegler theory from Eqs. (3.344) to (3.347) and also in our present case, the secondary is treated as a *"second primary"*, but the light direction is reversed because of $n'_1 = -1$. In the Z-form of the Schiefspiegler this brings a partial compensation of the coma, i.e. the signs of the coma introduced by the two primaries are opposite. This is particularly obvious in the case of the Czerny-Turner Z-form arrangement, where there is complete compensation of the two terms in Eq. (3.339), and a compensatory (subtractive) effect is obvious from the two halves of the system working as "telescopes" in opposition in one light direction. By contrast, in the so-called U-form of layout, the two components of the coma are *additive*, so this form should be avoided in practical systems.

If the sign reversal of the field angle is not introduced in the Schiefspiegler (Z-form) treatment, although it is, in fact, necessary and logical, the *additive* law of third order aberrations $\sum S_{II} = (S_{II})_1 + (S_{II})_2$ would have to be changed to a *subtractive* law with negative sign. But this would not be correct, since the third order treatment for the *aperture* parameters can be taken over without change from the normal centered case.

which is converted into angular coma in arcsec by multiplying by $(206\,265/f')$. Now $(Coma_t)_{\delta,S}$ gives the coma for the case of the small field angle u_{pr1} which is not the strict case of lateral decentering in the normal sense, shown in Fig. 3.97. Here, the principal ray coincident with the axis of the primary $(u_{pr1} = 0)$ is reflected back on itself, is incident on the decentered secondary at a small angle and reflected off as shown.

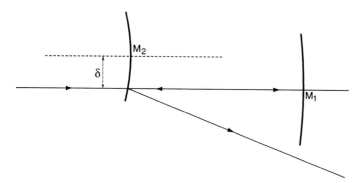

Fig. 3.97. Strict case of lateral decenter in a 2-mirror telescope

The difference between the cases shown in Figs. 3.96 and 3.97 is as follows. The Schiefspiegler case of Fig. 3.96 has been calculated from the recursion formulae with the "field angle" u_{pr1} strictly for an image point on this reflected "axis": no "field" for the system is considered and the pupil can be considered at both primary and secondary. The case of strict lateral decenter shown in Fig. 3.97 involves a rotation of the beam relative to the mirror system clockwise, in a negative cartesian sense, i.e. through the angle $-u_{pr1}$. This introduces an additional coma corresponding to the field coma of the system for a field angle $-u_{pr1}$. Thus, the field coma corresponding to the field angle u_{pr1} must be *subtracted* from the result above of Eq. (3.353) and this field coma will depend on the form of the telescope. If the form is aplanatic (RC in the case shown with a convex secondary), the field coma is zero and the decentering coma for the Schiefspiegler case of Fig. 3.96 and Eq. (3.353) gives the correct result; otherwise the field coma corresponding to u_{pr1} must be subtracted.

The general formula for the field coma for a 2-mirror telescope is, from Table 3.5

$$\left(\sum S_{II}\right)_f = -\left(\frac{y_1}{f'}\right)^3 \left[d_1\xi + \frac{1}{2}f' + \frac{s_{pr1}}{f'}\left(-f'\zeta + L\xi\right)\right]u_{pr1}$$

with

$$\xi = \frac{(m_2+1)^3}{4}\left[\left(\frac{m_2-1}{m_2+1}\right)^2 + b_{s2}\right]$$

If the telescope is corrected for spherical aberration, the last term is zero, giving

$$\left(\sum S_{II}\right)_{\delta,S,f} = -\left(\frac{y_1}{f'}\right)^3 \left[\frac{1}{2}f' + \frac{d_1}{4}(m_2+1)^3\left\{\left(\frac{m_2-1}{m_2+1}\right)^2 + b_{s2}\right\}\right]u_{pr1}$$

(3.354)

Replacing u_{pr1} by $-\delta/d_1$ and applying (3.196), we have for the lateral tangential field coma

$$(Coma_t)_{\delta,S,f} = -\frac{3}{2}\left(\frac{y_1}{f'}\right)^2\left[\frac{1}{2}\frac{f'}{d_1} + \frac{(m_2+1)^3}{4}\left\{\left(\frac{m_2-1}{m_2+1}\right)^2 + b_{s2}\right\}\right]\delta$$

(3.355)

This must be subtracted from $(Coma_t)_{\delta,S}$ of (3.353) to give the correct decentering coma for the general case of a laterally decentered 2-mirror telescope. The reduction is straightforward but somewhat tedious. The following paraxial relations are required:

$$\frac{f'}{d_1} = \frac{m_2}{1-R_A} \quad \text{from (2.72)}$$

$$\left(\frac{y_1}{r_1}\right)^2 = \frac{m_2^2}{16N^2} \quad \text{where } N \text{ is the f/no.}$$

$$1 - \frac{2d_1}{r_1} = \frac{y_2}{y_1} = R_A \quad \text{from (3.336) and (2.72)}$$

$$\mu_2^2 - 1 = \frac{1-m_2^2}{m_2^2} \quad \text{by definition}$$

$$\frac{y_1}{f'} = \frac{1}{2N}$$

Eq. (3.353) reduces to the term A_δ with

$$A_\delta = (Coma_t)_{\delta,S} = -\frac{3}{16}\frac{m_2}{N^2}\left[\frac{m_2^2 + R_A\left(1-m_2^2\right)}{1-R_A}\right]\delta$$

(3.356)

If we divide the field coma of (3.355) into two terms $B_{\delta 1}$ and $B_{\delta 2}$, then $B_{\delta 1}$ gives the field coma of the classical telescope as

$$B_{\delta 1} = -\frac{3}{2}\left(\frac{y_1}{f'}\right)^2\left(\frac{1}{2}\frac{f'}{d_1}\right)\delta = -\frac{3}{16}\frac{m_2}{N^2}\left(\frac{1}{1-R_A}\right)\delta$$

(3.357)

The lateral decentering coma of a *classical telescope* with primary of parabolic form is then given by

$$(Coma_t)_{\delta,cl} = A_\delta - B_{\delta 1}$$

giving

$$(Coma_t)_{\delta,cl} = -\frac{3}{16}\frac{m_2}{N^2}\left(m_2^2 - 1\right)\delta \,, \tag{3.358}$$

or, in arcsec from Eq. (3.198)

$$\left[(\delta u_p')_{Coma_t}\right]_{cl} = -\frac{3}{16}\frac{m_2}{N^2}\left(m_2^2 - 1\right)\frac{\delta}{f'}(206\,265)\ \text{arcsec} \tag{3.359}$$

The field coma effect is, in practice, relatively modest. With the normalized data of Table 3.2, the basic Schiefspiegler term of Eq. (3.356) gives $+48.871\delta$, corresponding to the aplanatic telescope, while Eq. (3.358) for the classical telescope gives $+45\delta$. This is a measure of the seriousness of decentering coma, the basic decentering effect being, in the configuration of Table 3.2, over twelve times as serious as the field coma of a classical telescope associated with the same beam angle u_{pr1}.

For the *general case of a 2-mirror telescope* with secondary defined by b_{s2} and corrected for spherical aberration, we have for the lateral decentering tangential coma

$$(Coma_t)_\delta = A_\delta - B_{\delta 1} - B_{\delta 2} = (Coma_t)_{\delta,cl} - B_{\delta 2} \tag{3.360}$$

From Eq. (3.358) and the second term of (3.355), we have

$$(Coma_t)_\delta = -\frac{3}{16}\frac{m_2}{N^2}(m_2^2 - 1)\delta + \frac{3}{2}\frac{1}{4N^2}\left[\frac{(m_2+1)^3}{4}\left\{\left(\frac{m_2-1}{m_2+1}\right)^2 + b_{s2}\right\}\right]\delta \tag{3.361}$$

This reduces to the simple general form

$$(Coma_t)_\delta = -\frac{3}{32}\frac{(m_2+1)^2}{N^2}\left[(m_2 - 1) - (m_2 + 1)b_{s2}\right]\delta \tag{3.362}$$

or the angular form

$$\left[(\delta u_p')_{Coma_t}\right]_\delta = -\frac{3}{32}\frac{(m_2+1)^2}{N^2}\left[(m_2 - 1) - (m_2 + 1)b_{s2}\right]\frac{\delta}{f'}(206\,265)\ \text{arcsec} \tag{3.363}$$

Since decentering coma is the principal curse, in practice, of the conventional Cassegrain telescope, the above formulae are extremely important. If the forms of b_{s2} corresponding to a classical telescope (Eq. (3.93)) or an aplanatic (RC) telescope (Eq. (3.109)) are inserted in (3.362), we recover the results of (3.358) and (3.356) respectively. In fact, Eq. (3.356) can be transformed at once into the form

$$(Coma_t)_{\delta,Aplan} = -\frac{3}{16}\frac{m_2}{N^2}\left[(m_2^2 - 1) + \frac{1}{1 - R_A}\right]\delta \tag{3.364}$$

The first term represents the form for the classical telescope of Eq. (3.358) and the second term the aplanatic supplement to the decentering coma, the latter only 8.6% of the former for the geometry of Table 3.2, Cases 3 and 4 ($m_2 = -4$, $R_A = +0.225$).

The same equations, in a slightly different form and adapted to his sign convention, are given by Bahner [3.5]. Other treatments have been given by Conrady [3.109], Maréchal [3.110], Slevogt [3.111], Baranne [3.112] and Schroeder [3.22(d)].

In the thirties, Ritchey built the first major RC *telescope*, with an aperture of 1 m, for the US Naval Observatory in Washington. A short account is given by Riekher [3.39(h)] and the basic data by Bahner [3.5]. This important telescope is discussed further in Chap. 5. Its sensitivity to decentering coma was higher than normal Cassegrain telescopes of the time, but this was due more to its optical geometry than its RC form. The general form for the *angular* decentering coma of Eq. (3.363) shows that, for any form of 2-mirror telescope, the most important parameter in producing decentering coma is the f/no. of the *primary*. This is particularly clear if $|m_2| \gg 1$, since the terms containing m_2 then vanish if N^2 is replaced by N_1^2 and f' by f_1'. (The dominant role of the primary, in the normal case where $|R_A| \ll 1$, is also clear from the basic Schiefspiegler formula (3.356) in which the first term from the primary is dominant). Ritchey's telescope had a primary of f/4.0 and the very fast final relative aperture of f/6.8 because he wished to have a fast photographic telescope. The corresponding m_2 was very low, only -1.7. The two terms in the bracket of Eq. (3.364) are then 1.89 and 1.67 respectively, with $R_A = 0.40$. With this low m_2, therefore, the aplanatic form did indeed almost double the decentering coma compared with the classical telescope form. However, for a given value of decenter δ, the aplanatic form only gave 33% more angular coma than a typical classical telescope would have given with the same primary and $m_2 = -4$.

Another interesting case is the *Dall-Kirkham (DK) telescope* with a value $(b_{s2})_{DK} = 0$. Eq. (3.362) then gives

$$(Coma_t)_{\delta,\,DK} = -\frac{3}{32}\frac{(m_2^2-1)(m_2+1)}{N^2}\,\delta \tag{3.365}$$

Comparison with (3.358) shows that the decentering coma of the DK form is $\frac{1}{2}(m_2+1)/(m_2)$ times that of the classical Cassegrain, a substantial gain.

Similarly, using (3.137) for $(b_{s2})_{SP}$, the decentering coma for a telescope with *a spherical primary* is derived from (3.362) as

$$(Coma_t)_{\delta,SP} = -\frac{3}{16}\frac{m_2}{N^2}\left((m_2^2-1)-\frac{m_2^2}{2R_A}\right)\delta\,, \tag{3.366}$$

in which the first term in the bracket corresponds to the classical Cassegrain case (3.358). With the values of Table 3.2 for m_2 and R_A (-4 and 0.225), the second term is dominated by R_A compared with the first, giving a $(Coma_t)_{\delta,SP}$ value -1.370 times that of the classical Cassegrain.

The angular tangential coma $\left[(\delta u_p')_{Coma_t}\right]_\delta$ due to transverse decentering according to Eq. (3.363) is given in arcsec in Table 3.20 for the 2-mirror telescopes listed in Table 3.2, using the same parameters $|m_2| = 4$, $|R_A| = 0.225$ and $|f'| = 1$. The formulae above are also valid for Gregory telescopes with

Table 3.20. Angular tangential coma produced by transverse decentering of the secondary in the 2-mirror telescopes of Table 3.2. $|m_2| = 4$; $|R_A| = 0.225$. The relative aperture at the final image is $N = 10$. The decenter is $|\delta/f'| = 10^{-4}$

Telescope type	Angular transverse decentering coma (arcsec)
Classical Cassegrain (C)	+2.320
Ritchey-Chrétien (RC)	+2.520
Dall-Kirkham (DK)	+0.870
Spherical Primary (SP)	−3.180
Classical Gregory (CG)	+2.320
Aplanatic Gregory (AG)	+2.447
DK Gregory (DKG)	+1.450
SP Gregory (SPG)	+7.821

the usual sign changes for these parameters. The values correspond to the relatively large decenter of $|\delta/f'| = 10^{-4}$ (+1 mm for $f' = \pm 10$ m) and for a final relative aperture $N = 10$. Only in the SP case is there a major difference between the Cassegrain and Gregory cases, provoked by the large second term of (3.366) and the sign change of R_A.

An important limit case is the *afocal* 2-mirror telescope, corrected for spherical aberration. The general formula (3.363) gives at once, for this limit case with $|m_2| \to \infty$

$$\left[(\delta u_p')_{Coma_t} \right]_{\delta, Afoc} = -\frac{3}{32} \frac{1}{N_1^2} (1 - b_{s2}) \frac{\delta}{f_1'} (206\,265) \text{ arcsec ,} \tag{3.367}$$

where N_1 and f_1' refer to the *primary*. We see that, in this limit case, the secondary only influences the decentering coma by its form b_{s2}, not by its geometry in the system. For the classical Mersenne form of the afocal telescope, with $b_{s2} = -1$, this gives the simple result

$$\left[(\delta u_p')_{Coma_t} \right]_{\delta, Afoc, cl} = -\frac{3}{16} \frac{1}{N_1^2} \frac{\delta}{f_1'} (206\,265) \text{ arcsec ,} \tag{3.368}$$

totally determined by the scale and speed of the primary. If we take the same primary as assumed in Table 3.20, with $N_1 = 2.5$ and $f_1' = -2500$ mm for $\delta = 1$ mm, then the angular lateral decentering coma in both classical Cassegrain and Gregory cases is +2.475 arcsec, slightly larger than for $|m_2| = 4$ in Table 3.20. The difference is small because $|m_2| = 4$ is quite a high value. As $|m_2| \to 1$, the limit case of a plane M_2, then the decentering coma vanishes according to (3.358) because M_2 has no optical power.

The afocal DK form gives, with $b_{s2} = 0$ in (3.367), for both Cassegrain and Gregory forms

$$\left[(\delta u_p')_{Coma_t} \right]_{\delta, Afoc, DK} = -\frac{3}{32} \frac{1}{N_1^2} \frac{\delta}{f_1'} (206\,265) \quad \text{arcsec} , \tag{3.369}$$

exactly half that of the classical Mersenne form, 1.238 arcsec. Finally, for the afocal SP form, $|m_2| \rightarrow \infty$ in (3.366) gives

$$\left[(\delta u_p')_{Coma_t} \right]_{\delta, Afoc, SP} = -\frac{3}{16} \frac{1}{N_1^2} \left(1 - \frac{1}{2R_A} \right) \frac{\delta}{f_1'} (206\,265) \quad \text{arcsec} , \tag{3.370}$$

again dominated by the second term. The values for the Cassegrain and Gregory cases with the primary defined above are -3.025 arcsec and $+7.975$ arcsec respectively. The large second term arises from the limit case of Eq. (3.137)

$$(b_{s2})_{Afoc, SP} = -1 + \frac{1}{R_A} , \tag{3.371}$$

since $f'/L = 1/R_A$ from (2.72).

All the above formulae are valid for the third order lateral decentering coma of a 2-mirror telescope *with any stop position* if the spherical aberration is corrected; but *only for a stop at the primary* if the spherical aberration is *not* corrected – see Eq. (3.354).

The formula (3.363) was derived above from the Schiefspiegler approach because the comparison is physically instructive and revealing. However, it can also be derived directly from the recursion formulae in the sense shown in Fig. 3.97. So far as the principal ray parameters are concerned, the secondary mirror is treated as a primary with a telecentric beam incident on it from infinity. Eqs. (3.336) are therefore applied to the principal ray. For the primary, $u_{pr1} = 0$ and its coma $(S_{II})_{1,\delta} = 0$. For the secondary with the telecentric incident beam

$$y_{pr2} = +\delta ,$$

the lateral decenter, and

$$\overline{A}_2 = \frac{y_{pr2}}{r_2}$$

$$(HE)_2 = \frac{y_{pr2}}{y_2} = \frac{y_{pr2}}{y_1 \left(1 - \frac{2d_1}{r_1} \right)}$$

Then, applying the recursion formulae (3.336) to the aperture ray as in the derivation of (3.352) from (3.342) and (3.343), we have, from (3.20)

$$(S_{II})_{2,\delta} = -\frac{2y_1^3}{r_1^2} \left(1 - \frac{2d_1}{r_1} \right) (\mu_2^2 - 1) \frac{y_{pr2}}{r_2} + \frac{y_{pr2}}{y_1 \left(1 - \frac{2d_1}{r_1} \right)} \tau_2 \tag{3.372}$$

This coma is the total coma since the primary has no contribution. Converted to $(Coma_t)$ from (3.196), it leads, using the definition of τ given above (3.19), to

$$(Coma_t)_\delta = \frac{3}{2}\frac{f'}{y_1}\left[2\left(\frac{y_1}{r_1}\right)^2\frac{y_2}{r_2}(\mu_2^2-1)+2b_{s2}\left(\frac{y_2}{r_2}\right)^3\right]y_{pr2} \qquad (3.373)$$

This can be easily reduced using the paraxial relations given above (3.356) and the further relation

$$\frac{y_2}{r_2} = \frac{(m_2+1)}{2}\frac{y_1}{f'}\ ,$$

derived from (2.90) and (2.72). Inserting also $y_{pr2} = \delta$, Eq. (3.373) then gives

$$(Coma_t)_\delta = -\frac{3}{32}\frac{(m_2+1)^2}{N^2}\left[(m_2-1)-(m_2+1)\,b_{s2}\right]\delta\ ,$$

identical with our general result above of Eq. (3.362).

This derivation of the lateral decentering coma formula using a telecentric beam on the secondary and the recursion formulae (3.336) is certainly one of the most direct and simple.

Lateral decentering coma has been dealt with in detail because of its capital importance in practical telescope technology. Technical aspects, connected with telescope alignment and maintenance, will be discussed in RTO II, Chap. 2 and 3. The theory given above is based on third order theory as expressed by Eqs. (3.20). It may legitimately be asked how accurate this theory is in practice. The answer is, in most cases, remarkably accurate. Except in extreme cases of aperture and field, which are rare in most applications of telescopes, the aberration function given in Table 3.1 is a rapidly converging function. This is particularly the case for lateral decentering coma in respect of field effects, defined by the decenter δ, which must be small in practice to give acceptable coma limits. It follows that only higher terms in aperture can normally be significant; but the convergence law will still apply.

Modern optical design programs permit decentering of elements both by lateral shifts and by rotations, enabling exact calculations of decentering aberrations. Coma will always be by far the most important term because of its third order linear dependence on field compared with the quadratic dependence of astigmatism. However, for modern telescopes using the edge of the field and particularly for anastigmatic telescopes, the effects of decentering astigmatism at the edge of the field may be significant. In such cases, freedom from decentering coma is no longer a sufficient condition. This is also true for modern, high quality, active telescopes like the ESO Very Large Telescope (VLT), for which the image analyser of the active optics system is near the edge of the field. Asymmetries in the astigmatic field produced by decentering astigmatism may then give wrong results for the astigmatism correction at the field center. In the VLT, therefore, decentering astigmatism is corrected resulting in virtual elimination of lateral decenter, even though this is not necessary for the correction of decentering coma. This whole matter is dealt with in detail in § 2.2.1 of RTO II.

It is sometimes asserted that higher order variations from strict conic sections defined by Eq. (3.11) will lead to more severe centering tolerances than those given from third order theory above. This could only be true in very extreme cases because such higher order form deformations are also converging functions following the general aberration function: the higher order wavefront effects are far lower than the third order effect of any powered optical element. The only clear exception to this general rule is the case of an element without optical power (plane mirror or plate) which is asB sandwich ... *only* with higher order terms.

3.7.2.2 Angular decenter of the secondary (rotation about its pole).

The contribution $(S_{II})_{\delta,S,2}$ leading to the second term in (3.352) of the Schiefspiegler analysis of decenter above gave directly the effect on coma of rotating the secondary through the angle u_{pr1}. This expression can be written for the general rotation angle u_{pr2} of the secondary, positive in the sense shown in Fig. 3.96, as

$$(S_{II})_{2,rot} = -2 \left(\frac{y_1}{r_1}\right)^2 y_2(\mu_2^2 - 1)u_{pr2} \tag{3.374}$$

With

$$(Coma_t) = -\frac{3}{2}\frac{f'}{y_1}S_{II}$$

from (3.196), we have

$$(Coma_t)_{2,rot} = 3f'\left(\frac{y_1}{r_1}\right)^2 \frac{y_2}{y_1}(\mu_2^2 - 1)u_{pr2}$$

With our paraxial substitutions given above for $\mu_2 = -\frac{1}{m_2}$, $y_2/y_1 = R_A$ and $(y_1/r_1)^2 = m_2^2/16N^2$, this reduces immediately to

$$(Coma_t)_{2,rot} = -\frac{3}{16}f'\frac{R_A(m_2^2 - 1)}{N^2}u_{pr2} \tag{3.375}$$

By definition in the derivation of this formula, the stop is considered as being at the secondary mirror, so the result of (3.375) is independent of b_{s2}, the form of the secondary. If u_{pr2} is expressed in radians, the angular coma in arcsec is given by $(206\,265/f')(Coma_t)_{2,rot}$. Expressing u_{pr2} in arcsec, we have, more directly

$$\left[(\delta u'_p)_{Coma_t}\right]_{2,rot} = -\frac{3}{16}\frac{R_A\left(m_2^2 - 1\right)}{N^2}(u_{pr2})_{arcsec} \quad \text{arcsec} \tag{3.376}$$

For the geometry of Table 3.2, $N = 10$ as in Table 3.20 and a rotation $u_{pr2} = 60$ arcsec, the angular coma according to (3.376) is only 0.380 arcsec, identical for the Cassegrain and Gregory forms apart from the sign (negative for the Cassegrain). Since 60 arcsec is quite a coarse rotation, causing a pointing change of

$$2u_{pr2}L/f' = 27 \text{ arcsec} ,$$

or almost two orders of magnitude more than the $Coma_t$ effect, it is clear that this source of decentering coma is far less sensitive than the lateral (translation) decentering coma given in Table 3.20, values corresponding to $\delta = 1$ mm for a 1 m telescope at f/10.

3.7.2.3 Coma-free (neutral) points. For any 2-mirror telescope system, there will be a point somewhere on its axis for which the translation coma $(Coma_t)_\delta$ and the rotation coma $(Coma_t)_{2,rot}$ will cancel out. This is called the coma-free point or neutral point for coma. Another important neutral point, independent of the shape of the secondary, is the neutral point for pointing, which is always at the centre of curvature of the secondary.

If the distance *behind* the secondary (towards the light) to the coma-free point is denoted by $-z_{CFP}$, then the translation δ of the pole of the secondary by a rotation through u_{pr2} is

$$\delta = -z_{CFP} u_{pr2}$$

Inserting this in Eq. (3.362) and combining the resulting relation with the rotation coma of Eq. (3.375), we have for the total coma in the general case

$$(Coma_t)_{tot} = \left[+\frac{3}{32}\frac{(m_2+1)^2}{N^2}\left\{(m_2-1)-(m_2+1)b_{s2}\right\}z_{CFP} \right. \\ \left. -\frac{3}{16}f'\frac{R_A(m_2^2-1)}{N^2}\right]u_{pr2}$$

$$(3.377)$$

For the CFP, this expression must be zero.

For the *classical Cassegrain telescope* from (3.358)

$$(Coma_t)_{tot,cl} = \left[+\frac{3}{16}\frac{m_2(m_2^2-1)}{N^2}z_{CFP} - \frac{3}{16}f'R_A\frac{(m_2^2-1)}{N^2}\right]u_{pr2} = 0$$

for the CFP. This gives with Eq. (2.72)

$$(z_{CFP})_{cl} = \frac{f'}{m_2}R_A = f_1'R_A = s_2 \qquad (3.378)$$

In other words, for the classical Cassegrain, the CFP lies at a distance s_2 behind the secondary, i.e. at the *prime focus*. Of course, this result is also valid for the classical Gregory.

For the *Dall-Kirkham (DK) telescope*, we have for the CFP from (3.377) setting $b_{s2} = 0$

$$(m_2+1)(z_{CFP})_{DK} = 2f'R_A$$

or

$$(z_{CFP})_{DK} = \frac{2f'R_A}{m_2+1}$$

Now, from (2.90) and (2.72)

$$r_2 = \frac{2L}{m_2 + 1} \ , \quad L = R_A f' = m_2 s_2 \ , \quad s_2 = R_A f'_1 \ ,$$

giving

$$(z_{CFP})_{DK} = r_2 = s_2 \left(\frac{2m_2}{m_2 + 1} \right) \tag{3.379}$$

For the Dall-Kirkham (DK), therefore, the CFP lies at the centre of curvature of the spherical secondary, the same as the neutral point for pointing, as must be the case for a spherical secondary. This is not necessarily an advantage since it precludes the possibility of influencing the coma without affecting the pointing (tracking). We shall return to this point later in connection with the active control of decentering coma. Conversely, field scanning without coma for observation in the IR is not possible for the DK form.

For an *aplanatic telescope* we use (3.364) with (3.375) to give from (3.377)

$$\left. \begin{array}{l} (Coma_t)_{tot,Aplan} = \left[\dfrac{3}{16} \dfrac{m_2}{N^2} \left\{ (m_2^2 - 1) + \left(\dfrac{1}{1 - R_A} \right) \right\} (z_{CFP})_{Aplan} \right. \\[3mm] \left. - \dfrac{3}{16} f' \dfrac{R_A(m_2^2 - 1)}{N^2} \right] u_{pr2} = 0 \end{array} \right\}$$

for the CFP. Using, from (2.72)

$$R_A = \frac{s_2}{f'_1} = \frac{s_2 m_2}{f'} \ ,$$

this reduces to the condition

$$(z_{CFP})_{Aplan} = s_2 \left[\frac{(m_2^2 - 1)(1 - R_A)}{1 + (m_2^2 - 1)(1 - R_A)} \right] \tag{3.380}$$

With our values from Table 3.2, the RC case with $m_2 = -4$ and $R_A = +0.225$ gives 0.921 for the factor in the bracket. With $m_2 \to \infty$, it becomes unity, giving the same result as the classical telescope because the forms are identical in the afocal case. In the focal case, the CFP is slightly nearer to the secondary than is the prime focus. The same situation applies to the position of the exit pupil ExP if the entrance pupil is at the primary. Applying (2.94) and Fig. 2.13, we can derive, making use of the paraxial relations (2.90), (2.55) and (2.72), the relation

$$z_{ExP} = s_2 \left[\frac{m_2(1 - R_A)}{m_2 R_A + (m_2 + 1)(1 - R_A)} \right] \tag{3.381}$$

for the position of the exit pupil if the entrance pupil is at the primary. This condition is *not* the same as (3.380), but the difference is small, the values $m_2 = -4$ and $R_A = +0.225$ giving 0.961 for the factor in the bracket of (3.381). In an important paper on decentering properties of Cassegrain telescopes, Meinel and Meinel [3.113] point out that the CFP in an RC telescope

is close to the exit pupil and attribute the discrepancy to aberrations arising from non-paraxial conditions. While the aberration of the pupil certainly
plays a role, the analysis below indicates that the situation is more complex
and that the agreement of CFP and ExP depends on the axial obstruction
ratio R_A and hence on the final image position. However, the agreement is
close in the RC case with normal geometries, as we see from the example
above.

It is instructive to analyse further the relation of the pupil position to
the position of $(CFP)_{Aplan}$ from the Schiefspiegler case of Fig. 3.96. We saw
that, because the aplanatic telescope has no field coma, there is no effect
in rotating the entrance beam about the pole of the primary, the entrance
pupil, to produce the situation of Fig. 3.97. If we were to rotate *the whole
system* about the exit pupil, we would have a similar effect. But this is not
the same as the CFP, about which only the secondary is rotated, not the
primary. Nevertheless, there is a remarkably close coincidence between the
two positions CFP and ExP for practical RC telescopes. It is easy to deduce
why this is the case. If the bracket terms of (3.380) and (3.381) are equated,
we can derive at once the resulting condition for $(R_A)_P$ as

$$(R_A)_P = -\frac{(m_2^2 - m_2 - 1)}{(m_2^2 - 1)(m_2 - 1)} \tag{3.382}$$

The numerator has the same form as Eq. (3.129), derived in connection with
the form of the primary in a certain simple limiting geometry of the DK
telescope. The roots of this numerator are the *magic number* $\tau = +1.618\,034$
in the case of m_2 positive (Gregory), or $-1/\tau = -(\tau - 1) = -0.618\,034$ in
the case of m_2 negative (Cassegrain). With $m_2 = \tau$ or $-1/\tau$ in the Gregory
or Cassegrain cases respectively, Eq. (3.382) gives $(R_A)_P = 0$, corresponding
to a telescope in which the secondary has shrunk to a point at the prime
focus. In the Gregory case, the range $0 < m_2 < \tau$ gives only virtual solutions
with R_A positive; while real solutions exist for $\tau < m_2 < \infty$. This function
has a minimum of $(R_A)_P = -0.357$ at $m_2 = +2.2695$. In the Cassegrain
case, there are two ranges with real solutions: the range $-\sqrt{2} > m_2 > -\infty$
covering normal Cassegrain telescopes and the range $0 > m_2 > -\tau/2$ covering
Schwarzschild-Couder types with a concave secondary without intermediate
image.

If the paraxial relations referred to above are substituted in (3.382), one
can derive the relation

$$L_P = -d_1 \left(\frac{m_2^2 - m_2 - 1}{m_2^2 - 2}\right) \tag{3.383}$$

for the condition that the final image distance L_P from the secondary shall
give agreement between the CFP and the exit pupil for a given value of m_2.
The numerator is again the magic number equation. In the Cassegrain case,
with m_2 negative, it is clear that the bracket of (3.383) is greater than unity
for the normal range of Cassegrain solutions: the final image must be *behind*

the primary. This is, of course, the case for normal Cassegrain telescopes. For example, if $m_2 = -4$, then $L_P = -1.357d_1$ and $(R_A)_P = +0.253$. These are quite typical values and validate the assumption that the CFP for RC telescopes is near the exit pupil. However, for $m_2 = -2$, the values are $L_P = -2.5d_1$ and $(R_A)_P = +0.556$; while for $m_2 \to -\infty$, the values are $L_P \to -d_1$ and $(R_A)_P \to 0$.

In the aplanatic Gregory case, Eq. (3.383) gives the factor of the bracket *less than* unity, resulting in positions of the final image *inside* the primary for real solutions. With the normal image position *behind* the primary, there is therefore a poorer agreement between the CFP and the exit pupil than in the Cassegrain RC case.

Finally, we consider the case of the *spherical primary (SP) telescope*. Combining (3.366) and (3.375) with (3.377) gives

$$
(Coma_t)_{tot,SP} = \left[\frac{3}{16} \frac{m_2^2}{N^2} \left\{ (m_2^2 - 1) - \frac{m_2^2}{2R_A} \right\} (z_{CFP})_{SP} \right.
$$
$$
\left. - \frac{3}{16} f' R_A \frac{(m_2^2 - 1)}{N^2} \right] u_{pr2} = 0
$$

for the CFP. Again with

$$
f' R_A = m_2 s_2
$$

from (2.72), this reduces to

$$
(z_{CFP})_{SP} = s_2 \left[\frac{2R_A(m_2^2 - 1)}{2R_A(m_2^2 - 1) - m_2^2} \right] \tag{3.384}
$$

In the Gregory case, the bracket term is always positive and less than unity because R_A is negative. For the Cassegrain, with R_A positive, the bracket term is *negative* in all practical cases: it can only be positive if R_A is high (>0.5) with a fairly large $|m_2|$, giving an image position far behind M_1. This means that, in practical Cassegrain cases, the CFP lies *in front of* the secondary. This sign reversal, compared with the other cases, arises because of the large *positive* value of b_{s2} (oblate spheroid) in (3.377) as distinct from the negative values in the classical and RC cases. In the afocal Cassegrain case, the CFP moves to ∞ if $R_A = 0.5$ and is behind the secondary if $R_A > 0.5$. With our standard values of Table 3.2 ($m_2 = -4$, $R_A = 0.225$), we have $(z_{CFP})_{SP,Cass} = -0.730s_2$, i.e. in front of the secondary.

The above specific cases show a general trend in the position of the CFP. Since the rotation decentering coma of M_2 is independent of the mirror form, the position of the CFP reflects the value of the lateral (translation) decentering coma of (3.362). The general case for the CFP is given by setting (3.377) to zero, which leads with simple reduction to

$$
z_{CFP} = s_2 \left/ \left[\left(\frac{m_2 + 1}{2m_2} \right) \left\{ 1 - \left(\frac{m_2 + 1}{m_2 - 1} \right) b_{s2} \right\} \right] \right. \tag{3.385}
$$

For a 2-mirror telescope of Cassegrain form, corrected for spherical aberration, large negative values of b_{s2} give a CFP behind the secondary, but close to it. As $|b_{s2}|$ reduces to the RC value, the CFP moves backwards almost to the prime focus, to a point normally very close to the exit pupil if the entrance pupil is at the primary. With further reduction of $|b_{s2}|$ to the value for a classical Cassegrain, the CFP is exactly at the prime focus. For lower values, it moves further back, reaching the centre of curvature of the secondary when $b_{s2} = 0$ for the DK form. After this, b_{s2} becomes positive (oblate spheroid) and the CFP moves further away until it is at $-\infty$ when the denominator in (3.385) becomes zero. Beyond this, the CFP is in front of the secondary, such cases including the SP form.

The telescope form giving $z_{CFP} = \infty$ is defined from (3.385) by

$$(b_{s2})_{TC,0} = + \left(\frac{m_2 - 1}{m_2 + 1} \right) , \qquad (3.386)$$

defining a 2-mirror telescope free from lateral (translation) decentering coma. This form can have practical application where *stability* of the image in the presence of lateral decentering is primordial. For spherical aberration correction, Table 3.5 and (2.72) give

$$(b_{s1})_{TC,0} = -1 + R_A \left(\frac{m_2 + 1}{m_2} \right)^3 \left[\left(\frac{m_2 - 1}{m_2 + 1} \right)^2 + \left(\frac{m_2 - 1}{m_2 + 1} \right) \right] ,$$

which reduces to

$$(b_{s1})_{TC,0} = -1 + 2R_A \frac{(m_2^2 - 1)}{m_2^2} , \qquad (3.387)$$

corresponding, in the Cassegrain case, to an ellipse with $(b_{s1})_{TC,0} = -0.578$ with our standard values of Table 3.2 with $m_2 = -4$, $R_A = 0.225$. With these values, the secondary has the form $(b_{s2})_{TC,0} = +1.667$, a fairly steep oblate spheroid, but not nearly as steep as the form required for the SP telescope of Case 6 in Table 3.2. The equivalent Gregory case with $m_2 = +4$ and $R_A = -0.225$ gives $(b_{s1})_{TC,0} = -1.422$ and $(b_{s2})_{TC,0} = +0.600$.

The field coma of such a TC,0 telescope is given from Table 3.5 as

$$\left(\sum S_{II} \right)_{TC,0} = - \left(\frac{y_1}{f'} \right)^3 \left(d_1 \xi + \frac{f'}{2} \right) u_{pr1}$$

Substitution for ξ and

$$d_1 = f' \frac{(1 - R_A)}{m_2}$$

from (2.72) gives

$$\left(\sum S_{II}\right)_{TC,0} = -\left(\frac{y_1}{f'}\right)^3 f'\left[\left(1 - R_A\right)\frac{(m_2+1)^3}{4m_2}\right.$$
$$\left.\left\{\left(\frac{m_2-1}{m_2+1}\right)^2 + \left(\frac{m_2-1}{m_2+1}\right)\right\} + \frac{1}{2}\right] u_{pr1}\Big\},$$

which reduces to

$$\left(\sum S_{II}\right)_{TC,0} = -\left(\frac{y_1}{f'}\right)^3 f'\left[\left(1 - R_A\right)\frac{(m_2^2-1)}{2} + \frac{1}{2}\right] u_{pr1} \qquad (3.388)$$

For the normalized case of Table 3.2 and the values $m_2 = -4$ and $R_A = 0.225$, we have

$$\left(\sum S_{II}\right)_{TC,0} = -(5.8125 + 0.5) = -6.3125 ,$$

the second term being the field coma of the classical Cassegrain. The total field coma is therefore 12.62 times worse than for the classical Cassegrain, compared with factors of 8.27 and 28.56 for the DK and SP solutions respectively. The field coma is therefore only some 50% worse than that of the DK solution.

It is possible to combine the condition for *freedom from lateral decentering coma* for the form of the secondary given by (3.386) with the condition for a *spherical primary*. The corresponding primary is spherical if

$$(b_{s1})_{TC,0} = 0$$

requiring from (3.387)

$$R_A = \pm\frac{1}{2}\left(\frac{m_2^2}{m_2^2-1}\right) , \qquad (3.389)$$

the $\pm$ sign being required because $R_A = L/f'$ and f' is positive for the Cassegrain, negative for the Gregory case. With $|m_2| = \infty$ for the afocal form, $R_A = \pm 1/2$. For finite values of m_2, R_A grows slowly: for $|m_2| = 4$, $R_A = \pm 0.533$; for $|m_2| = 2$, $R_A = \pm 0.666$. So this combined solution of zero lateral decentering coma with a spherical primary is only possible with high obstruction ratios R_A. In the Cassegrain case this reduces the field coma according to (3.388).

An interesting variant [3.114] is a "Schiefspiegler" using off-axis segments, shown in Fig. 3.98, which maintains the advantages while overcoming the obstruction problem. The insensitivity to lateral decenter means that the secondary is insensitive to shifts perpendicular to the axis so far as coma in the axial image I_2' is concerned. Of course, the strong field coma of this centered solution remains, but the whole field remains constant so far as coma is concerned. Tilt of the off-axis segment of the secondary about its centre is also relatively insensitive, in the sense that the point in the field corresponding to coma freedom wanders from the centre field point to a point in the field where the field coma is compensated by the rotation coma. This

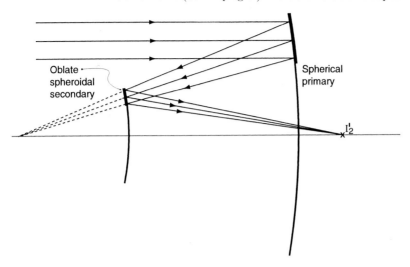

Fig. 3.98. Schiefspiegler with spherical primary and insensitive to lateral decenter [3.114]

type of compensation occurs in any 2-mirror telescope for which the field coma is not corrected, i.e. all forms except the RC and aplanatic Gregory.

The system of Fig. 3.98 is a special case of the general excentric system of Fig. 3.95.

The TC,0 form, free from lateral decentering coma, is simply a limit case of a Schiefspiegler with zero field angle at the primary and lateral shear of the secondary. As a Schiefspiegler, such a limit solution is normally illusory, as it does nothing to reduce the central obstruction; but the freedom from translation (lateral) decentering coma has considerable merit for special cases.

For 2-mirror telescopes with solutions $b_{s2} > +(m_2 - 1)/(m_2 + 1)$, we see from (3.385) that the bracket becomes negative and the CFP lies in front of the secondary in the Cassegrain case. The shorter the distance z_{CFP} from the secondary, the worse the lateral decentering coma. The SP solutions with normally acceptable values of R_A have shorter distances from (3.384) than the corresponding values for the RC, classical and, above all, DK solutions. With normal obstruction values, therefore, the SP solutions are not only the worst for field coma, but also for lateral decentering coma. As we have seen, only with $|R_A| > 0.5$ can this situation be remedied.

In general, the position of the CFP is of great importance in the design of systems for active correction of decentering coma. This will be dealt with in RTO II, Chap. 3.

3.7.2.4 Schiefspiegler with three or more mirrors. An analysis of 3-mirror Schiefspiegler and excentric uniaxial systems for telecommunication purposes has been given by Sand [3.115], for which the central obstruction of centered systems is a serious disadvantage (see § 3.10). Sand concludes that the 3-mirror Schiefspiegler of the Kutter form does not yield the necessary quality, although it is not clear why this is the case for the "axial" (centre field) image. He gives a 3-mirror excentric (uniaxial) solution equivalent to Fig. 3.95, in which the four conditions S_I, S_{II}, S_{III} and S_{IV} are fulfilled as in the systems of Paul-Baker, Korsch and Robb discussed above in § 3.6.5. The MTF (see § 3.10) is virtually perfect over a field of 1° or more because of the removal of central obstruction. System parameters are less sensitive than in the 2-mirror case. The baffling situation is favourable because of a fourth plane mirror giving the brachy-form.

Schiefspiegler with three or more mirrors clearly offer an enormous variety of solutions. Robb and Mertz [3.116] have investigated a scheme for an Arecibo type fixed spherical primary with a Gregory secondary and a corrector consisting of two oblique powered mirrors. For the steep primary considered, this design was not adequate because of higher order aberrations, although it represented a good third order solution. The authors reverted to a centered corrector.

The above system of Sand is an off-axis system *in aperture*. More recently, an off-axis version *in field* of the 3-mirror Paul telescope (Fig. 3.73) has been proposed by Stevick [3.166]. Since the Paul telescope has excellent field correction (coma and astigmatism corrected to the third order, residual field curvature which may be of little consequence for amateur use), the Stevick–Paul telescope yields very good imagery while avoiding the serious contrast loss (see Fig. 3.111) due to the central obstruction of a normal, centered Paul telescope.

In § 3.6.5.1 new work by Rakich [3.162] [3.163] was described, whereby an automatic program investigates the whole parametric space of 3-mirror solutions, of which two are spherical and one aspheric. Most of the interesting solutions revealed and discussed are normal centered, or folded, systems. However, Rakich also gives 4 new flat-field anastigmatic Schiefspiegler solutions. One of these is a flat-field modification of the Stevick–Paul telescope, giving excellent performance over a 0.5° circular field. If the primary is modified to become an ellipsoid ($b_s = -0.95$) as well as the liberation $c_2 \neq c_3$ to achieve the flat field, the system is called by Rakich a flat-field Paul–Rumsey arrangement, giving further improved quality. Other interesting flat-field systems given have the aspheric on the secondary ($f/7.9$), an all-spherical mirror system ($f/12.25$), a further system with aspheric on the secondary ($f/8.22$), and finally a Schiefspiegler of a type attributed to L.G. Cook (1987), for which the primary is an off-axis portion of an ellipsoid ($f/13$). A second version is shown with two folding flats to reduce the length.

A most interesting analysis of 4-mirror Schiefspiegler solutions has been performed by Schafer [3.117]. All the solutions are anastigmatic, free from obstruction, and contain only *spherical* surfaces. The starting basis for the designs is an inverted form of the Burch anastigmatic telescope [3.118], a modification using two concentric spheres of Schwarzschild's original design of Fig. 3.8.

Such proposals have in the past met with little interest. With modern progress in improvement of the reflectivity of larger mirrors, together with the stimulus to 3-mirror Schiefspiegler systems by the recent proposals of Stevick and Rakich above, there may well be considerable scope for further design work and realisation of 4-mirror Schiefspiegler solutions (see also § 3.6.5.1).

3.8 Despace effects in 2-mirror telescopes

3.8.1 Axial despace effects

Excellent treatments are given by Bahner [3.5] and Schroeder [3.22(e)].

3.8.1.1 Gaussian changes. There are three basic (independent) parameters of a 2-mirror telescope which are subject to error: f_1', f_2' and d_1. These errors will be determined by manufacturing and test procedures and tolerances, or, in the case of d_1, by the mechanical assembly conditions. The signs are as defined in Table 2.2 (and Table 2.3 for further derived quantities). For the axial despace characteristics, we require some relations from Chap. 2 (see Figs. 2.11 and 2.12). From (2.54) and (2.55):

$$f' = \frac{f_1' f_2'}{f_1' - f_2' - d_1} = m_2 f_1' \tag{3.390}$$

From (2.72):

$$L = f' - m_2 d_1 = m_2(f_1' - d_1) \tag{3.391}$$

From (2.72) and (2.90):

$$L = \frac{y_2}{y_1} f' = (m_2 + 1) f_2' \tag{3.392}$$

From (3.390) and (3.391):

$$L = \frac{(f_1' - d_1) f_2'}{f_1' - f_2' - d_1} \tag{3.393}$$

Note that f', L, y_2 and m_2 are *dependent* parameters. Their dependence on d_1 is particularly important since d_1 is used in all 2-mirror telescopes as the parameter for focusing the telescope correctly at a chosen image position.

With f_1' and f_2' constant, differentiation of (3.393) gives

$$dL = m_2^2 dd_1 \tag{3.394}$$

This is simply the magnification law for longitudinal image shifts [3.3] [3.6]. Similarly, with d_1 and f_2' constant, we derive

$$dL = -m_2^2 df_1' \,, \tag{3.395}$$

and with d_1 and f'_1 constant

$$dL = (m_2 + 1)^2 df'_2 \tag{3.396}$$

The despace sensitivity dm_2/dd_1 is derived by differentiating (3.391) with respect to d_1, in which f'_1 is constant and L, d_1 and m_2 are variables, and using (3.394), giving

$$dm_2 = \frac{m_2^2(m_2 + 1)}{L} dd_1 \tag{3.397}$$

The absolute shift of the image db/dd_1 is derived from (3.394) and (since the secondary and the image move in the same direction)

$$db = dL + dd_1 \; ,$$

giving

$$db = (m_2^2 + 1)dd_1 \tag{3.398}$$

Finally, the sensitivity df'/dd_1 is derived directly from (3.390) as

$$df' = \left(\frac{f'}{f'_1 - f'_2 - d_1} \right) dd_1 \tag{3.399}$$

3.8.1.2 Change in spherical aberration $\sum S_I$. A very important despace effect is the change of *spherical aberration* in the axial image (and therefore uniformly over the whole field) when d_1 is varied and the image shifts according to (3.398). The third order contribution $(S_I)_2$ of the secondary is easily derived from the recursion formulae (3.336) and (3.337) or directly from (3.39) and (2.72) as

$$(S_I)_2 = \left(\frac{y_1}{f'} \right)^3 y_2 \xi \tag{3.400}$$

with

$$\xi = \frac{(m_2 + 1)^3}{4} \left[\left(\frac{m_2 - 1}{m_2 + 1} \right)^2 + b_{s2} \right]$$

from (3.41). Then the required change in $\sum S_I$ is given by

$$\frac{d(S_I)_2}{dd_1} = \frac{d}{dd_1} \left[\left(\frac{y_1}{f'} \right)^3 y_2 \xi \right] = \left(\frac{y_1}{f'_1} \right)^3 \frac{d}{dd_1} \left[\frac{y_2}{m_2^3} \xi \right]$$

or

$$\frac{d(S_I)_2}{dd_1} = \left(\frac{y_1}{f'_1} \right)^3 \left[\frac{d}{dd_1} \left(\frac{y_2}{m_2^3} \right) \xi + \frac{y_2}{m_2^3} \frac{d\xi}{dd_1} \right] \tag{3.401}$$

From (3.391) and (3.392)

$$y_2 = y_1 \left(1 - \frac{d_1}{f_1'}\right) ,$$

giving

$$\frac{\mathrm{d}y_2}{\mathrm{d}d_1} = -m_2 \frac{y_1}{f'}$$

Using this and (3.397), the two terms of Eq. (3.401) reduce to

$$\frac{\mathrm{d}(S_I)_2}{\mathrm{d}d_1} = \left(\frac{y_1}{f'}\right)^4 \left[-m_2 \left(3m_2 + 4\right) \xi + m_2^2 \left(m_2^2 - 1 + 3\xi\right)\right]$$

or

$$\mathrm{d}(S_I)_2 = \left(\frac{y_1}{f'}\right)^4 \left[-4m_2\xi + m_2^2(m_2^2 - 1)\right]\mathrm{d}d_1 , \tag{3.402}$$

giving from (3.41)

$$\mathrm{d}(S_I)_2 = \left(\frac{y_1}{f'}\right)^4 \left[m_2^2 \left(m_2^2 - 1\right) - m_2(m_2 + 1)^3 \left\{\left(\frac{m_2 - 1}{m_2 + 1}\right)^2 + b_{s2}\right\}\right]\mathrm{d}d_1 \tag{3.403}$$

for the general case of a 2-mirror telescope.[10] Eq. (3.403) can also be written, introducing $f_1' = f'/m_2$, as

$$\mathrm{d}(S_I)_2 = \left(\frac{y_1}{f_1'}\right)^3 y_1 \left[\left(\frac{m_2^2 - 1}{m_2^2}\right) - \left(\frac{m_2 + 1}{m_2}\right)^3 \left\{\left(\frac{m_2 - 1}{m_2 + 1}\right)^2 + b_{s2}\right\}\right]\frac{\mathrm{d}d_1}{f_1'} , \tag{3.404}$$

in which $\mathrm{d}d_1$ is expressed as a proportion of the primary focal length f_1'. In the limit case of the afocal telescope with $m_2 = \infty$, this reduces to the simple form

$$\left[\mathrm{d}(S_I)_2\right]_{Afoc} = -\left(\frac{y_1}{f_1'}\right)^4 b_{s2}\mathrm{d}d_1 = -\left(\frac{y_1}{f_1'}\right)^3 y_1 b_{s2}\frac{\mathrm{d}d_1}{f_1'} , \tag{3.405}$$

which depends only on the aspheric form of the secondary.

---

[10] In a private communication (1997), Dr. Bahner indicated a form of equation equivalent to Eq. (3.403) which he considered appreciably simpler and therefore preferable. The difference arises simply because he introduced the parameters f' and L into the term in square brackets. Because of the innumerable relationships possible for the paraxial parameters indicated by Eqs. (2.54) to (2.90), almost all the formulae in this book can be expressed in many forms. Which form one chooses is a matter of individual preference. The form of Eq. (3.403) has been chosen here because all the terms in the square bracket contain only the fundamental dimensionless numerical parameters m_2 and b_{s2}. It also reduces easily to the simple forms of Eqs. (3.405) to (3.408) for the most important special cases.

In the case of the *classical telescope* (*cl*) form, $\xi_{cl} = 0$, giving from (3.402)

$$\left[d(S_I)_2 \right]_{cl} = \left(\frac{y_1}{f'} \right)^4 m_2^2 (m_2^2 - 1) dd_1 \qquad (3.406)$$

For the *aplanatic* form

$$(b_{s2})_{Aplan} = (b_{s2})_{cl} - \frac{2f'}{d_1(m_2 + 1)^3}$$

from (3.110), in which

$$(b_{s2})_{cl} = - \left(\frac{m_2 - 1}{m_2 + 1} \right)^2$$

Inserting this in (3.403) gives

$$\left[d(S_I)_2 \right]_{Aplan} = \left(\frac{y_1}{f'} \right)^4 \left[2m_2 \frac{f'}{d_1} + m_2^2 (m_2^2 - 1) \right] dd_1 \qquad (3.407)$$

Taking the typical Cassegrain parameters of Table 3.2 with $m_2 = -4$, $f' = 1$ and $d_1 = -0.19375$, Eq. (3.407) gives

$$\left[d(S_I)_2 \right]_{RC} = \left(\frac{y_1}{f'} \right)^4 \left[41.29 + 240 \right] dd_1 \;,$$

in which the second term is the value for a classical Cassegrain. The increase in sensitivity due to the RC form is modest, 17.2% for the above values. The increase in $(b_{s2})_{RC}$ over $(b_{s2})_{cl}$ is 13.8%. For large departures from the classical form, b_{s2} dominates over the fixed terms in (3.403) and the effect on the spherical aberration becomes more nearly proportional to $b_{s2} - (b_{s2})_{cl}$.

For the *Dall-Kirkham (DK)* form, $(b_{s2})_{DK} = 0$ and (3.403) reduces to

$$\left[d(S_I)_2 \right]_{DK} = \left(\frac{y_1}{f'} \right)^4 m_2 \left(m_2^2 - 1 \right) dd_1 \;, \qquad (3.408)$$

the spherical aberration change having reversed sign and, with $m_2 = -4$, only one quarter of that of a classical Cassegrain. The DK form is, therefore, not only favourable for transverse decenter but also for axial despace. The Gregory DK form (DKG) has the same value, but with positive sign with $m_2 = +4$. The favourable nature of the DK solution is particularly clear in the *afocal* form, when, from (3.405), the sensitivity becomes zero with $b_{s2} = 0$. The physical explanation is instructive. If the secondary is at the prime focus, it introduces zero $(S_I)_2$ because y_2 is zero in Eqs. (3.20). As the secondary is moved towards the primary, the first (spherical) term of $(S_I)_2$ increases, reaching a maximum in the afocal position, if r_2 is kept constant. It reduces as $|d_1|$ is further reduced, because A_2 reduces although y_2 increases, reaching zero when the centre of curvature of the secondary is at the prime focus. Hence the differential of $(S_I)_2$ is zero in the afocal case. By contrast, the aspheric term τ_2 increases monotonically with y_2^4, leading to Eq. (3.405).

From the conversion formulae (3.190), we can convert the above wavefront aberration differentials $d(S_I)_2$ into angular aberration. At the *best focus* (BF), the angular spherical aberration *in radians* corresponding to the *diameter* of the image (100% geometrical energy) is, from (3.404),

$$
\begin{aligned}
d(\delta u_p')_{BF,S_I} = & -\frac{1}{32 N_1^3} \left[\left(\frac{m_2^2 - 1}{m_2^2} \right) \right. \\
& \left. - \left(\frac{m_2 + 1}{m_2} \right)^3 \left\{ \left(\frac{m_2 - 1}{m_2 + 1} \right)^2 + b_{s2} \right\} \right] \frac{dd_1}{f_1'} \; \text{rad}
\end{aligned}
\tag{3.409}
$$

where N_1 is the f/no of the primary of focal length f_1'. At the best focus (BF), the disk of least confusion is one quarter of $d(\delta u_p')_{GF,S_I}$ at the Gaussian focus. A classical telescope with $N = 10$, $m_2 = \pm 4$ and $dd_1/f' = 10^{-3}$ or $dd_1/f_1' = 4 \cdot 10^{-3}$ gives from (3.409) with the second term of the square bracket zero

$$
\left[d(\delta u_p')_{GF} \right]_{cl} = -6.188 \text{ arcsec}
$$

and

$$
\left[d(\delta u_p')_{BF} \right]_{cl} = -1.547 \text{ arcsec}
$$

These values correspond to a despace of 10 mm in a 1 m telescope with $f' = 10$ m, the corresponding image shift being 170 mm according to (3.398).

If Eq. (3.406) is expressed in terms of the fixed parameter f_1' while f' varies with m_2, then

$$
\left[d(S_I)_2 \right]_{cl} = \left(\frac{y_1}{f_1'} \right)^4 \frac{(m_2^2 - 1)}{m_2^2} dd_1
\tag{3.410}
$$

and the change of wavefront aberration is virtually independent of m_2 if $m_2 \gg 1$. This is physically obvious if we consider that the change in beam diameter at the secondary depends only on $(y_1/f_1')dd_1$ and that the aspheric form of the secondary converges on the parabola as $m_2 \to \infty$.

We have seen above that the DK form is free from despace spherical aberration in the *afocal* case. A final case of interest is the focal case of a 2-mirror telescope, corrected for spherical aberration and with zero despace sensitivity (DE,0). Setting the bracket of (3.409) to zero gives the condition

$$
(b_{s2})_{DE,0} = \frac{m_2 - 1}{(m_2 + 1)^2}
\tag{3.411}
$$

With $m_2 = -4$, $(b_{s2})_{DE,0} = -0.556$, an ellipse. This is the axial equivalent of the TC,0 form of Eq. (3.386), which gave zero lateral decentering coma. But that condition requires an oblate spheroid.

The effect of axial despace error on the third order spherical aberration, expressed as the angular image diameter at *best focus* (BF), is given in Table 3.21 for the various 2-mirror telescope forms normalized as in Table 3.2.

Table 3.21. Angular despace spherical aberration at best focus (BF) and angular despace field coma for 2-mirror telescopes defined as in Table 3.2: $|m_2| = 4$, $|R_A| = 0.225$, $|N| = 10$ and the despace is $|dd_1/f'| = 10^{-3}$, or $dd_1/f_1' = -4 \cdot 10^{-3}$, ten times the decenter δ of Table 3.20. The semi-field angle for the coma is $u_{pr1} = 15$ arcmin

Telescope type	Despace angular spherical aberration at best focus $d(\delta u_p')_{BF}$ (arcsec)	Despace angular field coma $d(\delta u_p')_{Coma_t}$ (arcsec)
Classical Cassegrain (C)	$- 1.547$	$- 0.878$
Ritchey-Chrétien (RC)	$- 1.813$	$- 0.959$
Dall-Kirkham (DK)	$+ 0.387$	$- 0.287$
Spherical Primary (SP)	$+ 5.787$	$+ 1.362$
Classical Gregory (CG)	$- 1.547$	$+ 0.802$
Aplanatic Gregory (AG)	$- 1.715$	$+ 0.898$
DK Gregory (DKG)	$- 0.387$	$+ 0.144$
SP Gregory (SPG)	$- 8.880$	$+ 4.962$

The angular values are comparable with those of Table 3.20 although the axial change dd_1 is ten times larger than the lateral decenter δ. The right-hand column of Table 3.21 gives the axial despace effect on the field coma $\sum S_{II}$, which will be derived in the next section.

3.8.1.3 Change in field coma $\sum S_{II}$. This effect can be important, above all in aplanatic telescopes with extreme image requirements (such as space applications); or in cases where an image shift is required and the spherical aberration is compensated by other means.

We require here the contribution of the secondary mirror to $\sum S_{II}$, i.e. $(S_{II})_2$. This can be derived from the recursion formulae (3.336) and (3.337) or taken directly from (3.52):

$$(S_{II})_2 = \left(\frac{y_1}{f'}\right)^3 \left[-d_1\xi + \frac{f'}{2}(m_2^2 - 1) - \frac{s_{pr1}}{f'}L\xi\right] u_{pr1}$$

We will suppose the entrance pupil is at the primary, giving $s_{pr1} = 0$ and

$$(S_{II})_2 = -\left(\frac{y_1}{f_1'}\right)^3 \left[\frac{d_1\xi}{m_2^3} - \frac{f_1'}{2}\frac{(m_2^2 - 1)}{m_2^2}\right] u_{pr1} , \qquad (3.412)$$

in which f', a dependent variable, is replaced by the independent variable f_1'. Differentiating with respect to d_1 gives

$$\frac{d(S_{II})_2}{dd_1} = \frac{d}{dd_1}\left[-\left(\frac{y_1}{f_1'}\right)^3 \left(\frac{d_1\xi}{m_2^3} - \frac{(m_2^2 - 1)}{m_2^2}\frac{f_1'}{2}\right)\right] u_{pr1} \qquad (3.413)$$

With

$$\frac{d}{dd_1}\left[\frac{d_1\xi}{m_2^3}\right] = \frac{1}{m_2^3}\left[d_1\frac{d\xi}{dd_1} + \xi\right] - \frac{1}{m_2^4}\left[3d_1\xi\frac{dm_2}{dd_1}\right]$$

$$\frac{d\xi}{dd_1} = \frac{3}{4}\frac{dm_2}{dd_1}(m_2+1)^2\left[\frac{2}{3}\left(\frac{m_2-1}{m_2+1}\right) + \frac{1}{3}\left(\frac{m_2-1}{m_2+1}\right)^2 + b_{s2}\right]$$

and from (3.397) and (2.72)

$$\frac{dm_2}{dd_1} = \frac{m_2^2(m_2+1)}{L} = \frac{m_2(m_2+1)}{f_1'R_A},$$

we can derive from (3.413), after reduction using the paraxial relations of (2.72),

$$\frac{d(S_{II})_2}{dd_1} = -\left(\frac{y_1}{f'}\right)^3\frac{1}{R_A}\left[(4R_A-3)\xi \right.$$
$$\left. +(1-R_A)m_2(m_2^2-1) - m_2(m_2+1)\right]u_{pr1} \right\} \tag{3.414}$$

as the general form of axial despace field coma for any 2-mirror telescope with the stop at the primary, ξ being defined as in Table 3.5.

Converting to *angular tangential coma* with (3.198) and $n' = +1$, if dd_1 is referred to $f_1' = f'/m_2$, gives

$$d(\delta u_p')_{Coma_t} = \frac{3}{8N^2}\frac{1}{R_A}\left[\left(\frac{(4R_A-3)}{m_2}\right)\xi \right.$$
$$\left. +(1-R_A)(m_2^2-1) - (m_2+1)\right]u_{pr1}\frac{dd_1}{f_1'} \right\} \tag{3.415}$$

If we substitute for ξ and insert $N_1 = N/m_2$, this can be written as

$$d(\delta u_p')_{Coma_t} = \frac{3}{8N_1^2}\frac{1}{R_A}\left[\left(\frac{(4R_A-3)}{4}\right)\left(\frac{m_2+1}{m_2}\right)^3\left\{\left(\frac{m_2-1}{m_2+1}\right)^2 + b_{s2}\right\}\right.$$
$$\left. +(1-R_A)\left(\frac{m_2^2-1}{m_2^2}\right) - \left(\frac{m_2+1}{m_2^2}\right)\right]u_{pr1}\frac{dd_1}{f_1'} \right\}$$
$$\tag{3.416}$$

This form can also be applied to the *afocal* case with $m_2 \to \infty$, giving

$$\left[d(\delta u_p')_{Coma_t}\right]_{Afoc} = \frac{3}{8N_1^2}\frac{1}{R_A}\left[\left(\frac{(4R_A-3)}{4}\right)(1+b_{s2}) + (1-R_A)\right]u_{pr1}\frac{dd_1}{f_1'}$$
$$\tag{3.417}$$

For the classical afocal Mersenne telescopes with a parabolic secondary, this reduces to the simple form

$$\left[d(\delta u_p')_{Coma_t}\right]_{Afoc,cl} = \frac{3}{8N_1^2}\frac{1}{R_A}(1-R_A)u_{pr1}\frac{dd_1}{f_1'} \tag{3.418}$$

In the normal *focal* case, we can apply Eq. (3.415) to the various telescope forms of Table 3.2. For the *classical telescope*, ξ_{cl} is zero, giving

$$\left[d(\delta u_p')_{Coma_t}\right]_{cl} = \frac{3}{8N^2}\frac{1}{R_A}\left[(1-R_A)(m_2^2-1)-(m_2+1)\right]u_{pr1}\frac{dd_1}{f_1'} \tag{3.419}$$

For an *aplanatic telescope*, we have from (3.108)

$$\xi_{Aplan} = -\frac{f'}{2d_1} = -\frac{m_2}{2(1-R_A)}$$

from (2.72), giving

$$\left.\begin{array}{l}\left[d(\delta u_p')_{Coma_t}\right]_{Aplan} = \frac{3}{8N^2}\frac{1}{R_A}\left[-\left(\frac{(4R_A-3)}{2(1-R_A)}\right)\right.\\[2mm] \left.+(1-R_A)(m_2^2-1)-(m_2+1)\right]u_{pr1}\frac{dd_1}{f_1'}\end{array}\right\} \tag{3.420}$$

Following Schroeder [3.22(e)], it is possible to transform the above formulae by replacing the parameter R_A by $\bar{b} = b/f_1'$. From (2.65) and (2.72) one can deduce

$$R_A = \frac{\bar{b}-1}{m_2-1}, \tag{3.421}$$

a useful relation, since b is often a defining parameter in the layout of a telescope. In general, Eq. (3.415) shows that the despace coma will increase, for a fixed image position, if $|m_2|$ increases, since $|R_A|$ will be reduced. This is simply another example demonstrating the increased sensitivity of small secondaries with high magnification $|m_2|$.

Equation (3.420) is similar but not identical with the corresponding RC equation given by Schroeder [3.22(e)]. If this is transformed into the parameters of (3.420), the difference is in the first term of the square bracket i.e. $(3-4R_A)/2(1-R_A)$. Schroeder's form is $(3-R_A)/2$. With $R_A = 0.25$, these two forms give 1.333 and 1.375 respectively. Since, in the RC case, this term involving ξ is small compared with the residual terms corresponding to the classical Cassegrain if $|m_2| \gg 1$, the difference in $\left[d(\delta u_p')_{Coma_t}\right]_{RC}$ is very small in practical cases. For a case given by Schroeder with $N_1 = -2.5$, $N = 10$ (giving $m_2 = -4$), $\bar{b} = -0.25$ (giving $R_A = 0.25$), $u_{pr1} = 18$ arcmin and $dd_1 = -0.001f_1'$, $\left[d(\delta u_p')_{Coma_t}\right]_{RC}$ is -0.25245 arcsec from Eq. (3.420) and -0.25312 arcsec from Schroeder's formula. A ray-tracing check performed with the highly versatile ACCOS V optical design program and using the above parameters, to give an exact calculation of the third order field coma arising from despace, gives agreement with Eq. (3.420) to better than 0.00005 arcsec.

Table 3.21 gives the despace angular field coma for the different telescope forms under the same conditions as for the despace spherical aberration with $dd_1/f_1' = -4 \cdot 10^{-3}$. The semi-field angle in Eq. (3.415) is $u_{pr1} = 15$ arcmin, a typical field for RC telescopes. As is expected from the large field coma, the SP and SPG forms are by far the worst, the SPG value being very large. This arises not only because the SPG supplement term in (3.415) is larger, but also because it is additive to the classical telescope terms, whereas in the SP case it is subtractive. The Cassegrain SP form is, for all aberration effects, superior.

One reason for this is that the Gregory secondary, for the normalized geometry of Table 3.2, is markedly stronger than the Cassegrain secondary. However, this normalization is somewhat unfair to the Gregory. Since the Cassegrain and Gregory telescopes of Table 3.2 have the same values of $|R_A|$ and $|m_2|$, the final image position is not the same, because $L = |R_A|$ in the normalized system and d_1 is not the same in the two cases. From (3.421) we have $\bar{b} = -0.125$ for the Cassegrain systems with $m_2 = -4$ and $+0.325$ for the Gregory systems. For an aplanatic Gregory (AG) with $\bar{b} = -0.125$, the value of R_A is -0.375. Since R_A enters into the denominator of (3.420) linearly, this larger axial obstruction gives a more favourable value of $\left[d(\delta u_p')_{Coma_t}\right]_{AG}$ for the same image position as the Cassegrain, namely $+0.621$ arcsec compared with the value in Table 3.21 of $+0.898$ arcsec with $R_A = -0.225$. This is a further illustration of the price paid in sensitivity with small obstruction, high magnification secondaries.

In practice, the despace sensitivity to field coma is of little significance in non-aplanatic 2-mirror telescopes, since the field coma is not corrected. However, this is no longer true if field correctors are added which are also intended to correct the field coma (see Chap. 4).

3.8.2 Transverse despace effects

The important effect of introducing decentering coma, uniform over the field, was treated in detail in § 3.7.2. This adds vectorially to the field coma in non-aplanatic telescopes, giving a transverse shift of the image point in the field which is free from coma. For aplanatic telescopes, the decentering coma is not compensated at any field point and the coma-free property of the system is lost. Thus decenter is particularly serious for aplanatic telescopes, even though the formal sensitivity to decentering coma may not be significantly more than with classical forms.

Lateral decenter of the secondary (see Fig. 3.97) also produces a Gaussian *lateral shift of the image*. The change in the direction of the beam is twice the slope of the secondary corresponding to the height of the decenter δ. Now, from (3.2)

$$z_2 = \frac{y_2^2}{2r_2} = \frac{\delta^2}{2r_2} \quad \text{and} \quad \frac{dz_2}{dy} = \frac{\delta}{r_2}$$

From (2.90)

$$r_2 = \frac{2L}{m_2 + 1}$$

Since the image plane is at a distance L from the secondary, the lateral shift is

$$(\Delta\eta')_{dec} = L\left(2\frac{dz_2}{dy}\right) = (m_2 + 1)\delta, \tag{3.422}$$

in the opposite sense of the movement δ for a Cassegrain telescope where m_2 is negative. The *angular* lateral shift of the image related to the scale of the telescope in object space (pointing change) is

$$(\Delta u_p')_{dec} = (m_2 + 1)\frac{\delta}{f'} \text{ rad} = (m_2 + 1)\frac{\delta}{f'} \text{ 206 265 arcsec} \tag{3.423}$$

Such pointing changes are very large compared with angular decentering coma.

Rotation of the secondary through the angle u_{pr2}, using the notation of Eq. (3.375), produces an angular shift of the exit beam of $2u_{pr2}$. The linear image shift is then

$$(\Delta\eta')_{rot} = 2Lu_{pr2} = 2f'R_A u_{pr2} \tag{3.424}$$

The corresponding *angular* shift for the scale of the telescope (pointing change) is

$$(\Delta u_p')_{rot} = 2R_A(u_{pr2})_{arcsec} \text{ arcsec} \tag{3.425}$$

3.9 Zernike polynomials

The third order theory of telescope systems given in this chapter is based on the classical "Characteristic Function" of Hamilton discussed in § 3.2. Virtually all optical design programs provide analysis in these terms, but optimization of systems is usually performed using ray trace data which embraces *all* orders however they are defined. Nevertheless, most optical designers still think in terms of the "classical aberrations" because of the clear physical significance and the advantages of the simple third order terms. However, the Hamilton functions have the major disadvantage of being not only non-linear, but also of being (in general) non-orthogonal. The algorithms of optical design optimization routines have to take account of this.

In 1934 Zernike [3.119], in a classic paper, introduced the so-called Zernike circle polynomials. These have two very important advantages over the Hamilton terms. First, they are *orthogonal* functions and can be treated or corrected independently; second, they represent individually an optimum least square fit to the data within the accuracy of the degree of the polynomial.

The standard treatment of Zernike polynomials has been given by Wolf [3.120(a)] based on the derivation by Bhatia and Wolf [3.121] from the requirement of orthogonality and invariance. We shall reproduce this here in a condensed form. They show that there exists an infinity of complete sets of polynomials in two real variables x, y which are orthogonal within the unit circle (in our case, the pupil with radius normalized to unity), i.e. satisfying the orthogonality condition

$$\iint_{x^2+y^2\leq 1} V^*_{(\alpha)}(x,y)V_{(\beta)}(x,y)\mathrm{d}x\,\mathrm{d}y = A_{\alpha\beta}\delta_{\alpha\beta} , \qquad (3.426)$$

in which V_α and V_β denote two typical polynomials of the set, V^* being the complex conjugate, $\delta_{\alpha\beta}$ is the Kronecker symbol [11] and $A_{\alpha\beta}$ are normalization constants to be chosen. The circle polynomials of Zernike are distinguished from other sets by certain simple invariance properties, i.e. an invariance of form with respect to rotations of the axes about the origin. This has a resemblance to the Hamilton approach from the axial symmetry of a centered optical system. The function must then have the general type of solution

$$V(\rho \cos \phi, \rho \sin \phi) = R(\rho)e^{il\phi} \qquad (3.427)$$

in which ρ and ϕ have the same significance as in § 3.2, the radius and rotation angle in the pupil, and ρ is normalized to unity at the edge. In (3.427) $R(\rho) = V(\rho, 0)$ is a function only of ρ. The function $e^{il\phi}$ is now expanded in powers of $\cos\phi$ and $\sin\phi$. If V is a polynomial of *degree* n in the variables $x = \rho \cos \phi$ and $y = \rho \sin \phi$, it follows from (3.427) that $R(\rho)$ is a polynomial in ρ of degree n and *contains no power of ρ of degree lower than $|l|$. $R(\rho)$ is an even or odd polynomial depending on whether l is even or odd. The set of Zernike circle polynomials* is distinguished from all other possible sets by the property that it contains a polynomial for each pair of the permissible values of n (degree) and l (rotation angle in the pupil), *i.e. for integral values of n and l*, such that

$$n \geq 0 ; \quad l \underset{<}{\overset{>}{=}} 0 ; \quad n \geq |l| ; \quad (n - |l|) \text{ is an even number.} \qquad (3.428)$$

Wolfthen shows that the function may be normalized and expressed as

$$V_n^{\pm m}(\rho \cos \phi, \rho \sin \phi) = R_n^m(\rho)e^{\pm i\,m\,\phi} , \qquad (3.429)$$

in which $R_n^m(\rho)$ are the *radial polynomials* and $m = |l|$, a positive integer or zero.

The set of circle polynomials contains $\frac{1}{2}(n + 1)(n + 2)$ linearly independent polynomials of degree $\leq n$. It follows that every monomial function $x^i y^j$ where i and j are integers with $i \geq 0$, $j \geq 0$, *and consequently every polynomial in x,y*, may be expressed as a linear combination of a finite number of the circle polynomials V_n^l.

To derive explicit expressions for the radial polynomials they are orthogonalized by applying a special weighting factor related to the more general

---

one used in defining Jacobi polynomials. The Zernike normalization is then applied so that for all values of n and m

$$R_n^{\pm m}(1) = 1 \qquad (3.430)$$

for the edge of the unit circle, the pupil. Wolf then derives the explicit form for the radial polynomials as

$$R_n^{\pm m}(\rho) = \sum_{s=0}^{\frac{1}{2}(n-m)} (-1)^s \frac{(n-s)!}{s! \left(\frac{n+m}{2} - s\right)! \left(\frac{n-m}{2} - s\right)!} \rho^{(n-2s)} , \qquad (3.431)$$

in which

$$0 \leq s \leq \frac{n-m}{2} \qquad (3.432)$$

and bearing in mind that $0! = 1$. Wolf shows that the general generating function of the radial polynomials reduces in the case of $m = 0$ (axisymmetrical functions) to the generating function for the Legendre polynomials of argument $(2\rho^2 - 1)$, so that

$$R_{2n}^0(\rho) = P_n(2\rho^2 - 1) \qquad (3.433)$$

Table 3.22 gives the radial polynomials derived from (3.431) up to degree 8. It should be remembered that each radial polynomial with $m > 0$ has *two* components depending on $\rho \cos \phi$ and $\rho \sin \phi$, giving the number of linearly independent polynomials of degree $\leq n$ as $\frac{1}{2}(n+1)(n+2)$, as indicated above.

Resolving in this way, we derive from Table 3.22 the systematic ordering of radial polynomials given, following Dierickx [3.122], in Table 3.23. This comprises the terms in a triangle of Table 3.22 with base extending from $n = 1$ to $n = 10$ and apex at $n = m = 5$. By analogy with the Characteristic Function of § 3.2, the three lowest terms are the Gaussian effects. Terms with $m = 0$ and $n \geq 4$ are termed spherical aberration; with $m = 1$ and $n \geq 3$ coma; with $m = 2$ and $n \geq 2$ astigmatism; $m = 3$ triangular; $m = 4$ quadratic; $m = 5$ five-fold. The classical order numbers are given by $(n + m - 1)$.

From the normalization of (3.430), all functions are unity if $\rho = 1$. Integration of the functions confirms the orthogonality condition, that the integral be zero [3.123], is fulfilled. This is physically obvious for all the terms in Table 3.23 with $m \neq 0$ since they depend on $\cos m\phi$ or $\sin m\phi$. Setting $U = m\phi$, then $\int \cos m\phi \, d\phi = \frac{1}{m} \sin U$, giving a zero integral as in the case of the basic polynomial $\rho \cos \phi$. Similarly, it is easily shown that the radial polynomials represent a balance of classical terms up to the defined degree which minimizes the rms wavefront value [12]. Thus the focus term R_2^0 (No. 3) balances

---

[12] The rms wavefront error is given by

$$rms = \left[(\overline{W^2}) - (\overline{W})^2 \right]^{1/2} ,$$

i.e. the square root of the [mean of the square of the wavefront aberration minus the square of the mean wavefront aberration] (see § 3.10).

Table 3.22. The Zernike radial polynomials $R_n^m(\rho)$ up to degree 8 (after Born-Wolf [3.120(a)])

n / m	0	1	2	3	4	5	6	7	8
0	1		$2\rho^2 - 1$		$6\rho^4 - 6\rho^2 + 1$		$20\rho^6 - 30\rho^4 + 12\rho^2 - 1$		$70\rho^8 - 140\rho^6 + 90\rho^4 - 20\rho^2 + 1$
1		ρ		$3\rho^3 - 2\rho$		$10\rho^5 - 12\rho^3 + 3\rho$		$35\rho^7 - 60\rho^5 + 30\rho^3 - 4\rho$	
2			ρ^2		$4\rho^4 - 3\rho^2$		$15\rho^6 - 20\rho^4 + 6\rho^2$		$56\rho^8 - 105\rho^6 + 60\rho^4 - 10\rho^2$
3				ρ^3		$5\rho^5 - 4\rho^3$		$21\rho^7 - 30\rho^5 + 10\rho^3$	
4					ρ^4		$6\rho^6 - 5\rho^4$		$28\rho^8 - 42\rho^6 + 15\rho^4$
5						ρ^5		$7\rho^7 - 6\rho^5$	
6							ρ^6		$8\rho^8 - 7\rho^6$
7								ρ^7	
8									ρ^8

Table 3.23. Zernike polynomials resolved in the x, y directions. This table gives all terms up to R_5^5 and subsequent terms up to $n = 10$ with $n + m \leq 10$ (after Dierickx [3.122])

No. of Radial Polynomial	Mathematical expression	Classical interpretation of wavefront effect	$\underline{n}$	$\underline{m}$	$\underline{(n+m)}$
0	1	constant	0	0	0
1	$\rho \cos \phi$	tilt	1	1	
2	$\rho \sin \phi$	tilt	1	1	2
3	$2\rho^2 - 1$	focus	2	0	---
4	$\rho^2 \cos 2\phi$	astigmatism 3rd order	2	2	
5	$\rho^2 \sin 2\phi$	astigmatism 3rd order	2	2	
6	$(3\rho^2 - 2)\rho \cos \phi$	coma 3rd order	3	1	4
7	$(3\rho^2 - 2)\rho \sin \phi$	coma 3rd order	3	1	
8	$6\rho^4 - 6\rho^2 + 1$	spherical 3rd order	4	0	---
9	$\rho^3 \cos 3\phi$	triangular 5th order	3	3	
10	$\rho^3 \sin 3\phi$	triangular 5th order	3	3	
11	$(4\rho^2 - 3)\rho^2 \cos 2\phi$	astigmatism 5th order	4	2	6
12	$(4\rho^2 - 3)\rho^2 \sin 2\phi$	astigmatism 5th order	4	2	
13	$(10\rho^4 - 12\rho^2 + 3)\rho \cos \phi$	coma 5th order	5	1	
14	$(10\rho^4 - 12\rho^2 + 3)\rho \sin \phi$	coma 5th order	5	1	
15	$20\rho^6 - 30\rho^4 + 12\rho^2 - 1$	spherical 5th order	6	0	---
16	$\rho^4 \cos 4\phi$	quadratic 7th order	4	4	
17	$\rho^4 \sin 4\phi$	quadratic 7th order	4	4	
18	$(5\rho^2 - 4)\rho^3 \cos 3\phi$	triangular 7th order	5	3	
19	$(5\rho^2 - 4)\rho^3 \sin 3\phi$	triangular 7th order	5	3	
20	$(15\rho^4 - 20\rho^2 + 6)\rho^2 \cos 2\phi$	astigmatism 7th order	6	2	8
21	$(15\rho^4 - 20\rho^2 + 6)\rho^2 \sin 2\phi$	astigmatism 7th order	6	2	
22	$(35\rho^6 - 60\rho^4 + 30\rho^2 - 4)\rho \cos \phi$	coma 7th order	7	1	
23	$(35\rho^6 - 60\rho^4 + 30\rho^2 - 4)\rho \sin \phi$	coma 7th order	7	1	
24	$70\rho^8 - 140\rho^6 + 90\rho^4 - 20\rho^2 + 1$	spherical 7th order	8	0	---
25	$\rho^5 \cos 5\phi$	5-fold 9th order	5	5	
26	$\rho^5 \sin 5\phi$	5-fold 9th order	5	5	
27	$(6\rho^2 - 5)\rho^4 \cos 4\phi$	quadratic 9th order	6	4	
28	$(6\rho^2 - 5)\rho^4 \sin 4\phi$	quadratic 9th order	6	4	
29	$(21\rho^4 - 30\rho^2 + 10)\rho^3 \cos 3\phi$	triangular 9th order	7	3	10
30	$(21\rho^4 - 30\rho^2 + 10)\rho^3 \sin 3\phi$	triangular 9th order	7	3	
31	$(56\rho^6 - 105\rho^4 + 60\rho^2 - 10)\rho^2 \cos 2\phi$	astigmatism 9th order	8	2	
32	$(56\rho^6 - 105\rho^4 + 60\rho^2 - 10)\rho^2 \sin 2\phi$	astigmatism 9th order	8	2	
33	$(126\rho^8 - 280\rho^6 + 210\rho^4 - 60\rho^2 + 5)\rho \cos \phi$	coma 9th order	9	1	
34	$(126\rho^8 - 280\rho^6 + 210\rho^4 - 60\rho^2 + 5)\rho \sin \phi$	coma 9th order	9	1	
35	$252\rho^{10} - 630\rho^8 + 560\rho^6 - 210\rho^4 + 30\rho^2 - 1$	spherical 9th order	10	0	---
⋮	$924\rho^{12} - 2772\rho^{10} + 3150\rho^8 - 1680\rho^6$ $+420\rho^4 - 42\rho^2 + 1$	spherical 11th order	12	0	12

the Gaussian focus shift by adding the constant term which shifts the origin to minimize the rms; the term R_4^0 (No. 8) balances the third order spherical aberration against focus and a constant. Such balancing is, of course, well known from optical design using classical Hamilton terms.

Another excellent treatment of Zernike polynomials is given by Malacara [3.124]. He tabulates some comparative properties of a number of one-dimensional orthogonal polynomials: Legendre, Tschebyscheff, Jacobi, Laguerre and Hermite. In connection with *active optics* we shall introduce a further important type with great practical significance, namely *natural vibration modes*. Malacara also gives formulae and tables, based on work by Sumita [3.125], for conversion from Zernike polynomials to monomials and vice-versa.

3.10 Diffraction theory and its relation to aberrations

3.10.1 The Point Spread Function (PSF) due to diffraction at a rectangular aperture

Our treatment up to now has been entirely on the basis of *geometrical optics* for which *rays* are defined as the normals to a geometrical wavefront undisturbed by the physical boundary limiting its size.

Diffraction is a phenomenon arising from the wave theory of light. For a full account, the reader is referred to the standard work by Born-Wolf [3.120]. Many other excellent treatments with various emphases are available, e.g. Bahner [3.5], Welford [3.6], Maréchal and Françon [3.26] and Schroeder [3.22]. Schroeder's treatment in his Chap. 10 and 11, with particular reference to the Hubble Space Telescope, not only gives the basic theory but also an admirable practical treatment of the diffraction theory of aberrations.

Practical diffraction theory is based on certain important approximations: that the divergence angles of all rays from the object to the pupil, and the convergence angles from the pupil to the image, are small; that the pupil is large compared with the wavelength of the incident light; that polarisation effects are neglected. Normally, it is also assumed that the light transmission is uniform over the pupil: if a departure from this condition is introduced, then a procedure known as *apodisation* is effected. The first approximation above concerning *small* divergence and convergence angles leads to the neglect of all quadratic terms in the expression of the optical path length in terms of the pupil and field coordinates. This distinguishes *Fraunhofer* and *Fresnel* diffraction: in Fraunhofer diffraction the quadratic terms are ignored; in Fresnel diffraction they are taken into account. Fortunately, for most problems of optical instruments and virtually all cases of telescopes, the Fraunhofer approximations give a very good accuracy. However, for special cases of very high relative apertures or fields, the small angle approximation must be borne in mind.

For consistency with our previous notation in which a point in the pupil defined by ρ, ϕ was resolved into coordinates x, y (z being the axial direction) and the image height denoted by η, we shall reverse the notation used by Born-Wolf and Schroeder and use that shown in Fig. 3.99 for the case of a rectangular aperture (or slit). We assume a perfectly aberration-free spherical wavefront is converging from the exit pupil E' to the image point Q'_0. Following Born-Wolf [3.120(b)], the Fraunhofer diffraction integral giving the complex amplitude $U(Q')$ at point Q' near to Q'_0 is

$$U(Q') = C \int_{-x_m}^{+x_m} \int_{-y_m}^{+y_m} e^{-ik(px+qy)} \, dx \, dy = C \int_{-x_m}^{+x_m} e^{-ikpx} dx \int_{-y_m}^{+y_m} e^{-ikqy} dy$$

$$(3.434)$$

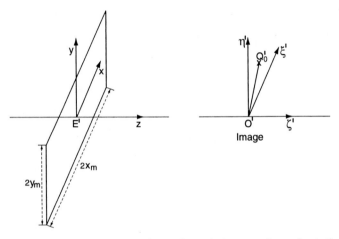

Fig. 3.99. Exit pupil (x, y, z) and image plane (η, ξ, ζ) coordinate systems

In this equation, $k = 2\pi/\lambda$ is defined from the basic differential wave equations, while the quantities p and q are defined by

$$p = l - l_o , \quad q = m - m_o ,$$

where l_o, m_o and l, m are the first two direction cosines of the incident and diffracted waves respectively. It is part of the definition of the approximation of Fraunhofer diffraction that the four direction cosines only enter linearly in the above combinations, implying that the diffraction effect is unchanged if the pupil aperture is shifted in its own plane. Now

$$\int_{-x_m}^{+x_m} e^{-ikpx} dx = -\frac{1}{ikp} \left[e^{-ikpx_m} - e^{ikpx_m} \right] = 2\frac{\sin(kpx_m)}{kp}$$

from the Euler relation, and similarly for the other integral in y. The *intensity* I at point Q' in the image plane is therefore

$$I(Q') = |U(Q')|^2 = \left(\frac{\sin kpx_m}{kpx_m}\right)^2 \left(\frac{\sin kqy_m}{kqy_m}\right)^2 I_0 , \qquad (3.435)$$

with $I_0 = C^2 A^2 = \tilde{E}A/\lambda^2$ is the intensity at the central point Q'_0 of the diffraction pattern, $\tilde{E}$ being the total energy falling on the pupil and $A = 4x_m y_m$ being its area. According to geometrical optics, the image at Q'_0 of an infinitesimal conjugate object point Q_0 would be an infinitesimal point, whereas diffraction theory gives the intensity distribution of (3.435).

From the definition of p and q above in terms of the differences of the direction cosines, which represent the angular shifts of Q' from Q'_0 in the x and y planes respectively, we have for an object at infinity in the normal telescopic case

$$\delta\eta' = (m - m_o)f' , \quad \delta u'_y = m - m_o = \frac{\delta\eta'}{f'}$$

as the linear and angular separations of Q' from Q'_0 in the y-plane. Then

$$\delta\eta' = qf' = \frac{\delta\bar{u}'_y}{ky_m} f' ,$$

if we define $\delta\bar{u}'_y$ as the normalized angle

$$\delta\bar{u}'_y \equiv kqy_m$$

appearing in (3.435). Substituting for k gives

$$\delta\bar{u}'_y = \pi\frac{2y_m}{\lambda}\frac{\delta\eta'}{f'} = \frac{\pi}{N_y}\frac{\delta\eta'}{\lambda} , \qquad (3.436)$$

in which N_y is the f/no of the rectangular aperture in the y-direction. The function

$$I = \left[\frac{\sin(\delta\bar{u}')}{\delta\bar{u}'}\right]^2 \qquad (3.437)$$

is shown in Fig. 3.100. It has a central peak of value $I_0 = 1$ at $\delta\bar{u}' = 0$ and zero minima at $\pm\pi, \pm2\pi, \dots.$ The secondary maxima with values of $(\delta\bar{u}')$ are given by the roots of the equation $\tan(\delta\bar{u}') - (\delta\bar{u}') = 0$, as shown in Table 3.24. The first minimum corresponds to

$$\frac{\delta\eta'}{f'} = \frac{\lambda}{2y_m} ,$$

so that an aperture with width n wavelengths gives a first minimum $1/n$ rad away from the central maximum.

The appearance of the diffraction pattern arising from a rectangular aperture is shown in Fig. 3.101, reproduced from Born-Wolf and Lipson, Taylor and Thompson.

If the aperture becomes a slit by virtue of making $2y_m \gg 2x_m$, the diffraction pattern in the y direction shrinks to negligible proportions and the one-dimensional pattern for a long slit results.

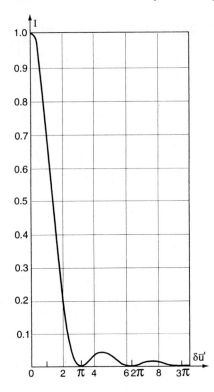

Fig. 3.100. Fraunhofer diffraction at a rectangular aperture showing the function $I = \left[\frac{\sin(\delta \overline{u}')}{\delta \overline{u}'}\right]^2$ (after Born-Wolf [3.120(b)])

Table 3.24. The first five maxima of the rectangular aperture function $I = \left[\frac{\sin(\delta \overline{u}')}{\delta \overline{u}'}\right]^2$ (after Born-Wolf [3.120(b)])

$\delta \overline{u}'$	$I = \left[\frac{\sin(\delta \overline{u}')}{\delta \overline{u}'}\right]^2$
0	1
$1.430\pi = 4.493$	0.04718
$2.459\pi = 7.725$	0.01694
$3.470\pi = 10.90$	0.00834
$4.479\pi = 14.07$	0.00503

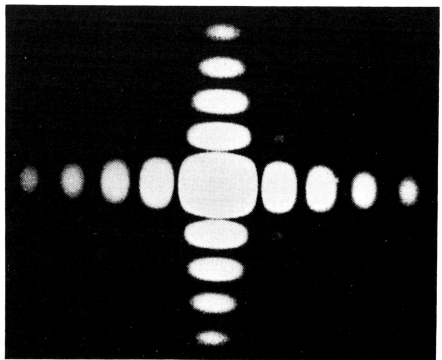

Fig. 3.101. Fraunhofer diffraction pattern of a rectangular aperture 8 mm × 7 mm, magnification 50×, $\lambda = 579$ nm. The centre was deliberately overexposed to reveal the secondary maxima (after Born-Wolf [3.120(b)] and Lipson, Taylor and Thompson, courtesy Brian Thompson)

3.10.2 Coherence

The above formulae are concerned with a point source. For an extended source, the result can be derived by integration, but here we must introduce the concept of *coherence*. If two beams originate from the same physical source, the phase fluctuations are normally fully correlated and the beams are mutually *coherent*. In beams from different sources, the phase fluctuations are uncorrelated and the beams are mutually *incoherent*. An intermediate situation can occur, that of *partial coherence*. The concept of the "degree of coherence" was introduced by Zernike [3.126] and applied further to image formation by Hopkins [3.127] [3.128]. For a coherent extended source, the complex amplitude must be integrated; for an incoherent extended source, the intensity. For a partially coherent extended source, the integration must take account of the degree of coherence between different elements of the source. Born-Wolf [3.120(b)] treats the important case of an incoherent luminous wire of infinite length diffracted by a narrow slit aperture parallel to the source, taken to be in the y-direction. Now $q = m - m_o$, where m_o is the position of a point source, so the intensity due to the line source is obtained by integrating (3.435) with respect to q. Born-Wolf show that the intensity I' is characterized by a similar function and is given by

$$I' = \left(\frac{\sin kp x_m}{kp x_m}\right)^2 I_0' , \tag{3.438}$$

where I_0' is the intensity at the centre of the pattern with $p = 0$.

3.10.3 The Point Spread Function (PSF) due to diffraction at a circular aperture

The derivation of the PSF in the case of the circular aperture is, in principle, similar to that for the rectangular aperture except that polar coordinates give a more natural formulation. Then, again following Born-Wolf [3.120(b)],

$$\rho \cos\phi = x , \quad \rho \sin\phi = y$$

for the pupil and

$$w \cos\psi = p , \quad w \sin\psi = q$$

for the image plane, where $w = (p^2 + q^2)^{1/2}$ is the angular separation of a point Q' in the image plane with direction cosines p, q relative to the central point of symmetry Q_0' of the pattern. The diffraction integral (3.434) becomes

$$U(Q') = C \int_0^{\rho_m} \int_0^{2\pi} e^{-ik\rho w \cos(\phi - \psi)} \rho \, d\rho d\phi , \tag{3.439}$$

which reduces to

$$U(Q') = 2\pi C \int_0^{\rho_m} J_0(k\rho w)\rho \, d\rho \tag{3.440}$$

and, finally

$$U(Q') = C\pi\rho_m^2 \left[\frac{2J_1(k\rho_m w)}{k\rho_m w}\right] , \tag{3.441}$$

where J_0 and J_1 are the Bessel functions of zero and first orders. The intensity I at Q' is therefore

$$I(Q') = |U(Q')|^2 = \left[\frac{2J_1(k\rho_m w)}{k\rho_m w}\right]^2 I_0 , \tag{3.442}$$

where $I_0 = C^2 A^2 = \tilde{E}A/\lambda^2$, with $A = \pi\rho_m^2$ and $\tilde{E}$ the total energy incident on the aperture, is the intensity at the central point of the pattern. The form of (3.442) is therefore similar to the result (3.438) for the slit, with the important difference that the sine is replaced by the Bessel function J_1. Figure 3.102 shows the function

$$I = \left[\frac{2J_1(\overline{w})}{\overline{w}}\right]^2 \tag{3.443}$$

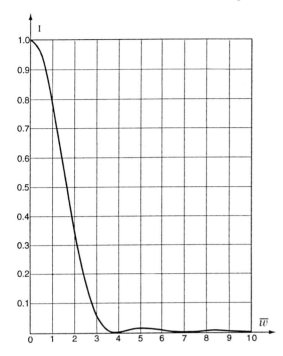

Fig. 3.102. Fraunhofer diffraction at a circular aperture showing the function $I = \left[\frac{2J_1(\overline{w})}{\overline{w}}\right]^2$ (after Born-Wolf [3.120(b)])

in which the normalized angle $\overline{w}$ is defined by

$$\overline{w} = k\rho_m w \tag{3.444}$$

The function is similar to that for a slit (Fig. 3.100), but the secondary maxima are relatively weaker. The positions of the zero intensity minima and the maxima are determined by the function J_1 and are given in Table 3.25. The minima are no longer equally spaced, although the separation tends towards the value π of the slit case if $\overline{w}$ becomes large. The appearance is shown in Fig. 3.103, in which the central peak has been overexposed in order to reveal the subsidiary maxima better. In a small telescope, well-corrected for spherical aberration, it is not easy to see even the first subsidiary maximum as its intensity from Table 3.25 is only 1.75% of the central maximum. When it is readily seen, it is normally a sign that aberrations have displaced energy into the subsidiary maxima from the central peak, as we shall see below.

The Eq. (3.442) was first derived in a different form by Airy [3.129] in a classical paper defining the so-called *Airy disk* (the effective diameter of the central maximum) as the diameter of the first minimum of the intensity function.

With $k = 2\pi/\lambda$, we have

$$\overline{w} = \frac{2\pi\rho_m w}{\lambda} \tag{3.445}$$

and the *radius* of the first dark ring is, from Table 3.25

Table 3.25. The first 3 subsidiary minima and maxima of the function $\left[\frac{2J_1(\overline{w})}{\overline{w}}\right]^2$ (after Born-Wolf [3.120(b)])

Case	$\overline{w}$	$\left[\frac{2J_1(\overline{w})}{\overline{w}}\right]^2$
Central max.	0	1
First min.	$1.220\pi = 3.832$	0
Max.	$1.635\pi = 5.136$	0.0175
Second min.	$2.233\pi = 7.016$	0
Max.	$2.679\pi = 8.417$	0.0042
Third min.	$3.238\pi = 10.174$	0
Max.	$3.699\pi = 11.620$	0.0016

$$w_0 = 0.610\frac{\lambda}{\rho_m} = 1.220\frac{\lambda}{D} \text{ rad} , \tag{3.446}$$

where $D = 2\rho_m$ is the diameter of the aperture. The *diameter* of the *Airy disk* is therefore

$$2w_0 = 2.440\frac{\lambda}{D} \text{ rad} \tag{3.447}$$

For $\lambda = 500$ nm and an aperture D of 5 m diameter, the diameter $2w_0$ is 0.0503 arcsec. This value is small compared with classical values of "good

Fig. 3.103. Fraunhofer diffraction at a circular aperture 6 mm in diameter, magnification 50×, $\lambda = 579$ nm. The central maximum has been overexposed to reveal the weak subsidiary maxima. (After Born-Wolf [3.120(b)] and Lipson, Taylor and Thompson, courtesy Brian Thompson)

seeing" due to atmospheric turbulence of about 1 arcsec. However, the potentialities of excellent modern sites, *if extreme care is taken with regard to local air conditions,* are such that diffraction must be taken into account in scientific specifications for the optical quality of modern telescopes for ground-based use. This will be discussed further in RTO II, Chap. 4. For space telescopes, the diffraction limit will anyway be the reference for the optical specification.

Another important criterion is the *encircled energy* as the fraction of the total energy contained within a given diameter. Expressing this fraction as L_w within the radius w, this is given by

$$L_w = 1 - J_0^2(k\rho_m w) - J_1^2(k\rho_m w) \,, \tag{3.448}$$

a formula originally derived by Rayleigh [3.130]. Figure 3.104 shows the function (3.448), in which 1, 2, 3 mark the positions of the dark minima from Table 3.25.

Rayleigh defined the resolving power or resolution limit of a telescope observing two incoherent point objects close to each other (e.g. a double star) *by the condition that the central maximum of the first diffraction pattern coincides with the first minimum of the second pattern (the Rayleigh criterion)* [3.131], originally introduced by Rayleigh as a criterion for the resolution of spectral lines whose intensity distribution follows the slit function of Eq. (3.438). In that case, the Rayleigh criterion gives a minimum intensity between the two peaks of 81% of the maximum. In the case of the circular aperture, the minimum intensity between the two peaks is 74%, more favourable. There is no special physical basis for considering this to be the resolution limit. Rayleigh considered experimental results agreed with an intensity minimum of about 80%. The Strehl criterion, considered below, is

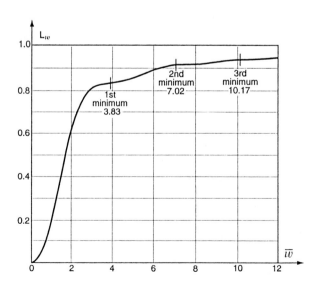

Fig. 3.104. Energy encircled in the radius w of the pattern due to Fraunhofer diffraction at a circular aperture. The fraction of energy is given by $L_w = 1 - J_0^2(\overline{w}) - J_1^2(\overline{w})$ with $\overline{w} = k\rho_m w$. (After Born-Wolf [3.120(b)])

a more scientific definition. Nevertheless, the Rayleigh criterion does correspond roughly with practice; though the real detection limits with modern detectors will depend on the signal/noise ratio in the observing mode used.

The *angular resolution* according to Rayleigh is therefore, from (3.446)

$$w_R = 1.220\frac{\lambda}{D} \text{ rad} , \qquad (3.449)$$

and the *linear resolution*

$$\delta\eta_R = f' w_R = 1.220 N\lambda , \qquad (3.450)$$

where N is the f/no. If the angular resolution limit of the human eye is taken as 1 arcmin, then the minimum magnification of a visual telescope to make use of its diffraction limited resolution is given by

$$m_{min} = \frac{0.000\,291 D}{1.220\lambda} = 2.38\frac{D}{\lambda}10^{-4} \qquad (3.451)$$

In practice, a magnification 2 or 3 times higher can be useful to relax the conditions for the eye: beyond this, the magnification is "empty". Taking $\lambda=580$ nm for visual use, we get

$$m_{min} = (0.410)D_{(mm)} \qquad (3.452)$$

Of course, this formula takes no account of the geometrical requirements of the eye pupil relative to the Ramsden disk discussed in § 2.2.4.

3.10.4 The Point Spread Function (PSF) due to diffraction at an annular aperture

The case of an annular aperture is of great importance for reflecting telescopes since all optically centered forms have a central obstruction. In the Cassegrain form the secondary is circular and its presence must lead to an annular aperture. More commonly, the baffle system of a Cassegrain telescope, designed to prevent light outside the field of view reaching the detector, forms the determinant central obstruction. The light distribution in the Fraunhofer diffraction pattern is then given by an integral of the form of Eq. (3.439), but the integration over the pupil radius ρ now extends over the domain limited by $\varepsilon\rho_m \leq \rho \leq \rho_m$, where ε is the obscuration factor, a positive number < 1. Eq. (3.441) then becomes

$$U(Q') = C\pi\rho_m^2 \left[\frac{2J_1(k\rho_m w)}{k\rho_m w}\right] - C\pi\varepsilon^2\rho_m^2 \left[\frac{2J_1(k\varepsilon\rho_m w)}{k\varepsilon\rho_m w}\right] \qquad (3.453)$$

The intensity is then given by

$$I(Q') = |U(Q')|^2 = \frac{1}{(1-\varepsilon^2)^2} \left[\left(\frac{2J_1(k\rho_m w)}{k\rho_m w}\right) - \varepsilon^2\left(\frac{2J_1(k\varepsilon\rho_m w)}{k\varepsilon\rho_m w}\right)\right]^2 I_0 , \qquad (3.454)$$

where

$$I_0 = |C|^2 \pi^2 \rho_m^4 (1 - \varepsilon^2)^2$$

is the intensity at the centre $w = 0$ of the diffraction pattern.

The positions of the minima (zeros) of this intensity function are given by the roots of the equation

$$J_1(k\rho_m w) - \varepsilon J_1(k\rho_m w) = 0 \tag{3.455}$$

for $w \neq 0$. The positions of the maxima are given by the roots of

$$J_2(k\rho_m w) - \varepsilon J_2(k\rho_m w) = 0 \tag{3.456}$$

If $\varepsilon = 0$, Eq. (3.455) gives the result for the unobstructed aperture of Table 3.25: the first minimum is at $\overline{w} = k\rho_m w = 3.832$, or at $w = 0.61\lambda/\rho_m$. As ε increases, the first root of (3.455) *decreases*. With $\varepsilon = 1/2$, it is at $\overline{w} = 3.15$ or $w = 0.50\lambda/\rho_m$. It is shown in Born-Wolf [3.120(c)] that

$$I_{(\varepsilon \to 1)} \to J_0^2(k\rho_m w)I_0 , \tag{3.457}$$

corresponding to $\overline{w} = 2.40$ or $w = 0.38\lambda/\rho_m$ for the first dark ring. The central obstruction therefore *increases* the resolving power in the Rayleigh sense. However, this gain in sharpness of the central maximum is paid for

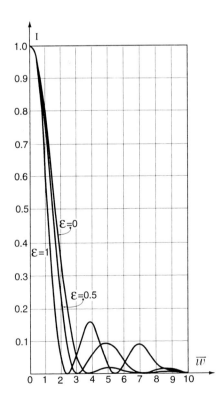

Fig. 3.105. The diffraction PSF at an annular aperture showing the effect of increasing the central obscuration factor ε (after Born-Wolf [3.120(c)] and G.C. Steward [3.132])

not only by the price of reduced integral energy in the image, but also by the fact that energy is transferred from the core into the secondary maxima, as shown in Fig. 3.105. As we shall see below in connection with the optical transfer function, this results in a loss of contrast in the image and gives the justification for Schiefspiegler in amateur telescope sizes.

If the transmission function over the pupil is no longer uniform but modified in a systematic way, the intensity distribution in the secondary maxima can be modified. This process is called *apodisation*. The general possibilities are treated by Maréchal and Françon [3.26(a)]. Jacquinot and Roizen-Dossier [3.133] exhaustively analysed the possibilities of 1-dimensional screening and this was extended by Lansraux to 2 dimensions [3.134]. Ref. [3.133] gives a general review.

3.10.5 The diffraction PSF in the presence of small aberrations

The effect of small aberrations on the diffraction PSF is of great importance in establishing specifications and tolerances for telescope optics (see RTO II, Chap. 4). We shall again essentially follow the treatment of Born-Wolf [3.120(d)]; other excellent accounts with different emphases are given by Welford [3.6], Maréchal and Françon [3.26(b)] and Schroeder [3.22(f)].

Born and Wolf show that the intensity at the point Q in the diffraction image associated with the Gaussian image point Q_0 [13] is given by

$$I(Q) = \frac{1}{\pi^2} \left| \int_0^1 \int_0^{2\pi} e^{i[kW - \tilde{v}\rho\cos(\phi-\psi) - (1/2)\tilde{u}\rho^2]} \rho \, d\rho \, d\phi \right|^2 I_0 \,, \qquad (3.458)$$

where I_0 is the intensity in the absence of aberrations, i.e. $W = 0$. ϕ and ψ are the azimuth angles in the pupil and image respectively, as defined in § 3.10.3, while $\tilde{u}$ and $\tilde{v}$ are the "optical coordinates" of Q:

$$\tilde{u} = k\left(\frac{\rho_m}{R}\right)^2 r \,, \qquad \tilde{v} = k\left(\frac{\rho_m}{R}\right)(p^2 + q^2)^{1/2} \,,$$

where p, q, r are the coordinates in the image plane, R is the distance from the exit pupil to the image Q_0 and $k = 2\pi/\lambda$.

In the absence of aberrations, the intensity maximum is at the Gaussian focus. If aberrations are present, there will be an intensity maximum at the *diffraction focus*, which may not be uniquely defined if the aberrations are large. However, if the *aberrations are small*, there will be a unique diffraction focus with a clearly defined intensity maximum. In this case, the ratio $I(Q)/I_0$ given by Eq. (3.458) can be used as a measure of the effect of the small aberrations on the image quality. This important concept was introduced by Strehl in 1902 [3.135]. Strehl called the ratio $I(Q)/I_0$ the "Definitionshelligkeit", literally "definition brightness". The commonly accepted term in English is the *Strehl Intensity Ratio*.

---

[13] For simplicity in the following formulation, we will drop the primes of Q', Q'_0 and W' indicating an image as distinct from an object in Fig. 3.99.

An important simplification of Eq. (3.458) is achieved by applying the so-called *displacement theorem*: the addition of terms in ρ^2, $\rho \sin \phi$, $\rho \cos \phi$ and ρ^0 (constant) to the aberration function have no effect on the intensity distribution of the diffraction image. Such terms refer to changes of the reference sphere, in radius or in tilt. They are Gaussian (first order), as shown in Table 3.1, and are concerned only with the definition of the *position* of the image, not its quality. This theorem is formally proven in Born-Wolf.

It follows that Eq. (3.458) can be simplified as

$$I(Q) = \frac{1}{\pi^2} \left| \int_0^1 \int_0^{2\pi} e^{ikW_Q} \rho \, d\rho \, d\phi \right|^2 I_0 \tag{3.459}$$

or

$$I(Q) = \frac{1}{\pi^2} \left| \int_0^1 \int_0^{2\pi} \left[1 + ikW_Q + \frac{1}{2}(ikW_Q)^2 + \ldots \right] \rho \, d\rho \, d\phi \right|^2 I_0 \tag{3.460}$$

Let $\overline{W}_Q^n$ denote the average value of the nth power of W_Q, so that

$$\overline{W}_Q^n = \frac{\int_0^1 \int_0^{2\pi} W_Q^n \rho \, d\rho \, d\phi}{\int_0^1 \int_0^{2\pi} \rho \, d\rho \, d\phi} = \frac{1}{\pi} \int_0^1 \int_0^{2\pi} W_Q^n \rho \, d\rho \, d\phi \tag{3.461}$$

We now make the approximation which is the normal one for the application of the Strehl Intensity Ratio: *that the aberrations are sufficiently small that the powers of (ikW_Q) in (3.460) above the second power may be neglected.* With this approximation, we can write from (3.461)

$$I(Q) = \left| 1 + ik\overline{W}_Q - \frac{1}{2}k^2 \overline{W}_Q^2 \right|^2 I_0$$

or

$$I(Q) = \left\{ 1 - \left(\frac{2\pi}{\lambda} \right)^2 \left[(\overline{W}_Q^2) - (\overline{W}_Q)^2 \right] \right\} I_0 \tag{3.462}$$

Now the "mean-square deformation" or "variance" of the wavefront is defined by

$$(\Delta W_Q)^2 = \frac{\int_0^1 \int_0^{2\pi} (W_Q - \overline{W}_Q)^2 \rho \, d\rho \, d\phi}{\int_0^1 \int_0^{2\pi} \rho \, d\rho \, d\phi}, \tag{3.463}$$

the rms deviation being ΔW_Q. Since, from the displacement theorem, the origin of measurement of W_Q is arbitrary, we also have for the variance

$$(\Delta W_Q)^2 = (\overline{W}_Q^2) - (\overline{W}_Q)^2, \tag{3.464}$$

the quantity in square brackets of (3.462). *Therefore, the relative loss in intensity of the central peak is only a function of the variance of the wavefront aberration, provided the aberrations are small enough to neglect powers above the second as indicated above.* This is a very important conclusion in specifying telescope optics. Eq. (3.462) can be written in the final simple form

$$\frac{I(Q)}{I_0} = 1 - \left(\frac{2\pi}{\lambda}\right)^2 (\Delta W_Q)^2 \tag{3.465}$$

Following Welford [3.6], it is sometimes useful to express (3.462) in terms of Cartesian coordinates in the form

$$\frac{I(Q)}{I_0} = 1 - \left(\frac{2\pi}{\lambda}\right)^2 \left(\frac{\int_0^1 \int_0^1 [W(x,y)]^2 dx\,dy}{A'} - \left[\frac{\int_0^1 \int_0^1 W(x,y)dx\,dy}{A'}\right]^2\right) , \tag{3.466}$$

where A' is the area of the pupil over which the integrations are taken. This will have to take account of any vignetting in the system disturbing the circular (or annular) form of the pupil.

Bearing in mind the approximation of "small" aberrations involved, we must now consider what values the ratio $I(Q)/I_0$ can assume. The Rayleigh criterion for resolution [3.131] discussed in § 3.10.3 above led to an intensity reduction between the images of a double star of about 20%. If we arbitrarily apply the same reduction to the intensity in the central maximum due to small aberrations, we have

$$\frac{I(Q)}{I_0} = 0.8 , \tag{3.467}$$

giving from (3.465)

$$\left(\frac{2\pi}{\lambda}\right)^2 (\Delta W_Q)^2 = 0.2 \tag{3.468}$$

or

$$(\Delta W_Q)^2 = \lambda^2/197.4 \tag{3.469}$$

Maréchal and Françon [3.26(c)] give $\lambda^2/180$, a value derived with the approximation $\pi \simeq 3$. Following Welford [3.6], we shall use the more accurate value of (3.469) which leads to the conveniently close approximation

$$(W_Q)_{rms} \equiv \Delta W_Q \simeq \lambda/14 \tag{3.470}$$

Eqs. (3.469) and (3.470) refer to the total aberration according to (3.463) or (3.466). In practice, the modal concept of the aberration function is usually more instructive than numerical integration.

Following Maréchal and Françon [3.26(c)], we shall apply Eq. (3.465) to the following aberration polynomial:

$$W_Q = k_d\rho^2 + k_{S1}\rho^4 + k_{S2}\rho^6 + (k_T\rho + k_C\rho^3)\cos\phi + k_A\rho^2\cos 2\phi \tag{3.471}$$

From § 3.2.1, the terms can be identified as defocus, third and fifth order spherical aberration, wavefront tilt and third order coma, and third order astigmatism. Since the denominator of (3.463) with zero vignetting is π, we have from (3.463) and (3.464)

$$(\Delta W_Q)^2 = (\overline{W_Q^2}) - (\overline{W}_Q)^2$$

or

$$(\Delta W_Q)^2 = \left[\frac{1}{\pi} \int_0^1 \int_0^{2\pi} W_Q^2 \rho \, d\rho \, d\phi\right] - \left[\frac{1}{\pi} \int_0^1 \int_0^{2\pi} W_Q \rho \, d\rho \, d\phi\right]^2 \quad (3.472)$$

We will apply this, as an example, to the case of the first three terms in (3.471) and deduce the aberration coefficients k_d, k_{S1} and k_{S2} giving optimum tolerances according to the criterion of (3.468). Then in (3.472) for this case

$$W_Q = k_d \rho^2 + k_{S1} \rho^4 + k_{S2} \rho^6 , \quad (3.473)$$

with ρ normalized to 1 at the edge of the pupil. Eq. (3.472) gives

$$(\Delta W_Q)^2 = \frac{k_d^2}{12} + \frac{4}{45} k_{S1}^2 + \frac{9}{112} k_{S2}^2 + \frac{k_d k_{S1}}{6} + \frac{3}{20} k_d k_{S2} + \frac{k_{S1} k_{S2}}{6} \quad (3.474)$$

If k_{S2} is considered as a constant and k_d and k_{S1} are varied to minimize the function, giving maximum tolerances, then

$$\frac{\partial (\Delta W_Q)^2}{\partial k_d} = \frac{\partial (\Delta W_Q)^2}{\partial k_{S1}} = 0 ,$$

giving

$$\frac{k_d}{6} + \frac{k_{S1}}{6} + \frac{3}{20} k_{S2} = 0$$

$$\frac{k_d}{6} + \frac{8}{45} k_{S1} + \frac{k_{S2}}{6} = 0$$

The solution gives the simple result

$$k_d = +\frac{3}{5} k_{S2} , \quad k_{S1} = -\frac{3}{2} k_{S2} \quad (3.475)$$

Substituting back in (3.474) gives

$$(\Delta W_P)^2 = \frac{k_{S2}^2}{2800} \quad (3.476)$$

If this result is applied to the tolerance criterion (3.468), we get

$$k_{S2} = \frac{\sqrt{140}}{\pi} \lambda = 3.77\lambda \quad (3.477)$$

for the amount of fifth order spherical aberration leading to a drop of 20% in the Strehl Ratio if it is optimally balanced by

$$k_d = 2.26\lambda , \quad k_{S1} = -5.65\lambda$$

from (3.475).

The three terms of (3.473) can also be taken individually or in the combination of k_d and k_{S1}. Other groups are k_T and k_C, individually or combined, and k_A. The complete variance function of all the terms in (3.471) is

$$(\Delta W_Q)^2 = \left(\frac{k_d^2}{12} + \frac{4}{45}k_{S1}^2 + \frac{9}{112}k_{S2}^2 + \frac{k_d k_{S1}}{6} + \frac{3}{20}k_d k_{S2} + \frac{k_{S1} k_{S2}}{6} \right)$$
$$+ \left(\frac{k_T^2}{4} + \frac{k_T k_C}{3} + \frac{k_C^2}{8} \right) + \left(\frac{k_A^2}{6} \right) \tag{3.478}$$

Applying (3.465) to the various groupings leads to the results of Table 3.26. It should be noted that all these cases refer to *incoherent illumination* for point sources, the normal case for astronomical telescopes. Other cases are dealt with by Maréchal and Françon [3.26(d)]. If the third order astigmatism is defined as $k_A \rho^2 \cos^2 \phi$ (Welford [3.6] and Schroeder [3.22(g)]), the transformation $\cos^2 \phi = \frac{1}{2}(1 + \cos 2\phi)$ produces a shift of the mean focus giving different values for $(\Delta W_Q)^2$ and k_A.

It is noteworthy that the definition of $I(Q)/I_0 = 0.8$ leads to coefficients for defocus and third order spherical aberration, considered individually, which are about $\lambda/4$. Rayleigh [3.131] deduced this tolerance for defocus on an experimental basis and suggested it as a general rule. Table 3.26 confirms that it gives a reasonable approximation for the individual aberrations, but the Strehl Intensity Ratio puts the matter on a more rigorous physical basis.

The combination of defocus with third order spherical aberration shows that, as in the geometrical-optical case treated in §3.3.1, a multiplication of the tolerance by a factor of 4 takes place. However, in the present diffraction case, the optimum focus is *not* at the disk-of-least-confusion of the geometrical-optical case: Table 3.26 shows it is *half way* from the paraxial to the marginal focus, not three quarters of the way. It is easily shown that the reference sphere cuts the wavefront at the edge of the pupil.

The case of the balance of defocus with spherical aberration of both third and fifth orders is instructive in showing the large individual coefficients which can appear when non-orthogonal terms are balanced.

The final case shown in Table 3.26 is a departure from the normal aberration polynomial but represents, in idealized form, a very important practical error in the figure of large optics: ripple. Ripple is a succession of concentric zones and arises from resonance effects in the motion of the polishing tool. The steeper the aspheric function, the greater the danger of ripple, if conventional figuring methods are used. Figure 3.106 shows an idealized sinusoidal ripple with the wavefront function

$$W_Q = k_Z \cos\left(\frac{\rho}{\lambda_Z} \right) , \tag{3.479}$$

in which λ_Z is the normalized ripple wavelength and k_Z is the amplitude. Then

$$\frac{\rho_m}{\lambda_Z} = \frac{1}{\lambda_Z} = n_Z 2\pi ,$$

where n_Z is the number of complete zones over the radius. Our treatment assumes n_Z is any whole positive integer.

Table 3.26. Coefficients of various aberrations giving a Strehl Intensity Ratio of 0.8 (incoherent light, point source, circular unvignetted pupil)

Aberration	Aberration polynomial	$(\text{rms})^2$ $\equiv$ $(\Delta W_Q)^2$	Coefficients for $I(Q)/I_0 = 0.80$
Defocus	$k_d\rho^2$	$\frac{k_d^2}{12}$	$k_d = 0.25\lambda$
Third order Spherical Aberration alone	$k_{S1}\rho^4$	$\frac{4}{45}k_{S1}^2$	$k_{S1} = 0.24\lambda$
Third order Spherical Aberration with optimum focus	$k_d\rho^2 + k_{S1}\rho^4$	$\frac{k_d^2}{12} + \frac{4}{45}k_{S1}^2$ $+\frac{k_d k_{S1}}{6}$	$k_{S1} = 0.95\lambda$ $k_d = -0.95\lambda$
Third order Coma alone	$k_C\rho^3\cos\phi$	$\frac{k_C^2}{8}$	$k_C = 0.20\lambda$
Third order Coma with optimum Tilt	$(k_T\rho + k_C\rho^3)\cos\phi$	$\frac{k_T^2}{4} + \frac{k_C^2}{8}$ $+\frac{k_T k_C}{3}$	$k_C = 0.60\lambda$ $k_T = -0.40\lambda$
Third order Astigmatism	$k_A\rho^2\cos 2\phi$	$\frac{k_A^2}{6}$	$k_A = 0.17\lambda$
Ripple (concentric sinusoidal zones)	$k_Z\cos\left(\frac{\rho}{\lambda_Z}\right)$	$\frac{k_Z^2}{2}$	$k_Z = 0.10\lambda$

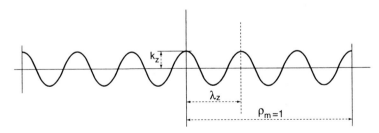

Fig. 3.106. Idealized sinusoidal ripple showing 3 complete wavelengths of ripple in the pupil radius

Applying (3.472) to (3.479) gives

$$(\Delta W_Q)^2 = \overline{W}_Q^2 = \frac{k_Z^2}{2} ,$$
(3.480)

since $(\overline{W}_Q)^2 = 0$. This result is independent of n_Z or λ_Z. Relative to the square of its coefficient k_Z^2, it has easily the highest variance of the aberrations listed, giving the single error tolerance for $I_Q/I_0 = 0.80$ of 0.10λ. This is a measure of the high significance of ripple and justifies the complex measures taken to avoid zones and keep their amplitude low.

3.10.6 The diffraction PSF in the presence of small aberrations and an annular aperture

Table 3.26 supposes the full circular pupil is operative. In practice, the normal case in reflecting telescopes is the *annular* aperture, whose diffraction PSF was considered in § 3.10.4. Following Schroeder [3.22(g)], we must now consider the effect of an obscuration factor ε in a pupil normalized to $\rho_m = 1$. Eq. (3.472) can be taken over directly with the changed limits of integration, as

$$(\Delta W_Q)^2 = \left. \begin{array}{c} \left(\dfrac{1}{\pi(1 - \varepsilon^2)} \displaystyle\int_\varepsilon^1 \int_0^{2\pi} W_Q^2 \, \rho \, d\rho \, d\phi \right) \\[14pt] - \left(\dfrac{1}{\pi(1 - \varepsilon^2)} \displaystyle\int_\varepsilon^1 \int_0^{2\pi} W_Q \, \rho \, d\rho \, d\phi \right)^2 \end{array} \right\} ,$$
(3.481)

since the denominator of (3.463) is now $\pi(1-\varepsilon^2)$. In the case of defocus alone, as a typical case, (3.481) gives for the variance

$$(\Delta W_Q)^2 = \frac{k_d^2}{3} \left(\frac{1 - \varepsilon^6}{1 - \varepsilon^2} \right) - \frac{k_d^2}{4} \left(\frac{1 - \varepsilon^4}{1 - \varepsilon^2} \right)$$
(3.482)

In all cases, we arrive at functions of the form

$$F(\varepsilon^2) = \frac{1 - \varepsilon^{2\tilde{n}}}{1 - \varepsilon^2}$$

which reduces to

$$F(\varepsilon^2) = 1 + \varepsilon^2 + \varepsilon^4 + \varepsilon^6 + \ldots + \varepsilon^{2(\tilde{n}-1)} ,$$

with $\tilde{n} = 1, 2, 3, \ldots$ Table 3.27 gives the results for the variances in the cases given in Table 3.26 and the ratio of the peak-to-valley (ptv) aberration to the rms aberration with $\varepsilon = 0$. Note that the ptv wavefront aberration is the same as the wavefront aberration coefficient in the case of the symmetrical aberrations defocus and spherical aberration, but *twice* as large in all other cases because of the ± 1 limits of the $\cos n\phi$ function.

In Table 3.27, the variance in the case of ripple is given as independent of ε. This is true under the same condition as in the unobscured case, namely,

Table 3.27. The effect of the obscuration factor ε on the wavefront variance and the ratio (ptv/rms) with $\varepsilon = 0$

Aberration	Wavefront function	Variance with obscuration factor ε	ptv/rms $(\varepsilon = 0)$
Defocus (d)	$k_d\rho^2$	$\frac{k_d^2}{12}(1-\varepsilon^2)^2$	$\sqrt{12}$ $= 3.46$
Third order Spherical Aberration (S_1) alone	$k_{S1}\rho^4$	$\frac{4}{45}k_{S1}^2\left(1 - \frac{\varepsilon^2}{4} - \frac{3}{2}\varepsilon^4 - \frac{\varepsilon^6}{4} + \varepsilon^8\right)$	$\sqrt{11.25}$ $= 3.35$
$S_1 + d$ (optimum focus)	$k_d\rho^2 + k_{S1}\rho^4$	$\frac{k_d^2}{12}(1-\varepsilon^2)^2$ $+ \frac{k_d k_{S1}}{6}(1 - \varepsilon^2 - \varepsilon^4 + \varepsilon^6)$ $+ \frac{4}{45}k_{S1}^2\left(1 - \frac{\varepsilon^2}{4} - \frac{3}{2}\varepsilon^4 - \frac{\varepsilon^6}{4} + \varepsilon^8\right)$	$\sqrt{\frac{180}{16}}$ $= 3.35$
Tilt (T)	$k_T\rho\cos\phi$	$\frac{k_T^2}{4}(1+\varepsilon^2)$	$2\sqrt{4}$ $= 4.00$
Third order Coma (C) alone	$k_C\rho^3\cos\phi$	$\frac{k_C^2}{8}(1+\varepsilon^2+\varepsilon^4+\varepsilon^6)$	$2\sqrt{8}$ $= 5.66$
$C_1 + T$ (optimum Tilt)	$(k_T\rho + k_C\rho^3)\cos\phi$	$\frac{k_T^2}{4}(1+\varepsilon^2)$ $+ \frac{k_T k_C}{3}(1 + \varepsilon^2 + \varepsilon^4)$ $+ \frac{k_C^2}{8}(1+\varepsilon^2+\varepsilon^4+\varepsilon^6)$	$2\sqrt{72}\left(\frac{1}{3}\right)$ $= 5.66$
Third order Astigmatism (A)	$k_A\rho^2\cos 2\phi$	$\frac{k_A^2}{6}(1+\varepsilon^2+\varepsilon^4)$	$2\sqrt{6}$ $= 4.90$
Ripple - sinusoidal with $(n_Z - n_Z')$ complete wavelengths over annular pupil	$k_Z\cos\left(\frac{\rho}{\lambda_Z}\right)$	$\frac{k_Z^2}{2}(1)$	$2\sqrt{2}$ $= 2.83$

that the annular radius contains a complete integral number $(n_Z - n'_Z)$ of ripple (zonal) wavelengths, i.e. both n_Z and n'_Z are integers, where n'_Z is the number of obscured ripple waves.

As in Table 3.26, the result for astigmatism in Table 3.27 differs from that given by Schroeder because of the difference in the wavefront functions with $\cos 2\phi$ and $\cos^2 \phi$ respectively. Both formulations are valid and can readily be transformed into each other.

For the combined function of spherical aberration and defocus

$$k_d \rho^2 + k_{S1} \rho^4$$

Table 3.27 shows that the ratio ptv/rms is unchanged by the optimum focus combination with $k_d = -k_{S1}$, since both quantities are reduced by the factor 4 (ptv = zonal aberration in this case). Similarly for the coma-tilt combination

$$(k_T \rho + k_C \rho^3) \cos \phi$$

the ratio ptv/rms is also unchanged for the optimum tilt $k_T = -\frac{2}{3}k_C$. The maximum aberration here is not the zonal aberration but the aberration at the edge of the pupil $(W_Q)_{\rho=1} = +\frac{1}{3}k_C$, giving ptv $= \frac{2}{3}k_C$. With rms $= k_C/\sqrt{72}$, the ratio remains $2\sqrt{8} = 5.66$. These two cases of invariance to changes of the reference sphere are illustrations of the *displacement theorem* discussed above in the derivation of Eq. (3.459).

3.10.7 The diffraction PSF in the presence of larger aberrations: the Optical Transfer Function (OTF)

The effect of aberrations in optical systems may broadly be classified in three main groups: first, aberrations near the diffraction limit; second, aberrations above the diffraction limit but acceptable for the detector; third, very large aberrations relative to the diffraction limit. The *first* group has been dealt with in §§ 3.10.5 and 3.10.6 above. It assumes more and more importance with the improvement in quality of ground-based telescopes and sites, and the steadily increasing advance of space telescopes. The *third* group represents the domain where diffraction effects become negligible and geometrical optical image sizes give a good description of the image. Until recently, accepted values for external "seeing" (non-local air turbulence) were so far above the diffraction limit, that geometrical-optical interpretations of the image quality were reasonable, above all, if the detector resolution is much lower than the diffraction image size given by (3.447). This is no longer the case if accurate assessment is desired. The *second* group is the most complex situation: neither the series approximations used in the first group nor the asymptotic developments of the third are adequate. The best tool for this group is the *Optical Transfer Function* (OTF).

The theory of the OTF, essentially a systematic application of Fourier theory to optical imagery, was initiated by Duffieux in 1946 [3.136]. The application to television, with analogy to filter theory in communication systems, was a powerful impetus. A very extensive literature exists. General treatments of varying depth, all excellent within the aims of the work in question, are given by Maréchal-Françon [3.26(e)], Welford [3.6], Schroeder [3.22(h)] (particularly well adapted to the case of telescopes) and Wetherell [3.137]. For complete treatments of the Fourier theory required, see Maréchal-Françon [3.26(f)] or Goodman [3.138]. Here, we shall confine ourselves to the essential aspects applicable to normal use of telescopes, i.e. the case of incoherent light.

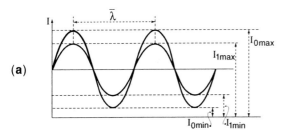

(a)

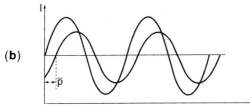

(b)

Fig. 3.107. Transfer of a sinusoidal wave with reduced contrast through an optical system **(a)** without phase shift, **(b)** with phase shift $\bar{p}$

The OTF is based on the concept of *spatial frequency* of the intensity in the object plane, represented as the sum of a continuous infinity of sinusoidal components. Each component is transmitted by the optical system with reduced contrast, the contrast being defined by

$$C_0 = \frac{I_0\,\text{max} - I_0\,\text{min}}{I_0\,\text{max} + I_0\,\text{min}} \tag{3.483}$$

for the object and by C_I for the image, as shown in Fig. 3.107. If we consider an object consisting of only one sinusoidal frequency in the t-section, then the *Modulation Transfer Function* (MTF) for this frequency s is given simply by

$$T_C(s) = \frac{C_I}{C_0} \tag{3.484}$$

This ignores the phase shift $\bar{p}$.

Following Welford [3.6], we take coordinates in the object plane as ξ_0 and η_0 and consider these as transmitted to the image plane with the magnification factor m, then for the image $\xi = m\xi_0$, $\eta = m\eta_0$. For a sinusoidal object whose "lines" are parallel to the η_0-axis, we can write

$$I_0(\xi_0) = a_0(1 + \cos 2\pi s \xi_0) \tag{3.485}$$

for the function of the object intensity (*incoherently* illuminated), in which $s = 1/\overline{\lambda}$ is the spatial frequency in mm^{-1}, if ξ_0 is expressed in mm, and a_0 is the normalized amplitude. According to (3.483), the contrast of the object $C_0 = 1$. The transfer through the system then gives at the image

$$I_I(\xi) = a_I\left[1 + T_C \cos(2\pi s \xi + \overline{p})\right] , \tag{3.486}$$

where the factor a_I/a_0 represents the constant intensity loss due to absorption etc.

As with the diffraction integral, it is most convenient in the general formulation to use the complex form from the Euler relation. This general formulation is given, for example, by Maréchal-Françon [3.26]. For a simple frequency s, (3.486) then becomes

$$I_I(\xi) = a_I\left[1 + \underline{R}\left(\underline{L}(s)e^{2\pi i s \xi}\right)\right] \tag{3.487}$$

where $\underline{R}$ is the real part of the complex quantity $\underline{L}$. It follows that $T_C = \underline{R}$, the real part of the OTF $\underline{L}$, is the MTF and the phase shift is $\arg \underline{L}(s)$.

In the general case, we have a two-dimensional situation depending not only on s but on the position in the field and the aberrations affecting that image point. The OTF will depend on whether the intensity function is in the t- or s-section. For symmetrical aberrations, the phase shift $\overline{p}$ will be zero; for coma it is *not* zero, though it may be small in practice. An important condition for the generalisation is that the aberration function be invariant over the range of the lateral aberration at any point: this is the *isoplanatism condition*. It will only break down if aberrations become very large, in which case the optical system is useless in practice.

If the wavefront aberration in the pupil is $W(x, y)$, then as for (3.459) the complex amplitude in the exit pupil induced by a point object is proportional to $e^{ikW(x,y)}$ with $k = 2\pi/\lambda$ as before. We define

$$F(x, y) = e^{ikW(x,y)} \tag{3.488}$$

as the *pupil function*, having the form of (3.488) over the area of the pupil and zero outside it. The diffraction integral (3.439) can be written to give the complex amplitude at the image point ξ, η in terms of rectangular coordinates, if constants are ignored, as

$$U(\xi, \eta) = \int_{-\infty}^{+\infty} \int_{-\infty}^{+\infty} F(x, y)e^{-i\,(k/R)(\xi x + \eta y)}\,\mathrm{d}x\,\mathrm{d}y , \tag{3.489}$$

where R is the radius of curvature of the reference sphere. The integration here is taken over infinite limits although in (3.439) it was taken over the area of the pupil. This change is made to simplify the application of Fourier transform theory and is validated, for example, in Maréchal-Françon [3.26] and Welford [3.6]. If $\xi/\lambda R$ and $\eta/\lambda R$ are considered as single variables as in Eq. (3.434), then Eq. (3.489) expresses $U(\xi, \eta)$ as the two-dimensional

Fourier transform of $F(\xi, \eta)$ with the appropriate re-scaling (normalization) of ξ, η. This statement is generally true for any shape or size of pupil and any aberration function $W(x, y)$. If the transmission over the pupil is not uniform (apodisation), this can also be introduced into the function $F(\xi, \eta)$. The constant factors omitted are given by Born-Wolf [3.120(e)] or Maréchal-Françon [3.26] and disappear in the subsequent normalization process. The intensity point spread function is given by

$$I(\xi, \eta) = |U(\xi, \eta)|^2 \qquad (3.490)$$

Again following the treatments of Welford [3.6] and Maréchal-Françon [3.26], we now generalize from the expression (3.487) for two dimensions to give the OTF as the normalized inverse Fourier transform of the intensity PSF as

$$\underline{L}(s, t) = \frac{\int_{-\infty}^{+\infty} \int_{-\infty}^{+\infty} I(\xi, \eta) e^{2\pi i(s\xi + t\eta)} \mathrm{d}\xi \, \mathrm{d}\eta}{\int_{-\infty}^{+\infty} \int_{-\infty}^{+\infty} I(\xi, \eta) \mathrm{d}\xi \, \mathrm{d}\eta}, \qquad (3.491)$$

where s and t are the spatial frequencies in the ξ and η directions respectively. They are expressed as lines/mm (or line-pairs/mm in television usage) if linear dimensions are in mm. $\underline{L}(s)$ in (3.487) is simply the one-dimensional equivalent and is equal to $\underline{L}(s, 0)$. From (3.491), the OTF in the ξ-direction only is

$$\underline{L}(s, 0) = \frac{\int_{-\infty}^{+\infty} I_L(\xi) e^{2\pi i s \xi} \mathrm{d}\xi}{\int_{-\infty}^{+\infty} \int_{-\infty}^{+\infty} I(\xi, \eta) \mathrm{d}\xi \, \mathrm{d}\eta} \qquad (3.492)$$

with

$$I_L(\xi) = \int_{-\infty}^{+\infty} I(\xi, \eta) \mathrm{d}\eta \qquad (3.493)$$

The function $I_L(\xi)$ is the line spread function (LSF), the image of an infinitely narrow line source illuminated incoherently. Eq. (3.492) states that the one-dimensional OTF is the Fourier transform of the LSF. The denominator in (3.492) is the normalizing factor which gives $\underline{L}(0, 0) = 1$, a condition expressed in (3.485) above giving $C_0 = 1$ for the object with no loss of contrast.

It is shown by Maréchal-Françon [3.26(g)] that the expression (3.491) for the OTF can be converted into an alternative form which has an elegant physical interpretation. This form uses the autocorrelation function of the pupil function:

$$\underline{L}(s, t) = \frac{\int_{-\infty}^{+\infty} \int_{-\infty}^{+\infty} F(x + \lambda Rs, y + \lambda Rt) F^*(x, y) \mathrm{d}x \, \mathrm{d}y}{\int_{-\infty}^{+\infty} \int_{-\infty}^{+\infty} |F(x, y)|^2 \mathrm{d}x \, \mathrm{d}y} \qquad (3.494)$$

As before, the denominator normalizes to make $\underline{L}(0, 0) = 1$. Applying the theorem of Parseval, it is shown in Maréchal-Françon [3.26(h)] that the numerator of (3.494) can be transformed into the symmetrically equivalent form

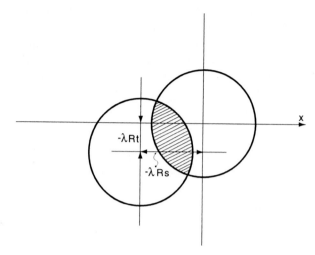

Fig. 3.108. Shearing of a circular pupil corresponding to the calculation of the OTF from the autocorrelation function for shears of λRs, λRt

$$\int_{-\infty}^{+\infty} \int_{-\infty}^{+\infty} F(x,y)F^*(x - \lambda Rs, y - \lambda Rt)\mathrm{d}x\,\mathrm{d}y \; , \tag{3.495}$$

in which $F(x,y)$ is the pupil function defined in (3.488). Although formally the range of integration is infinite, the pupil function is, in fact, unity (normalized) within the area of the pupil and zero outside it. The autocorrelation function of (3.494) corresponds to a shear of the pupil of the amounts $-\lambda Rs$, $-\lambda Rt$ in the directions x and y respectively, as shown in Fig. 3.108. Although the pupil is represented as a complete circle, the shear principle is equally valid for a pupil with vignetting or central obstruction. Clearly, if the vector shear in Fig. 3.108 reaches the diameter of the pupil, there is no common area and no signal at higher frequencies can be transmitted through the system, i.e. the contrast according to (3.483) is zero and the optical system is the equivalent of a low bandpass filter. In Fig. 3.109, the shear situation is represented in one dimension. The shear is λRs. Clearly, the limit frequency is given by

$$\lambda Rs_{max} = D \; ,$$

where D is the diameter of the pupil, so that

$$s_{max} = \frac{D}{\lambda R} = \frac{2u'_{max}}{\lambda} = \frac{1}{N\lambda} \; , \tag{3.496}$$

where u'_{max} is the semi-angular aperture of the exit pupil and $2u'_{max} = 1/N$, the reciprocal of the f/no. Now $s_{max} = 1/\overline{\lambda}_{min}$, where $\overline{\lambda}_{min}$ is the minimum wavelength of transmission of a sinusoidal spatial intensity function. Then

$$\overline{\lambda}_{min} = 1/s_{max} = N\lambda \tag{3.497}$$

This result can be compared directly with the linear resolution based on the Rayleigh criterion of the radius of the first diffraction minimum given

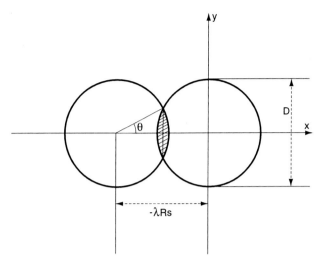

Fig. 3.109. One-dimensional pupil shear to demonstrate the low bandpass filter function of an optical system

in (3.450). The results are identical except for the factor 1.220 in the latter case, which referred to the PSF for a circular aperture. Here we are dealing with the Line Spread Function, LSF. Referring to the case of the PSF at a rectangular aperture, Eq. (3.436) shows that the linear resolution corresponding to the radius of the first minimum is identical to (3.497).

Referring again to Fig. 3.109, we have

$$\cos\theta = \frac{\lambda Rs}{D} \tag{3.498}$$

The common area A of the sheared pupils, normalized to the area of the pupil, is

$$\frac{A}{A_0} = \frac{2}{\pi}(\theta - \sin\theta\,\cos\theta) , \tag{3.499}$$

which, combined with (3.498), gives for the OTF of an aberration-free circular pupil of diameter D

$$\underline{L}(s) = \frac{A}{A_0} = \frac{2}{\pi}\cos^{-1}\left(\frac{\lambda Rs}{D}\right) - \frac{2\lambda Rs}{\pi D}\left(1 - \frac{\lambda^2 R^2 s^2}{D^2}\right)^{1/2} \tag{3.500}$$

In terms of the f/no $N = R/D$, we have

$$\underline{L}(s) = \frac{2}{\pi}\left[\cos^{-1}(N\lambda s) - N\lambda s\sqrt{1 - (N\lambda s)^2}\right] \tag{3.501}$$

If λ is given in mm, then s is in mm^{-1}. This equation gives, strictly, only the real part of the complex function, i.e. $\underline{R}$ in Eq. (3.487), the MTF. We have in general

$$\underline{L}(s) = \underline{R} + i\underline{P} \tag{3.502}$$

where $\underline{P}$ is the phase function. The autocorrelation function (3.494) must be multiplied by the cosine of the argument of (3.502) in terms of the pupil function for the real part A and the sine for the imaginary part B. For details the reader is referred to Wetherell [3.137]. The two components are given by

$$\text{MTF} = \underline{R} = (A^2 + B^2)^{1/2}, \quad \text{PTF} = \tan^{-1}(B/A) \tag{3.503}$$

Figure 3.110 shows the function $\underline{L}(s)$ from (3.501). It was one of the basic conditions (see §3.10.1) of the diffraction integral that the relative aperture of the optical system should not be too large. This condition applies equally to the above derivation of the MTF. However, this theory is applicable to large field angles. This introduces a cosine term into (3.491) in which $I(\xi, \eta)$ becomes $I(\xi, \eta \cos u'_{pr})$ if the field angle is in the t-section. Welford [3.6] shows that the cosine term disappears in the autocorrelation function.

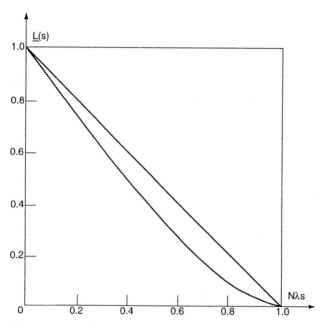

Fig. 3.110. The MTF for a circular pupil free from aberrations and obstruction, corresponding to the diffraction PSF with incoherent illumination

Some important properties of the Fourier transform relations must be emphasized. If $I_I(\xi, \eta)$ is the intensity distribution in the image, $I_0(\xi_0, \eta_0)$ is the intensity distribution in the object (*incoherently* illuminated) and $I_Q(\xi - \xi_0, \eta - \eta_0)$ is the PSF for some image point Q related to its object point by the magnification m, then

$$I_I(\xi, \eta) = \int \int I_0(\xi_0, \eta_0) I_Q(\xi - \xi_0, \eta - \eta_0) \, d\xi_0 \, d\eta_0 \qquad (3.504)$$

It is shown (for example in Maréchal-Françon [3.26(i)]) that it follows from the theorem of Parseval that the Fourier transforms of the above quantities are related by the simple equation

$$i_I(s, t) = i_0(s, t) i_Q(s, t) , \qquad (3.505)$$

i.e. the Fourier transform of the image is simply the product of the Fourier transform of the object and the Fourier transform of the PSF, the latter being $i_Q(s, t) = $ OTF. It is clearly far simpler to operate with (3.505) than with (3.504), a convolution process usually written in the form

$$I_I = I_0 \otimes I_Q$$

The simple multiplicative form of (3.505) also has the immense advantage that a chain of factors affecting the final image quality can be handled by multiplying their individual OTFs. This can be applied not only to a succession of individual optical systems or elements but also to detectors and disturbing factors such as image motion and manufacturing errors.

An important modification is introduced into the MTF function shown in Fig. 3.110 if a central obstruction is introduced. This problem was first solved in analytical form by O'Neill [3.139]. The form of the MTF function for various values of obstruction ratio ε is shown in Fig. 3.111. The essential feature is that the contrast is *increased* near the transmission limit, but *reduced* at low frequencies. This result is exactly what would be expected from the intensity distribution of Steward [3.132] in the diffraction PSF for an annular

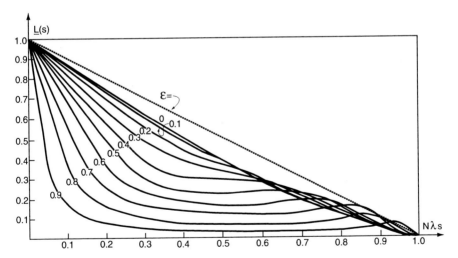

Fig. 3.111. MTF for a circular pupil free of aberration with central obstruction ε (after Lloyd [3.140])

aperture (Fig. 3.105). The narrowing of the central maximum corresponds to the increased contrast at the limit frequencies of resolution, while the rapid increase of intensity in the diffraction rings corresponds to the contrast reduction at low frequencies. O'Neill [3.139] points out that Rayleigh [3.141] comments on experiments of Herschel in improving resolution of double stars by introducing a central obstruction. O'Neill also points out the interesting limit case of an infinitely narrow annular aperture as $\varepsilon \to 1$, which gives simply a sharp peak near the limit of resolution and is the two-dimensional analogue of a Young's double slit aperture with the peak occurring at the spatial frequency of the fringes.

The complete formula for the functions of Fig. 3.111 is also given by Wetherell [3.137] including correction of an error in the original O'Neill paper.

The reduction in contrast for intermediate frequencies of about 0.3 is the principal justification for the Schiefspiegler of Kutter and other forms, discussed in § 3.7 above. For an amateur telescope with $D = 30$ cm, the diameter of the Airy disk with $\lambda = 500$ nm is 0.83 arcsec. The resolution limit in the Rayleigh sense is 0.42 arcsec, corresponding roughly to our normalized limit frequency 1. If the best atmospheric "seeing" is 1 arcsec, the limit frequency observable in practice is about 0.4 on the normalized scale. With a central obstruction $\varepsilon = 0.35$, the loss of contrast from the curves in Fig. 3.111 is of the order of 30%. This loss can be serious for critical planetary observation. For constant atmospheric seeing, the larger the telescope, the more the normalized atmospheric limit frequency is pushed leftwards to small values where the relative contrast loss becomes much less. This is the main reason the Schiefspiegler is of much less interest for large professional telescopes.

The results given in Figs. 3.110 and 3.111 are for the MTF of *aberration-free systems* affected only by diffraction. If the aberration in the pupil function F in Eq. (3.494) is non-zero, then the contrast will be reduced. Pioneer work in the solution of (3.494) for basic aberrations with incoherent illumination was done by Hopkins [3.142]. If, for example, the function W in (3.488) represents pure defocus, then we insert

$$W = k_d \rho^2$$

Results for defocus, third order spherical aberration combined with defocus and third order astigmatism were given by Hopkins [3.142], Black and Linfoot [3.143] and De [3.144] respectively, and are quoted by Born-Wolf [3.120(f)]. Figure 3.112 shows the results for defocus. Unlike the Strehl Intensity Ratio, which is only a useful measure for small aberrations, the MTF calculations are valid for large aberrations. The parameter shown on the curves measures the defocus effect and is defined by

$$\underline{m} = \frac{\pi}{2\lambda} \left(\frac{D}{2R} \right)^2 z = 2\pi N^2 z , \qquad (3.506)$$

where z is the distance of the defined image plane from the Gaussian focus. Values of $\underline{m} > 2$ lead to *negative* contrasts (contrast inversion) at quite low frequencies. The appearance is shown by Maréchal-Françon [3.26(j)].

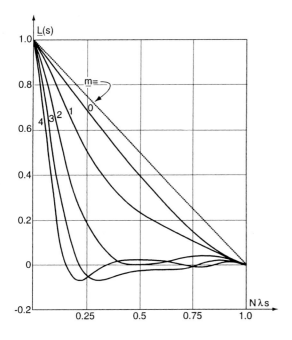

Fig. 3.112. The MTF for a non-obstructed circular aperture with incoherent illumination in the presence of pure defocus aberration (after Hopkins [3.142])

Hopkins [3.145] proved the important principle that the introduction of aberration into a system will reduce the contrast *for all frequencies* compared with the non-aberrated diffraction MTF.

Further examples of the degradation of MTF by various combinations of aberrations are given by Wetherell [3.137].

It can be shown (see Wetherell [3.137]) from Eqs. (3.466), (3.491) and (3.494) that the Strehl Intensity Ratio is simply the area under the MTF function, normalized such that the area under the diffraction MTF in the absence of aberrations is defined as unity. O'Neill [3.146] first pointed out this important relation as a consequence of the fact that $I(Q)$ and $\underline{L}(s,t)$ are Fourier transform pairs. This gives an excellent understanding of the physical link between the two approaches. For small aberrations, the MTF function is independent of the nature of the aberrations, only depending on their variance. But for large aberrations, both the form of the MTF and the diffraction image depend strongly on the nature of the aberrations. The integral under the function, i.e. the Strehl Intensity Ratio, cannot take account of these differences and is therefore an inadequate measure of the effect of strong aberrations.

The OTF may also be calculated as an approximation of geometrical optics: the larger the aberrations, the better the approximation will be. The formulation is given by Welford [3.6]. As was mentioned in § 3.3, the formulae of Nijboer [3.27] express the Fourier transform process of the rays from the pupil to the image. The intensity function I_g of the geometrical optical PSF is given by

$$I_g(\xi, \eta) = C \frac{\partial(x, y)}{\partial(\xi, \eta)} \tag{3.507}$$

giving

$$\underline{L}_g(s, t) = K \int_{-\infty}^{+\infty} \int_{-\infty}^{+\infty} \frac{\partial(x, y)}{\partial(\xi, \eta)} e^{2\pi i(\xi s + \eta t)} \mathrm{d}\xi \, \mathrm{d}\eta \tag{3.508}$$

Modern, sophisticated optical design programs can determine the geometrical PSF as a spot-diagram. The intensity function can then be used directly to give the geometrical optical approximation of the OTF; or combined with the diffraction integral to give the combined effect.

We shall return to these considerations in RTO II, Chap. 4, in connection with the optical specification of telescopes.

3.10.8 Diffraction effects at obstructions in the pupil other than axial central obstruction

The commonest obstruction is the "spider", or supporting cross, holding the secondary in place in a Cassegrain. Such effects are best handled by applying the Babinet principle of complementary screens – see Born-Wolf [3.120(g)] or Ditchburn [3.147]. The two screens are complementary in the sense that the opaque and transmitting parts are inverted. If the complex amplitude induced at a point is U if no screen is present and U_1 and U_2 when the screens are present, then the Babinet principle states that

$$U_1 + U_2 = U \tag{3.509}$$

If $U = 0$, then $U_2 = -U_1$. The principle was first enunciated by Babinet in 1837 [3.148]. The effect of a supporting (obstructing) rectangular spider is therefore the complement of two slits at right-angles placed over the pupil. The case of a rectangular aperture was treated in § 3.10.1.

Dimitroff and Baker [3.23(e)] give results from Scheiner and Hirayama [3.149] for diffraction effects at a large number of obstructed apertures, including a single-arm spider and the normal double-arm form shown in Fig. 3.113. Each arm produces a diffraction "spike" at right-angles to its own direction, the combination then producing the well-known "diffraction cross" present in most astronomical photographs with bright stars in the field. Figure 3.114 shows this in a typical plate taken with the ESO (MPIA) 2.2 m telescope.

The avoidance of such diffraction spikes is the second advantage of the Schiefspiegler. Alternatively, the spikes can be modified by other forms of

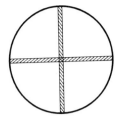

Fig. 3.113. Normal form of supporting spider for secondary mirrors shown here without central obstruction of the secondary

masking or support. An excellent resumé of the possibilities and diffraction images produced is given by Cox [3.150]. He shows the evolution of the diffraction image with a normal spider with reduction of the spider projected area. The results of masking with 4, 5 and 16 round holes over the pupil are also shown. A 4-hole mask removes the spikes but affects the intensity distribution in a more symmetrical way. Any obstruction will cause loss of energy in the central maximum and corresponding loss of contrast. Only when an obstruction becomes small compared with the wavelength of light will its presence become undetectable. Hence, if non-linear detectors such as photo-

Fig. 3.114. Typical astronomical photograph of a star field where diffraction spikes appear on the bright star images. The galaxy is NGC 253, photographed with the 2.2 m telescope at La Silla with 40 m exposure. (Courtesy ESO)

graphic plates are operating with bright stars in a strongly saturated regime, even fine wires across the pupil can produce diffraction effects comparable with normal spider supports. This is also illustrated by the complementary character of the normal supporting spider and two slits at right angles referred to above from the Babinet principle. A spider arm with significant thickness is then complementary to a rectangular aperture, the case treated in § 3.10.1 above. Fig. 3.100 shows the minima of the diffraction pattern for one dimension at the values $\pi, 2\pi, 3\pi, \ldots$ of the function $\delta \bar{u}'$. This function is defined for the y-direction by Eq. (3.436). For the first minimum at π, it follows that

$$\frac{2y_m}{\lambda} \frac{\delta \eta'}{f'} = 1 \,, \tag{3.510}$$

where $2y_m$ is the aperture (slit) width and $\delta \eta'$ is the linear distance in the image plane of the first minimum from the centre of the diffraction pattern at $\delta \bar{u}' = 0$. Now if y_m is doubled, then $\delta \eta'$ must be halved for the first minimum if Eq. (3.510) is maintained. In other words, a doubling of the slit width halves the length of the diffraction effect. For the complementary spider arm, this means that the length of the diffraction spike is inversely proportional to the thickness of the spider arm responsible. This inverse linear law is the same as that governing the case of diffraction at a circular aperture (see § 3.10.3) leading to the well-known Airy disk formula of Eq. (3.447) and the classical definition of the resolving power of a telescope given in Eq. (3.449).

4 Field correctors and focal reducers or extenders

4.1 Introduction

The term *"field corrector"* in telescope systems implies some system placed inside the image plane, but relatively close to it, whose primary function is to correct the field aberrations of the mirror system. If the mirror system does not correct the axial image at the focus in question (e.g. the prime focus of an RC telescope), the field corrector will also be required to achieve this. Normally, field correctors are substantially afocal, i.e. they only have a minor effect on the f/no of the incident beam from the mirror system.

The term *"focal reducer"* refers to a transfer system (which may, in principle, be before or after the real telescope image) whose primary function is to reduce the f/no of the mirror system, e.g. convert an f/8 Cassegrain image to f/3. A focal reducer will normally have the task of *maintaining* the quality of imagery over a larger angular field. Inverse focal reducers are also of practical importance, i.e. *focal extenders* which reduce the angular field and increase the focal length. In most practical cases, the focal reducers are closely related to the optical systems of spectrographs.

Field correctors fall, in principle, into two groups: those designed to work with an *existing* (fixed) mirror system and those optimized with one or more free parameters of the mirror system. In the latter case, there is no real distinction between an optimized telescope with field corrector and the optimized telescopes of Chap. 3 in which additional plates, mirrors or lenses were added to the 1- or 2-mirror basic system. Since the theory is closely linked, the two field corrector groups will be considered together.

The first reference to field correctors was apparently made by Sampson [4.1] [4.2]. In two remarkable papers based on Schwarzschild's theory, he laid down some of the basic principles of field correctors with lenses, both for Cassegrain and Newton foci, the Newton case being optically the same as the prime focus (PF). This work was extended in 1922 by Violette [4.3] to the field correction of the newly invented RC telescope of Chrétien. This was followed in 1935 by a classic paper by Paul [4.4], who systematically analysed the possibilities of lens correctors for both Cassegrain and prime foci and introduced the concept of aspheric plate correctors as the theoretical equivalent of a deformed plane mirror. He also treated the general case of 3 aspheric mirrors leading to the Paul telescope discussed in Chap. 3.

The period from 1945 to about 1975 saw intensive activity in the design of field correctors to give larger fields for photographic plates. An excellent review was given by Wynne [4.5]. Their significance declined with the replacement in normal telescopes of the photographic plate by modern electronic detectors, above all the CCD. These detectors had – and still have – relatively small fields compared with photographic plates. However, arrays of detectors are becoming possible with larger fields, and techniques such as multi-image spectroscopy at the Cassegrain focus using fibres also place higher demands on field correction.

Compared with 3- or 4-mirror solutions, refracting field correctors are usually more light-efficient but suffer from ghost reflections and chromatic aberrations. Aspheric plates are, in general, superior for ghost reflections because they are essentially similar to a flat, thin filter glass: the ghost image of a bright object is, for the small angular fields of telescopes, close to its primary image and normally swamped by the latter's over-exposure (Fig. 4.1). For a plane-parallel plate in air, the ghost image displacement in the image

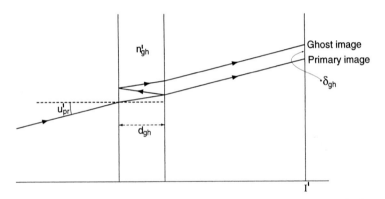

Fig. 4.1. Ghost images through 2 reflections at a plane parallel plate

plane δ_{gh} is given for a small principal ray angle u'_{pr} by

$$\delta_{gh} = \frac{2u'_{pr}}{n'_{gh}} d_{gh} \; , \tag{4.1}$$

where d_{gh} is the plate thickness and n'_{gh} its refractive index. In a Cassegrain telescope u'_{pr} is magnified by the telephoto effect typically by a factor of the order of 4 compared with the object field angle. For a semi-field of 0.25°, this gives $u'_{pr} \sim 1°$, say 0.02 rad. For $d_{gh} = 10$ mm and $n'_{gh} = 1.5$

$$\delta_{gh} = 0.267 \text{ mm} \; ,$$

which is only of the order of 2 arcsec or less with the plate scale of a typical 3.5 m telescope in the Cassegrain focus. By contrast, a lens can produce a ghost image widely separated from the primary image. It will also, in general, be strongly defocused whereas the ghost image of the plate is only defocused by $-2d_{gh}/n'_{gh}$ relative to the primary image.

4.2 Aspheric plate correctors

4.2.1 Prime focus (PF) correctors using aspheric plates

Paul [4.4] was the first to recognize, on the basis of Schwarzschild-Chrétien theory, the correction possibilities of aspheric plates for both the PF and Cassegrain foci.

For the prime focus, he considered first the case of a parabolic primary, showing that a single plate could correct the field coma at the cost of introducing spherical aberration. With 2 separated plates the second condition of field astigmatism could be corrected, again at the cost of introducing spherical aberration which Paul proposed to correct by changing the form of the primary into a hyperbola. Implicitly, the complete theory of correctors using aspheric plates was laid out in this work by Paul. However, the practical significance was apparently not appreciated until the definitive work of Meinel [4.6] and Gascoigne [4.7] [4.8] [4.9], the two latter papers giving perhaps the best review of modern telescope optics available.

The theory of aspheric plates was given in § 3.4, the conditions for the correction of the first three Seidel aberrations, S_I, S_{II} and S_{III}, being given by Eq. (3.220) for a 2-mirror telescope and aspheric plate near the pupil:

$$\left.\begin{array}{rcl} \sum S_I & = & \left(\dfrac{y_1}{f'}\right)^4 \left[-f'\zeta + L\xi + \delta S_I^*\right] = 0 \\[3mm] \sum S_{II} & = & \left(\dfrac{y_1}{f'}\right)^3 \left[-d_1\xi - \dfrac{f'}{2} + \dfrac{s_{pl}}{f'}\delta S_I^*\right] u_{pr1} = 0 \\[3mm] \sum S_{III} & = & \left(\dfrac{y_1}{f'}\right)^2 \left[\dfrac{f'}{L}(f' + d_1) + \dfrac{d_1^2}{L}\xi + \left(\dfrac{s_{pl}}{f'}\right)^2 \delta S_I^*\right] u_{pr1}^2 = 0 \end{array}\right\}$$

We can use this 2-mirror formulation also for the PF case by taking the limit case of a flat secondary in Cassegrain geometry, giving a positive focal length f'. This gives the same result as Eqs. (3.219).

In these equations, δS_I^* is the contribution of the aspheric plate to $\sum S_I$ and s_{pl} is the distance of the plate from the primary *in object space*, a positive s_{pl} corresponding to a virtual plate position to the right of the primary with the light incident from the left. It was shown by Burch [4.10] that the above formulation can be extended to any number of plates with $(\delta S_I^*)_1, (\delta S_I^*)_2 \ldots$ and $(s_{pl}/f')_1, (s_{pl}/f')_2 \ldots$, the plate effects being additive. Furthermore, the effect of an aspheric plate in the convergent beam forming the image (Fig. 4.2) is the same as a plate at the conjugate point at s_{pl} in object space whose

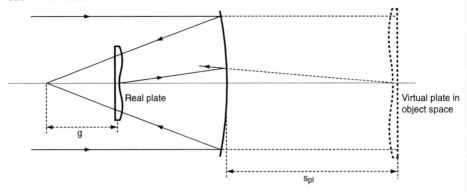

Fig. 4.2. Transformation of a real corrector plate to a virtual plate in object space

aspheric constant a is defined by Eq. (3.221) as

$$\delta S_I^* = 8(n' - n) a\, y^4 \ ,$$

which is reduced by the factor $(g/f')^4$, where g is the distance of the plate from the image. The wavefront aberration contributions of the real and virtual plates are identical over the respective axial beam widths.

In the PF case, the paraxial transformation for the virtual plate is simply

$$s_{pl} = f' \left(\frac{f' - g}{g} \right) \ , \tag{4.2}$$

in which $f' = |f_1'|$, the focal length of the primary, and g is defined as a positive quantity with $g < |f_1'|$. If, for our PF case, we place the fictitious plane secondary in contact with the primary, we can substitute in Eqs. (3.220) above

$$m_2 = -1, \quad L = f', \quad d_1 = 0, \quad \xi = 0, \quad u_{pr1} = 1$$

giving, with our usual normalization[1] of $y_1 = f'$, the conditions

$$\left. \begin{array}{rll} \sum S_I &= -f'\zeta + \delta S_I^* &= 0 \\[2mm] \sum S_{II} &= -\dfrac{1}{2} f' + \dfrac{s_{pl}}{f'}\, \delta S_I^* &= 0 \\[2mm] \sum S_{III} &= f' + \left(\dfrac{s_{pl}}{f'}\right)^2 \delta S_I^* &= 0 \end{array} \right\} \tag{4.3}$$

---

[1] It should be noted that this normalization with $y_1 = f'$, thereby eliminating the factors (y_1/f') in various powers in Eq. (3.220), is only valid in the general case if the conditions are corrected to zero, as in Eq. (4.3). This must be borne in mind for cases where a telescope is deliberately *not* corrected for spherical aberration ($\sum S_I$) because of desired compensation of the aberration in a subsequent instrument.

As explained in §3.4, if $\sum S_I = \sum S_{II} = 0$, the stop position in the system has no effect on the three Seidel aberrations and may be considered to be at its normal position, the primary. Then, with our normalization

$$f' = y_1 = u_{pr1} = 1 \,,$$

it follows at once that

$$\frac{s_{pl}}{f'} = \frac{y_{pr}}{y}$$

As in §3.4, we will define

$$\frac{s_{pl}}{f'} = E = \frac{f' - g}{g} \tag{4.4}$$

from (4.2) and simplify the notation with

$$(\delta S_I^*)_\nu = S_\nu$$

for aspheric plate number ν in a multiple plate corrector. Then (4.3) becomes for a 3-plate corrector:

$$\left.\begin{aligned} S_1 + S_2 + S_3 &= +f'\zeta \\ E_1 S_1 + E_2 S_2 + E_3 S_3 &= +f'/2 \\ E_1^2 S_1 + E_2^2 S_2 + E_3^2 S_3 &= -f' \end{aligned}\right\} \tag{4.5}$$

The quantity ζ defines the asphericity of the primary according to Table 3.5 as

$$\zeta = \frac{m_2^3}{4}(1 + b_{s1}) \,,$$

in which $m_2 = -1$ in our PF case. If $\zeta = 0$ in (4.5), we have the conditions for the correction of a parabolic primary, essentially the same as those given by Wynne [4.5] with a slightly different normalization.

4.2.1.1 One corrector plate. The first two equations of (4.5) immediately enable us to prove the first fundamental property given by Paul [4.4]. For a parabolic primary (i.e. a conventional Newton telescope) and a single corrector plate:

$$S_1 = 0$$
$$ES_1 = f'/2$$

The only solution requires $E = \infty$ or, from (4.4), $g = 0$. The plate is in the image plane and would require infinite asphericity. So a single plate at a finite distance g can correct the coma of the parabolic primary but introduces, from (4.4), spherical aberration

$$\sum S_I = S_1 = \frac{f'}{2}\left(\frac{g}{f' - g}\right) \tag{4.6}$$

With $g = f'/10$, this gives $\sum S_I = (0.05556)f'$.

If the primary is allowed to assume *a non-parabolic form* giving a real solution in the case of a single plate corrector, we have from (4.5)

$$\left.\begin{array}{rl} S_1 &= +f'\zeta \\ ES_1 &= +f'/2 \end{array}\right\} , \tag{4.7}$$

giving

$$\zeta = \frac{1}{2}\left(\frac{g}{f'-g}\right) = -\frac{1}{4}(1+b_{s1}) \tag{4.8}$$

or

$$(1+b_{s1}) = -2\left(\frac{g}{f'-g}\right) , \tag{4.9}$$

which is negative since f' and g are positive. The required form of the primary is thus *hyperbolic* and the aspheric form of the plate is the *opposite* of a Schmidt plate correcting a spherical primary. If b_{s1} has the value for an RC primary, then we have the *Gascoigne PF aspheric corrector* [4.7] [4.9]. The necessary position of the plate is given at once by transforming (4.9):

$$g = -f'\left(\frac{1+b_{s1}}{1-b_{s1}}\right) \tag{4.10}$$

This relation was given by Gascoigne [4.9]. Taking the value of $(b_{s1})_{RC}$ from Table 3.2 as -1.03629, (4.10) gives

$$g = (0.017\,822)f' ,$$

only about 1.8% of $|f_1'|$. The single Gascoigne plate can correct the first two conditions of (4.5) by its two parameters of position and asphericity; but it cannot correct the third condition for astigmatism. In fact, it makes this far worse, but is still a most useful form of corrector as we shall now prove.

For a single plate, the astigmatism is from the third equation of (4.5), and from (4.7) and (4.8):

$$\sum S_{III} = f' + E_1^2 S_1$$

$$= f' + \left(\frac{f'-g}{g}\right)^2\left[-\frac{f'}{4}(1+b_{s1})\right]$$

Eliminating g from (4.10) gives

$$\sum S_{III} = f'\left(\frac{b_{s1}}{1+b_{s1}}\right) , \tag{4.11}$$

the same result as in Gascoigne [4.9] with different normalization. This result can be written as

$$\sum S_{III} = f'\left[1 - \left(\frac{1}{1+b_{s1}}\right)\right] ,$$

the first term being the contribution of the primary mirror and the second that of the plate. For the RC primary of Table 3.2 with $b_{s1} = -1.03629$, we have $\sum S_{III} = f'(1 + 27.56)$, so the astigmatism of the primary is increased by a factor about 29 times. In terms of g, (4.9) gives

$$\sum S_{III} = f' \left(\frac{f' + g}{2g} \right) \tag{4.12}$$

Clearly, the larger the negative value of $(1 + b_{s1})$ and, as a result g, the more favourable the correction of the Gascoigne plate at the PF.

For RC telescopes with a lower value of $|m_2|$ than that in Table 3.2 ($|m_2| = 4$), the primary is more eccentric and thus more favourable. We shall see later that certain correctors at the Cassegrain focus favour quasi-RC solutions for which the eccentricity of the primary is of the order of 30% higher than for the strict RC solution. This favours the Gascoigne plate corrector for the PF. Such a telescope is the ESO 3.6 m at La Silla. The Gascoigne plate corrector is provided with an additional field-flattening lens which flattens the mean astigmatic field. For an angular aberration of 1 arcsec at the mean astigmatic focus for the edge of the field, a field diameter of *16 arcmin* is possible in this case. Such a Gascoigne plate corrector with field flattener is an excellent solution for modern telescopes equipped with a CCD detector at the PF: it is robust, favourable for ghost images as we saw above, and also for chromatic effects as we shall show below. Let us consider first the field lens required.

The mean field curvature is given as in Table 3.3 and from Eq. (3.206) as

$$\text{MFC} = 2 \sum S_{III} + \sum S_{IV}$$

Now $S_{IV} = -f'$ for the mirror and is zero for the plate. From (4.12)

$$\text{MFC} = \frac{f'^2}{g} = -H^2 P_C , \tag{4.13}$$

from (3.20), where H is the Lagrange Invariant and P_C the Petzval curvature. Now $H = n'u'\eta'$ and, with our normalization and the fictitious plane secondary, $n' = 1$, $u' = -1$, and $\eta' = f'$ in image space, so that $H = -f'$. The wavefront aberration MFC of the mirror-plate combination is *positive* according to (4.13), with P_C negative, so that the mean field curvature is *concave* towards the primary, i.e. the sign is reversed compared with the much weaker field curvature of the primary alone. Then (4.13) gives

$$(P_C)_{MFC} = -1/g \tag{4.14}$$

From the definitions for Eqs. (3.20), we have for a surface ν of the field flattener (FF)

$$(P_C)_{FF,\nu} = \frac{1}{(r_{FF})_\nu} \left(\frac{1}{(n'_{FF})_\nu} - \frac{1}{(n_{FF})_\nu} \right) ,$$

where $(n_{FF})_1 = n' = 1$ for air in the image space and $(n'_{FF})_1$ is the index of the lens FF. For compensation, we require

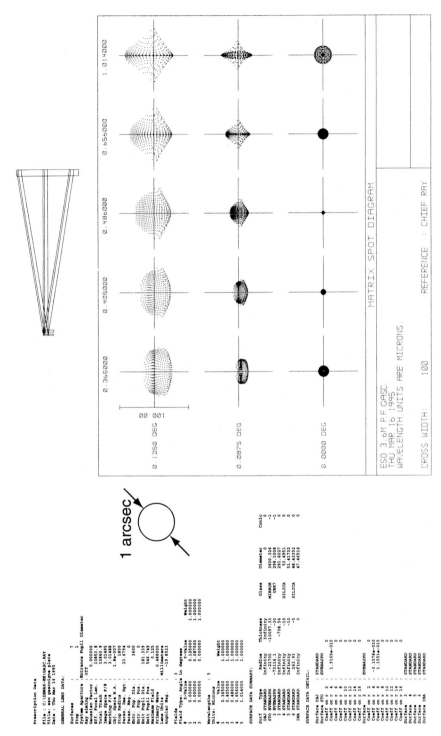

Fig. 4.3. Spot-diagrams for the PF Gascoigne plate-field flattener corrector (with filter) of the ESO 3.6 m telescope on La Silla. The Schwarzschild constant of the primary is $b_{s1} = -1.1567$ for a quasi-RC solution

$$(P_C)_{FF} = -(P_C)_{MFC} \, ,$$

giving for the first surface of FF if its second surface is plane

$$(r_{FF})_1 = -\frac{1}{(P_C)_{MFC}} \left(\frac{1}{(n'_{FF})_1} - \frac{1}{(n_{FF})_1} \right)$$

This gives from (4.14)

$$(r_{FF})_1 = g \left(\frac{1}{(n'_{FF})_1} - \frac{1}{(n_{FF})_1} \right) = -g \left(\frac{(n'_{FF})_1 - 1}{(n'_{FF})_1} \right) \tag{4.15}$$

Since $(n'_{FF})_1 > 1$ and g is positive, the lens is a plano-concave field flattener with its first radius negative.

From (3.221) and (3.20), a single plate in air has a chromatic variation of each aberration directly proportional to the chromatic variation of $(n'_{pl} - 1)$, about 10% for crown glass from 1000 nm to 300 nm. Over most of the range it is far less and chromatism can be further reduced by supplying two Gascoigne plates for different wavelength regions.

The finite thickness of the plate adds a *plane-parallel plate* which, in the converging beam, also produces aberrations. As windows, filters, etc., plane-parallel plates are common elements in the converging beams of telescopes. The aberrations introduced by such a plate are given in the works of Hopkins and Welford cited in Chap. 3 (refs. [3.3] and [3.6]) and are as follows (see Eqs. (3.254))

$$
\left.
\begin{aligned}
(S_I)_{PP} &= -\left(\frac{n'^2_{pp} - 1}{n'^3_{pp}} \right) u^4 \, d_{pp} \\[2mm]
(S_{II})_{PP} &= \left(\frac{u_{pr}}{u} \right) S_I \\[2mm]
(S_{III})_{PP} &= \left(\frac{u_{pr}}{u} \right)^2 S_I
\end{aligned}
\right\}
\tag{4.16}
$$

where d_{pp} is the plate thickness, $u = \frac{1}{2N}$ the semi-aperture angle and u_{pr} the principal ray angle which is the incidence angle of the beam on the plate.

The final optical design of a Gascoigne plate with field flattener for the PF will take account of all such effects as well as balancing of higher order effects. In practice, such aspheric plates will also have a balancing term in y^2, as in Schmidt plates, to reduce their asphericity. Figure 4.3 gives spot-diagrams for the Gascoigne plate, field flattener corrector of the PF of the ESO 3.6 m telescope.

4.2.1.2 Two corrector plates. Paul [4.4] investigated the case of a 2-plate corrector for a *parabolic* primary. His aim was the correction of coma and astigmatism. From (4.5) we have, then, the conditions

$$S_1 E_1 + S_2 E_2 = \frac{1}{2} f'$$

$$S_1 E_1^2 + S_2 E_2^2 = -f'$$

giving

$$S_1 = \frac{1}{2}f' \frac{(E_2 + 2)}{E_1(E_2 - E_1)} , \quad S_2 = \frac{1}{2}f' \frac{(E_1 + 2)}{E_2(E_1 - E_2)}$$

From (4.4), with

$$E_1 = \frac{f' - g_1}{g_1} , \quad E_2 = \frac{f' - g_2}{g_2} ,$$

we can derive

$$\sum S_I = S_1 + S_2 = \frac{1}{2}f'^2 \left[\frac{g_1 + g_2}{(f' - g_1)(f' - g_2)} \right] , \tag{4.17}$$

which is the same as the form given by Gascoigne [4.9] apart from a different normalization. Since f' and g are finite positive quantities and $0 < g < f'$, it follows that it is impossible to avoid introducing spherical aberration with a parabolic primary and two aspheric plates correcting coma and astigmatism. With $g_1 = f'/10$ and $g_2 = f'/20$, (4.17) gives

$$\sum S_I = \frac{f'}{4} \left(\frac{20}{57} \right) = (0.08772)f' ,$$

whereas a spherical primary mirror alone would give $f'/4$. The spherical aberration is therefore of the order of $1/3$ of the aberration of a spherical primary. This is over one and half times the spherical aberration of Eq. (4.6) for the single plate correcting coma alone and with the same g as the larger plate of the 2-plate corrector. This illustrates an important consequence of multi-plate systems: because of compensations, the individual plates are stronger with corresponding effects on any uncorrected aberrations. The contribution S_1 of the larger plate is

$$S_1 = \frac{1}{2} \left[\frac{g_1^2(f' + g_2)}{(g_1 - g_2)(f' - g_1)} \right] , \tag{4.18}$$

which is almost exactly twice that of a single plate, from Eq. (4.6), correcting coma alone at the same distance g_1.

Paul [4.4] drew the logical conclusion that a *2-plate corrector* should be used with a *hyperbolic* primary to correct $\sum S_I$. The general condition of (4.5) can then be written:

$$\left. \begin{aligned} S_1 + S_2 \quad &= +f'\zeta = -\frac{f'}{4}(1 + b_{s1}) \\ S_1 E_1 + S_2 E_2 &= +f'/2 \\ S_1 E_1^2 + S_2 E_2^2 &= -f' \end{aligned} \right\} \tag{4.19}$$

The solution of the second and third equations is the same as above giving

$$\left(\sum S \right)_{Plates} = \frac{1}{2}f'^2 \left[\frac{(g_1 + g_2)}{(f' - g_1)(f' - g_2)} \right]$$

The condition for correction of spherical aberration by the primary is therefore

$$\sum S = \frac{1}{2} f'^2 \left[\frac{(g_1 + g_2)}{(f' - g_1)(f' - g_2)} \right] + \frac{f'}{4}(1 + b_{s1}) = 0 , \qquad (4.20)$$

or

$$1 + b_{s1} = -2f' \left[\frac{(g_1 + g_2)}{(f' - g_1)(f' - g_2)} \right] \qquad (4.21)$$

This should be compared with Eq. (4.9), the equivalent expression for the single (Gascoigne) plate correcting a hyperboloid for spherical aberration and coma. Eq. (4.21) for the 2-plate corrector gives

$$(1 + b_{s1})_{2pl} = -0.351$$

with $g_1 = f'/10$, $g_2 = f'/20$. The Gascoigne plate with $g = f'/10$ gives from (4.9)

$$(1 + b_{s1})_{1pl} = -0.222$$

The 2-plate corrector correcting $\sum S_I$, $\sum S_{II}$ and $\sum S_{III}$ therefore requires a much stronger hyperbola for the spherical aberration correction than the Gascoigne plate and this places it normally outside the useful range of RC or quasi-RC telescopes. In other words, its use in the PF requires a mirror combination unfavourable for field correction at the Cassegrain focus.

There remains one interesting possibility: a *2-plate corrector* with a *parabolic* or *hyperbolic* primary correcting $\sum S_I$ and $\sum S_{II}$ only, without $\sum S_{III}$. From (4.5) for the general case with a hyperbolic primary, we have the conditions

$$\left. \begin{array}{rl} S_1 + S_2 & = +f'\zeta \\ S_1 E_1 + S_2 E_2 & = +f'/2 \end{array} \right\} \qquad (4.22)$$

giving

$$S_1 = \frac{f' \left(\frac{1}{2} - E_2 \zeta \right)}{(E_1 - E_2)} , \quad S_2 = \frac{-f' \left(\frac{1}{2} - E_1 \zeta \right)}{(E_1 - E_2)} \qquad (4.23)$$

Clearly a positive value of ζ (i.e. $b_{s1} < -1$ with $m_2 = -1$) reduces the power of the plates compared with the case of a parabolic primary with $\zeta = 0$.

From the third equation of (4.5) the astigmatism is given by

$$\sum S_{III} = f' + S_1 E_1^2 + S_2 E_2^2$$

or, from (4.23)

$$\sum S_{III} = f' + \frac{f'}{2} \left(E_1 + E_2 \right) - f'\zeta E_1 E_2 \qquad (4.24)$$

Substituting for E from (4.4) gives

$$\sum S_{III} = \frac{1}{2}f'^2\frac{(g_1+g_2)}{g_1 g_2} - f'\zeta\frac{(f'-g_1)(f'-g_2)}{g_1 g_2} \tag{4.25}$$

Finally, substituting for ζ from Table 3.5 with $m_2 = -1$ gives the form equivalent to that given by Gascoigne [4.9]:

$$\sum S_{III} = \frac{1}{2}f'^2\frac{(g_1+g_2)}{g_1 g_2} + \frac{f'}{4}(1+b_{s1})\frac{(f'-g_1)(f'-g_2)}{g_1 g_2} \tag{4.26}$$

Gascoigne gives an approximate form for the second term on the reasonable assumption from practical application that $g_1, g_2 \ll f'$. Simplifying in this way, the second term of (4.26) reduces to

$$\frac{f'}{4}(1+b_{s1})\frac{(f'-g_1)(f'-g_2)}{g_1 g_2} \simeq \frac{f'^3}{4}\frac{(1+b_{s1})}{g_1 g_2}$$

With this approximation and taking into account a factor for different normalization, the second term of Gascoigne should have a factor 8 in the denominator which is missing in [4.9]. This is important for the general conclusions of the potential of correctors consisting of two aspheric plates. Taking $g_1 = f'/10$, $g_2 = f'/20$ and $b_{s1} = -1.03$ as a typical RC value, the approximate form of (4.26) gives

$$\sum S_{III} \simeq 15f' - 1.5f' \ ,$$

i.e. the compensating second term is only $1/10$ of the first term, which gives the astigmatism for such a system with a parabolic primary. This modest compensation is a consequence of the small departure from the parabola of RC primaries and confirms the conclusion from (4.21) that a much larger departure from the paraboloid ($b_{s1} = -1.351$) is required to correct all three aberrations with a 2-plate corrector with the above values of g_1, g_2. Without the factor 8 in the denominator of his second term, Gascoigne's formula gives 80% compensation by the second term which would indicate the 2-plate corrector can give excellent correction of all three aberrations with modern RC primaries. Unfortunately, this is not the case.

With the above values of g_1, g_2, the first term of (4.26) gives $\sum S_{III} = 15f'$ for a 2-plate corrector and a parabolic primary correcting $\sum S_I = \sum S_{II} = 0$. Using (4.12) we can compare this with the astigmatism of a single plate corrector of similar mean size ($g = 0.075f'$) correcting a hyperboloid, giving $\sum S_{III} = 7.17f'$, less than half that of the 2-plate corrector with parabolic primary. However, from (4.9), the corresponding eccentricity of the primary for $g = 0.075f'$ in the singlet case is $b_{s1} = -1.162$, much more favourable than a normal RC case. If $b_{s1} = -1.03$ for a typical RC, the singlet Gascoigne plate requires from (4.10) $g = 0.0148f'$ and gives $\sum S_{III} = 34.33f'$ from (4.12), over twice that of the parabola, 2-plate case using much larger plates.

The general conclusion is that the PF corrector with 2 aspheric plates is only of practical interest for a special PF telescope with hyperbolic primary of high eccentricity ($b_{s1} \sim -1.35$). This was the suggestion of Paul [4.4]. It is not a practical solution for parabolic or normal RC primaries, even if larger plates are used, because of the residual astigmatism.

4.2.1.3 Three corrector plates. We can apply the 3 equations of (4.5)

$$\left.\begin{array}{rl} S_I + S_2 + S_3 & = +f'\zeta = -\dfrac{f'}{4}(1 + b_{s1}) \\[2mm] E_1 S_1 + E_2 S_2 + E_3 S_3 & = +f'/2 \\[2mm] E_1^2 S_1 + E_2^2 S_2 + E_3^2 S_3 & = -f' \end{array}\right\}$$

directly to correct the three conditions $\sum S_1 = \sum S_{II} = \sum S_{III} = 0$ *for any value of* ζ. Meinel [4.6] and Wynne [4.5] give the solution for the parabolic primary and subsequent work, which we shall consider in more detail, has extended the application to RC and quasi-RC primaries. We can derive the general solution from (4.5) which is

$$S_1 = \frac{f' \left[\{ \frac{1}{2}(E_2 + E_3) + 1 \} - E_2 E_3 \zeta \right]}{(E_1 - E_3)(E_2 - E_1)}, \tag{4.27}$$

with equivalent expressions for S_2 and S_3. Substituting for E from (4.4) reduces (4.27) to

$$S_1 = \frac{g_1^2}{f'} \left[\frac{\frac{1}{2}f'(g_2 + g_3) - (f' - g_2)(f' - g_3)\zeta}{(g_3 - g_1)(g_1 - g_2)} \right] \tag{4.28}$$

If the second term of the numerator is set to zero for a parabolic primary, (4.28) is the same as the result given by Wynne [4.5] apart from small differences of normalization. If the suffixes 1,2,3 are taken in the direction of the light, then 1 refers to the first (largest) plate with the largest g. The denominator in (4.28) is therefore *negative*. For S_2 it is *positive* and for S_3 *negative*. Since the numerator is positive in all cases, these are the signs of the aspherics on the three plates: 1 and 3 as for a Schmidt plate, 2 the opposite like a Gascoigne plate for a hyperbolic primary. Figure 4.4 shows this situation schematically for 1-, 2- and 3-plate correctors. For 2-plate correctors the sign of the front plate will depend on whether the conditions $\sum S_{II} = \sum S_{III} = 0$ are corrected for a strongly hyperbolic primary (positive front plate); or $\sum S_I = \sum S_{II} = 0$ for a parabolic or RC primary (negative front plate).

Theoretically, a solution of (4.5) exists for any plate spacing and any asphericity of the primary; but small values of g and small differences in the 3 values will lead to very high plate asphericities. Meinel suggests a maximum g of the order of $0.15f'$. Let us take, as an example, $g_1 = 0.15f'$, $g_2 = 0.10f'$, $g_3 = 0.05f'$ and $b_{s1} = -1.04$ for a typical quasi-RC giving $\zeta = +0.01$ from (4.8). Eq. (4.28) then gives

$$S_1 = f' \frac{9}{400} \left[\frac{0.075 - 0.00855}{-0.005} \right] = -f'(0.299\,025)$$

(a) **(b)**

(c) **(d)**

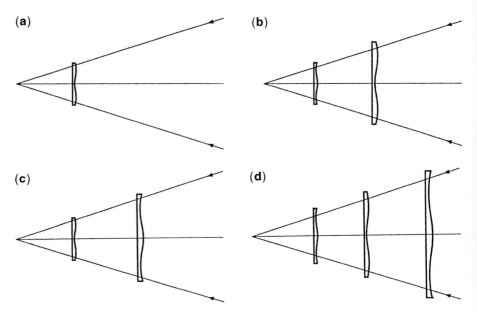

Fig. 4.4. Schematic appearance of aspherics on PF plate correctors: (**a**) Gascoigne plate (singlet) corrector for RC hyperbolic primaries; (**b**) 2-plate corrector for strongly hyperbolic primaries correcting $\sum S_I = \sum S_{II} = \sum S_{III} = 0$; (**c**) 2-plate corrector for parabolic or RC primaries correcting $\sum S_I = \sum S_{II} = 0$; (**d**) 3-plate corrector for parabolic or RC primaries correcting $\sum S_I = \sum S_{II} = \sum S_{III} = 0$

Transposing suffixes 1 and 3 gives for the third plate

$$S_3 = f' \frac{1}{400} \left[\frac{0.125 - 0.00765}{-0.005} \right] = -f'(0.058\,675)$$

and similarly for the second plate, transposing suffixes 1 and 2,

$$S_2 = f' \frac{1}{100} \left[\frac{0.10 - 0.008\,075}{0.0025} \right] = +f'(0.367\,700)$$

Then $\sum S = 0.01\,f' = +f'\zeta$ for the primary, as required by the first equation of (4.5). The balance of terms in this example is instructive. The second term in ζ for the typical case chosen brings a reduction of only 11.4% for the first plate and only 6.1% for the third plate, relatively insignificant relaxations compared with the plates for a parabolic primary. The individual plate contributions are respectively about 30, 37 and 6 times higher than the contribution of the quasi-RC primary. Since a *spherical* primary would give a contribution $0.25f'$, the contributions of plates 1 and 2 are even higher than this! Furthermore, these contributions measure the required asphericity of the plates over the axial beam width. But the axial free aperture of the plates must be multiplied by $(1 + E_R)$ to avoid vignetting in the field, where E_R is the E-value for the real field and aperture. Since the asphericities increase with y^4 compensated by a y^2 term, the field surplus on the plate diameter rapidly increases the total asphericity for small g-values. We see, therefore, that the elegance of a multi-element corrector in correcting several conditions also exacts its price: the individual elements are very strong and this leads to higher order effects which limit the performance, as we shall see with practical examples.

Multiplying (4.28) by E_1 and E_1^2 respectively, gives the coma and astigmatism contributions of the first plate as

$$(S_{II})_1 = \frac{g_1}{f'} \left[\frac{\frac{1}{2}f'(f'-g_1)(g_2+g_3) - (f'-g_1)(f'-g_2)(f'-g_3)\zeta}{(g_3-g_1)(g_1-g_2)} \right] \quad (4.29)$$

$$(S_{III})_1 = \frac{1}{f'} \left[\frac{\frac{1}{2}f'(f'-g_1)^2(g_2+g_3) - (f'-g_1)^2(f'-g_2)(f'-g_3)\zeta}{(g_3-g_1)(g_1-g_2)} \right], \quad (4.30)$$

with similar expressions for the other two plates. With the values given above, the contributions of the *third* plate to coma and astigmatism are $(S_{II})_3 = f'(-1.260)$ and $(S_{III})_3 = f'(-23.94)$. The latter is thus 24 times larger than the astigmatism of the primary mirror we are correcting, again a measure of the sensitivity of the system and the fine tolerances required.

The *chromatic variations* of the aberrations S_I, S_{II}, S_{III} will be proportional to the contributions of the individual plates but, in third order theory, will balance out as for the monochromatic aberrations themselves. From this theory, then, a set of plates of one glass will simply have the chromatic variation of the mirror contributions on the right-hand side of Eqs. (4.5). As stated above, this will only be about 10% for crown glass and the full spectral range of 1000 nm to 300 nm. Unfortunately, in practice, the situation is much less favourable because of higher order aberrations, both monochromatic, and above all, chromatic. Third order theory assumes the paraxial *heights* of the rays at the different elements are respected in all cases. This is not the case with strong plates: the largest plate disperses the rays to different heights on the other plates, above all the principal ray affecting E which is very sensitive for the coma and astigmatism contributions. Such 3-plate correctors are therefore essentially limited by the higher order chromatic aberrations, above all chromatic differences of coma and, to a lesser extent, spherochromatism and astigmatism. These effects are mitigated by the balancing y^2 term in the plate form, but are still determinant. The only way to eliminate these chromatic effects would be to achromatise *each plate* individually as an achromatic plate using 2 glasses. In practice, this is barely possible since the individual plates of the achromatic combinations would be many times stronger than their combination, also producing higher order effects apart from the technical difficulty and cost.

We shall return to this issue in comparing lens and plate correctors.

Wynne [4.5] stated in 1972 that the manufacture of such 3-plate correctors had not apparently been envisaged for the correction of a *parabolic* primary. I think this is still true today. The manufacture was seriously envisaged for RC and quasi-RC telescopes but, I believe, has never been carried out.

Schulte [4.11] investigated such a 3-plate corrector on the basis of Meinel's work [4.6] for the Kitt Peak 3.8 m, f/2.8 RC primary. A detailed investigation was carried out by Köhler [4.12] [4.13] and colleagues for the ESO 3.6 m, f/3.0 quasi-RC primary. This system has been re-calculated by Wynne [4.5] [4.14] for comparison with lens correctors. The original system (Wilson [4.15]) had 25% vignetting. Spot diagrams for an optimized form without vignetting and with extended spectral range were given for this 3-plate corrector plus field lens by Cao and Wilson [4.16]. These are reproduced in Fig. 4.14 in a comparison with a 3-lens corrector. Since the latter gives superior chromatic performance, the 3-plate system was abandoned at the time. The relative merits are further discussed later in comparison with the lens corrector.

4.2.2 Cassegrain or Gregory focus correctors using aspheric plates

This possibility was referred to briefly in § 3.4 in connection with the addition of full-size aspheric plates in the object space of 2-mirror telescopes. We are concerned now with a real aspheric plate (or plates) near the Cassegrain (or Gregory) image which is then treated as a virtual plate in object space, as in the PF case above. For a single plate, we have from Eq. (3.220) in our simplified notation above for the fulfilment of the three aberration conditions $\sum S_I = \sum S_{II} = \sum S_{III} = 0$

$$\left.\begin{array}{rcl} -f'\zeta + L\xi + S & = & 0 \\[2mm] -d_1\xi - \frac{f'}{2} + ES & = & 0 \\[2mm] f'\dfrac{(f'+d_1)}{L} + \dfrac{d_1^2}{L}\xi + E^2 S & = & 0 \end{array}\right\}, \tag{4.31}$$

with $E = s_{pl}/f'$ for the virtual plate in object space at a distance s_{pl} from the primary, as shown in Fig. 4.5. A paraxial ray must be traced back from the real plate at distance g inside the secondary image to the conjugate point in the object space. If the paraxial parameters r_1, r_2 and d_1 are first eliminated to give an expression in f', L, m_2 and g (g being defined as positive and the other quantities having the signs given in Chap. 2 for the Cassegrain and Gregory cases), one can derive

$$s_{pl} = f'\left\{\frac{f'}{g} - \frac{f'}{L}\left[1 + \frac{1}{m_2}\left(1 - \frac{L}{f'}\right)\right]\right\} \tag{4.32}$$

The second term of this expression can be simplified by eliminating m_2 in terms of d_1, a negative quantity in both cases, giving

$$\frac{f'}{L}\left[1 + \frac{1}{m_2}\left(1 - \frac{L}{f'}\right)\right] = \frac{f'+d_1}{L} = \frac{d_1 + f_2'}{f_2'}, \tag{4.33}$$

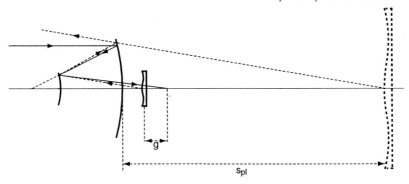

Fig. 4.5. Conjugate virtual plate in object space for a real aspheric plate at distance g in front of the Cassegrain focus

with f_2' defined as negative for a Cassegrain telescope. Therefore

$$E = \frac{s_{pl}}{f'} = \frac{f'}{g} - \left(\frac{f' + d_1}{L}\right) = \frac{f'}{g} - \left(\frac{d_1 + f_2'}{f_2'}\right) \tag{4.34}$$

This expression in terms of f_2' is the same as that given by Burch [4.10] and Gascoigne [4.9]. Eq. (4.34) is valid for both Cassegrain and Gregory forms: since f' is positive for Cassegrain and negative for Gregory, E also has these signs.

4.2.2.1 Strict aplanatic telescope. If such a plate is applied to a *strict aplanatic (normally RC) focus*, corrected for $\sum S_I = \sum S_{II} = 0$, then the first equation of (4.31) requires $S = 0$ if $\sum S_I = 0$ with the plate. Since $S \neq 0$ if $g > 0$, it follows that such a plate changes all three aberrations. Clearly, it can correct the astigmatism of an RC telescope at the cost of introducing spherical aberration and coma. From Eq. (3.108) or (4.31), the RC condition for the secondary is

$$\xi_{RC} = -\frac{f'}{2d_1} \,,$$

which, by substitution in the third equation of (4.31), gives the condition for correction of the RC astigmatism by the plate as

$$\sum S_{III} = \frac{f'}{L} \left(f' + \frac{d_1}{2}\right) + E^2 S = 0 \tag{4.35}$$

or, from (4.34)

$$\sum S_{III} = f' \left(\frac{f' + \frac{d_1}{2}}{L}\right) + \left[\frac{f'}{g} - \left(\frac{f' + d_1}{L}\right)\right]^2 S = 0 \tag{4.36}$$

Gascoigne [4.9] expresses the first term in the equivalent form with his normalization of $f_1'(2f' + d_1)/2(f_1' - d_1)$, to which it can be converted using $L = m_2(f_1' - d_1)$ from (2.75).

The necessary form for a *Gascoigne plate* correcting the astigmatism of an RC telescope is then given from (4.36) by

$$S = -f' \left[\frac{f' + \frac{d_1}{2}}{L} \right] \Big/ \left[\frac{f'}{g} - \left(\frac{f' + d_1}{L} \right) \right]^2 \tag{4.37}$$

and the equivalent plate profile equation from Eq. (3.221)

$$S = 8(n' - n)a\, y^4 \tag{4.38}$$

Gascoigne [4.9] points out that the second term in the denominator of (4.37) is small compared with the first. For the RC values of Table 3.2, the second term is 3.58. Gascoigne [4.8] gives $g = 250$ mm for a 1 m telescope. If we take $f'/g = 35$ as a reasonable minimum, then the error in neglecting the second term is $<$ ca. 10% and we can write, following Gascoigne, from (4.37)

$$S \approx -\frac{g^2}{f'} \left(\frac{f' + \frac{d_1}{2}}{L} \right), \tag{4.39}$$

showing that the strength of the plate is approximately proportional to g^2. In both the RC and aplanatic Gregory cases the plate is *negative* like a Schmidt plate, the opposite form from a Gascoigne plate in the PF.

Since the contribution of the aplanatic system to the spherical aberration and coma is zero, we have from (4.31)

$$\sum S_I = S = -f' \left[\frac{f' + \frac{d_1}{2}}{L} \right] \Big/ \left[\frac{f'}{g} - \left(\frac{f' + d_1}{L} \right) \right]^2 \tag{4.40}$$

$$\sum S_{II} = ES = -f' \left[\frac{f' + \frac{d_1}{2}}{L} \right] \Big/ \left[\frac{f'}{g} - \left(\frac{f' + d_1}{L} \right) \right] \tag{4.41}$$

With the approximation neglecting the second term of the denominator as above

$$\sum S_{II} \approx -g \left(\frac{f' + \frac{d_1}{2}}{L} \right), \tag{4.42}$$

i.e. the coma introduced is roughly proportional to g.

Taking the RC values of Table 3.2, we have $f' = 1$, $d_1 = -0.19375$, $L = 0.225$. Then with $g = f'/35$ we have from (4.40) for the normalized telescope with f/0.5:

$$S = -(0.004\,066\,7)f'$$

With $f' = 35000$ mm for a 3.5 m telescope and setting the relative aperture at f/10, we have from Eq. (3.181)

$$(W'_I)_{GF} = \frac{1}{8} \left(\frac{y_1}{y_{m1}} \right)^4 S = -\frac{1}{8} \frac{1}{(20)^4} (0.004\,066\,7) 35000 \text{ mm}$$

$$= -0.000\,111\,2 \text{ mm} = -0.1112\ \mu\text{m}$$

At the best focus, Eq. (3.190) then gives the angular aberration with $y_1 = 1750$ mm as

$$(\delta u_p')_{BF} = -\frac{2}{y_1}(W_I')_{GF}(206\,265) \text{ arcsec}$$

$$= +0.02621 \text{ arcsec}$$

The effect on the spherical aberration is therefore negligible in practice.

Similarly, from (4.41) and Eq. (3.181) we can derive the coma as

$$W_{II}' = \frac{1}{2}\left(\frac{y_1}{y_{m1}}\right)^3 \left(\frac{u_{pr}}{u_{prm}}\right) \sum S_{II} ,$$

in which u_{pr}/u_{prm} gives the real semi-field compared with the normalized value u_{prm} of 1 rad. Taking $u_{pr} = 15$ arcmin gives

$$(W_{II}')_{GF} = -0.001\,219\,5 \text{ mm} ,$$

which from Eq. (3.198) gives the angular tangential coma as

$$(\delta u_p')_{Coma_t} = -\frac{3(W_{II}')_{GF}}{y_1}(206\,265) \text{ arcsec}$$

$$= +0.4312 \text{ arcsec}$$

This coma is not negligible for best quality but may be acceptable if we compare it with the residual astigmatism of the RC telescope without the corrector. We have from (4.36)

$$\left(\sum S_{III}\right)_{RC} = -E^2 S = -E\sum S_{II}$$

and from (3.181) and (3.208)

$$(\delta u_p')_{ast,m} = -\frac{\sum S_{III}}{y_1}(206\,265) \text{ arcsec}$$

for the angular astigmatism at the mean focus. For $u_{pr} = 15$ arcmin, we deduce for the naked RC telescope

$$(\delta u_p')_{ast,m} = -0.788 \text{ arcsec}$$

Since this declines with the square of the field and the coma only linearly, the use of such a singlet Gascoigne plate corrector with these parameters would be marginal: the astigmatism correction is paid for by a coma half as large at a semi-field of 15 arcmin. A more attractive solution would be to halve g to give $f'/g = 70$. From (4.42) the coma would be roughly halved. But the asphericity of the plate is increased, according to (4.38), although S is reduced.

4.2.2.2 Quasi-RC (aplanatic) telescope. A more interesting possibility, in practice, is the optimization of the whole telescope system to fulfil all three conditions $\sum S_I = \sum S_{II} = \sum S_{III} = 0$ from (4.31) using all three aspheric parameters ζ, ξ and S. We can re-write these equations in the form

$$
\left.\begin{aligned}
-f'\zeta + L\xi + S &= 0 \\
U - d_1\xi + ES &= 0 \\
V + \frac{d_1^2}{L}\xi + E^2 S &= 0
\end{aligned}\right\} , \tag{4.43}
$$

where

$$
\left.\begin{aligned}
U &= -\frac{1}{2}f' \\
V &= f'\frac{(f' + d_1)}{L} \\
E &= \frac{f'}{g} - \frac{(f' + d_1)}{L}
\end{aligned}\right\} \tag{4.44}
$$

are constants for a given telescope geometry, with g positive and the other quantities with the signs given in Chap. 2.

The solution of (4.43) is

$$
\left.\begin{aligned}
S &= -\frac{(LV + d_1 U)}{E(d_1 + LE)} \\
\zeta &= \frac{LEU - LV - d_1 U}{d_1 f' E} \\
\xi &= \frac{f'\zeta - S}{L}
\end{aligned}\right\} \tag{4.45}
$$

Taking again the data of Table 3.2 and $f'/g = 35$, the values are

$$
d_1 = -0.19375 f', \quad L = +0.225 f'
$$

$$
U = -0.5 f', \quad V = +3.583\,333 f', \quad E = +31.4167
$$

giving

$$
\begin{aligned}
S &= -0.004\,181\,33 f' \\
\zeta &= +0.729\,015 \\
\xi &= +3.258\,650
\end{aligned}
$$

From Table 3.5 with $m_2 = -4$:

$$
\begin{aligned}
b_{s1} &= -1.045\,563 \\
b_{s2} &= -3.260\,541
\end{aligned}
$$

Comparing with the values for the RC system of Table 3.2, we see that b_{s1} has increased by 0.89% which is negligible from a manufacturing viewpoint. However, the increase in $(1 + b_{s1})$ is 25.6% which is very advantageous for prime focus correctors. The increase in b_{s2} is 3.18%, again negligible from a manufacturing viewpoint.

We now note a very important property which is common to most Cassegrain correctors: the value of S above must be similar to that calculated above for a Gascoigne plate for a strict RC system, in fact it is 2.82% larger than that value. This effect on the spherical aberration was only 0.026 arcsec, completely negligible. *So the corrector can be removed without any effect on the axial image quality.* However, it *does* affect the coma: this was the reason for liberating the mirror constants. We have for the contribution of the plate

$$(S_{II})_{pl} = ES = -0.13136f'$$

and E has the same value taken for the strict RC case. So the coma on removing the plate is also 2.82% more than the value calculated above, which gives a tangential coma for our quasi-RC on removing the plate of 0.4434 arcsec at a semi-field of 15 arcmin.

Since all three conditions are fulfilled, this quasi-RC system with an aspheric plate is the theoretical equivalent of a system of 3 aspheric mirrors with the third (plane) mirror near the focus. Such systems have been studied in detail by Schulte [4.17] and Bowen and Vaughan [4.18]. According to Gascoigne [4.9], several such telescopes had been built by 1973. Schulte [4.17] discusses the corrector for the 1.52 m quasi-RC design at Cerro Tololo, including a field-flattener, and gives a quality of less than 0.5 arcsec for a semi-field of $u_{pr} = 0.75°$ for the wavelength range 660 nm to 340 nm. The plate is of fused silica and has a diameter of 380 mm. Gascoigne [4.9] also pointed out that a flat field of 3° diameter was realised for a 1 m telescope by using equal radii on primary and secondary and correcting with such a plate to a *monochromatic* quality within 0.2 arcsec.

As in the PF case, the limits will be set here by the higher order chromatic aberrations, above all chromatic differences of coma and astigmatism. These can be considerably reduced if large amounts of lateral chromatic aberration (C_2) and chromatic difference of distortion are tolerated; but this is only acceptable if the use is confined to narrow spectral bands with interference filters. Normally, all of these aberrations must be reasonably balanced against each other. Figure 4.6 gives spot-diagrams for such a system based on the Cassegrain geometry of the ESO 3.6 m telescope and including a field flattener.

In view of the favourable ghost images of such correctors (see above in the introduction to this chapter) and the robust and favourable position tolerances of plate correctors, there is no doubt that such a plate corrector with a quasi-RC solution represents a very attractive design. Furthermore, the increased eccentricity of the primary in the quasi-RC solution favours the design of PF correctors in general and the Gascoigne plate for the PF in particular.

Although the effect on the spherical aberration is negligible, this does *not* mean that the asphericity over the whole plate diameter is small. The

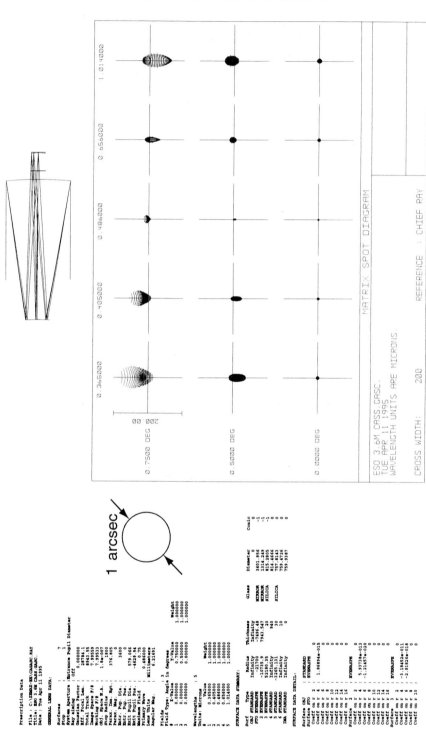

Fig. 4.6. Spot-diagrams for a quasi-RC telescope with Gascoigne plate corrector and field flattener based on the geometry of the ESO 3.6 m telescope (f/3 - f/8)

large values of E, such as 31.4167 above, mean that the principal ray is E-times higher than the aperture ray for the normalized telescope with f/0.5 and $u_{pr} = 1$. For a real telescope with f/10 and a field diameter of 1.5°, the real value $E_{real} = 8.225$. The aspheric form in Eq. (3.221) defining the plate depends on y^4 and the aspheric constant is determined by S over the axial beam width. So S_{real} must be multiplied by $(E_{real})^4 = 4577$ for $E_{real} = 8.225$. This gives an asphericity of about 1 mm from Eq. (3.221) for $n' = 1.5$ and a 3.5 m telescope at f/10. The compensating y^2 term, as in the Schmidt plate, reduces this to about one quarter, i.e. ≈ 250 μm. We see that a 3° diameter field would require an asphericity 16 times as high, becoming prohibitive from a manufacturing viewpoint. In practice, such plate strengths will anyway lead to unacceptable chromatic effects. Schulte [4.17] gives the maximum departure from the nearest sphere as ± 64 μm for the corrector plate of the 1.52 m telescope for Cerro Tololo. Scaling the telescope size from 3.5 m and allowing for his use of fused silica with $n' \sim 1.46$, this is in excellent agreement with the value above of about 250 μm for a 3.5 m telescope. Chromatic effects are worse for Cassegrain telescopes with a high m_2 because the angle u'_{pr} of the emergent principal ray is magnified by the same order. But this is the angle of incidence of the principal ray on the plate as shown in Fig. 4.1 and the chromatic differences of astigmatism and coma will increase with it.

4.2.2.3 Cassegrain correctors consisting of 2 or more aspheric plates.
It is clear that similar considerations apply to the Cassegrain focus as to the prime focus, but with the important difference that the relative aperture is far lower for the Cassegrain and, in consequence, the factor E is much higher.

A classical Cassegrain presents an identical situation to a parabolic primary of the same relative aperture, whereas an RC combination has no PF equivalent. With 2 plates at the Cassegrain, 2 conditions can be fulfilled. As in the PF case, the only interest (in principle) would be the correction to achieve $\sum S_{II} = \sum S_{III} = 0$ with free parameters for ζ or ξ, or both. The correction of the astigmatism of a strict RC, maintaining the other two conditions, would require a 3-plate corrector.

We have seen above that the limitations of the single Gascoigne plate in the Cassegrain *with a quasi-RC solution* are set only by the plate strength due to high values of E and the corresponding chromatic aberrations of higher order. We have also seen that multi-plate correctors, because of the nature of the aberration compensations in the solution matrix, must have far higher individual plate strengths than is the case for single-plate correctors, leading to increased higher order chromatic aberrations. For these reasons, although the 3-plate case for an RC telescope does not appear to have been formally designed for a practical case, it seems very unlikely that it would be a viable solution.

Nevertheless, for special cases, 2- or 3-plate correctors may be of considerable interest. One such case, considered briefly in §3.4, was the correction of spherical aberration in the *Hubble Space Telescope* (HST). In that case, since the error was on the primary at the pupil, there was no effect on coma or astigmatism. Since the RC astigmatism was corrected at the individual instruments the correction required is only in $\sum S_I$. This is the opposite case from a 3-plate corrector mentioned above to correct astigmatism in an RC telescope, but the principle is the same. An excellent optical solution was possible, but the logistics of its mounting in space would have presented major technical problems and dangers of failure, as was also the case for other technically interesting solutions. For these reasons, a 2-mirror corrector (COSTAR) was preferred, which also has the advantage of avoiding spherochromatism and other higher order chromatic effects [4.19]. More details are given in RTO II, Chap. 3.

4.3 Correctors using lenses

4.3.1 Prime focus (PF) correctors using lenses

4.3.1.1 Theory of basic solutions. Reference has been made above to the early work of Sampson [4.1]. Gascoigne [4.7] [4.8] and Wynne [4.5] give excellent reviews of this and later work.

In 1913 Sampson was concerned with a corrector for a Newton telescope with an f/5 *parabolic* primary. Following his work on a 3-lens corrector for a Cassegrain telescope (see below), he investigated the possibilities of a PF corrector consisting of 3 thin lenses of a single glass, effectively afocal, the largest being at a distance 0.215 f' from the focus, the smallest at 0.175 f', in front of the Newton flat. The lens spacings were therefore quite small and arbitrarily selected. Sampson set up the three third order equations for $\sum S_I$, $\sum S_{II}$ and the "curvature" combination of $\sum S_{III}$ and $\sum S_{IV}$ and realised he could not, with practicable lens shapes, satisfy the condition $\sum S_I = 0$ as well as the other two conditions with the parabolic primary; or indeed meet the conditions $\sum S_I = \sum S_{II} = 0$. He proposed, therefore, to abandon the parabolic form, replacing it by a hyperboloid "nearly as far beyond the paraboloid as the paraboloid is beyond the sphere" ($b_{s1} = -1.944$ in our notation). This permitted correction of the coma and the curvature conditions. The central lens had opposite sign from the two outer ones, a principle applied in later successful systems.

Apart from establishing the fundamental difficulty of correcting spherical aberrationand the field aberrations with a parabolic primary, Sampson also pointed out the well-known advantage of eliminating secondary spectrum by the use of a single glass in an effectively afocal corrector system.

In 1933 Ross [4.20] published results of astrometric investigations with the 60-inch Mt. Wilson reflector, to which a prime focus corrector of his design had been added to the parabolic primary. This seems to be the first PF corrector to be made and applied in practice. It consists of a compact (effectively "thin") doublet system which is of a single glass and roughly afocal. In the above paper, Ross refers to the problem of residual spherical aberration and believed at that time that an aspheric surface could improve this.

In 1935 he wrote a second paper [4.21] with a complete optical analysis of the system. He drew attention to the important fact, well-known from lens theory, that it is impossible to correct both first order chromatic aberrations, longitudinal and transverse, with separated lenses, whether or not the corrector is afocal. This is easily proved from the theory given in Eq. (3.222) where the two first order chromatic coefficients were termed C_1 and C_2. From Hopkins [4.22] for two lenses

$$\left.\begin{array}{rcl} \sum C_1 & = & (C_1)_1 + (C_1)_2 \\ \sum C_2 & = & \left[E_1(C_1)_1 + E_2(C_1)_2 \right] H \end{array}\right\} , \tag{4.46}$$

in which the Lagrange Invariant $H = 1$ for our usual normalization and

$$E = \frac{1}{H^2} \frac{y_{pr}}{y}$$

as before. If the longitudinal chromatism $\sum C_1$ is to be corrected, then

$$(C_1)_2 = -(C_1)_1$$

and

$$\sum C_2 = (C_1)_1 (E_1 - E_2)$$

This can only be zero if $E_1 = E_2$ ("thin" system without spacing) or $(C_1)_1 = 0$ (lenses of zero power).

This theorem has important consequences for lens correctors in general. On account of it, Ross started with a 2-lens afocal corrector with nominally zero spacing and zero thickness. The theory of this case is strictly analogous to that given above for aspheric plates, except that the aberrations of the corrector are generated by the powers and bendings of the lenses which may or may not, at this stage, have aspheric surfaces. From Eqs. (4.3), (4.4), (4.5) and (3.213) we have

$$\left.\begin{array}{rcl} \sum S_I & = & -f'\zeta + (S_I)_{cor} \\ \sum S_{II} & = & -\frac{1}{2}f' + (S_{II})_{cor} + E(S_I)_{cor} \\ \sum S_{III} & = & f' + (S_{III})_{cor} + 2E(S_{II})_{cor} + E^2(S_I)_{cor} \end{array}\right\} , \tag{4.47}$$

in which $(S_{II})_{cor}$ and $(S_{III})_{cor}$ are the "central" contributions, i.e. the contributions of the corrector if the stop were in its plane. Eq. (4.4) is the same as for the aspheric plate since the lens corrector is afocal:

$$E = \frac{f' - g}{g} \tag{4.48}$$

For a parabolic primary, $\zeta = 0$ giving

$$(S_I)_{cor} = 0 \ \text{ if } \ \sum S_I = 0$$

Since [4.22] for a "thin", afocal corrector $S_{III} = H^2(K_1 + K_2) = 0$, the equations give for $\sum S_{II} = \sum S_{III} = 0$

$$(S_{II})_{cor} = \frac{1}{2}f'$$

and

$$f' + Ef' = 0$$

or, from (4.48)

$$\frac{f' - g}{g} = -1$$

Since, in our formulation using a dummy flat secondary in § 4.2.1, both f' and g are defined as positive quantities with $g < f'$, this condition cannot be fulfilled. In other words, such a thin, afocal doublet cannot fulfil all three conditions with a parabolic primary.

If we accept that $\sum S_I \neq 0$, and use the second and the third equations of (4.47) to achieve $\sum S_{II} = \sum S_{III} = 0$, we can derive at once

$$(S_I)_{cor} = \sum S_I$$

with the parabolic primary as

$$(S_I)_{cor} = f'\left(\frac{1 + E}{E^2}\right) \tag{4.49}$$

or, substituting (4.48)

$$(S_I)_{cor} = g\left(\frac{f'}{f' - g}\right)^2 \tag{4.50}$$

In his derivation, Ross [4.21] assumed the astigmatism of the primary was negligible and simply required the corrector to be free from astigmatism. (It should be noted here that the field curvature of the primary is even weaker and can be considered as a small compensation of its astigmatism. In the PF case, it is not necessary to introduce it as a separate condition in Eqs. (4.47), since we shall mainly be dealing with afocal correctors for reasons of achromatism). This approximation gives in our notation

$$(S_I)_{cor} \simeq \frac{f'}{E} \simeq g\left(\frac{f'}{f' - g}\right), \tag{4.51}$$

the form also given by Gascoigne [4.9] with a different normalization factor. This result given by Ross was of fundamental importance in corrector development since it confirmed in explicit form the implicit conclusions of Sampson. It enabled Ross to conclude that the correction of coma and astigmatism with a thin, afocal lens system inevitably leaves finite spherical aberration whose amount depends (roughly linearly) only on the parameter g, i.e. the distance from the focal plane relative to the focal length. The powers and shapes of the lenses are immaterial. By implication, it must also have been clear to Ross that the result implied independence from asphericity on the corrector as well, in contradiction with his first paper [4.20]. However, this important corollary was stated neither by Ross [4.21] nor by Paul [4.4] and was first clearly enunciated in a classic paper by Wynne [4.23] in 1949 in a general analytical treatment of field correctors for parabolic mirrors. We shall formally prove this below.

Since Ross was interested in correcting existing parabolic primaries, the possibility of correcting the spherical aberration at the primary, as proposed by Sampson, was not acceptable. Instead, he fixed limits on the amount of $\sum S_I$. Converting (4.51) to the disk of least confusion in arcsec (see Eq. (3.184) and Eq. (3.190)), Ross postulated a mean value of $g/f' = 0.04$, giving an image spread of 1.7 arcsec for an f/5 paraboloid and 5.8 arcsec for f/3.3 (Palomar). The value of g/f' was a delicate compromise: if made too small, the powers and/or bendings of the lenses will be so strong that higher order field aberrations will dominate.

Following Hopkins [4.22], the "central" contributions of a thin lens (stop at the lens) can be expressed as

$$
\left.
\begin{aligned}
(S_I)_L &= \tfrac{1}{4}y^4 K_L^3\left[\frac{(n'+2)}{n'(n'-1)^2}X_L^2 - \frac{4(n'+1)}{n'(n'-1)}X_L Y_L \right. \\
&\qquad \left. + \frac{(3n'+2)}{n'}Y_L^2 + \frac{n'^2}{(n'-1)^2}\right] \\
(S_{II})_L &= \frac{1}{2}Hy^2 K_L^2\left[\frac{(n'+1)}{n'(n'-1)}X_L - \frac{(2n'+1)}{n'}Y_L\right] \\
(S_{III})_L &= H^2 K_L \\
(S_{IV})_L &= H^2(K_L/n')
\end{aligned}
\right\}, \tag{4.52}
$$

where n' is the refractive index, K_L the power, X_L a dimensionless "shape factor" and Y_L a dimensionless "magnification factor" [2]. The paraxial ray height y is proportional to g. Y_L is largely determined by the converging ray bundle from the primary mirror; but K_L and X_L are free parameters in the equations for $(S_I)_L$ and $(S_{II})_L$. There are therefore, as Ross pointed out, an infinity of solutions to be chosen between powers and bendings.

Ross noted that the spherical aberration could be reduced by introducing a small separation – but this inevitably introduces chromatic aberration from (4.46). The finite thicknesses of the lenses also disturbed the achromatism. The transverse chromatic condition C_2 then requires glasses with slightly different dispersions (7% with the thicknesses Ross chose). He preferred to

---

[2] It should be noted that Eq. (4.52) is simply a more general form, using different parameters, of Eq. (3.271).

correct the longitudinal condition C_1 using *identical glasses*, giving a small departure from afocality and a slight transverse chromatism. The secondary spectrum is, with identical glasses, zero irrespective of the total power of the system.

Ross proved that the Seidel distortion $(S_V)_{cor}$ of the corrector is also uniquely fixed by the parameter g by the relation

$$(S_V)_{cor} \propto \left(\frac{f'-g}{g}\right)^2 ,$$

(4.53)

thereby *increasing* rapidly with reduction in g and in conflict with the spherical aberration condition of Eq. (4.50). Distortion has been neglected in our general treatment of telescope systems because it is usually small enough to be of no consequence when calibrated. Field correctors, however, can introduce appreciable distortion: even when calibrated, significant distortion may lead to objectionable photometric effects.

In his pioneer work, Ross investigated many doublet arrangements in detail. He concluded that the best solutions put the negative lens at the front (nearer the mirror), particularly for small values of g/f', say < 0.05. For larger g-values, viable solutions were also found with the positive lens at the front, but spherical aberration and distortion values were less good than in the reverse order. In all cases, the total power of the corrector was negative. His original system for the 60-inch, f/5 Mt. Wilson telescope primary had an 8-inch aperture and $g = 0.05f'$.

An elegant general theorem concerning the impossibility of correcting $\sum S_I$ as well as $\sum S_{II}$ and $\sum S_{III}$ for a parabolic primary and a corrector consisting of any number of thin lenses in contact was given by Wynne [4.23], who showed that correction of all three conditions is only possible in the limit case where the total power of the system, including the primary, is zero. Such an afocal total system is the strict equivalent of an afocal Mersenne beam compressor using two confocal paraboloids, which is also anastigmatic. Such a beam compressor was used by Paul [4.4] (see Fig. 3.73) as the basis for his 3-mirror telescope proposal. A similar possibility would exist, in principle, with the above afocal system with lenses. But, in practice, as Wynne points out, the secondary spectrum would preclude its use. The Paul telescope would be preferable in every way, including obstruction aspects.

In general, one could say that, although Ross's work effectively introduced practical PF correctors, the spherical aberration of his doublet solutions for paraboloids would not be acceptable today. He mentions [4.21] that "a system consisting of three lenses will be discussed in another paper". Apparently this paper never appeared, but Wynne [4.5] [4.24] has given details of such a design realised for the 200-inch, f/3.3 Mt. Palomar primary – see Fig. 4.7 taken from Wynne's paper [4.5]. Ross used his compact doublet, with negative lens at the front in combination with a thin, strongly curved meniscus placed before the doublet at about twice the g-value. This gave good correction on axis, but according to Wynne [4.24] about 3 arcsec of monochromatic comatic image

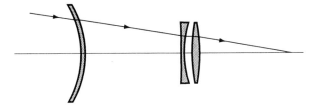

Fig. 4.7. The 3-lens Ross corrector for the Mt. Palomar 5 m, f/3.3 parabolic primary (schematic, after Wynne [4.5] [4.24])

spread at 10 arcmin off-axis, apart from chromatic difference of coma. Wynne gives a modified design, with thicker meniscus and inverted powers on the doublet, giving much improved field performance. The Ross 3-lens designs were more satisfactory for less steep paraboloids such as the 120-inch, f/5 Lick primary.

Apart from his pioneer work on aspheric plate correctors, Paul [4.4] carried out in 1935 a detailed analysis of lens correctors, based only on the earlier paper of Ross [4.20]. Paul considered first the case of a single lens, pointing out that it was only of theoretical interest because of the inevitable chromatic aberration. The equations for the three conditions $\sum S_I, \sum S_{II}, \sum S_{III}$ are identical to those given for an afocal doublet in Eqs. (4.47), except that the definition of E given in (4.48) has to be modified to take account of $f' \neq |f'_1|$, which changes the length metric in our normalized system. However, this has no effect on the validity of Eqs. (4.47). The proof that it is impossible to correct all three conditions with a parabolic primary therefore remains valid. Paul gives solutions either for the correction of $\sum S_I = \sum S_{II} = 0$ with finite $\sum S_{III}$; or, following Ross, with $\sum S_{II} = \sum S_{III} = 0$ with finite $\sum S_I$. In principle, the two available parameters K_L and X_L from (4.52) are sufficient to satisfy the two conditions; but Paul realised that with the necessary small distance from the focus, a practical solution led to extreme bendings X_L. He therefore introduced a third parameter, an aspheric surface, to relax the requirements, enabling an equiconvex lens to be used. Taking $g = 0.20|f'_1|$, he derived a residual $\sum S_I$, for the system with $\sum S_{II} = \sum S_{III} = 0$, equivalent to a disk of least confusion of 1.7 arcsec for the 60-inch, f/5 paraboloid of Mt. Wilson.

Although Paul does not state this explicitly through an equation equivalent to (4.49) for the inevitable residual spherical aberration resulting from the paraboloid-lens combination, it is clear that he understood this limitation because he states that the solution of all three conditions requires $\zeta \neq 0$ in (4.47) and that the primary must have hyperbolic form. The subsequent aspheric on the lens was solely to relax the shape factor X_L, giving a better practical solution for $\sum S_{II} = \sum S_{III} = 0$ for a parabolic primary, not an attempt to correct $\sum S_I$ which Paul knew was inevitable.

Paul then treated the case of an afocal doublet of a single glass, following Sampson [4.2] and Ross [4.20]. Again, Eqs. (4.47) express the requirement for the correction of the three conditions and Paul explicitly suggests correcting

$\sum S_I$ by a hyperbolic form of the primary. With $g/f' = 0.05$, he deduces the necessary eccentricity $b_{s1} = -1.22$ for the primary from a formula equivalent to (4.50) for $(S_I)_{cor}$, dependent only on g/f'. Paul noted that this departure from the parabola is much less than that proposed by Sampson [4.2] ($b_{s1} = -1.944$ – see above) and that such a solution would be applicable to the first RC telescope, then just completed by Ritchey. This telescope has been referred to in §§ 3.2.6.3 and 3.7.2 above. The constructional data are given in Table 3 of Bahner's book [4.25]. With a primary of f/4.0 and an RC focus of f/6.85, the secondary magnification $m_2 = -1.71$. This value is far lower than those typical of modern RC telescopes and led, from Eq. (3.114) or (3.115) to a much more eccentric primary. This favoured at that time Paul's proposal for such correctors for RC primaries, a basic idea which was to be very fruitful 30 years later.

We shall now consider several specific cases, starting with the proof of the important theorem of Wynne [4.23] *that an aspheric surface brings no advantage for a thin, afocal corrector for a parabolic primary, irrespective of the number of lenses.* Adding an aspheric extends the Eqs. (4.47) to

$$\left. \begin{aligned} \sum S_I &= -f'\zeta + (S_I)_{cor} + S_P \\ \sum S_{II} &= -\tfrac{1}{2}f' + (S_{II})_{cor} + E(S_I)_{cor} + ES_P \\ \sum S_{III} &= f' + (S_{III})_{cor} + 2E(S_{II})_{cor} + E^2(S_I)_{cor} + E^2 S_P \end{aligned} \right\} \quad (4.54)$$

For the parabolic primary, $\zeta = 0$ and $(S_I)_{cor} + S_P = 0$ if $\sum S_I = 0$. This can be written $(S_I)^*_{cor} = 0$. The last two equations of (4.54) can then be written

$$\left. \begin{aligned} \sum S_{II} &= -\tfrac{1}{2}f' + (S_{II})_{cor} + E(S_I)^*_{cor} \\ \sum S_{III} &= f' + (S_{III})_{cor} + 2E(S_{II})_{cor} + E^2(S_I)^*_{cor} \end{aligned} \right\}, \quad (4.55)$$

which have exactly the same form as the equivalent equations of (4.47) and lead to exactly the same spherical aberration residual of (4.49). The only advantage of an aspheric surface is, in the sense used by Paul, to relax the bending requirements for the lenses to achieve coma and astigmatism correction.

As stated above, the Eqs. (4.47) apply also to a single lens with power, as discussed by Paul. If an aspheric term S_P is added as in Eqs. (4.54), the conclusion is identical. *Therefore, an aspheric surface cannot influence the residual spherical aberration of a "thin" corrector satisfying $\sum S_{II} = \sum S_{III} = 0$ for a parabolic primary, irrespective of whether the corrector is afocal or not: the property comes from its "thin" nature, whereby all constructional parameters have the same value of E.*

Let us consider now the possibility of a thin, afocal doublet combined with a parabolic primary to satisfy $\sum S_I = \sum S_{II} = 0$ with $\sum S_{III} \neq 0$. Setting the first two equations of (4.47) to zero with $\zeta = 0$ gives for the astigmatism

$$\sum S_{III} = f'(1 + E), \quad (4.56)$$

bearing in mind that the central astigmatism $(S_{III})_{cor} = 0$ because the doublet is afocal. From (4.48)

$$\sum S_{III} = \frac{f'^2}{g} = f'\left(\frac{f'}{g}\right) \tag{4.57}$$

Again g may be chosen at will, as for the case of a corrector for $\sum S_{II} = \sum S_{III} = 0$. With $g = 0.05f'$, it follows from (4.57) that the astigmatism of the primary mirror, which is f', is increased by a factor 20 times in this case. Gascoigne [4.9] implies this solution is useless in practice because the astigmatism is excessive. It is instructive to compare it with the astigmatism of the Gascoigne plate as a PF corrector for hyperbolic primaries which was given in Eq. (4.12) as

$$\left(\sum S_{III}\right)_{GP} = f'\left(\frac{f'+g}{2g}\right)$$

For small values of g, the astigmatism of the thin doublet from (4.57) is about twice that of the Gascoigne plate with the same value of g. However, in the Gascoigne plate case, g is determined by the eccentricity of the primary, whereas the doublet corrector for a parabolic primary allows a free choice of g. With g-values $\sim 0.05f'$, the astigmatism residual may be more favourable than that given by Gascoigne plates for typical RC primaries.

We will now consider more closely the proposal of Paul [4.4] to use a compact, afocal doublet with *a hyperbolic primary* to correct $\sum S_I$. We have from (4.47) and (4.50) with $\sum S_{II} = \sum S_{III} = 0$

$$\begin{aligned}\sum S_I &= -f'\zeta + (S_I)_{cor} &&= 0 \\ &= -f'\zeta + g\left(\frac{f'}{f'-g}\right)^2 &&= 0\end{aligned}$$

From (3.71), setting $m_2 = -1$ for a Cassegrain with plane secondary, equivalent to the PF

$$\zeta = -\frac{1}{4}(1+b_{s1}),$$

giving

$$b_{s1} = -1 - 4\frac{g}{f'}\left(\frac{f'}{f'-g}\right)^2 \tag{4.58}$$

Since f' and g are defined as positive, this proves Paul's statement that correction of all three conditions with a thin, afocal doublet requires a hyperbolic primary. Setting $g = 0.05f'$ gives $b_{s1} = -1.222$, in agreement with Paul. Eq. (4.58) gives the approximate relation quoted by Gascoigne [4.9]

$$g \simeq -\frac{(1+b_{s1})}{4}f', \tag{4.59}$$

a good approximation because g must be small for modern RC telescopes. For the RC system of Table 3.2 with $m_2 = -4$, Eq. (4.59) gives $g \simeq 0.0091f'$ for such a PF corrector. By comparison, the Gascoigne plate required from (4.9)

$$(b_{s1})_{GP} = -1 - 2\left(\frac{g}{f' - g}\right),$$

giving for small values of g/f' a distance g from the focus about twice that of the doublet with correspondingly larger diameter. This is an advantage for higher order aberrations, but we must remember that the Gascoigne plate does not correct $\sum S_{III}$ – indeed it is vastly increased.

Equation (4.59) effectively expresses the *scaling law* first enunciated in a major paper on RC telescopes and correctors by Wynne [4.14]. Since $g \ll f'$, the spherical aberration of the corrector will increase roughly linearly with g/f'. But if the corrector is scaled in the same proportion, this metric change has no effect on its angular contributions to coma and astigmatism. (This can easily be formally confirmed from Eqs. (4.47), (4.48) and (4.51)). This property is precisely what is required for different eccentricities of different RC telescope primaries of constant focal length and relative aperture, for which the coma and astigmatism contributions are constant. Only minor changes will need to be made for real lens thicknesses. Wynne shows that the same principle applies to a more complex corrector such as one consisting of three lenses.

Consider now the case of a *thin, afocal doublet combined with an aspheric plate separated from the doublet*. The E-values are now different, Eqs. (4.47) giving

$$\left.\begin{array}{l} \sum S_I \ \ = -f'\zeta + (S_I)_{cor} + S_P \\ \sum S_{II} \ = -\frac{1}{2}f' + (S_{II})_{cor} + E_{cor}(S_I)_{cor} + E_P S_P \\ \sum S_{III} = f' + (S_{III})_{cor} + 2E_{cor}(S_{II})_{cor} + E_{cor}^2(S_I)_{cor} + E_P^2 S_P \end{array}\right\} \quad (4.60)$$

For a *parabolic primary*, $\zeta = 0$ gives $S_P = -(S_I)_{cor}$ if $\sum S_I = 0$. Reduction leads to the requirement, if all three conditions are to be fulfilled:

$$(S_I)_{cor} = f'\frac{(E_{cor} + 1)}{(E_{cor} - E_P)^2} = \frac{g_P^2 \, g_{cor}}{(g_P - g_{cor})^2}, \quad (4.61)$$

confirming the previous conclusion that no solution exists for $E_{cor} = E_P$. But with $E_{cor} \neq E_P$, a solution exists. This leads to the important general conclusion that aspheric surfaces on lenses, or separate plates, are useful parameters in combination with other lenses, or lens groups, from which they are axially separated. (This conclusion is, of course, identical with the basis of multi-plate correctors dealt with in § 4.2.1.) In other words, an aspheric on a lens is not useful for the correction of spherical aberration in its own function; but may be useful in combination with other elements with different E, the utility being principally in the relaxation of bending requirements as recognised by Paul [4.4]. The solution for a parabolic primary corrected by a thin afocal doublet and separated aspheric plate was already treated by Wynne in 1949 [4.23]. It was proposed also by Schulte [4.11] for an RC-telescope. According to Wynne [4.5], with the spherical aberration correction

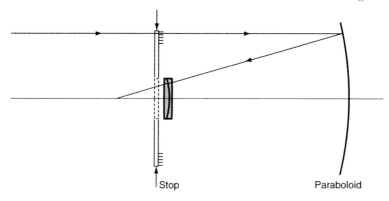

Fig. 4.8. Reflector-corrector due to Baker [4.27]

liberated by the hyperbolic primary, it offers little advantage over the doublet-only type corrector with a hyperboloid proposed by Paul [4.4] and taken up in 1961 by Rosin [4.26], using a two-glass doublet design.

At this stage, we must introduce the *reflector-corrector* due to Baker [4.27]. This solves the problem of spherical aberration with a Ross-type doublet correcting a parabolic primary by adding into the incident beam, in roughly the same plane as the doublet, a Schmidt-type corrector plate – Fig. 4.8. This system is also discussed by Wynne [4.5]. The doublet corrects coma and astigmatism for the front stop and also field curvatureby appropriate residual positive power, giving somewhat more spherical aberration than that of a normal Ross doublet. This is corrected by the Schmidt-type plate, rather as in a Wright-Väisälä camera (see § 3.6.4.1), except that this latter cannot correct astigmatism and field curvature and works with a primary of oblate-spheroidal form. Since the stop is at the plate, it has no effect on other aberrations than $\sum S_I$. This system has only half the length of the Schmidt telescope and gives in addition a flat field. The reflector-corrector can also be removed for naked use of the paraboloid. If this convertibility is of no consequence, the reflector-corrector can be used with a spherical primary as a compact modification of the Schmidt system.

We will now consider the case of an afocal corrector consisting of 2 *spaced lenses*, *i.e.* with different *E*-values. In fact, the same treatment applies to *2 spaced afocal doublets*, since, as we saw above, the same formulae apply except for the precise form of the *E*-parameters in terms of physical dimensions in the system. For this case, Eqs. (4.47) can be written

$$
\begin{aligned}
\sum S_I &= -f'\zeta + (S_I)_{cor\,1} + (S_I)_{cor\,2} \\
\sum S_{II} &= -\tfrac{1}{2}f' + (S_{II})_{cor\,1} + (S_{II})_{cor\,2} + E_1(S_I)_{cor\,1} + E_2(S_I)_{cor\,2} \\
\sum S_{III} &= f' + (S_{III})_{cor\,1} + (S_{III})_{cor\,2} + 2E_1(S_{II})_{cor\,1} + 2E_2(S_{II})_{cor\,2} \\
&\quad + E_1^2(S_I)_{cor\,1} + E_2^2(S_I)_{cor\,2}
\end{aligned}
\right\}(4.62)
$$

If the conditions $\sum S_I = \sum S_{II} = \sum S_{III} = 0$ are to be fulfilled with $\zeta = 0$ for a parabolic primary, these equations give:

$$\left.\begin{aligned}
(S_I)_{cor\,2} &= -(S_I)_{cor\,1} \\
(S_{II})_{cor\,2} &= \tfrac{1}{2}f' - (E_1 - E_2)(S_I)_{cor\,1} - (S_{II})_{cor\,1} \\
(S_{III})_{cor\,2} &= -\big[\,(S_{III})_{cor\,1} + 2(E_1 - E_2)(S_{II})_{cor\,1} \\
&\quad + (E_1 - E_2)^2(S_I)_{cor\,1} + f'(1 + E_2)\,\big]
\end{aligned}\right\} \tag{4.63}$$

For *2 separated single lenses*, the transverse chromatic condition C_2 cannot be fulfilled as well as the longitudinal condition C_1, as was discussed above. Otherwise, the four conditions $\sum S_I$, $\sum S_{II}$, $\sum S_{III}$ and C_1 can, in principle, be fulfilled by the parameters: 2 powers, 2 bendings, E_2 and $(E_1 - E_2)$, as was implicit in the work of Paul [4.4]. For *2 separated thin afocal doublets*, 2 more parameters are available, 2 bendings. The two extra powers are not free parameters because the thin doublets are each defined as afocal, correcting both C_1 and C_2 with a single glass without secondary spectrum. This parametric situation is more generous than that of a single afocal doublet with a separated plate which gave the solution of Eq. (4.61) for given E-values. With 2 thin afocal doublets, there are, in principle, an infinite number of solutions for given E-values. Such a 4-lens system is, therefore, clearly promising as a corrector for paraboloids. An intermediate solution would be a 3-lens system consisting of a single afocal doublet and a single afocal meniscus with significantly different E-values, as was chosen by Ross (Fig. 4.7).

If the primary is hyperbolic, this will further relax these systems.

4.3.1.2 Modern correctors for parabolic primaries.

Wynne's work [4.24] on improvements to the three-lens designs of Ross has been mentioned above. Although this brought considerable improvement, the limitation from chromatic difference of coma was still larger than desirable. Accordingly, Wynne [4.28] investigated 4-lens correctors for paraboloids. He started from the concept above of two separated, thin, afocal doublets of a single glass (UBK7). Above all, the coma correction was distributed evenly between the two doublets, whereas the 3-lens Ross corrector of Fig. 4.7 achieved the coma correction mainly at one surface of the doublet. The spherical aberration contributions of the doublets are equal and opposite and the astigmatism contributions largely so, giving a sum $\sum(S_{III})_{cor}$ balancing that of the primary. The scaling law, mentioned above, can be applied. For the initial design, the ratio g_1/g_2 of the distances from the focus was taken to be about 3. Since the finite thicknesses disturb the chromatic conditions, the order of powers in the two doublets is reversed to allow compensations. Individual powers were given initially with numerical apertures of about ± 0.3, but this is not critical. This design corrects $\sum C_1$, $\sum C_2$, $\sum S_I$, $\sum S_{II}$, $\sum S_{III}$ and $\sum S_{IV}$. The system was then optimized for real thicknesses and separations. This leads to separations of the two doublets, above all of the second one. Wynne gave details of such a design for the Palomar 5 m, f/3.34 paraboloid.

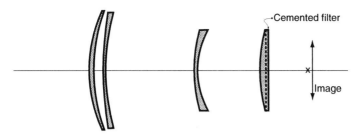

Fig. 4.9. Wynne design for a 4-lens corrector of the Palomar 5 m, f/3.34 paraboloid. The cross shows the focus of the naked primary (after Wynne [4.28])

Only spherical surfaces were used. The final corrector had a small negative power, giving f/3.52 in the final image. Figure 4.9 shows the section through the system and Fig. 4.10 the spot-diagrams, both taken from Wynne's paper [4.28]. Later, Wynne showed a similar design for the Isaac Newton 2.5 m, f/3.0 primary with somewhat thicker front lenses. Thinner lenses give better performance in such designs, but can give sag problems. Also, the larger the system, the better its theoretical performance.

The spot-diagrams clearly reveal the limitation of such correctors by chromatic difference of coma, particularly due to the reversal of sign of the residual coma at the edge of the field of 25 arcmin diameter. The spot-diagrams fall within 0.5 arcsec over the whole field for the mean focal position except at the two extreme blue wavelengths.

The higher order chromatic aberrations, mainly chromatic difference of coma and astigmatism, limit all practical correctors. This is inevitable with separated, powered lenses because of the dispersion of ray heights on subsequent elements. The situation is alleviated by inverting powers of successive lenses, as Wynne has done. This limitation could only be removed by achromatising *each* lens. Not only would this lead to unacceptable absorption, but also much increased individual powers with serious monochromatic higher order effects as well as secondary spectrum. Such systems would not be practical, so the limitation of higher order chromatic effects will remain.

In the above analysis of the possible correction of all three third order conditions for paraboloids, a solution with 3 lenses (an afocal doublet and a meniscus) was referred to, as attempted by Ross and improved by Wynne [4.24]. Since ghost images are a significant problem with correctors [4.29], a 3-lens solution is of great interest if satisfactory quality can be achieved. Applying the argument discussed above, which had its origins in the work of Paul [4.4], that an aspheric surface can relax a system if it has a different E-value from those of other elements, Faulde and Wilson [4.30] designed a 3-lens corrector for paraboloids with an aspheric surface on the concave side of the central negative lens. The aim was to achieve a similar quality *for the paraboloid* as was achieved by the 3-lens corrector for *RC primaries* designed

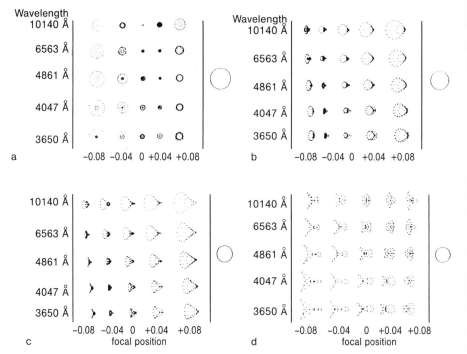

Fig. 4.10. Spot-diagrams for the Wynne design of Fig. 4.9: (**a**) on axis, (**b**) 6 arcmin off-axis, (**c**) 9 arcmin off-axis, (**d**) 12.5 arcmin off-axis. The circle has a diameter of 0.5 arcsec. (After Wynne [4.28])

by Wynne [4.14], which we shall discuss in the next section. The design is shown in section in Fig. 4.11 and the spot-diagrams in Fig. 4.12. This work was suggested by Elsässer and Bahner in the framework of studies for the 3.5 m MPIA telescope. A paraboloid of 3.5 m with f/3.0 and a well corrected, flat field of ±0.5° were requirements using the single glass UBK7.

Since the field is ±0.49° compared with ±0.25° for the Wynne system of Fig. 4.10, and the paraboloid somewhat steeper, the quality of this corrector is not within 0.5 arcsec. The aim was spot-diagrams within 1.0 arcsec over the whole field and for all wavelengths, which was just about achieved.

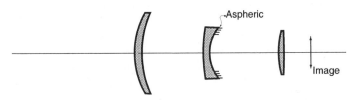

Fig. 4.11. 3-lens corrector for paraboloids using one aspheric by Faulde and Wilson [4.30]

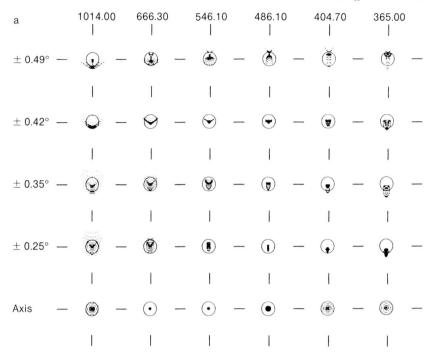

Fig. 4.12a. Spot-diagrams for the 3-lens corrector of Fig. 4.11. The circle represents 0.5 arcsec. (**a**) Basic focus

As we shall see, the quality is fully comparable with the 3-lens RC primary corrector (without an aspheric) of Wynne, calculated for a similar field and mirror geometry. Faulde and Wilson concluded that the aspheric surface in the paraboloid corrector can replace one lens in the all-spherical 4-lens solution. The aspheric is a moderate ellipse with a Schwarzschild constant from Eq. (3.11) of -0.0619, but the higher order terms show appreciable departures from the series corresponding to the above ellipse. Relative to a sphere with the same curvature at the pole, the asphericity at the edge of the lens is about 0.22 mm, but this would be much reduced by another reference sphere. The aspheric form emerges above all from the control of fifth order coma. Whether some exactly elliptical form would be adequate would require further investigation. In view of aspheric testing with null-systems, it does not follow that an exact ellipse is the best form for manufacture. For ease of testing, it was considered essential to apply the aspheric to a *concave* surface. This aspheric assumes a quite different role from that of the overcorrection of an RC primary because the relative pupil position (E-value) and significant incidence angle of the principal ray cause it to influence the field aberrations as well as the axial spherical aberration.

   It should be noted that the front lens of the Faulde-Wilson 3-lens design is larger, relative to the primary, than that of the Wynne 4-lens system. For

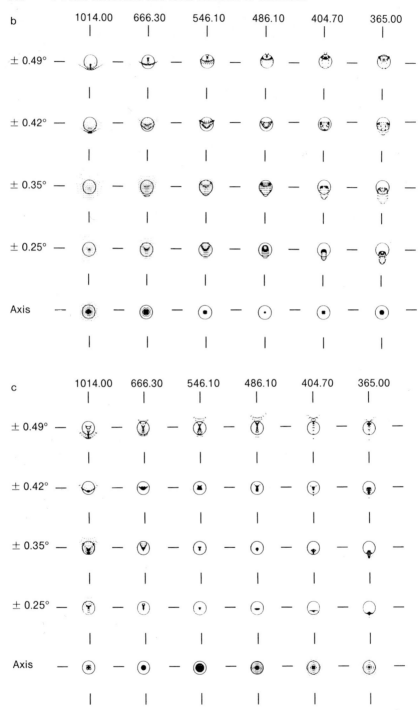

Fig. 4.12b, c. Spot-diagrams for the 3-lens corrector of Fig. 4.11. The circle represents 0.5 arcsec. (**b**) Focus shift +0.05 mm, (**c**) focus shift −0.05 mm

the 3.5 m MPIA primary and an unvignetted field diameter of 0.982°, the free diameter of the front lens was 471.2 mm. However, its thickness of about 50 mm gives it a robust form. The corrector has slight negative power, the case with most such correctors: it converts the relative aperture from f/3.0 to f/3.35.

For the same reasons given for the Wynne 4-lens corrector, this corrector is also limited by the higher order chromatic aberrations. The role of the aspheric, relative to the size of the corrector, is further discussed in the next section.

4.3.1.3 Modern correctors for RC primaries. The invention of Paul [4.4] to use a Ross-type afocal doublet and correct the spherical aberration with the RC primary is still valid today as a modern solution for fields up to the order of 30 arcmin diameter. Reference was made above to Rosin's design [4.26] with a hyperboloid, who used a doublet of two glasses to reduce the lateral chromatic aberration C_2 in the sense discussed by Ross (see above). Small longitudinal chromatic aberration C_1 had to be accepted. Rosin aimed to achieve 0.1 arcsec resolution over a field of 1° diameter (or more!). He claimed this quality was attained over the 1° diameter field for the wavelengths 589, 656 and 486 nm. Unfortunately, he did not give spot-diagrams in the conventional form. Extension to extreme wavelengths 1014 and 365 nm places much greater chromatic demands on the system. More significantly, he assumed a primary of 2 m at f/12.5 and with a Schwarzschild constant $b_{s1} = -2.49$! This has little relevance for modern RC telescopes and must be seen as a special telescope of exceptional length.

In general, doublet correctors for RC primaries will be of more interest if of the single glass type. An excellent example is the doublet corrector of two quartz lenses designed by Wynne [4.14] [4.5] for the 105-inch, f/4 RC primary of the McDonald telescope with $m_2 = -2.25$. Spot-diagrams are given for a maximum field diameter of 28 arcmin. They are within 0.5 arcsec for the wavelength range 770 to 365 nm except for the lateral chromatic aberration C_2. If the axial chromatism C_1 is corrected, it is the lateral chromatism that limits these doublet systems. This is a consequence of the fact that no compensation is possible by a further element or elements such as can take place with triplet correctors.

For many purposes, the corrected field with the doublet will be satisfactory; but for classical photography or multi-fibre spectrographs wider fields are desirable. In the 1960's, when the fundamental work was done, PF photography with baked plates was still a significant demand.

The classical work on the application of *3-lens correctors* to RC primaries was performed by Wynne [4.24] [4.14] [4.5]. In 1965, Wynne [4.24] investigated such a corrector for the Kitt Peak 3.8 m RC telescope with an f/2.8 primary. Five different designs were considered, the best consisting of a front doublet with leading negative lens and a rear cemented triplet which was

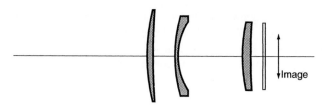

Fig. 4.13. 3-lens corrector by Wynne [4.14] [4.5] for the Kitt Peak 3.8 m, f/2.8 - f/8 RC telescope. All three lenses are of UBK7. (After Wynne)

monochromatically effectively a single lens, but whose centre lens was of a high dispersion glass. The front lenses were also in different glasses. The spot-diagrams are shown for a field diameter of 30 arcmin and are mainly within 0.5 arcsec. Secondary spectrum effects of the lateral chromatic aberration C_2, due to the different glasses, are significant at the edge of the field.

In his 1968 paper [4.14], Wynne described a new design for the Kitt Peak telescope for a field of 50 arcmin diameter. The magnification had been slightly reduced compared with the original proposal (f/2.8 to f/8 instead of f/9). This increased the eccentricity of the primary and favoured the corrector design. It is a three-lens system of one glass as shown in Fig. 4.13 and *has become the standard basic design for RC primaries*. In view of the inevitable limitations of higher order chromatic aberrations, the spectral range was split over 2 interchangeable correctors in order to increase the field coverage at high quality. Spot-diagrams are within 1 arcsec for the field diameter of 50 arcmin and a spectral range of 486 to 405 nm.

Wynne applied the same basic design to a corrector for the ESO 3.6 m, f/3-f/8 quasi-RC telescope. Because the primary of this telescope had been optimized for use with a singlet corrector at the Cassegrain focus (see below), its eccentricity was higher than the strict RC form would give, favouring the corrector. Even the strict RC form would have been more favourable than the Kitt Peak telescope because $m_2 = -2.67$ instead of -2.86 for Kitt Peak, apart from the advantage of the somewhat smaller relative aperture. Such a higher eccentricity would still not be sufficient to compensate the spherical aberration of a Ross-type doublet placed similarly to the front lens of these 3-lens designs; but it relaxes the amount of spherical aberration the front lenses have to compensate. The rear positive lens compensates the astigmatism of the front pair, as well as the lateral chromatic aberration generated by their separation. Data for this corrector are given in [4.14]. Wynne [4.5] compares this 3-lens corrector with the design using 3 aspheric plates and a field flattener given by Köhler [4.12] [4.13], discussed in § 4.2.1.3 above. Wynne's comparative spot-diagrams show that his lens system gives superior performance over the entire field and spectral range specified. This is confirmed by the comparison of Cao and Wilson [4.16], where the plate and lens systems are compared without vignetting for the wide spectral range from 334 - 1014 nm. This comparison is shown in Fig. 4.14. Above all, the 3-lens system is superior in the higher order chromatic aberrations, particularly chromatic difference of coma. This arises because the lenses have 2 parameters, power and bending, compared with essentially 1 parameter

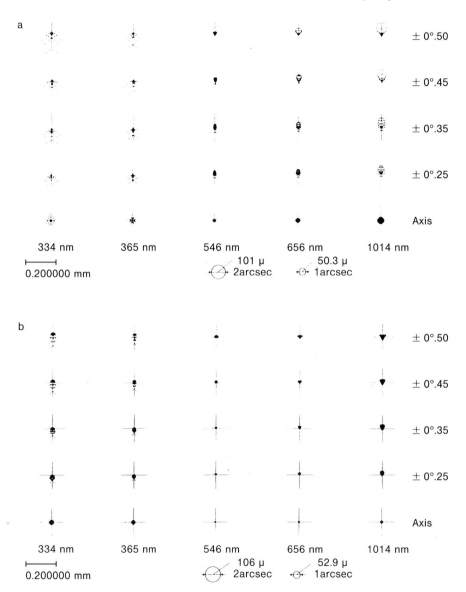

Fig. 4.14. Spot-diagrams for two correctors for the ESO 3.6 m quasi-RC telescope: (a) the basic plate system of Köhler with field flattener, recalculated without vignetting; (b) the Wynne-type lens corrector also recalculated without vignetting. (After Cao and Wilson [4.16])

for the plates, the asphericity. Even the monochromatic performance of the plate system is slightly inferior, although this is not the limitation in either system. The only advantages of the plate system would be its ghost images, as discussed in § 4.1, and the insensitivity of plates to flexure through sag effects.

In connection with the ESO 3.6 m design, a 3-lens corrector with opposite power distribution to that of Wynne, namely negative-positive-negative, was investigated by Baranne [4.31] [4.32], using one aspheric surface. Its performance was less good than the Wynne form, which appears to be the definitive solution.

Wynne [4.5] [4.14] also gives results for a 4-lens corrector, similar to his system for a paraboloid shown in Fig. 4.9. He reported a modest reduction of the largest skew aberrations to about 75%. A similar level of improvement was obtained by adding an aspheric to each of the three lenses of the triplet corrector, with 2 powers of figuring. Little improvement was obtained with less than three aspherics. Wynne concluded that these gains were too modest to justify the additional complication. Wilson [4.15] approached this aspect from the opposite direction by introducing weak lens power with one plane surface into the 3-aspheric plate solution, discussed above. This produced considerable improvement, but the final system of 3 weak aspheric lenses was no better than the Wynne all-spherical 3-lens corrector.

This matter was investigated by Cao and Wilson [4.16]. Experience with the Wynne-type 3-lens corrector for the ESO 3.6 m quasi-RC telescope had revealed flexure problems for the first two lenses. The chromatic performance improves with reduced thicknesses, so these had been held as small as possible. The authors' aim was to thicken these elements and to attempt an improvement of the higher order chromatic aberrations by introducing aspheric surfaces and, possibly, a fourth lens. This work essentially confirmed the trends reported by Wynne. The best system had 3 aspherics (one on each lens) with 2 powers of figuring plus a fourth lens to flatten the field. The front thicknesses were significantly increased: the second lens was certainly more resistant to flexure, the first lens probably so, but this required more investigation. The results confirmed conclusions drawn by Richardson, Harmer and Grundmann [4.33] [4.34], who also achieved improved performance through bigger correctors with an aspheric. Allowing a front lens some 62% larger relative to the primary compared with the Wynne-type corrector of the ESO 3.6 m telescope and using an aspheric on the second lens, Richardson et al. [4.33] attained significantly improved performance in a corrector for the 3.9 m Anglo-Australian telescope (AAT). A similar system was designed by Henneberg at Carl Zeiss for the 3.5 m MPIA telescope [4.35]. Richardson et al. [4.34] also showed that successful designs are possible with fast primaries of f/2.0. An example is given for the proposed 7.6 m telescope of the University of Texas – Fig. 4.15.

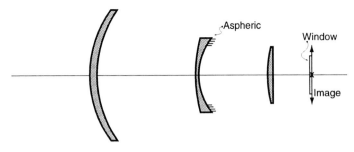

Fig. 4.15. 3-lens prime focus corrector designed by Richardson et al. [4.34] for the then proposed 7.6 m, f/2 primary of University of Texas. (After Richardson et al.)

The front meniscus lens of Fig. 4.15 is about the same size as the same authors' design for the 3.9 m AAT, i.e. the size relative to the primary (g-value above) is much larger in the latter case. The AAT primary has f/3.25 compared with f/2.0 proposed for Texas. As a result, while the first two lenses of the AAT design are excellently robust, those for the f/2.0 Texas primary are markedly more curved, above all the front lens. With the diameter proposed, 613 mm, it seems possible that flexure problems would appear, as discussed by Cao and Wilson [4.16]. This system was never manufactured as the telescope project did not come to fruition.

The work of Richardson et al. showed, above all, the gain achievable by increasing the size of the corrector. This was pointed out by Wynne [4.28] in connection with correctors for paraboloids. A significant increase in size of the front lens was also applied in the 3-lens paraboloid corrector, using an aspheric on the second lens, by Faulde and Wilson [4.30]. Wynne's scaling law with ($b_{s1}+1$) corresponds, according to Eq. (3.115) to a proportionality to the inverse cube of the secondary magnification ($1/m_2^3$). Modern RC telescopes tend to steeper primaries with larger values of m_2. For one Texas design by Richardson et al. [4.34], $m_2 = -6.75$ giving b_{s1}-values hardly departing from the parabola. In contrast with the early RC designs discussed by Paul [4.4], the modern RC form of primary hardly contributes any overcorrection of spherical aberration. This is the reason why the aspheric on the concave surface of the second lens, as proposed by Faulde and Wilson [4.30] for the paraboloid, is increasingly valuable for modern RC correctors. However, as Richardson et al. [4.34] correctly point out, the larger the system, the less the aspheric is needed. Faulde and Wilson allowed a front lens diameter of 471 mm with a 3.5 m primary, compared with 360 mm for the equivalent RC corrector. The resulting aspheric was, to a third order, quite weak. Conversely, for the smaller correctors used earlier by Wynne [4.14] and considered with only modest increase in diameter by Cao and Wilson [4.16] for less extreme m_2-values such as -2.7 for the ESO 3.6 m telescope, three aspheric surfaces are required to obtain a significant gain. This supports the design direction of Richardson et al. to go to the limit in corrector size. In such cases, though, flexure problems may set the practical limits [4.16].

As discussed above in connection with paraboloid correctors, the only theoretically complete solution to the limitations of higher order chromatic aberrations would be achromatism of each lens with 2 glasses. Apart from problems of secondary spectrum, this is impossible in practice for the reasons given above. Furthermore, the larger systems become, the less practicable such achromatism becomes. In this sense, no further major advance over existing systems seems possible with refracting elements.

Much work on PF correctors for recent telescopes, often with extremely steep primaries, has been done by Epps and collaborators. A good example is the design for the 10 m Keck telescope segmented primary (at that time still an RC design) by Epps, Angel and Anderson [4.36]. Figure 4.16 is reproduced from their paper and shows that the front meniscus is used with an interchangeable "blue" or "red" unit which contains the two smaller lenses of the 3-lens corrector and the flat window of the detector. The blue unit also contains an atmospheric dispersion corrector (ADC), a subject we shall deal with in § 4.4. The primary is very steep with f/1.75 and correspondingly only slightly hyperbolic ($b_{s1} = -1.0038$). Very complete spot-diagrams are given for four spectral ranges, with refocusing for the blue system. Figure 4.17 gives results for 2 of the 4 spectral bands shown. The middle and rear lenses each have an aspheric. The form of the spot-diagrams off-axis is complex, no doubt a result of the high relative aperture of the primary and the

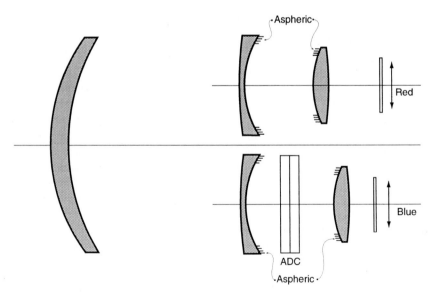

Fig. 4.16. 3-lens PF corrector for the 10 m Keck primary (after Epps et al. [4.36])

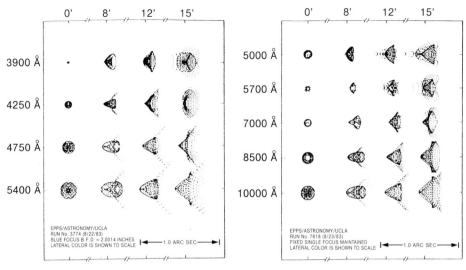

Fig. 4.17. Spot-diagrams for the system of Fig. 4.16 for two of the four spectral bands given (after Epps et al. [4.36])

effects of the aspherics. The worst spot-diagrams have a spread of the order of $\frac{2}{3}$ arcsec, which the authors consider, in tentative conclusions, to fall short of requirements.

Earlier in 1982, the same authors [4.37] [4.38] had favoured a Paul-Baker reflecting type corrector. This, as was shown in Chap. 3, can give superb imagery and is completely achromatic. However, as Richardson and Morbey [4.39] pointed out, the corrector has anyway to work with refracting elements such as wide-band filters, detector faceplates and cold-box windows whose refractive effects cannot be compensated in purely reflective correctors. This is a major advantage of refractive correctors, in spite of their limitations.

Brodie and Epps [4.40] have also designed improved correctors for an older telescope, the 120-inch Shane telescope of the Lick Observatory. This has a parabolic f/5.0 primary. Depending on geometrical space constraints, 3-lens (quartz) solutions with spherical surfaces were set up for fields of 30, 40, 50 and 60 arcmin diameter. Other solutions for 60 arcmin field diameter used larger elements (front lens 24.7 inch diameter) with spherical surfaces, and "normal" sized elements (front lens 19.3 inch diameter) with one aspheric on the second lens. The aspheric gives improved performance, for lenses of this size, over the equivalent all-spherical solution, in agreement with the conclusions above.

Modern designs confirm that the optimum type of corrector will depend on the Schwarzschild (conic) constant and f/no of the primary, and on the permissible size of the front lens relative to that of the primary. This latter may well be limited by flexure, above all if the primary has a high relative aperture. Depending on the above parameters, an aspheric surface (or surfaces) may or may not be worthwhile. Supplementary conditions can greatly complicate the problem. For example, Epps (private communication) has attacked the problem of maintaining a *telecentric* output for use of the corrector with fibres for a multi-object spectrograph. This condition is in conflict with a *flat-field* requirement. For the Shane telescope, he obtained a good design with all spherical surfaces if a strongly concave, hyperbolic field is accepted. This forced the use of a strong aspheric. Other constraints he mentions in retrofit designs for existing telescopes are parfocality with the naked focus, limited allowed change in f/ratio and long back focus.

The Wynne design for the RC prime focus correctors goes back to 1968 and has set the standard for many years. This (and virtually all subsequent variants discussed above) used only one glass, normally Schott UBK7, for cost reasons and with optimum transmission for materials available at that time in the required diameters. This situation was determinant for the design, above all the relatively wide spacing of the 3 lenses. As discussed above, since the lenses are not themselves achromatized, the separations lead to serious higher order chromatic effects, notably chromatic difference of coma and astigmatism. Thus two correctors were necessary to cover the spectral range from 334 nm to 1014 nm with a quality more or less meeting the 1 arcsec requirement over a 1° field. The finite thicknesses of the first two lenses also impair the chromatic performance, so these thicknesses were held to what was deemed a reasonable minimum. In fact, the image quality of the ESO Wynne-type corrector has been constantly impaired by flexure astigmatism of these two elements, as discussed above in connection with the work of Cao and Wilson.

More recently (1996) the whole problem has been approached with different premises by Delabre [4.75]. Special glasses are now available in adequate diameters, notably fused silica and Schott FK5. Figure 4.18 shows the Delabre system with its performance and data in the standardized format of Fig. 4.6. In fact, it is no longer a triplet corrector, but a *4-lens corrector*. However, this does not involve an extra lens since the fourth lens is simply the cryostat window, which is given some optical power, of the CCD detector. The figure shows that the first three lenses are in a compact group of axial length 152 mm instead of 490 mm in the old Wynne-type corrector. The front lens is about 19% smaller than the old front lens. More significantly, the front lens is nearly plano-convex and much thicker relative to its diameter, while the second lens no longer has the flexure-sensitive meniscus form and is also much thicker. The new corrector is therefore completely robust against flexure-induced astigmatic images. The chromatic correction is so much better, above all because of the design possibilities provided by the fourth lens (window), that a single corrector can operate over the whole spectral range from 320 nm to 1100 nm. The system delivers a final image of $f/3.02$ and has an average optical quality with 80% energy concentration

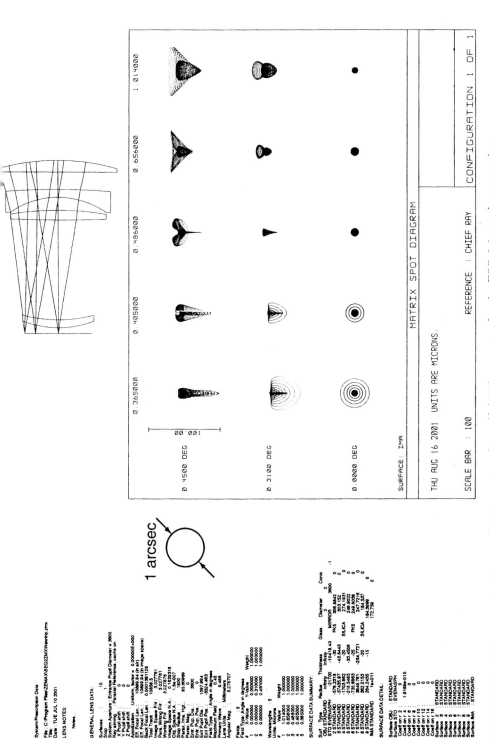

Fig. 4.18. Spot-diagrams for the new Delabre prime focus (f / 3.0) corrector for the ESO 3.6 m telescope using two glasses, Schott FK5 and fused silica (1996)

within 0.4 arcsec in the range 330 nm to 1000 nm. The spot-diagrams of Fig. 4.18 may be compared with those shown in Fig. 4.14(b) for the original Wynne-type corrector, for which the field performance in the blue part of the spectrum is much inferior to the 1 arcsec specification. It should be emphasized that the new corrector, like its predecessor, only uses spherical surfaces. The circular field covered is marginally smaller, 0.9° diameter instead of 1.0°. However, this field size is solely determined by the detector: a larger field, at least 1°, would be perfectly feasible. The limitation in all prime focus correctors lies not with the monochromatic aberrations, but with the chromatic effects. A balance between the lateral chromatic aberration and the higher order chromatic aberrations of coma and astigmatism presents the usual limitation.

Delabre has also designed a similar 4-lens PF-corrector for a 3.6 m telescope with a parabolic primary of $f/3.8$ [4.76]. The lack of overcorrection of spherical aberration given by an RC primary is a disadvantage that is compensated by the lower relative aperture. The field covered is 1.4° circular (1.0° by 1.0° square). Both the monochromatic performance and the chromatic performance are comparable with the system of Fig. 4.18.

4.3.2 Secondary focus correctors using lenses

4.3.2.1 Origins. Secondary focus correctors normally refer to Cassegrain foci but are, of course, also relevant to Gregory telescopes. The normal application is to Cassegrain foci, either in the classical form or in RC or quasi-RC form.

Sampson's first paper in 1913 [4.1] was the first proposal to be made for a corrector which was substantially afocal with a view to correcting field aberrations. Sampson's design used an f/5 primary whose asphericity was allowed to vary to correct the spherical aberration, the final form being an ellipse close to the parabola. The Cassegrain secondary, which he called the "reverser", was of a Mangin-type with a spherical back reflecting surface and weak negative lens effect. A spaced, nearly afocal doublet was then placed about 2/3 of the distance to the primary. Sampson calculated an image diameter, strictly circular, within 2.2 arcsec for a field diameter of 2°. The final f/ratio was f/14.05.

In modern terms, Sampson's design was effectively a Dall-Kirkham telescope with a Mangin secondary and afocal doublet corrector. (We recall from Chap. 3 that the Dall-Kirkham telescope with spherical secondary also has an elliptical primary fairly near the parabola). In this sense, it was *not* a corrector for a classical Cassegrain telescope, but a new – and remarkable – telescope design. His aim was similar to Schwarzschild's telescope, but in a more practical form.

Sampson's work was followed in 1922 by the analysis of Violette [4.3], immediately succeeding the invention of the RC telescope and closely following the approach of Chrétien discussed in Chap. 3. Although Violette's point of departure was the RC telescope, his approach was, in fact, the same as Sampson's in that he designed a complete system with a doublet corrector in which the forms of *both mirrors* were allowed to depart from the RC asphericities. His aim, then, was a telescope fulfilling all four conditions $\sum S_I = \sum S_{II} = \sum S_{III} = \sum S_{IV} = 0$. Since the RC system did not satisfy the last two, Violette's corrector required finite negative power to correct the Petzval sum. He used two different glasses for achromatism, supposing thin lenses, and noted the inevitable secondary spectrum. He applied thin lens theory for the lenses and the Eikonal approach of Chrétien (following Schwarzschild) for the mirrors. He noted a spare degree of freedom for limiting distortion. Another solution, following Sampson, to eliminate the secondary spectrum, was proposed for a Cassegrain with equal radii on the two mirrors, giving a mirror system with $\sum S_{IV} = 0$. This permitted an afocal doublet with a single glass. We have considered a Cassegrain with $\sum S_{IV} = 0$ in Chap. 3: its major disadvantage is the large obstruction ratio R_A for acceptable image positions.

Surprisingly, the pioneer work of Sampson and Violette attracted little attention until Bowen [4.41] in 1961 proposed Cassegrain foci of the order of f/8 with f/3 primaries in order to have acceptable "speed" at the Cassegrain with photographic plates. This led to proposals for RC telescopes to extend the Cassegrain field and to corrector systems for further extension for various telescope forms.

4.3.2.2 Third order theory for classical Cassegrain and RC telescopes. We can apply the same thin lens theory to the Cassegrain (or secondary focus in general) that we have applied above to the PF corrector case. Wynne [4.23] analysed the situation in a similar way already in 1949. While the same theory is applicable, there is an important difference. The PF case was dominated by considerations of spherical aberration and coma, arising from the appreciable relative aperture; astigmatism and field curvature of the primary only played a minor role. In the Cassegrain case, we have an opposite situation: since the relative aperture is weak and the telescope is often aplanatic or near to that state, the corrector's role is dominated by astigmatism and field curvature, above all because the field curvature concave to the incident light of the secondary is usually far stronger than the opposite field curvature of the primary.

Let us consider first the general case of a Cassegrain telescope with correction by a thin lens. Combining Eqs. (4.31), omitting asphericities, with Eqs. (4.47) for lenses and adding the field curvature condition from Table 3.5, we have:

$$\sum S_I = -f'\zeta + L\xi + (S_I)_{cor}$$

$$\sum S_{II} = -\tfrac{1}{2}f' - d_1\xi + (S_{II})_{cor} + E(S_I)_{cor}$$

$$\sum S_{III} = f'\left(\frac{f'+d_1}{L}\right) + \frac{d_1^2}{L}\xi + (S_{III})_{cor} + 2E(S_{II})_{cor} + E^2(S_I)_{cor}$$

$$\sum S_{IV} = (S_{IV})_{Tel} + (S_{IV})_{cor}$$

$$(4.64)$$

As before, the quantities f' and L are defined as positive quantities in a Cassegrain telescope with normalization to $f' = 1$, while d_1 is negative. The quantity E, expressing the position of the corrector relative to the stop at the primary, was given for an aspheric plate corrector by (4.34) as

$$E = \frac{f'}{g} - \left(\frac{f'+d_1}{L}\right) ,$$

in which g is the distance from the Cassegrain focal plane, again defined as positive. Because E was originally defined by

$$E = \frac{1}{H}\frac{y_{pr}}{y} ,$$

in which H is the Lagrange-Helmholtz Invariant, the power of a thin lens only changes the metric of the equations by a scaling factor of the focal length. Since this affects all linear quantities equally, this has no effect on the relations derived below.

We can now consider the conditions under which all four aberrations of Eqs. (4.64) can be corrected. We note first of all that, while from (4.52) the expressions $(S_I)_L$ and $(S_{II})_L$ for a thin lens are relatively complex, those for $(S_{III})_L$ and $(S_{IV})_L$ are very simple:

$$(S_{III})_L = H^2 K_L = f'^2 K_L \tag{4.65}$$

in our normalized system; and

$$(S_{IV})_L = H^2\left(\frac{K_L}{n'}\right) = f'^2\left(\frac{K_L}{n'}\right) , \tag{4.66}$$

where K_L is the power of the lens and n' the refractive index of the glass. (It should be emphasized that these are the "central" aberrations arising if the stop is placed at the thin lens). Furthermore, from Table 3.5

$$(S_{IV})_{Tel} = H^2\left[\frac{m_2}{f'} - \left(\frac{m_2+1}{L}\right)\right] ,$$

in which we have again

$$H^2 = f'^2$$

from the normalization, giving

$$(S_{IV})_{Tel} = f'^2\left[\frac{m_2}{f'} - \left(\frac{m_2+1}{L}\right)\right] \tag{4.67}$$

We now apply (4.64), (4.65), (4.66) and (4.67) to the case of a *classical Cassegrain* telescope with a parabolic primary. Since ζ and ξ are zero, the correction of all 4 aberrations requires

$$
\left.
\begin{aligned}
\sum S_I &= (S_I)_{cor} &&= 0 \\
\sum S_{II} &= -\tfrac{1}{2}f' + (S_{II})_{cor} &&= 0 \\
\sum S_{III} &= f'\left(\frac{f'+d_1}{L}\right) + (S_{III})_{cor} + 2E(S_{II})_{cor} &&= 0 \\
\sum S_{IV} &= (S_{IV})_{Tel} + (S_{IV})_{cor} &&= 0
\end{aligned}
\right\} \tag{4.68}
$$

This gives from $\sum S_I = \sum S_{II} = 0$

$$
(S_{III})_{cor} = -f'\left(\frac{f'+d_1}{L}\right) - Ef' \,,
$$

and, substituting for E from (4.34)

$$
(S_{III})_{cor} = -\frac{f'^2}{g} \tag{4.69}
$$

From (4.65), if the corrector is a single lens

$$
(S_{III})_{cor} = f'^2 K_L \,,
$$

giving

$$
K_L = -\frac{1}{g} \tag{4.70}
$$

for the condition for correction of astigmatism in a classical Cassegrain. For the correction of field curvature, (4.66), (4.67) and (4.68) give

$$
K_L = -n'\left[\frac{m_2}{f'} - \left(\frac{m_2+1}{L}\right)\right] \tag{4.71}
$$

Combining (4.70) and (4.71) gives the condition for correcting all four aberrations in a classical Cassegrain with such a single thin lens:

$$
\frac{1}{g} = n'\left[\frac{m_2}{f'} - \left(\frac{m_2+1}{L}\right)\right] \tag{4.72}
$$

This assumes, of course, that it is possible for the lens to contribute $(S_I)_{cor}=0$ and $(S_{II})_{cor} = \tfrac{1}{2}f'$ from (4.68). In fact, it is normally not possible, as was recognised in 1949 by Wynne [4.23] – we shall return to this point later. Consider first the condition (4.72). If we take $g = f'/35$, as assumed above for an aspheric plate corrector, and the values $f' = +1$, $m_2 = -4$ and $L = +0.225$ from Table 3.2, then (4.72) gives

$$
n' = \frac{35}{9.333} = 3.75 \,,
$$

impossible in practice. This result means that the correction of astigmatism requires a lens which, with a real glass with $n' \sim 1.5$ and in this position, would over-compensate the field curvature. Of the other conditions, $(S_{II})_{cor} = \frac{1}{2}f'$ can be met by suitable bending from (4.52); but $(S_I)_{cor} = 0$ cannot be formally met if $K \neq 0$, which is in conflict with the field curvature condition. However, the axial beam width is sufficiently small, with small g-values, that the residual $(S_I)_{cor}$ may be acceptable.

The same formulation applies to any lens corrector which is *"thin"*, whatever the number of lenses involved. In particular, it applies to a *"thin"* doublet, substantially afocal except for the balance of power to correct the field curvature, if a *single glass* is used. This is clear if we write for the doublet from (4.65) and (4.66)

$$\left. \begin{array}{rcl} (S_{III})_{cor} & = & f'^2(K_{L1} + K_{L2}) \\[2mm] (S_{IV})_{cor} & = & f'^2\left(\dfrac{K_{L1} + K_{L2}}{n'}\right) \end{array} \right\} , \qquad (4.73)$$

which lead to the same equation (4.72). However, the doublet will permit the formal correction of $(S_I)_{cor} = 0$ and $(S_{II})_{cor} = \frac{1}{2}f'$ because of its extra degrees of freedom. But the limitation of (4.72) remains. It can be met by *shifting the corrector about* $2\frac{1}{2}$ *times further from the image to give* $f'/g \sim 14$. This is, in principle, possible for a doublet (though it may cause problems in practice), but would lead to problems of spherical aberration with a singlet lens. There are also chromatic aberrations which we shall deal with later.

There are two other ways we may relax the situation: by introducing glasses with different refractive indices n' for the basic wavelength (different dispersions will help us with the chromatic conditions anyway, but at the cost of secondary spectrum); and by departing from the classical Cassegrain form with one or both of the mirrors.

Consider first *2 different glasses* in a doublet corrector. Eqs. (4.73) become

$$(S_{III})_{cor} = f'^2(K_{L1} + K_{L2})$$
$$(S_{IV})_{cor} = f'^2\left(\frac{K_{L1}}{n_1'} + \frac{K_{L2}}{n_2'}\right) ,$$

giving with (4.67)

$$K_{L1}\left(\frac{n_2'}{n_1'}\right) + K_{L2} = -n_2'\left[\frac{m_2}{f'} - \left(\frac{m_2 + 1}{L}\right)\right] \qquad (4.74)$$

as the condition for $\sum S_{IV} = 0$. This relaxes the requirement of (4.72), as will be shown in detail in the more important application to RC telescopes, below. The different dispersions in a "thin" system also permit the correction of both chromatic aberrations C_1 and C_2, at the cost of secondary spectrum.

The other possibility, to allow a variation of the aspheric constants on the mirrors, leads to a *quasi-classical Cassegrain*. Suppose this mirror system still corrects $(\sum S_I)_{Tel} = 0$, but introduces changes of coma ΔS_{II} and astigmatism ΔS_{III} which will be functions of ξ. (The corrector must be capable of producing $(S_I)_{cor} = 0$, which is possible with a doublet; with a singlet near the image the value would not be zero but would be small). Then (4.68) becomes with a "thin" corrector:

$$\left.\begin{aligned}
\sum S_I &= (S_I)_{cor} = 0 \\
\sum S_{II} &= -\tfrac{1}{2}f' + \Delta S_{II} + (S_{II})_{cor} \\
\sum S_{III} &= f'\left(\frac{f' + d_1}{L}\right) + \Delta S_{III} + (S_{III})_{cor} + 2E(S_{II})_{cor} \\
\sum S_{IV} &= (S_{IV})_{Tel} + (S_{IV})_{cor}
\end{aligned}\right\} \tag{4.75}$$

The equivalent condition of (4.72) is then

$$-\frac{1}{g} + \left(\frac{2E\Delta S_{II} - \Delta S_{III}}{f'^2}\right) = -n'\left[\frac{m_2}{f'} - \left(\frac{m_2 + 1}{L}\right)\right] \tag{4.76}$$

We will suppose that ΔS_{III} is small compared with $2E\Delta S_{II}$. Setting $E \approx f'/g$ from (4.34), Eq. (4.76) can be written approximately as

$$-\frac{1}{g} + \frac{1}{g}\left(\frac{2\Delta S_{II}}{f'}\right) \approx -n'\left[\frac{m_2}{f'} - \left(\frac{m_2 + 1}{L}\right)\right] \tag{4.77}$$

With the values used above and $n' = 1.5$, this gives

$$\Delta S_{II} \approx +0.300 f'$$

for the same value $f'/g = 35$ postulated before. This is an important result. It shows that a change of the mirror aspherics to correct about 60% of the coma of the classical Cassegrain completely liberates the condition for correcting $\sum S_{III} = \sum S_{IV} = 0$ of (4.72) by adding the term in ΔS_{II} appearing in (4.76) and (4.77). This is very sensitive because $E \approx f'/g$ is a huge multiplier of its effect – an illustration of the see-saw diagram interpretation of Eqs. (4.75) due to Burch [4.10]. If ΔS_{II} is a free parameter, the position g becomes completely uncritical.

Let us consider now the other most important case, the *strict RC telescope*. Here $(\sum S_I)_{Tel} = (\sum S_{II})_{Tel} = 0$, while $(\sum S_{III})_{Tel}$ is given by Eq. (3.119). Eqs. (4.64) can then be written for the correction of all four aberrations

$$\left.\begin{aligned}
\sum S_I &= (S_I)_{cor} & = 0 \\
\sum S_{II} &= (S_{II})_{cor} & = 0 \\
\sum S_{III} &= f'\left(\frac{f' + \frac{d_1}{2}}{L}\right) + (S_{III})_{cor} & = 0 \\
\sum S_{IV} &= (S_{IV})_{Tel} + (S_{IV})_{cor} & = 0
\end{aligned}\right\} \tag{4.78}$$

Proceeding as before, the condition $\sum S_{III} = \sum S_{IV} = 0$ is then simply

$$\left(\frac{f' + \frac{d_1}{2}}{f'L}\right) = n'\left[\frac{m_2}{f'} - \left(\frac{m_2 + 1}{L}\right)\right] \tag{4.79}$$

for the case of a *"thin"* corrector of one or more elements *of the same glass*. It is important to note that the stop-related quantity E is absent in (4.78) and (4.79) because both $(S_I)_{cor}$ and $(S_{II})_{cor}$ are zero. The values of Table 3.2, as used above, lead to

$$n' = +0.430$$

as the requirement for fulfilling this condition for the optical geometry of Table 3.2. This value of n' is physically impossible. It is the opposite situation from what we had with Eq. (4.72) for the classical Cassegrain with $f'/g = 35$. Now, the compensation of the astigmatism of the telescope alone, without a stop-induced supplement from a coma term, requires a corrector whose power is *too weak* to compensate the field curvature. A real solution can only exist for lower magnifications m_2 than the value -4 assumed: but this is not the modern trend. In the classical Cassegrain case of Eq. (4.72), a solution exists if g is increased; but in the RC case Eq. (4.79) is independent of g.

We now consider in detail the advantage of a *2-glass solution* for a thin doublet corrector of a strict RC telescope. The astigmatism condition can be written

$$K_{L1} + K_{L2} = -\left(\frac{f' + \frac{d_1}{2}}{f'L}\right) \tag{4.80}$$

from (4.73) and (4.78), while the field curvature condition is the same as for the Cassegrain, given in (4.74) as

$$K_{L1}\left(\frac{n_2'}{n_1'}\right) + K_{L2} = -n_2'\left[\frac{m_2}{f'} - \left(\frac{m_2 + 1}{L}\right)\right] \tag{4.81}$$

Subtracting (4.80) from (4.81) gives

$$K_{L1}\left(\frac{n_2'}{n_1'} - 1\right) = -n_2'\left[\frac{m_2}{f'} - \left(\frac{m_2 + 1}{L}\right)\right] + \left(\frac{f' + \frac{d_1}{2}}{f'L}\right) \tag{4.82}$$

Using again the values of Table 3.2 and setting

$$n_2' = 1.65 , \quad n_1' = 1.50$$

in (4.82) leads to

$$K_{L1} = -113.86$$

and from (4.80)

$$K_{L2} = +109.85$$

The difference of about 4% in the powers gives the required astigmatism while the absolute values combined with the refractive indices give the required Petzval sum from (4.81). (By contrast, with a single glass, the powers K_{L1} and K_{L2} could be freely chosen to correct $(S_I)_{cor} = (S_{II})_{cor} = 0$, only the sum being important for $(S_{III})_{cor}$ or $(S_{IV})_{cor}$, correction of both being impossible with one glass from (4.79) for a compact system). Although the above individual powers seem high, it must be remembered they are normalized to $f' = 1$ and are of a feasible order of magnitude for relatively small lenses near the Cassegrain image correcting field of the order of $1°$ or less in diameter.

Equation (4.82) demonstrates the important advantage of 2-*glass* doublet solutions in thin correctors for strict RC telescopes. We shall give examples confirming this. For a "thin" system, both chromatic conditions C_1 and C_2 can be fulfilled, an important advantage in view of the residual negative power required for the correction of astigmatism and field curvature. But we shall see the effects of secondary spectrum in C_2 become significant.

As in the case of the classical Cassegrain, a powerful means of relaxing the condition (4.79) for a *thin corrector* of *one glass type* is to vary the aspheric constants on the mirrors to produce a *quasi-RC telescope*. As in the quasi-classical Cassegrain case above, we will assume a solution of the 2-mirror system giving $(S_I)_{Tel} = 0$, but introducing finite changes of coma and astigmatism of ΔS_{II} and ΔS_{III} which will be functions of ξ from (4.64). By analogy with (4.75), we have:

$$\left.\begin{aligned}
\sum S_I &= (S_I)_{cor} &&= 0\\
\sum S_{II} &= \Delta S_{II} + (S_{II})_{cor} &&= 0\\
\sum S_{III} &= f'\left(\frac{f' + \frac{d_1}{2}}{L}\right) + \Delta S_{III} + (S_{III})_{cor} + 2E(S_{II})_{cor} &&= 0\\
\sum S_{IV} &= (S_{IV})_{Tel} + (S_{IV})_{cor} &&= 0
\end{aligned}\right\}$$

$$(4.83)$$

The first three conditions give

$$K_{cor} = -\left(\frac{f' + \frac{d_1}{2}}{f'L}\right) - \frac{\Delta S_{III}}{f'^2} + 2E\frac{\Delta S_{II}}{f'^2} \qquad (4.84)$$

Assume, as before, that $\Delta S_{III} \ll 2E\,\Delta S_{II}$ and that $E \approx f'/g = 35$. Then (4.84) leads, when combined with the $\sum S_{IV} = 0$ condition from (4.71), to

$$-\left(\frac{f' + \frac{d_1}{2}}{f'L}\right) + 2\frac{\Delta S_{II}}{f'g} \approx -n'\left[\frac{m_2}{f'} - \left(\frac{m_2 + 1}{L}\right)\right], \qquad (4.85)$$

in which the second term on the left has been added compared with (4.79) for the strict RC. Setting the same values from Table 3.2 as before with $n' = 1.5$ gives

$$\Delta S_{II} \approx -0.143 f',$$

which should be compared with the equivalent result of $\Delta S_{II} \approx +0.300 f'$ from (4.77) for the quasi-classical Cassegrain. Bearing in mind that the classical Cassegrain has $(\sum S_{II})_{Tel} = -0.5 f'$, we see that, within our rough approximations, the two solutions are about the same, being (from the point of view of coma) about one third of the way from the RC back to the classical Cassegrain. This, then, is the optimum amount of coma from a 2-mirror telescope to favour a thin one-glass corrector with $E \approx 35$. Practical designs will confirm this advantage.

Implicitly, the theory of Violette [4.3] contains both the above relaxation possibilities (2 different glasses *and* variation of the asphericities of the mirrors), although Violette introduced two glasses for reasons of achromatism. It was thus a very far-sighted analysis. He recognised, too, that he had more variables than necessary and proposed a corrector form minimising a further condition, distortion.

Up to now, we have considered solutions with a *"thin"* corrector in the sense that all its surfaces have effectively the same value of E. We must now consider the possibility of *separated* lenses, in particular a separated doublet. We will consider the case of a *strict RC* with two separated corrector lenses *of the same glass*. (An equivalent formulation and conclusion applies to the case of a classical Cassegrain). Then Eqs. (4.78) become:

$$
\left.
\begin{aligned}
\sum S_I &= (S_I)_{L1} + (S_I)_{L2} & = 0\\
\sum S_{II} &= (S_{II})_{L1} + (S_{II})_{L2} & = 0\\
\sum S_{III} &= f'\left(\frac{f' + \frac{d_1}{2}}{L}\right) + (S_{III})_{L1} + (S_{III})_{L2}\\
&\quad + 2E_1(S_{II})_{L1} + 2E_2(S_{II})_{L2} + E_1^2(S_I)_{L1} + E_2^2(S_I)_{L2} & = 0\\
\sum S_{IV} &= (S_{IV})_{Tel} + (S_{IV})_{cor} & = 0
\end{aligned}
\right\}
$$
$$(4.86)$$

Reduction as before leads to the condition for $\sum S_{III} = \sum S_{IV} = 0$:

$$
\left.
\begin{aligned}
&-\left(\frac{f' + \frac{d_1}{2}}{f'L}\right) + \frac{2(S_{II})_{L1}}{f'^2}(E_2 - E_1) + \frac{(S_I)_{L1}}{f'^2}(E_2^2 - E_1^2)\\
&= -n'\left[\frac{m_2}{f'} - \left(\frac{m_2 + 1}{L}\right)\right]
\end{aligned}
\right\}
,
$$
$$(4.87)$$

in which two additional terms appear compared with the case of a "thin" corrector represented by (4.79). Since E_1 and E_2 are large, we will again assume the approximation that the contribution $(S_I)_{L1}$ to $\sum S_I$ is negligible and ignore this term. Assume, as before, $E_2 = 35$ and that $E_1 = 20$ giving $(E_2 - E_1) = 15$. Substitution of the values of Table 3.2 as before then gives for $n' = 1.5$:

$$(S_{II})_{L1} \approx -0.333 f'$$
$$(S_{II})_{L2} \approx +0.333 f'$$

So, separation of the elements of a doublet of a single glass also converts (4.79) into a form with a real solution for a strict RC telescope. However, we must remember the law, first pointed out by Ross in connection with field correctors, that a *separated* doublet of a single glass cannot correct both chromatic conditions C_1 and C_2, even if they are afocal. In this case, they are not afocal so chromatism is anyway present with one glass.

The above theory can readily be extended to a corrector consisting of *three* separated thin lenses. By analogy with the 3-aspheric plate theory of § 4.2.2, a solution of the three conditions $\sum S_I = \sum S_{II} = \sum S_{III} = 0$ must also exist with thin lenses. Because of the greater parametric freedom of lenses (power and bending), $\sum S_{IV} = 0$ can also be corrected. This applies to both strict RC and classical Cassegrain telescopes. The third lens resolves, in principle, the conflict between $\sum S_I$ and $\sum S_{II}$ correction which exists for classical Cassegrain telescopes. But, as with plates, high individual lens powers may emerge from the solution which may lead, in practical cases, to limitations of higher order aberrations. Aspheric surfaces may help here, as indicated by Paul [4.4] or Wynne [4.23].

4.3.2.3 Summary of the results of the above theory. As Wynne [4.5] has pointed out, thin lens theory becomes less accurate for elements close to the image. Nevertheless, the above theory reveals clearly by Eq. (4.79) that, for a strict RC, a "thin" corrector of a single glass cannot provide a solution of all four conditions for a real glass, irrespective of the number of elements. For such a *strict RC telescope*, a real monochromatic solution can be achieved at once for 4 conditions by one of the following devices, using a *doublet* corrector:

a) Maintain the *strict RC form* and introduce *2 different glasses* into a *thin* doublet corrector with as large a difference of refractive index as reasonable. The chromatic aberrations can then also be corrected at the cost of secondary spectrum.
b) Establish a *quasi-RC form* of the 2-mirror system to deliver some coma to the *thin* doublet corrector consisting of *one glass*. Some chromatic aberration is inevitable but there is no secondary spectrum. This solution also applies to singlet lens correctors.
c) Maintain the *strict RC form* and the *single glass* doublet but introduce a finite *separation* between the lenses. With the single glass, the chromatic residuals will tend to be worse.

Of course, combinations of (a) and (b), or (a) and (c) are also possible. The former was the solution chosen by Violette [4.3].

The situation for a *classical Cassegrain* corrector is similar for the above relaxation possibilities. However, there is an important difference. A solution is theoretically possible for a thin, single glass doublet if the distance g from the image is chosen to give a feasible refractive index according to (4.72). But for modern values of m_2, this may be an unacceptably large value of g. Furthermore, as has been mentioned above and shown by Wynne [4.23]

[4.42], the bending of the doublet elements to produce the coma correction normally introduces too much spherical aberration. These problems can be resolved by using three separated lenses.

4.3.2.4 Third order theory for Cassegrain telescopes without correction of $(\sum S_I)_{Tel}$.

In the above theory, we have considered normal Cassegrain telescopes delivering an image from the 2-mirror system corrected for spherical aberration, i.e. $(\sum S_I)_{Tel} = 0$ or acceptably small. This also enables the mirror system to work *without* corrector. It will always be the case if a corrector is added to an existing telescope.

If a telescope is designed *always* to work with the corrector, this restriction is not necessary, giving the following general formulation for a thin, single-glass doublet corrector:

$$\left.\begin{aligned}
\sum S_I &= (S_I)_{Tel} + (S_I)_{cor} &= 0\\
\sum S_{II} &= (S_{II})_{Tel} + (S_{II})_{cor} + E(S_I)_{cor} &= 0\\
\sum S_{III} &= (S_{III})_{Tel} + (S_{III})_{cor} + 2E(S_{II})_{cor} + E^2(S_I)_{cor} &= 0\\
\sum S_{IV} &= (S_{IV})_{Tel} + (S_{IV})_{cor} &= 0
\end{aligned}\right\}$$

$$(4.88)$$

Here $(S_{II})_{Tel}$ will include a stop-shift term in $(S_I)_{Tel}$, and $(S_{III})_{Tel}$ includes stop-shift terms in $(S_{II})_{Tel}$ and $(S_I)_{Tel}$. The required corrector aberrations are

$$\left.\begin{aligned}
(S_I)_{cor} &= -(S_I)_{Tel}\\
(S_{II})_{cor} &= -\left[(S_{II})_{Tel} + E(S_I)_{cor}\right]\\
(S_{III})_{cor} &= -\left[(S_{III})_{Tel} + 2E(S_{II})_{cor} + E^2(S_I)_{cor}\right] = f'^2(K_{L1} + K_{L2})
\end{aligned}\right\}$$

$$(4.89)$$

$(S_I)_{Tel}$, $(S_{II})_{Tel}$ and $(S_{III})_{Tel}$ are given from the defined values of ζ and ξ from Table 3.5. Then, as before, the requirement for $\sum S_{III} = \sum S_{IV} = 0$ gives

$$-\frac{1}{f'^2}\left[(S_{III})_{Tel} + 2E(S_{II})_{cor} + E^2(S_I)_{cor}\right] = -n'\left[\frac{m_2}{f'} - \left(\frac{m_2 + 1}{L}\right)\right]$$

$$(4.90)$$

Equation (4.90) and the first two equations of (4.89) can give a real solution for a thin doublet in certain circumstances. The most interesting example is the proposal of Rosin [4.43], based on work by Wynne [4.23], to use a parabolic primary with a spherical secondary of the same radius and an afocal, single-glass doublet. Essentially the same system was proposed later by Harmer and Wynne [4.44]. We shall consider the results of this design in the next section.

4.3.2.5 Practical examples of secondary focus correctors using lenses

Field flatteners: A negative lens placed close to the image can flatten the field of an RC telescope in an optimum way for the mean astigmatic field curvature $2S_{III} + S_{IV}$. This solution achieves essentially the same result as bending a photographic plate except for small chromatic differences of C_2. Wynne [4.14] gives the gain for a curved photographic plate for typical RC telescopes: for a 3.5 m, f/3-f/8 RC, the semi-field for 0.5 arcsec image spread is increased by bending the plate optimally from 9.2 arcmin to 13.0 arcmin.

Single lens correctors: In spite of the inevitable chromatic aberrations, such simple correctors can be considered in certain circumstances. Köhler [4.12] [4.13] noted that the negative field-flattening lens of the RC telescope can also correct the astigmatism if it is shifted from the image and the aspheric constants of the mirror system are changed to compensate the coma inevitably introduced by the lens. This is a good practical example of the relaxation (b) in §4.3.2.3. The properties of this corrector, which was manufactured for the ESO 3.6 m quasi-RC telescope and determined the forms of the mirrors, were discussed by Wilson [4.15]. Figure 4.19 reproduces spot-diagrams given for the best focus of the mean wavelength (546 nm) over a

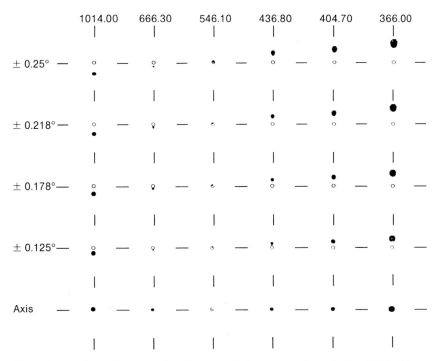

Fig. 4.19. Spot-diagrams for the singlet lens corrector of the ESO 3.6 m quasi-RC telescope (after Wilson [4.15]). Circle 0.18 arcsec

field of $\pm\,0.25°$. The monochromatic correction, as predicted by the third order theory above, is very good, within 0.18 arcsec at the field edge. The uncorrected C_1 is modest over a wide spectral band and can be refocused. With optimum focus, the higher order chromatic aberrations are also quite modest. However, the great weakness is the uncorrected C_2, the lateral chromatic aberration, which amounts to about 2 arcsec over the whole wavelength band 365 nm–1014 nm. The corrector must therefore be used over relatively narrow spectral bands, which also allows optimum focus. One advantage of this corrector is the reduction of ghost images compared with doublet correctors. Another is that its contribution to $\sum S_I$ is sufficiently low that the telescope can be used without the corrector, although it has some field coma as a *quasi*-RC system.

The aspheric plate plus field-flattener corrector of Fig. 4.6, as discussed by Schulte [4.17], is much superior in performance, but much more expensive to make.

Another possible single lens solution was mentioned by Wilson [4.45]. It is well known that a meniscus lens, roughly concentric to the image and placed near the image plane, will strongly affect astigmatism without much affecting the aperture aberrations $\sum S_I$ and $\sum S_{II}$. If its thickness is finite, it can also correct the field curvature. Because of an inevitable coma contribution, a modification of the mirror constants is also required here. On account of the thickness and bending, the chromatic aberrations are worse than in the thin Köhler singlet above. However, in an achromatised doublet form, the meniscus concept has been applied successfully by Rosin – see next section.

Two- and three-lens correctors for quasi-classical Cassegrain telescopes: In his pioneer paper of 1949, Wynne [4.23] investigated correctors for *classical Cassegrain* telescopes. Because of the problems referred to in §§ 4.3.2.2 and 4.3.2.3, he concluded that a modification of the asphericity of the secondary mirror was essential for the correction of spherical aberration. The problem of field curvature was solved by giving the secondary the same curvature as the primary – a solution which has only limited use, in practice, because of obstruction. This allowed an afocal doublet corrector of a single glass. For two examples given, the figuring on the secondary was 63% and 15%, respectively, of the classical hyperboloid for a doublet of diameter $\frac{1}{4}$ that of the primary. Of course, the mirror system without corrector is *not* corrected for spherical aberration. This design principle corresponds to the relaxation (b) of § 4.3.2.3.

Wynne [4.23] mentions the possibility of making the secondary *spherical*, but does not give an example. Rosin [4.43] was the first to apply this to a finished design. He followed Wynne's proposal of eliminating field curvature by using equal radii on primary and secondary, which leads to either high obstruction or unacceptable image position in most cases.

Harmer and Wynne [4.44] published a similar design with a *spherical secondary*, suggesting its greatest interest would be for modest-sized telescopes with fields up to 1.5° diameter. The corrector is a nearly afocal doublet in UBK7 glass. Spot-diagrams are shown for two designs, one with the positive lens leading and one with the negative lens leading. They are effectively within 0.5 arcsec for the whole field of ±0.75° and for most of the spectral range from 365-852 nm.

Wynne [4.42] has also given a more general treatment of the possibilities of doublet correctors in classical Cassegrain telescopes or for parabolic primaries with changed secondary asphericity. He also mentions [4.23] corrector designs for classical Cassegrain telescopes with 3 *separated* lenses, corresponding to the relaxation principle (c) of § 4.3.2.3. He points out that this can lead to problems of high individual powers – exactly the same problem as occurs with 3 aspheric plates, as discussed in § 4.2.

In general, 2-lens correctors will give good solutions for strict classical Cassegrain (or quasi-classical Cassegrain) telescopes with residual field curvature if one or more of the relaxation principles of § 4.3.2.3 is applied, depending on field, m_2 and the relative aperture of the primary. If the latter is very steep then the relative aperture of the secondary focus is also appreciable, even with a fairly large m_2.

Such extreme systems have been investigated by Epps et al. [4.36]. A strict classical Cassegrain with f/2.0 to f/5.28 has been taken as the basis for the calculation of a 3-lens corrector giving a final f/ratio of f/6.00. The lenses are of a single glass, both in this and similar designs and have an aspheric on the two rear lenses. All such designs (Fig. 4.20) have an ADC (atmospheric dispersion corrector) built in. In a later paper [4.46], Epps indicates the use of fused silica for the lenses for a similar design intended for the MMT conversion. The choice of the glass is uncritical for single-glass designs. Figure 4.21 reproduces the spot-diagrams for the design of Fig. 4.20.

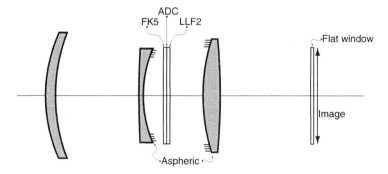

Fig. 4.20. Three-element corrector for an f/2.00 to f/5.28 classical Cassegrain designed by Epps et al. [4.36] for a 300-inch telescope, giving f/6.00 with the corrector

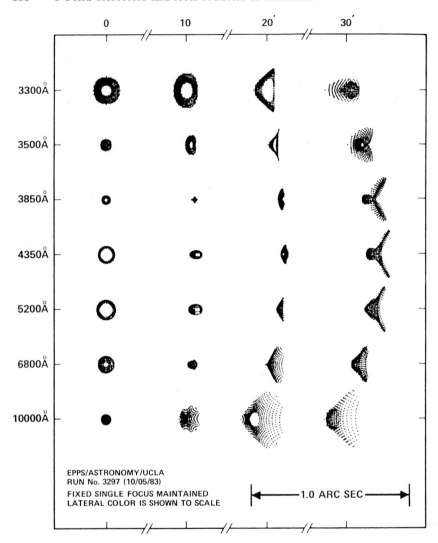

Fig. 4.21. Spot-diagrams for the design of Fig. 4.20 for a field of 1° diameter (Epps et al. [4.36])

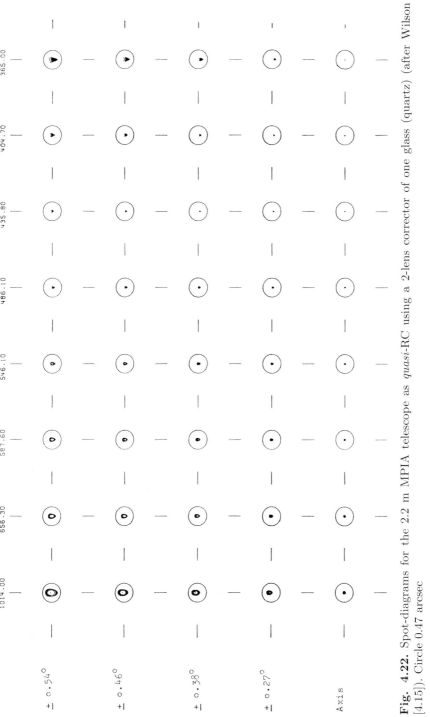

Fig. 4.22. Spot-diagrams for the 2.2 m MPIA telescope as *quasi*-RC using a 2-lens corrector of one glass (quartz) (after Wilson [4.15]). Circle 0.47 arcsec

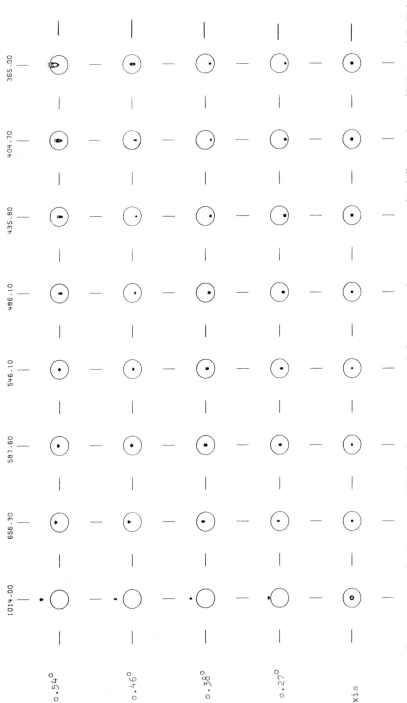

Fig. 4.23. Spot-diagrams for the 2.2 m MPIA telescope as *strict* RC using a 2-lens corrector of 2 different glasses (PK50 and BaF₃) (after Wilson [4.15]) Circle 0.48 arcsec

In [4.36] and [4.46], a large number of such designs are investigated for different f/ratios. Further such studies have been made for the Magellan 315-inch, f/1.20 parabolic primary [4.47] using all spherical surfaces and delivering a mildly curved field. Epps [4.48] has even applied such designs to a 256-inch, f/1.00 to f/3.962 naked classical Cassegrain to give f/4.50.

Lens correctors for RC and quasi-RC telescopes: Many designs have been published. Wynne [4.24] gave a design for the Kitt Peak 150-inch RC telescope. It consists of two separated lenses agreeing with the relaxation principle (c) of § 4.3.2.3. The field given was ± 15 arcmin, with correction within 0.2 arcsec. Later, Wynne [4.14] gave a similar design for a field of ±25 arcmin, the spot-diagrams somewhat exceeding 0.5 arcsec. Wilson [4.15] gives comparisons of a number of designs showing the advantage of relaxation with the principles of § 4.3.2.3. These become rapidly more significant as the field increases towards ±30 arcmin or beyond. Figs. 4.22 and 4.23 reproduce examples of designs for the 2.2 m MPIA telescopes. Out to a field of ±32.5 arcmin, the monochromatic spot-diagrams are within 0.1 arcsec. The chromatic spread is within 0.3 arcsec over the whole field and spectral range (365–1014 nm) in the first case, and even substantially better in the second case apart from secondary spectrum effects of lateral chromatic aberration (C_2). Wilson shows another design, for ±27 arcmin field, using a quartz-

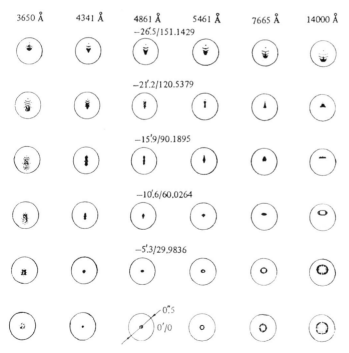

Fig. 4.24. Spot-diagrams for a doublet corrector using a single glass designed by Su, Zhou and Yu [4.51] for the Chinese 2.16 m, *strict* RC telescope with f/3 to f/9

LLF1 two-glass corrector in which all spot-diagrams are within 0.1 arcsec, the only significant errors being the secondary spectrum of C_2 which amounts to 0.57 arcsec for the edge of the field and extreme wavelength 1014 nm. The effect of this aberration depends, of course, on the bandpass of the filters.

Refsdal [4.49] gave a design for a *quasi*-RC telescope using a doublet of fused silica with one aspheric giving very good correction over a field of ±45 arcmin. An elegant design by Rosin [4.50] used two glasses (K3 and SK16) in a meniscus form with surfaces either concentric to the image or using the "aplanatic condition" ($\Delta\left(\frac{u}{n}\right) = 0$ in Eqs. (3.20)). This is a powerful extension of the singlet, thick meniscus design mentioned above, using relaxation principle (a) of § 4.3.2.3.

For a *strict* RC telescope, the best design of a doublet corrector of a single glass has been given by Su, Zhou and Yu [4.51] for the Chinese 2.16 m RC telescope, f/3 to f/9. Using a corrector consisting of 2 fused quartz lenses with quite large separation (about 20 cm for diameters of 32.6 and 30.1 cm), they attained the performance shown in Fig. 4.24 for a field of ±26.5 arcmin. This shows the power of the relaxation principle (c) of § 4.3.2.3 (separation of the elements). The spot-diagrams over the whole field and wavelength range from 365 to 1400 nm are within 0.32 arcsec. The lens separation produces small lateral chromatic aberration (C_2) effects which are well balanced between the orders.

This design, originally carried out in 1974, sets the standard for normal RC correctors.

Finally, reference must be made to a large number of designs by Epps and collaborators for RC systems. For example, Epps et al. [4.36] give several designs for hyperbolic primaries with small departures from the paraboloid typical of modern RC telescopes with high m_2. These are 3-lens designs which, as one would expect, are similar in appearance to that of Fig. 4.20 for a parabolic primary. They also use 2 aspherics and give similar performance to the spot-diagrams of Fig. 4.21. Another example using a much more eccentric primary ($b_{s1} = -1.1523$) gives an improvement of a factor of about 2 in the spot-diagrams over the whole field of ± 30 arcmin, as shown in Fig. 4.25, also designed for a 300-inch telescope. The improvement due to the higher eccentricity is another example of the gain through relaxation (b) of § 4.3.2.3. Further examples are given by Epps [4.47] for the Magellan and Columbus project designs. Since the primaries are very close to paraboloids, the designs are based on the paraboloid although the Columbus project envisaged an RC telescope. Since $m_2 = -12.5$ (f/1.20 to f/15), the departure from the classical Cassegrain is small. As with his PF designs, Epps also presents designs with reduced non-telecentricity for use with fibre-optics spectrographs. Typical errors of non-telecentricity are of the order of 1° - 2°.

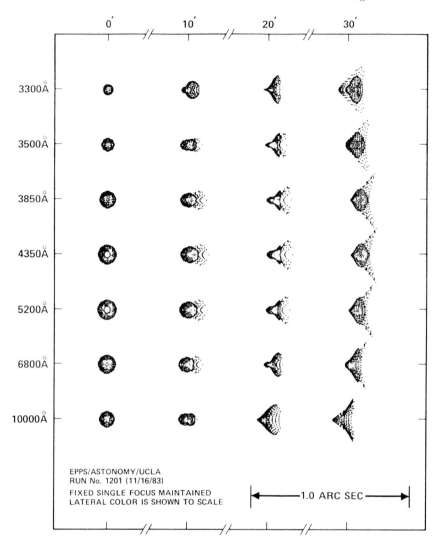

Fig. 4.25. Spot-diagrams for a three-element corrector designed by Epps et al. [4.36] for an f/1.80 – f/4.50 Cassegrain system using a significantly hyperbolic primary ($b_{s1} = -1.152\,310\,5$)

4.4 Atmospheric Dispersion Correctors (ADC)

We have referred above to the Epps designs [4.36] of sophisticated correctors including ADC. All of these included two plane-parallel plates, incorporated into each corrector, to provide material for later design of a pair of counter-rotating zero-deviation prisms serving as an ADC. No further information was given about their nature, but they were considered from an optical design viewpoint as plane-parallel plates. The glasses given were FK5 and LLF2.

The use of prisms for ADC is, according to Wallner and Wetherell [4.52], a concept that goes back to Airy [4.53] and was often used for special observational situations. But systematic analysis for general purpose correction has only recently been carried out.

The classical formula for atmospheric refraction, due to Comstock, is quoted in older books, e.g. Russell et al. [4.54]. In modern units it can be written as

$$\phi_{ref} = (16.114)\frac{b}{t} \tan Z \quad arcsec , \tag{4.91}$$

where Z is the zenith distance, b the barometric pressure in *mbar*, t the absolute temperature in $°K$ and ϕ_{ref} the mean refraction deviation in *arcsec*. This formula is based on a model treating the atmosphere as successive layers of uniform density and ignoring the curvature of the Earth. It is easily proven [4.55] that, to this approximation, the general form of (4.91) is

$$\phi_{ref} \simeq (n_a - 1) \tan Z \quad rad , \tag{4.92}$$

where n_a is the effective refractive index of the atmosphere and is a function primarily of b, t and λ, the wavelength. If the curvature of the Earth is taken into account, it was shown by Cassini [4.55] that the more accurate form is

$$\phi_{ref} = (n_a - 1)\left[\left(1 - \frac{h}{R}\right) \tan Z - \frac{h}{R} \tan^3 Z\right] rad , \tag{4.93}$$

in which h is the height of the equivalent homogeneous atmosphere and R the radius of the Earth. Since $\frac{h}{R}$ is a small quantity, the term in $\tan^3 Z$ can be ignored unless $\tan Z \gg 1$. For ADC purposes[3], Eqs. (4.91) and (4.92) are sufficient, so that

$$(206\,265)(n_a - 1) = (16.114)\frac{b}{t} ,$$

giving $(n_a - 1) = 0.000\,289\,8$ if $b = 1013.25$ mbar (760 Torr) and $t = 273.15°K$ ($0°C$). The more accurate formula of Cassini would give a value for $(n_a - 1)$ nearer to the measured value of the atmosphere for the visual sensitivity maximum at about 550 nm, *i.e.* $(n_a - 1) \simeq 0.000\,293\,3$. Values as given by von Hoerner and Schaifers [4.56] are reproduced in Table 4.1.

---

[3] A more general treatment of atmospheric refraction and dispersion is given in RTO II, Chap. 5.

Table 4.1. Refractive index values n_a as a function of wavelength λ, taken from von Hoerner and Schaifers [4.56], for 760 Torr and $0°C$

λ (nm)	280	300	400	500	600	700	800
$(n_a - 1)10^3$	0.3111	0.3077	0.2984	0.2944	0.2923	0.2910	0.2902

The above simple, approximate theory of atmospheric refraction is quite adequate for ADC systems because the match of real glasses to the atmospheric dispersion function is anyway only a rough approximation.

From (4.92), it follows immediately that the approximate formula for *atmospheric dispersion* for the wavelength range n_{a1} to n_{a2} is

$$\delta\phi_{ref1,2} \simeq \left(\frac{n_{a1} - n_{a2}}{n_a - 1}\right)\phi_{ref} \tag{4.94}$$

For the same conditions, we can calculate the approximate value of ϕ_{ref} from (4.91). $Z = 70°$ gives $\phi_{ref} = 164.2$ arcsec. For the range *300–800 nm*, (4.94) gives with $(n_a - 1) = 0.000\,289\,8$ as the effective value from (4.91) and (4.92) and the dispersion from Table 4.1

$$(\delta\phi_{ref})_{Z=70°} = 9.92 \text{ arcsec}$$

The corresponding values for $Z = 45°$ are $\phi_{ref} = 59.8$ arcsec and

$$(\delta\phi_{ref})_{Z=45°} = 3.61 \text{ arcsec}$$

For a more modest range *400–800 nm*, the equivalent dispersions are *4.65 arcsec* and *1.69 arcsec*. These values reveal how serious atmospheric dispersion is compared with modern image requirements of 0.5 arcsec or better, and the need for effective ADC.

Wallner and Wetherell [4.52] point out the practical value of *zero-deviation* solutions to avoid vignetting and pointing changes and suggest a Risley variable dispersion prism solution based on two zero-deviation prisms, each delivering finite dispersion from two different glasses, which are mutually rotated against each other to vary the dispersion and compensate the atmospheric dispersion for different values of Z. In fact, Wallner and Wetherell were not concerned with the general case of correcting a telescope image, but rather with the special case of *speckle cameras*. Such systems had already been described for this purpose by Beddoes et al. [4.57] and Breckenridge et al. [4.58]. It is possible for speckle purposes to mount the ADC in a parallel or nearly parallel beam, when prisms introduce no aberrations. The basic principles of ADC for a normal Cassegrain telescope, with a converging beam, were laid down in an admirable paper by Wynne [4.59].

Wallner and Wetherell investigated in detail the possibilities of glass pairs matching as closely as possible the dispersion – wavelength function of the atmosphere. Their aim was to achieve, for speckle purposes, diffraction limited ADC performance over a wide spectral band (350–1300 nm) for apertures up to 25 m! They achieve this for the Schott glass pairs LaF24/KF9 and (better) LaKN14/K11. The former pair gives serious absorption problems.

Wynne [4.59] demonstrates that the above approach has little relevance to the general problem of ADC at the Cassegrain focus. The converging beam will produce well-known aberrations due to the prism pairs as plane-parallel plates (see § 3.6.2.3 and references [3.3] and [3.6] of Chap. 3) and other, more complex decentering aberrations if the surfaces are not normal to the telescope axis. If the individual prisms have the same refractive index for the central wavelength, then the zero-deviation requirement will give identical angles for the unit prisms and each prism pair will be, *monochromatically*, simply a plane-parallel plate perpendicular to the axis. Two such pairs in op-

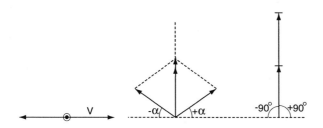

Fig. 4.26. Dispersion variation by opposite rotation of two prism pairs

posite directions will produce a zero resultant dispersion vector. If the pairs are rotated through an angle $\pm\alpha$ in opposite directions about the telescope axis, the resultant vectors increase giving a maximum with rotations of $\pm$ 90° (Fig. 4.26). The effect is a vertical dispersion vector, variable between 0 and $2V$ where V is the dispersion of an individual pair. The necessary direction of the resultant vector will depend on the mount and attitude of the telescope and is achieved by a supplementary rotation of the entire unit. Two air-glass surfaces can effectively be economised by using an optical immersion oil between the pairs, as used by Breckenridge et al. [4.58]. The dispersions of the two glasses should be as different as possible to reduce the prism angles, since the inclined faces still produce *chromatic* variations of aberrations for wavelengths away from the central wavelength. Similarly, the ADC should be as far from the image as possible since this distance is the lever arm for it to produce transverse compensating dispersion: the necessary prism angles reduce linearly with this focal distance. However, a greater focal distance increases the size and hence the necessary thickness of the prisms: their effect as plane-parallel plates increases. Wynne [4.59] points out that the longitudinal chromatic aberration C_1 is by far the most serious effect and that this can be entirely removed by arranging that the glasses in oiled contact have different dispersions and that the contact surfaces be given a slight curvature instead of being plane.

Wynne gives preliminary results for such an ADC for the William Herschel 4.2 m telescope (f/11). The (Schott) glasses chosen were UBK7 and LLF6 to give maximum dispersion difference with adequate transmission.

The ADC was placed 2 m from the focus. Correction up to $Z = 71.6°$, giving an uncorrected lateral chromatic aberration of 5.8 arcsec between 340 and 800 nm, required prism angles of 1.506° and 1.454°. With single prism elements of central thickness 17 mm the contact surfaces had a radius of curvature of 50 m to correct C_1. The residual aberrations over the whole spectral range are negligible, less than 0.05 arcsec, in the field centre, with very small changes over a field of 3.7 arcmin diameter. The limitation to optical quality is due to mismatch of the dispersion relative to the atmosphere, the problem considered by Wallner and Wetherell [4.52]. This amounts to about 7% of the atmospheric dispersion, i.e. about 0.4 arcsec for this spectral range at $Z = 71.6°$. This limitation is fundamental, unless the requirement of roughly equal central refractive indices is abandoned, as the work of Wallner and Wetherell showed, or more exotic materials are used (Fig. 4.27). Their preferred glass pairs had a very large index difference of about 0.2. It should be noted that $\tan 71.6° = 3$. From (4.92) it follows that this 7% dispersion discrepancy reduces to only about 0.14 arcsec at $Z = 45°$.

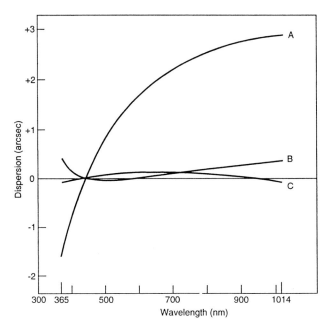

Fig. 4.27. Performance of the ADC designed by Wynne and Worswick for the William Herschel 4.2 m telescope with $Z = 70°$. Curve A shows the uncorrected atmospheric dispersion; curve B the correction achieved with the glasses used (UBK7 and LLF6); curve C what could be achieved with FK50 and Calcium Fluoride. (After Wynne and Worswick [4.60])

In a later paper, Wynne and Worswick [4.60] give the complete aberration theory of such ADC for the Cassegrain focus, also spot-diagrams of the precise performance of the ADC (within 0.1 arcsec) for the 4.2 m telescope. Figure 4.27, reproduced from their paper, shows the correction relative to the uncorrected atmospheric dispersion at $Z = 70°$ from 365-1014 nm.

As an independent *Cassegrain ADC*, the designs of Wynne and Worswick [4.60] appear to be the definitive solution. Their aberration analysis shows, moreover, that such simple systems are impracticable for typical PF focal ratios (f/3.5 or faster) because of unacceptable amounts of spherical aberration and coma arising from the thicknesses. Of course, this assumes the ADC is *added* to an existing telescope corrector. If the ADC is integrated in the corrector, as envisaged by Epps et al. [4.36], then compensations within the system may well be feasible.

Wynne [4.61] further analysed the possibilities of an *independent ADC for the PF case*. If the equivalent plane-parallel plates of the prism pairs are replaced by menisci, concentric to the focus, there will be no spherical aberration or longitudinal chromatic aberration. The sagittal field curvature is also zero but the tangential field curvature is twice the Petzval sum of the meniscus system, giving astigmatism. Wynne estimated this astigmatism to give about 0.5 arcsec aberration at the edge of a field diameter of 28 arcmin for the 3.6 m, f/3.2 PF of the Anglo-Australian Telescope (AAT). Logically, the ADC would be placed in front of the PF corrector. This aberration can easily be compensated in the corrector, with the disadvantage of limiting the corrector field if the ADC is removed for near-zenith observations. An oiled contact ADC has only two air-glass surfaces and may well be left in place. If a fully integrated design is envisaged, larger meniscus thicknesses become possible, so that sufficient dispersion may be available for a facility equivalent to an objective prism.

Wynne's work on *ADC for prime focus correctors (PF)* was reported more completely by Wynne and Worswick [4.62] with a detailed analysis of the aberration theory in this focus. Two cases were investigated, as shown in Fig. 4.28. It was found that the effective thick meniscus in front of a normal 3- or 4-lens PF corrector, if the whole system is re-designed, actually slightly

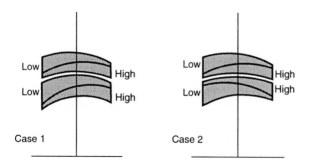

Fig. 4.28. Configurations of prisms for maximum dispersion setting, in cases 1 and 2 (angles exaggerated) for ADC in the prime focus (PF). (After Wynne and Worswick [4.62])

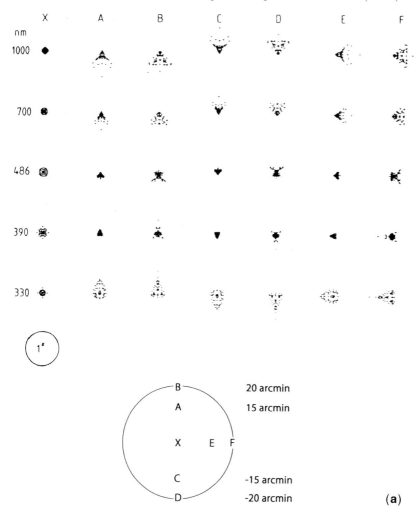

Fig. 4.29a. Spot-diagrams, reproduced from Wynne and Worswick [4.62], for their PF ADC for the 4.2 m, f/2.5 William Herschel Telescope. (**a**) Zero dispersion setting

improves the performance. This enables the use of a thicker meniscus with more powerful dispersions, giving an objective prism facility. In this case, the inclination angles are no longer negligible as they are for normal ADC, making Case 2 in Fig. 4.28 preferable. For the normal application limited to ADC, it is unimportant whether Case 1 or Case 2 is used.

Figure 4.29 gives the spot-diagrams for the ADC case of modest prism angles for 7 field positions and 5 wavelengths. (a) is for zero dispersion setting, (b) for ± 45° and (c) for maximum dispersion ± 90°. The field diameter is 40 arcmin.

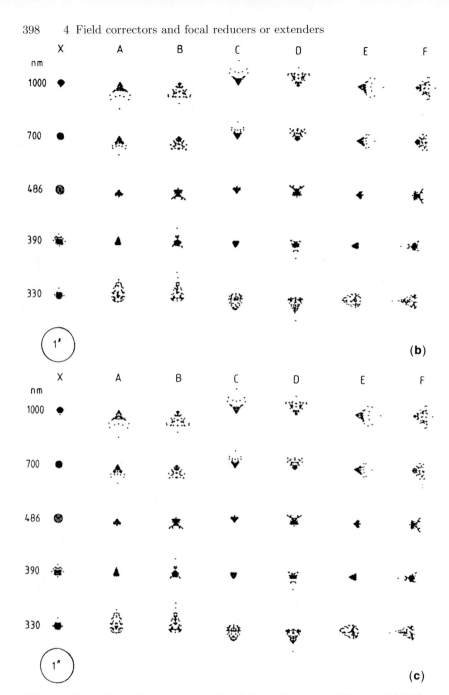

Fig. 4.29b, c. Spot-diagrams, reproduced from Wynne and Worswick [4.62], for their PF ADC for the 4.2 m, f/2.5 William Herschel Telescope. (**b**) At ± 45° and (**c**) maximum dispersion setting ± 90°

Bearing in mind the rotations of the spot-diagrams for different parts of the field, it is clear that the ADC causes negligible deterioration of the performance of the total corrector system. The glasses are UBK7 and LLF6, as in the Cassegrain case above. The authors also give spot-diagrams for a system some seven times more powerful for use as an objective prism. Inevitably, there is now appreciable effect of the prism system at $\pm 45°$, but very little at $\pm 0°$ and $\pm 90°$.

Another account of the above ADC-corrector systems has been given by Bingham [4.63], who shows the section through the finalised total system with an ADC prism arrangement as in Case 1 of Fig. 4.28. This is reproduced in Fig. 4.30. The ADC glasses are PSK3 and LLF1, and the contact layer is flat. The WHT primary is a paraboloid. Bingham also shows a similar ADC-corrector (the latter with 4 lenses) for an f/1.6 parabolic primary giving an f/2.0 final image. The image quality is similar over a somewhat reduced field of 30 arcmin diameter.

The PF systems described above use ADCs consisting of prisms with curved faces, that is combinations of lenses and prisms; or, putting it another way with *spherical* surfaces, off-axis sections of lenses.

In 1986 Su [4.64] proposed *Cassegrain correctors* in which the ADC feature is *integrated* into a doublet corrector. He called the lens-prism combinations "lensms". Figure 4.31 shows the form of his corrector, reproduced from his

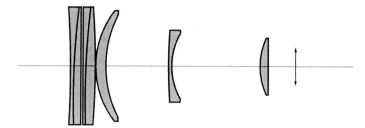

Fig. 4.30. Section through the complete ADC-corrector system for the 4.2 m, f/2.5 PF of the WHT, reproduced from Bingham [4.63]

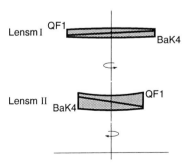

Fig. 4.31. "Lensm" design of Su [4.64] for an ADC integrated into a doublet corrector for a 5 m, f/2 to f/4.5 strict RC (Cassegrain) focus. The glasses are from the Chinese glass catalogue

paper. Su gives spot-diagrams for two dispersion rotations. The individual spot-diagrams are within 0.5 arcsec for a field diameter of 45 arcmin, most of them being much smaller. The largest total dispersion from 350 to 1014 nm is 7.6 arcsec, sufficient for $Z \simeq 70°$. Su also gives a diagram of secondary spectrum errors compared with the atmospheric dispersion function.

In 1990 this work was followed up by a detailed analysis by Wang and Su [4.65] of similar fully integrated ADC *in PF correctors*. All cases concerned correctors for a *7.5m, f/2 paraboloid*. Three types of lensm corrector were used, two of which are shown in Fig. 4.32. These are based on standard types of corrector of the sort proposed by Faulde and Wilson [4.30] and Wynne [4.28] respectively.

Figure 4.33 shows the field points for the calculation of spot-diagrams. Spot-diagrams are reproduced for lensm corrector type I in Fig. 4.34 for a field diameter of 45 arcmin and zero dispersion. The significance of the vertical scale of 6 arcsec is that this represents the maximum dispersion of the lensm corrector for the wavelength range 350 to 1014 nm. The mean diameters of the spot diagrams for 434 nm, 350 nm and 1014 nm respectively are 0.29,

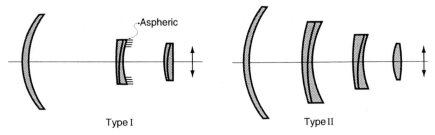

Type I   Type II

Fig. 4.32. Two types of lensm corrector designed by Wang and Su [4.65] for the PF of a 7.5 m, f/2 paraboloid

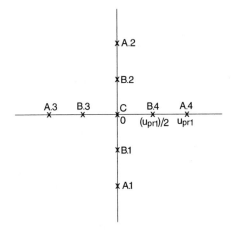

Fig. 4.33. Points in the field for the calculation of spot-diagrams, from Wang and Su [4.65]

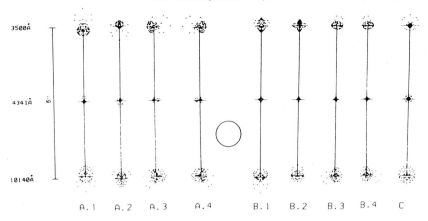

Fig. 4.34. Spot-diagrams for lensm corrector type I for a field diameter of 45 arc-min with a 7.5 m, f/2 paraboloid. Rotation angles of the lensms are 0°, 0°, i.e. zero dispersion. Circle diameter = 1 arcsec. Reproduced from Wang and Su [4.65]

0.76 and 0.85 arcsec, the maximum being 1.11 arcsec. The field covered is somewhat larger than that of Wynne and Worswick (Fig. 4.29) who are using one more "lens" in the total system and a less steep paraboloid. But these authors used no aspheric surface.

The variations in quality of the lensm correctors of Wang and Su with different rotations (dispersions) are very small.

Figure 4.35 shows another example with lensm corrector type II for the same field of 45 arcmin diameter and zero dispersion. The mean diameters of the spot-diagrams for 434, 350 and 1014 nm respectively are 0.35, 0.87 and 0.96 arcsec with a maximum of 1.35 arcsec. Adding an aspheric to the

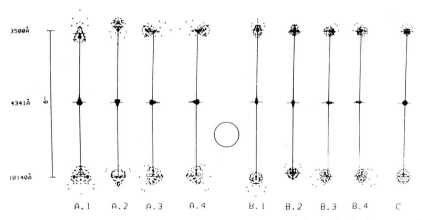

Fig. 4.35. Spot-diagrams for lensm corrector type II for a field diameter of 45 arcmin with a 7.5 m, f/2 paraboloid. Rotation angles of the lensms are 0°, 0°, i.e. zero dispersion. Circle diameter = 1 arcsec. Reproduced from Wang and Su [4.65]

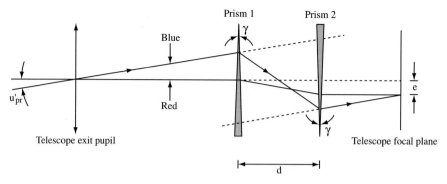

Fig. 4.36. Optical design of the LADC – schematic (after Avila, Rupprecht and Beckers[4.78])

first (larger) lensm (type III lensm corrector) improves these values to 0.33, 0.65 and 0.63 arcsec with a maximum of 0.97 arcsec. The same system at maximum dispersion gives 0.33, 0.66 and 0.65 arcsec with a maximum of 1.07 arcsec, a very minor increase.

The extra lens element of types II and III seems to bring little gain over type I. Since ADC in this system is achieved with the minimum number of normal corrector elements, it seems a most interesting solution. It should be noted that the front meniscus lens has a diameter only 9% of that of the primary and $g/|f_1'| = 0.08$. This relatively low value accounts for the effectiveness of the aspheric surface, as discussed above.

Finally, it should be said that ADC does not necessarily have to be introduced as part of the *telescope* optics. If slit or fibre instruments are used, it will then have to be done *before* the image plane, but such an ADC can still be seen as part of the auxiliary instrument and can be directly adapted to its field requirements. This solution was adopted, for example, in the ESO NTT.

This symbiosis of telescope and instrument is illustrated in another form by the more recent development for the ESO VLT unit telescopes when combined with the two FORS (Focal Reducer and Spectrograph) instruments [4.77] at the Cassegrain focus. For direct imagery, but above all for spectroscopy of extended sources and for multi-object spectroscopy where the slits cannot be aligned along the direction of the atmospheric dispersion, it is not possible to include the ADC in the instrument. The size of such telescopes of 8 m aperture makes the classical solutions discussed above, which were developed for telescopes of half this size or less, both difficult and expensive. Above all, the necessary size of the dispersive prisms was prohibitive for the elegant double-prism pair (zero deviation) system of Wynne. Apart from the high cost of the individual prisms with significant prism angles γ, the absorption of the two glasses with appreciable thicknesses would be a major problem. Furthermore, because cementing of the prism pairs in sizes of the order of 0.5 m would be too dangerous, oiled contact surfaces would be essential. But this would also be technically problematic.

A solution better adapted to such large telescopes was proposed at ESO by J. Beckers and designed in detail by G. Avila and G. Rupprecht [4.78]. This system has been called the "Linear Atmospheric Dispersion Corrector" (LADC). It consists of two single prisms, but with significant separation as shown in Fig. 4.36. The whole unit is about 900 mm in diameter and has a length of 1570 mm. It is part of the M_1 cell/M_3 tower construction. The figure arbitrarily defines the red principal ray emerging from the telescope exit pupil as being on the axis of the telescope. Since the two prism angles γ are identical and reversed, there is a deflection e of the axis, depending linearly on the separation, *but no pupil tilt*. The latter property was a fundamental requirement of the VLT system. The axis shift is automatically compensated by the guide probe. No rotation is required in an alt-az mounted telescope since the dispersion is always in the altitude direction. The blue field ray is shown with vastly exaggerated field angle u'_{pr}. Bearing in mind that the deviation of a small angled prism is $(n-1)\gamma$ if the incidence angle remains small, then the emerging rays from prism 2 are parallel to the incident rays on prism 1. The separation d is adjusted so that the emerging coloured rays focus on the telescope focal plane. In practice, prism 2 remains fixed at a suitable distance from the image plane, while prism 1 is moved along the axis. The larger the dispersion field angle, the larger d must be. Prism 1 produces the effective dispersion correction, part of which is removed by prism 2. The positive balance arises from the rule stated for the Wynne system above, that the compensating dispersion of a prism is proportional to its distance (lever arm) from the focal plane. The real role of prism 2 is therefore simply to remove image and pupil tilt. In the zenith, $d = 0$ and the two prisms form a parallel plate.

Since there is only one glass type, fused silica can be used giving optimum transmission, an important property bearing in mind that the system cannot be routinely removed from the telescope, even if it is not required for observations close to the zenith. A disadvantage, inherent in the system compared with the 2-glass doublet system of Wynne and Worswick, is that the matching of the atmospheric dispersion is inevitably less precise. However, in the most critical case of the high spatial resolution imaging mode with a field diameter of 3.4 arcmin, the authors show that, for the spectral range 350–850 nm and a zenith distance $Z = 50°$, the atmospheric dispersion is reduced from 1.00 arcsec to 0.12 arcsec at the field centre and from 1.14 arcsec to 0.18 arcsec at the field edge. The broad-band correction is therefore between 88% and 84% for a field diameter of 3.4 arcmin and the residues have, in practice, proved undetectable. Scaling factor variations of the field would be very serious for multi-object spectroscopy with FORS using slits of about 0.5 arcsec, but the variation produced by the LADC is less than 0.01 arcsec and therefore negligible. If the FORS instrument has to be moved to another unit telescope, the LADC unit must also be removed and remounted with it, since the presence of the LADC facility is considered essential for high quality FORS operation.

4.5 Focal reducers and extenders

4.5.1 Simple reducers and extenders in front of the image

In § 3.6.4.4, systems were discussed in which refracting elements were used in combinations with Mangin mirrors and lens objectives, e.g. Fig. 3.67. In this spirit, between 1828 and 1833, Barlow (see ref. [3.39] in Chap. 3) carried out experiments with liquid lenses and, as a by-product, introduced the negative *Barlow lens* as a device for increasing the focal length. This was a small-field telephoto system. In principle, a positive lens as a focal reducer is completely analogue; but as a "wide-angle" system for a given detector size it has problems of field aberrations that do not occur with the Barlow lens. In both cases, the system must function with a nearly telecentric pupil because of the position of the telescope exit pupil behind the secondary, far away from the Barlow or reducer (Shapley) lens.

A focal reducer or extender must operate *by definition* with an axially corrected telescope: otherwise it is a special form of corrector with positive or negative power.

Since a focal reducer (FR) or extender (FE) has significant power by definition, it must be achromatised in the classic sense with two different glasses. The simplest form is therefore a thin doublet achromat. This can be treated by the same thin lens theory as that given above in § 4.3.2.2 for quasi-afocal doublet correctors applied to classical Cassegrain and RC telescopes. However, the situation with FR and FE is much less favourable since the total power is finite and prescribed by the magnification. In the case of RC telescopes, the situation of Eqs. (4.78) applies, where stop-shift terms are absent and a perfect FR or FE would have $S_I = S_{II} = 0$ with compensation of the RC S_{III} and S_{IV} residues. Then (4.82) would apply, in principle, for the correction of $\sum S_{III}$ and $\sum S_{IV}$. For available values of n'_1 and n'_2, the condition can only be fulfilled with a small negative total power of the FR or FE. This is in fundamental conflict with the prescribed finite power determined by its magnification. The direction of the FE is more favourable than that of the FR since it compensates the RC residues. But significant power (magnification) is bound to overcompensate. The required power K_{FE} is dependent on the distance g_{FE} from the original telescope focus and is given from (2.8), for a "thin" FE, by

$$m_{FE} = \frac{1}{1 + K_{FE}g_{FE}} \, , \tag{4.95}$$

where g_{FE} and m_{FE}, the magnification, are positive quantities. Since the FE has $m_{FE} > 1$, it follows from (4.95) that K_{FE} must be *negative*. In the limit case with $g_{FE} = 0$, $m_{FE} = 1$ and the FE becomes a field lens. If $K_{FE}g_{FE} = -1$, then $g_{FE} = -f'_{FE}$ and $m_{FE} = \infty$, the limit case of an afocal Galilean telescope. From (4.52) we have the simple situation for a "thin" FE that $(S_{III})_{FE} \propto (\sum K)_{FE}$ and $(S_{IV})_{FE} \propto \left(\frac{K_1}{n'_1} + \frac{K_2}{n'_2} \right)_{FE}$, if there are no stop-shift terms affecting the astigmatism. The overcompensation by $(S_{IV})_{FE}$ can be mitigated for a given $(\sum K)_{FE}$ by choosing n'_1 and n'_2 to be as high as possible, but there is nothing to be done about the astigmatism except to make g_{FE} as large as possible to reduce $(\sum K)_{FE}$. The larger the m_{FE}, the worse the situation becomes. The requirement that $(S_I)_{FE} = (S_{II})_{FE} = 0$ is the same as that for a normal achromatic objective and can normally be met, provided g_{FE} is not too small and m_{FE} not too large. But the quasi-telecentric stop produces supplementary coma.

The situation for an FR with an RC telescope is the same except that the residual RC aberrations *increase* those of the FR instead of compensating them to some extent. Furthermore, the reduction in *linear* field means the *angular* field for a given detector size is increased, whereas it is reduced with the FE. For both these reasons, FE are more favourable for $m_{FE} = 1/m_{FR}$ than FR.

With a classical Cassegrain, the coma of the telescope produces a stop-shift term in astigmatism which introduces a further complication. As with the quasi-afocal doublets, the correction of the telescope coma may require a bending which prohibits correction of spherical aberration. For most practical cases, residual coma from the quasi-telecentric effects in the FE or FR is more serious.

Barlow lenses are very popular with amateur telescopes of small size, usually for visual use. An excellent description of the possibilities is given by Rutten and van Venrooij [4.66]. They give a design for a cemented achromat for a 200 mm, f/10 Schmidt-Cassegrain telescope (in theory, a similar situation to the RC case above) without precise indication of the glasses used. Normally, common glasses are used, but (following from the theory above) high index lanthanum glasses give better performance provided a consider-able dispersion difference (Abbe number) is available. The design of Rutten and van Venrooij has the positive (flint) lens towards the incident light in agreement with Hartshorn [4.67]: the reverse order gives worse field aberra-tions. The authors give spot-diagrams for the combination with the Schmidt-Cassegrain, the performance being acceptable (relative to the Airy disk) over a field diameter of 40 mm for a Barlow magnification of 2 and a final focal length of 4000 mm (about $\frac{1}{2}^\circ$). The field limitation is by coma and astigma-tism.

Rutten and van Venrooij also give an equivalent design for a focal reducer with magnification 0.55. Again the positive lens (crown) leads. The linear field diameter given is 20 mm (about 1°) with acceptable performance compared with a 25μm circle set by grain size for photography. Because of its "thin lens" nature, its field performance is limited mainly by astigmatism.

Of course, the design of such systems with a modern optical design program is a trivial operation. Separation of the elements gives additional design freedom (e.g. coma) at the cost of chromatic errors and increased air-glass surfaces. Such systems are more in the domain of powered correctors, dealt with above.

Simple Barlow (diverging) doublets are capable of correcting linear fields comparable with their own diameters [4.67] for moderate magnifications (≈ 2) on small telescopes, the diameters being of the order of a tenth of the primary (or objective) diameter. For larger telescopes, the linear field corrected for a given doublet size will be the same, but the angular field will decrease linearly with the telescope size with fixed f/ratio. If it is wished to correct a larger angular field, the doublet must be made larger and placed further from the focus, giving increased problems of chromatic aberrations (including secondary spectrum) and, possibly, spherical aberration. The performance deteriorates if the Barlow magnification increases because the total power, and hence that of the individual lenses, must increase. Similarly, if the relative aperture of the telescope beam incident on the Barlow doublet is increased, the problems of correction of spherical aberration and coma grow rapidly.

4.5.2 Wide-field focal reducers (FR) as a substitute for a prime focus

This was the subject of a study carried out at Carl Zeiss for the 3.5 m, f/3 to f/8 MPIA telescope. A résumé of the results was given by Wilson [4.15]. The following properties of an *ideal* focal reducer were listed:

a) It should optically replace the PF, providing a similar field with similar quality over the whole spectral range.
b) It should use the normal telescope secondary (RC or quasi-RC in the case considered).
c) It should yield a convenient position of the final focus.
d) The final focus should be so arranged that the image receiving apparatus (including IR) can be used without causing unacceptable obstruction.
e) It should not cause construction problems such as a long overhang.
f) It should have as small a length and weight as possible to reduce handling problems.
g) It should have as few optical elements as possible.
h) It should contain only UV-transmitting elements.

This is a formidable list and it may be doubted whether a full solution exists. Above all, the linear field size implies at once that this general problem is quite different from the simple small-field doublet systems discussed above. In the above 3.5 m telescope, the required 1° diameter field of the f/3 PF, with corrector, gives a linear field at the f/8 Cassegrain of nearly 490 mm diameter. The magnification is $3/8 = 1/2.67$, appreciably stronger than the small telescope examples given above.

4.5.2.1 Wide-field focal reducers (FR) *without* an intermediate image.

Such a system is the wide-field, large-scale Gaussian equivalent of the simple doublet solutions given in § 4.5.1. The biggest problem for the wide-field extension is the position of the exit pupil of a Cassegrain telescope far in front of the FR so that the FR is effectively working telecentrically. This situation is already a limitation with the small-field doublets discussed above, but becomes far more serious with large linear fields.

Space limitations will normally rule out mirror solutions for an FR without an intermediate image. With lenses, the diameter of the front lenses must be larger than that of the virtual image, i.e. over half a meter for the above 1° field requirement. Figure 4.37 shows schematically a basic design restricted to 5 lenses, the largest having a diameter of 654 mm, covering a field of 0.9° diameter. With such diameters, the choice of glasses is very limited. Spot-diagrams are shown in Fig. 4.38. The monochromatic correction is favoured by increasing the length and is within 0.5 arcsec. The lateral chromatic aberration C_2 and the higher order chromatic aberrations are extremely serious even at modest fields. The system would only be useable over narrow spectral ranges and, even then, over a modest total spectral range because of chromatic differences of coma and astigmatism. Such linear fields cannot be covered by a practical lens system for a significant spectral range. The basic requirement a) above cannot be met. Achromatisation of the last singlet lens in Fig. 4.37, which balances the astigmatism and field curvature, would produce an improvement, but the finite thickness and powers of the individual doublets are fatal for the higher order chromatic performance.

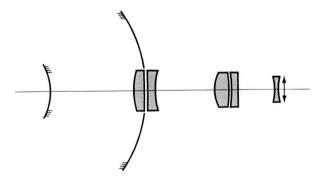

Fig. 4.37. Basic design (schematic) for an FR *without* intermediate image for a field of 0.9° diameter at the Cassegrain (RC) focus of the 3.5 m MPIA, f/3 to f/8 telescope [4.15]. $m_{FR} = 1/2.67$

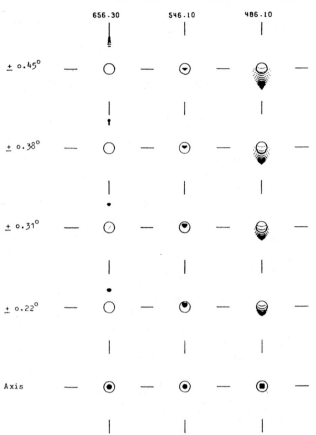

Fig. 4.38. Spot-diagrams for the FR system of Fig. 4.37 [4.15]. The circle is 0.98 arcsec = 50μm

Lens solutions for modest linear fields in large telescopes may be useful in special cases. A general solution equivalent to the PF with corrector for an FR *without* intermediate image seems impossible.

4.5.2.2 Wide-field focal reducers (FR) *with* an intermediate image. The importance of the quasi-telecentric stop for the performance of an FE or FR *without* intermediate image has been emphasised above: it is the fundamental factor limiting performance. If the telescope image is allowed to form and a field lens placed near it, the exit pupil of the telescope can, in principle, be re-imaged to an optimum point for an FR with its elements disposed round it like those of a photographic objective round a central stop.

Such a system is really no longer part of the telescope optics in the strict sense: it is effectively a supplementary instrument for direct imaging. In this sense, it is closely associated with spectrograph design: indeed, a normal spectrograph is simply an FR with dispersive means at the transferred pupil. Instrument optics is too large a subject to be treated in this book, but we will give a brief account of some of the work on FR solely considered for direct imaging.

In the study referred to above [4.15] for the MPIA, a lens system with field lens was investigated as shown in Fig. 4.39. The spot-diagrams are shown

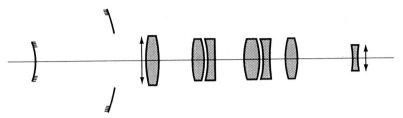

Fig. 4.39. Basic design (schematic) for an FR *with* intermediate image for a field of 0.9° diameter at the Cassegrain (RC) focus of the 3.5 m MPIA f/3 to f/8 telescope [4.15]. $m_{FR} = 1/2.67$

in Fig. 4.40. The monochromatic image suffers from appreciable coma which may be correctable with some design modification. Fundamental, however, is the large chromatic difference of coma. The elements of the FR must be disposed over a large axial distance, so that some of them are still a long way from the pupil. The lens thicknesses cause problems as with the system of Fig. 4.37. A further problem is the strong singlet field lens which produces a pupil image with large chromatic aberration. This means that the principal rays of different wavelengths traverse the elements of the FR at different heights, a fatal situation for higher order chromatic aberrations.

We may conclude that, for an FR replacing a PF facility with a 1° field diameter, lens systems, even with an intermediate image, have no chance of fulfilling the quality requirement a) above with regard to the spectral range. If this is to be achieved, the basic *power* of the FR must come from *mirrors*, lens elements being only quasi-afocal correctors. Mirrors solve the chromatic problems associated with the power at once, but lead to obstruction problems. In [4.15] a number of mirror solutions were investigated, all using the PF of a supplementary Schmidt-based system. The basic design arrangement is shown in Fig. 4.41. To minimise obstruction problems of the detector at the PF of the Schmidt mirror, a higher reduction factor of $m_{FR} = 1/4.71$ had to be used (f/8 to f/1.7). The optical correction would have been better with the previous reduction of f/8 to f/3. Figure 4.41 shows a 2-element corrector in front of the image. In some systems, this was simply a normal corrector for the Cassegrain focus; in others, it was effectively part of the FR and gave an uncorrected intermediate image. The systems considered were a singlet field lens with: a simple Schmidt system with shifted pupil; the same

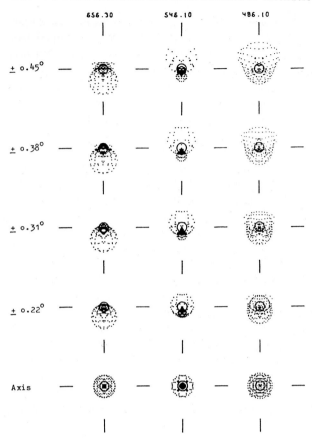

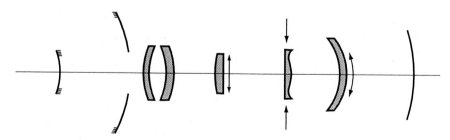

Fig. 4.40. Spot-diagrams for the FR system of Fig. 4.39 [4.15]. The circle is 0.98 arcsec

Fig. 4.41. Basic focal reducer geometry using a Schmidt-based mirror system for the 3.5 m MPIA telescope. Reduction is f/8 to f/1.7 ($m_{FR} = 1/4.71$) [4.15]

with a doublet corrector (*un*corrected); a Bouwers-Maksutov system with a doublet corrector (corrected); a Hawkins-Linfoot system with a doublet corrector (corrected); the same with a doublet corrector (*un*corrected); an extra field lens with a Baker-type system using 2 menisci and 1 aspheric plate and doublet corrector (*un*corrected). Finally, the field lens (or lenses) were replaced by a *field mirror* combined with the same Baker-type system with doublet corrector (*un*corrected) in order to establish the importance of the chromatic aberrations of the pupil. All of these telescope systems were discussed in some detail in § 3.6.4, the relevant literature references in Chap. 3 being [3.32] [3.38] [3.50] [3.54] [3.56] [3.57]. All results are for curved Schmidt fields. Here we shall reproduce two results, the best system with one field lens and doublet corrector (corrected intermediate image) and the best system with 2 field lenses and doublet corrector (*un*corrected intermediate image). Spot-diagrams of these systems are shown in Figs. 4.42 and 4.43 respectively, in each case for a field of 1° diameter and a circle of 0.5 arcsec. In both of these designs, the doublet corrector effectively removes the astigmatism of the telescope system. In the first system, it is a normal corrector correcting the intermediate image; in the second system, it produces an optimum balance of aberrations feeding into the FR.

In the Hawkins-Linfoot design, the chromatic aberration of the concentric meniscus is corrected by a weak afocal doublet in the transferred pupil, with an aspheric to correct the zonal spherical aberration. The secondary spectrum of the afocal doublet corrector gives lateral chromatic aberration curvature of the upper spot-diagram line (Fig. 4.42) and also the additive effect of spherochromatism and secondary spectrum focus error at 1014 nm.

In the Baker 2 meniscus – plate design, there is a meniscus on each side of the pupil and an aspheric plate in the pupil. All three elements are of quartz, enabling chromatic correction with no secondary spectrum. The field lens is split into two since this system is very sensitive to pupil aberration. The secondary spectrum of the transverse colour originates in the 2-glass corrector in front of the intermediate image. This could be suppressed with a single glass type with somewhat greater spacing.

With further design improvements, the system of Fig. 4.43 is capable of 0.5 arcsec performance over the whole field and wavelength range. For a field of 0.7° diameter, the quality is appreciably better. In this sense, the quality aim a) above has been attained. Although a single glass was not used in the above design, it would be possible to do this, eliminating secondary spectrum effects and giving optimum UV-transmission. The weak point is requirement g), that not too many optical elements be required. The system has a total of 2 field lenses, 4 corrector lenses (including 2 of a normal Cassegrain field corrector), an aspheric plate and a spherical mirror. By comparison, the equivalent PF corrector uses 3 lenses with spherical surfaces and suffers less from obstruction problems of the detector.

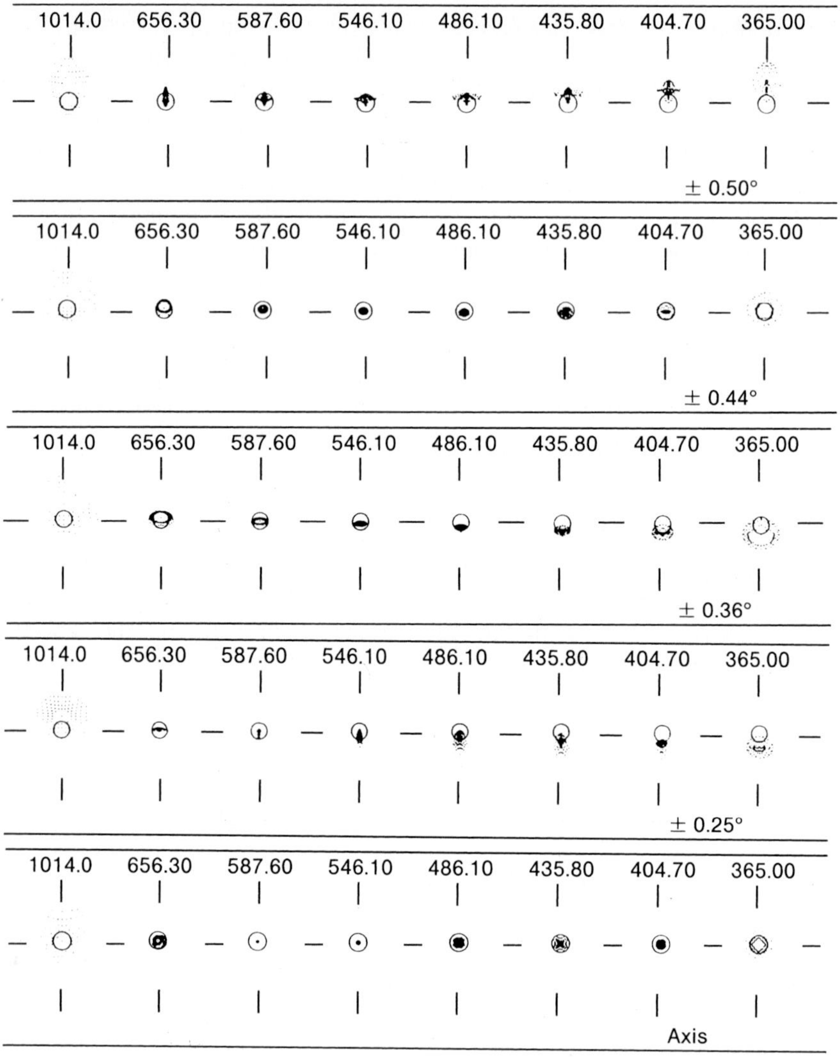

Fig. 4.42. Spot-diagrams for a focal reducer f/8 to f/1.7 designed for the 3.5 m MPIA telescope for a field diameter of 1°. Doublet corrector (corrected intermediate image), one field lens and a Hawkins-Linfoot camera. Circle = 0.50 arcsec = 14 μm. Image radius 1031 mm [4.15]

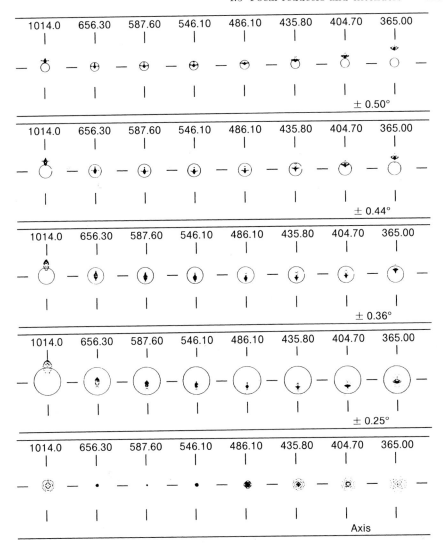

Fig. 4.43. Spot-diagrams for a focal reducer f/8 to f/1.7 designed for the 3.5 m MPIA telescope for a field diameter of 1°. Doublet corrector (*uncorrected intermediate image*), 2 field lenses and Baker-type camera with 2 menisci and one plate. Circle = 0.50 arcsec = 14 μm. Plot field for axial spot-diagrams = 10 μm. Image radius = 1052 mm [4.15]

It is clear that FR with large angular fields at the Cassegrain focus can only compete with PF correctors at considerable extra cost and effort in the number of optical elements involved and their arrangement with the detector. Concerning the latter, it should be added that a Schmidt-Cassegrain camera solution would also be quite feasible.

4.5.3 Other Cassegrain focal reducers.

For more modest fields and requirements, many proposals have been made. In 1956 Meinel [4.68] published results of an f/2 Cassegrain camera constructed with a field lens at the focus of the 82 inch McDonald telescope (field diameter 20 arcmin) and a standard Leitz Summicron photographic objective placed behind a Tessar type collimator with a filter between them. The pupil is imaged to the correct point in the Summicron. According to Meinel, the huge gain in speed with the camera was attained without image quality loss beyond the normal seeing limit.

An example of successful improvisation under time pressure is the Stockholm FR described by Jörsäter [4.69]. This converts a field of 43 mm diameter (5.4 arcmin) at the f/11 RC focus of the 2.5 m Nordic Optical Telescope (NOT) to an f/3.5 image. The field lens is a spectacle lens of 70 mm diameter and $f' = 600$ mm. The FR is an f/2.5 Konica camera objective. Images of FWHM of 1 arcsec are produced, giving 2 pixel sampling on a CCD detector. An important property is excellent freedom from ghost images.

More ambitious FR using mirrors in various forms have been proposed by Boulesteix et al. [4.70] for Fabry-Perot interferometry, and by Geyer and Nelles [4.71].

Very interesting work was done for the Texas 7.6 m telescope project with a view to replacing the PF by a Cassegrain FR with more modest fields than the 1° discussed above. Meinel et al. [4.72] described a system consisting of a field lens working with a 4-mirror FR to convert the f/13.5 Nasmyth focus to f/3.0, the field covered being 8 arcmin. Figure 4.44 shows the arrangement. The system consists of a modified "Bowen-type" camera using an inverted Cassegrain to increase the aperture, combined with a Gregory to avoid a refractive field flattener near the image. The exit pupil of the telescope is imaged on to the first, convex, mirror of the inverted Cassegrain, which in turn images it on to its Gregory secondary half. The mirrors are respectively spherical, nearly spherical, hyperbolic and elliptical. The great advantage of the system is its *compact* nature. The authors show spot-diagrams over the 8 arcmin diameter field which are *substantially within 0.2 arcsec*. There is some chromatic variation due to the field lens. For the modest field transmitted, there is no vignetting and only moderate central obstruction.

MacFarlane [4.73] has also proposed a number of Nasmyth FR designs for the same Texas 7.6 m telescope project. Most of them are based on the INCA (Inverted Cassegrain) type discussed by Rosin [4.74]. These are basically similar to the Meinel proposal of Fig. 4.44 but use only two reflections instead of

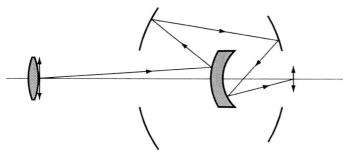

Fig. 4.44. Focal reducer designed by Meinel et al. [4.72] for the Texas 7.6 m telescope project. The f/13.5 Nasmyth focus is converted to f/3.0 over a field of 8 arcmin diameter

4 – see Fig. 4.45. Rosin gave a solution with spherical mirrors, corrected for spherical aberration if the magnification is correct. MacFarlane gives such a solution with mirrors of equal curvature and a magnification of 1/3.73. If both mirrors are made aspheric, a flat-field anastigmat corrected for all four aberrations is possible. In fact, this system was effectively invented by Schwarzschild (Ref. [3.1] in Chap. 3) for a parallel incident beam, as shown in Fig. 3.8, and abandoned as useless for a normal telescope. In the MacFarlane design, it was unclear whether a modification could retain anastigmatic imagery and also compensate the additive field curvature of the Cassegrain telescope and positive field lens. He gave a reduction from f/13.0 to f/6.0, using a configuration with the Cassegrain image and field lens slightly to the *right* of the concave mirror. A reduction to f/3.0 gave a much better performance over a field of 10 arcmin diameter, the spot-diagrams being within 0.2 arcsec. But no solution could approach the field performance of PF cor-

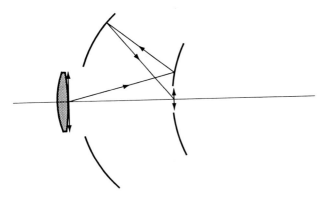

Fig. 4.45. INCA (Inverted Cassegrain) focal reducer proposed by MacFarlane [4.73]

rectors, confirming the conclusion of Wilson [4.15] that this is only possible with more complex systems.

MacFarlane [4.73] also considered the advantage of a field mirror to avoid the problems of chromatic aberrations of the pupil, which Wilson [4.15] had investigated. MacFarlane confirmed a significant advantage and proposed a practical field mirror arrangement as in Fig. 4.46. The Nasmyth M3 is inclined to the axis at about 40° instead of 45°. The concave field mirror is then inclined at about 5° to the axis, which should be acceptable as it is very near the image. The system of Fig. 4.46 shows a simple Schmidt camera as FR, but more complex systems would be possible. The Petzval sum of a concave field mirror is opposite to that of a positive field lens. This would be advantageous with an INCA type FR but Wilson [4.15] pointed out that a Schmidt type FR usually dominates the total field curvature and a field mirror adds further to its effect. So a field mirror is more interesting for INCA or straight Schmidt-Cassegrain FR solutions.

It is most instructive to compare the proposal of MacFarlane of Fig. 4.46 with some of the telescope solutions proposed in § 3.6.5.3. The system of Sasian (Fig. 3.90) achieves pupil transfer by a cylindrical "Nasmyth" mirror instead of a field mirror. But the final f/no is conventional because of the basic geometry. The system of Wilson and Delabre (Fig. 3.89), on the other hand, produces the pupil transfer with a normal "Nasmyth" flat by using an extra mirror. The fourth powered mirror (M_1 and M_2 being spherical) is replaced by a Cassegrain pair, *which is effectively an FR*, giving f/2.8. This gives a field quality within 0.2 arcsec over a field diameter of 30 arcmin, compared with MacFarlane's aim of 10 arcmin diameter, and also has the advantage of a steep spherical primary and a spherical secondary. This shows the advantage of building the FR into the basic design of the telescope. The total number

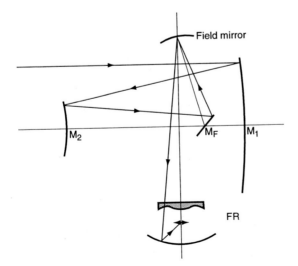

Fig. 4.46. FR concept using a field mirror proposed by MacFarlane [4.73]

of elements is the same as in the MacFarlane proposal, assuming he uses a simple Schmidt without further complication.

The general conclusion on FR, considered as an addition to a normal telescope, still seems to be that they cannot replace a PF corrector giving a field of the order of 1° diameter. They are much more interesting for more limited fields. CCD detectors favour such applications. The commonest FR are simply a non-spectral mode in a general spectrograph.

5 Major telescopes from Lord Rosse to about 1980

5.1 Major telescopes in the speculum mirror epoch to 1865

In Chap. 1 a brief account of the development of the reflector up to William Herschel was given. From 1800 to about 1840, the pendulum of rivalry between the reflector and the refractor swung back in favour of the refractor through the epoch-making work of Fraunhofer in the development of optical glass and its systematic application to refractors such as the Dorpat refractor of aperture 24.4 cm. The fact that refractors of relatively modest size were able, in spite of the limitation of secondary spectrum, to compete with or even excel reflectors with apertures up to 1.22 m (Herschel) was a measure of the two great weaknesses of the reflectors of the time: the poor efficiency through low reflectivity and the problems of mechanical manipulation in such sizes.

Excellent accounts of developments in the nineteenth and early twentieth centuries are given, above all, by Danjon and Couder [5.1], King [5.2] and Riekher [5.3]. Here the intention is only to establish the essential aspects leading to the modern reflector. As pointed out in Chap. 1, the theoretical basis until Schwarzschild in 1905 [5.4] was still the "classical" telescope with a parabolic primary as laid down by Descartes and applied in Newton, Herschel, Gregory or Cassegrain forms.

The further development of the reflector after William Herschel was, above all, due to *Lord Rosse* (William Parsons) and *William Lassell* in Great Britain. In the 1830's, Rosse systematically investigated the problems of casting large mirrors in speculum metal. A major problem up to then was the danger of crystallisation through slow cooling. This had forced Herschel to use a very high copper content for his largest mirror with negative consequences for the reflectivity. Some workers tried to solve the crystallisation problem by rapid solidification on a cold iron plate. Rosse showed that this was an illusion as it softened the material and led to an inferior polish [5.3]. By ingenious technical compromises in the cooling procedure, he was able to produce blanks up to 0.9 m in 1839 and 1.82 m (60 inch) in 1842 using an optimum alloy for polishing (68.2% Cu and 31.8% Sn). In parallel, Rosse performed pioneer work in a concept which has now again become ultra-

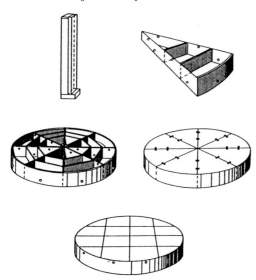

Fig. 5.1. 90 cm lightweighted, built-up blank made by Lord Rosse in 1839 (courtesy Rolf Riekher)

modern (see RTO II, Chap. 3), namely lightweighted *built-up* blanks [5.1] [5.3]. Fig. 5.1 is reproduced from Riekher [5.3] and shows the construction of a 90 cm blank made by Rosse. The ribs were of a Cu-Zn alloy with the same expansion coefficient as the speculum faceplate. The sectors were soldered together. Rosse could detect no difference in optical performance between the 90 cm massive cast blank and the built-up blank. However, he decided in favour of the massive cast approach for his 6-foot blank. Five blanks were cast, of which the first and last were polished, the other 3 breaking because of uneven temperature in cooling.

Rosse was also the first telescope maker to develop a polishing machine systematically.

Another important advance over the Herschel technology was the first use in a major telescope (1.82 m) of the *whiffle-tree support concept* invented by Thomas Grubb [5.5] (Fig. 5.2). Three plates support the mirror weight on universal joints at the centre of gravity of the 3 mirror sectors. These, in turn, each support 3 more plates on universal joints. With 4 stages, 81 supports were finally used at the back of the mirror [5.3]. All modern mirror support systems are based either on this principle due to Grubb or the astatic principle due to Lassell discussed below (see also RTO II, Chap. 3).

Rosse's largest 1.82 m primary had a focal length of 16.5 m giving f/9.0, only slightly shorter than Herschel's normal f/ratios. Now from Eq. 3.11 we see that the asphericity for a given conic section defined by the Schwarzschild constant b_s varies linearly with the size of a mirror and as the inverse cube of its f/no. *For a parabola*, the difference from the circle *with the same vertex curvature c* is simply

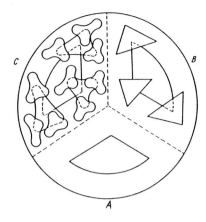

Fig. 5.2. Whiffle-tree support system in 4 stages designed by Thomas Grubb for the Rosse 6-foot reflector completed in 1845 (courtesy Rolf Riekher)

$$(\delta z)_{par} = \frac{c^3}{8}y^4 = \left(\frac{y}{r}\right)^3\frac{y}{8}$$

Inserting the f/no as $N = f'/2y$ gives with $f' = r/2$

$$(\delta z)_{par} = \frac{y}{512N^3} \tag{5.1}$$

The above values for the Rosse 6-foot telescope give from (5.1)

$$(\delta z)_{par} \simeq 2.4~\mu m$$

Rosse parabolised from the sphere of equal curvature c according to formula (3.11) by *flattening* the outer parts of the mirror. The modern method abandons the equal curvature reference sphere, so that the radial aspherising function can be freely chosen according to

$$\delta z = -\frac{(c_p - c_s)}{2}y^2 + \frac{c_s^3}{8}y^4, \tag{5.2}$$

where the suffixes s and p refer to the reference sphere and parabola respectively. For example, if the two terms in (5.2) are made equal for the full aperture y_m, then material must only be removed in a zonal operation. This is further discussed in RTO II, Chap. 1. The amount of material to be removed is reduced and the zonal position of the zero of the function can be chosen at will.

Rosse apparently was the first to introduce zonal testing using masks. He knew that the paraboloid no longer produced theoretically perfect geometrical optical images on axis if a test object were placed fairly near. This effect was first recognised by Herschel and its avoidance requires fulfilling the so-called Herschel condition (see ref. [3.3] and [3.6] of Chap. 3) which is rarely possible in practical optical systems. Rosse therefore calculated the theoretical focus shift for different zones and measured this with his zonal masks.

Rosse recognised the disadvantage of field coma introduced by the Herschel tilted single mirror form (Fig. 1.1) and reverted to the Newton form

Lord Rosse's Riesenteleskop.

Fig. 5.3. Lord Rosse's 6-foot (1.82 m) telescope completed in 1845 (courtesy Deutsches Museum, Munich)

with a second reflection, made possible essentially by the better composition and reflectivity of his speculum. Figure 5.3 shows the mounted telescope. Recognising the immense mechanical problems experienced by Herschel with his largest telescope, Rosse prudently renounced the possibility of a mounting permitting general sky coverage. Instead, the huge iron tube swung between walls on trunnion bearings through an angle of 160° along the meridian. By raising or lowering the bearings on each side with wedges, the tube could be inclined ± 12° giving an observation period of about $1\frac{1}{2}$ hours for an object near the equator.

This telescope represented a remarkable advance but was not yet a modern telescope. Using visual magnifications up to 1300, limited by the "seeing" in Ireland, Rosse was able to discover the spiral structure of external galaxies (Riekher gives a comparison drawing and photograph in [5.3]) some 75 years before their physical nature was proven. He also found that parts of some spiral nebulae could be resolved into stars [5.6].

About the same time as Rosse, *William Lassell* also made fundamental advances. His casting techniques and alloys were similar to those of Rosse and equally successful up to his largest size of 1.22 m (4 feet) with f/9.2. Lassell's largest telescope was set up in Malta in 1861. This telescope possessed three notable new features

- A primary mirror support system based on astatic levers (see RTO II, Chap. 3)
- An equatorial mount based on a fork-type equatorial axis
- An open slat tube to permit natural ventilation of the air in the tube

Figure 5.4 shows a reproduction of the original single astatic lever support invented for a 9-inch telescope in 1842 [5.7] and Fig. 5.5 the complete 1.22 m

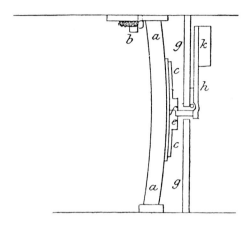

Fig. 5.4. Original single astatic counterweight described by Lassell in 1842 [5.7] (reproduced from Danjon and Couder [5.1])

telescope [5.8]. Figure 5.4 is reproduced from Danjon and Couder [5.1] who point out that Lassell was apparently not fully aware of the significance of his own invention. The 1.22 m telescope had a multi-lever astatic support and was, in this sense, the immediate precursor of the majority of support systems of subsequent telescopes. The development of such astatic supports will be discussed in RTO II, Chap. 3. Suffice it to say here that Lassell, underestimating his own invention for lack of proper scientific analysis, erroneously believed (ref. [5.1] p. 687) that the lever arms had to be maintained roughly horizontal. To this end, Lassell designed the telescope tube to be rotatable about its axis, an appreciable mechanical complication. Observational experience rapidly demonstrated to him that this rotation made no difference and that the levers worked equally well in all telescope attitudes. The essential reason for Lassell's error was the unfamiliarity with the new equatorial mount: with the old altitude-azimuth (*Alt-Az*) mount, lever arms arranged horizontally in the zenith position would have stayed so at all elevations. We shall see (RTO II, Chap. 3) that the recent reversion to the *Alt-Az* mount requires an inverse adaptation to its simplifications.

Lassell's rotating tube had, in fact, a second and highly practical purpose: to allow easy access from the observing tower (Fig. 5.5) to the Newton focus [5.3]. This is a problem of which amateurs making visual use of Newton telescopes rapidly become aware. If Lassell had used the Cassegrain form, this telescope could have rated, from its opto-mechanical *form*, as the first

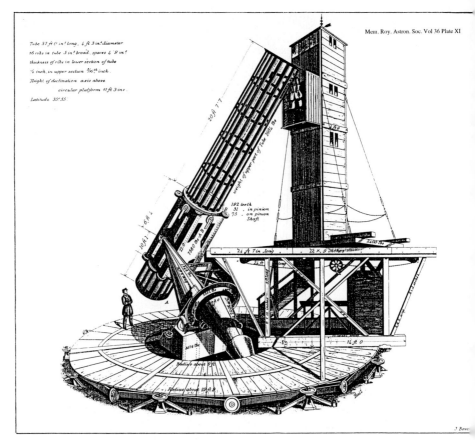

Fig. 5.5. Lassell's 1.22 m telescope set up in 1861 in Malta (reproduced from the original plate of [5.8])

modern reflector. But its speculum mirror still belonged to the pre-modern era.

Another important telescope of this epoch which must be mentioned was built by *James Nasmyth* in about 1845. It had an aperture of 20 inches (51 cm) and two most important features (Fig. 5.6). These were, firstly, the use of the *Cassegrain* optical form, although Fig. 5.6 shows that a Newton focus was also available. Secondly, Nasmyth's interest in the Cassegrain optical form was his modification of it to give the *Nasmyth focus* by adding a Newton-type flat to send the beam through the hollow altitude axis to a fixed focus position, if the observer turns with the azimuth turntable axis. This was a very important invention and makes this Nasmyth telescope the direct precursor of the ESO NTT. This 20-inch Nasmyth telescope, together with a 15-inch Newtonian-Cassegrain built by *Thomas Grubb* in 1835 for the Armagh Observatory, also represented the first successful manufacture of a

Fig. 5.6. James Nasmyth's 20-inch Cassegrain-Nasmyth telescope about 1845 (reproduced from King [5.2])

convex Cassegrain secondary for a major telescope, almost 200 years after its theoretical invention by Cassegrain! Unfortunately, little clear information concerning the optical quality in the Nasmyth focus compared with the Newton focus or other Newton telescopes seems to be available, since Nasmyth mostly observed the sun and moon rather than star fields.

The last, and potentially the most impressive telescope of the speculum mirror era was the *Melbourne reflector* of 48-inch aperture. The responsible committee, chaired by Robinson, included Rosse, John Herschel and Lassell, the latter offering his own 48-inch telescope operating in Malta. The latter was refused, to the chagrin of Lassell, because a "more manageable" instrument [5.2] was desired. The basis of this decision was the clear recognition of the advantage of the *Cassegrain form* because of its telephoto property: a 1.22 m primary with f/7.5 and $m_2 = -5.54$. It is notable that, for the first time, a significant reduction in primary f/no compared with W. Herschel's telescopes was attempted. This feature, combined with the Cassegrain form, presented a new dimension of optical manufacturing and testing difficulty. The contract was given to *Thomas Grubb* (working with his son Howard), who had invented the whiffle-tree support used by Rosse. He proposed a similar 27-pad support for the Melbourne reflector. No Newton focus was envisaged, which much simplified the visual access to the focus compared with the telescopes of Rosse and Lassell. Apart from the access advantage and that of the shorter, lighter tube, Robinson believed the convex secondary compensated the errors of the primary and produced a flatter image [5.2]. Table 3.3 shows that these statements are incorrect for a Newton and Cassegrain normalized to the same focal length, but the large telephoto effect was justification enough. The mounting, of the so-called English type as an equatorial, was an admirable concept (Fig. 5.7).[1] The tube was a lightweight structure allowing full ventilation and reducing windloading. Since observation was in the open, this was an important feature. Note that the telescope had *no fixed dome*, only a sliding-roof weather protection, again a direction which is ultra-modern today for some of the most advanced telescopes.

Had the committee taken the decision to equip the Melbourne reflector with the newly invented silver-on-glass mirrors, it would undoubtedly have been the first successful modern telescope. In fact, Robinson preferred the cautious approach of retaining speculum, believing that silver films of this size were too risky in the unfavourable climate of Melbourne [5.9]. This proved to be a fatally wrong decision.

Shortly before the first edition of this book was published, my attention was drawn by Peter Hingley, Librarian of the Royal Astronomical Society,

---

[1] S.C.B. Gascoigne has kindly pointed out that the well-known print of Fig. 5.7 does not, in fact, show the Melbourne telescope erected at its final destination in 1869, but a preceding trial erection near Dublin in 1867. This is clear from the angle of the polar axis to the horizontal, which is *more than* 45°, corresponding to the latitude of Dublin (53.3° N). Melbourne has a latitude of 37.8° S.

Fig. 5.7. The 4-foot (1.22 m) Melbourne reflector erected in 1869 (reproduced from King [5.2])

London, to the existence of a marvellous engraving showing the casting of the first 48-inch speculum blank. I am most grateful to him that this can now appear in the second edition. The original appeared in Vol. XII of the Strand Magazine (London) in 1896, in an article in which Howard Grubb was interviewed. He was responsible for the casting operations in Dublin in 1866, when the original engraving was produced. His account of the casting of the first blank is hair-raising. Two tons of metal were melted in a furnace fired by a mixture of coke and compressed peat. The dramatic nature of the operation is wonderfully shown in the engraving, reproduced in Fig. 5.8.

Fig. 5.8. The casting of the first 48-inch speculum blank for the Melbourne reflector (courtesy Royal Astronomical Society, through Peter Hingley, original engraving reproduced in the Strand Magazine, London, Vol. XII, 1896, p. 372)

Howard Grubb describes it thus: "The metal was poured in about six seconds and the extraordinary spectacle is depicted in the illustration on the next page. Every man wore a large apron and gauntlets of thick felt, with an uncanny-looking calico hood, soaked in alum, drawn completely over his head. This hood was provided with large, glistening talc eyes. These wierd figures flitted around in the ghastly light of the intense soda-flame that leapt from the great furnace, and the windows were filled with the eager faces of fascinated spectators". The whole operation in the terrible heat lasted 24 hours, but this blank was a failure. However, the experience enabled the successful casting of the second and third blanks, which were the ones figured and used in the telescope.

The early history of the application of chemical silvering on glass to telescope mirrors is given by King [5.2]. The first successful chemical silvering was displayed in 1851 at the Great Exhibition in London for decorative purposes, following a patent of Varnish and Mellish. This was refined in 1856 by the German chemist Liebig. The process was first applied to small telescopes by Steinheil in the same year, but independently and more strikingly by *Foucault* [5.10] to a 10 cm telescope (used as a test collimator) in 1857. Foucault then made a number of silver-on-glass reflectors up to 80 cm diameter, the larger ones profiting from his brilliant invention of the "knife-edge test" in 1859 (see RTO II, Chap. 1). The largest, with 80 cm aperture, was completed in 1862 and is shown in Fig. 5.9 [2].

Since the decision to use speculum in the Melbourne reflector was only taken in 1862 [5.13], five years after the successful completion of Foucault's first glass telescope, it appears now as an almost inexcusable blunder, particularly as Draper had also had successful results as early as 1858 [5.2]. The tarnishing problem was well-known in speculum and frequent re-polishing of the mirrors was required, an enormous complication which was completely avoided by silver-on-glass. The acceptance tests of the optics in 1868 claimed very high quality. For transport, the mirrors were protected by shellac. Two primaries, A and B, were supplied [5.13] and the incorrect removal of the shellac damaged the surface of primary A from the start. With primary B, the resolution was claimed to be fair and the reflectivity high. The climate of Melbourne, because of extreme temperature changes and damp, was very unfavourable to speculum; also the building gave inadequate dust and wind protection. Over 15 years a programme of observation of nebulae, following J. Herschel's catalogue, was pursued, entirely with hand drawings. A photographic attachment had been supplied but was useless at the very low Cassegrain relative aperture of f/41.6, above all with existing slow emulsions. The committee had failed to recognise the already well-known law in photography of unresolved extended objects whereby exposure time increases with

---

[2] I am most grateful to William Tobin for pointing out that King's Fig. 108 is misidentified as Foucault's 33 cm telescope. A complete account of Foucault's telescopes is given in two papers by Tobin [5.11] [5.12].

Fig. 5.9. Foucault's largest (80 cm) silver-on-glass reflector, completed in 1862 (reproduced from King [5.2])

N^2: the Cassegrain form was totally unsuitable for the astronomical programme pursued. Wind-shake was also a fatal problem with this long focal length.

In 1874, primary A was re-installed and then gave good images: the previous poor images were due to incorrect mounting in the cell, leading to constraints and astigmatic errors of various types. In 1877 the director, Ellery, announced that the mirrors were so tarnished that re-polishing was essential [5.2]. Returning the mirrors to the manufacturer, Grubb, in Ireland was considered impracticable. The technician accompanying the telescope, Le Sueur, who was supposed to deal with the optics, had left in disgust in 1870. Ellery attempted repolishing himself but lacked expertise, above all in testing. In 1890, he claimed success, but the optics never functioned reasonably again.

Much later (1953), the telescope was transferred to Mt. Stromlo where one of the primaries was dropped and smashed: the metal paid for a 50-inch glass blank to be worked into Schmidt-type optics. Subsequently, the remaining metal primary and the mounting were returned to Melbourne, intended to become a museum exhibit [5.13].

The story of the Melbourne reflector is correctly seen as one of the greatest tragedies in the history of the telescope. In 1904 Ritchey [5.2] [5.14] wrote: *"I consider the failure of the Melbourne reflector to have been one of the greatest calamities in the history of instrumental astronomy; for by destroying confidence in the usefulness of great reflecting telescopes, it has hindered the development of this type of instrument, so wonderfully efficient for photographic and spectroscopic work, for nearly a third of a century".*

Indeed, the lessons of the Melbourne reflector are fundamental and a warning for all subsequent telescope projects:

- The dangers of a design by a *committee* rather than dedicated individuals. All previous successful telescopes (above all those of Herschel, Rosse and Lassell) were designed and built by enthusiasts who themselves optimized and used them.
- The failure to involve the designer-manufacturer in the erection *and optimization* of his telescope, in function, on site.
- The failure to give authority and power to an astronomical director of sufficient enthusiasm, vision and astronomical and technical competence to ensure an astronomical programme suited to the nature of the telescope and the necessary technical expertise to maintain it.

5.2 Glass optics telescopes up to the Palomar 200-inch

As a result of the failure of the Melbourne reflector, the pendulum of the rivalry of reflector versus refractor swung back for the last time to the refractor, culminating in the 36-inch Lick and 40-inch Yerkes refractors of Alvan Clark. Nevertheless, the reflector continued to make steady progress without repeating the aperture achieved by Rosse (1.82 m) before the end of the nineteenth

century. In 1859 Foucault [5.15] invented the Foucault knife-edge test, the first
scientific test of telescope mirrors of high sensitivity (see RTO II, Chap. 1).
After 1865 silvered glass dominated mirror technology entirely. In spite of
tarnish by sulphur and moisture and relatively poor reflectivity in the far
blue, the ease of replacement and average high reflectivity compared with
speculum made *this one of the most important advances in the history of the
reflecting telescope.* It favoured the further development of the Cassegrain because 2 reflections were much more acceptable. It was also much lighter than
speculum, a heavy alloy. The further rapid development of photography also
favoured the reflector because of its complete absence of chromatism. Foucault, Draper, Brashear, H. Grubb, With, Calver, Martin, Eichens, Gautier
and Common were the principal successful manufacturers up to 1900. Calver
made a 36-inch (91 cm) silvered glass mirror for Common in 1879. Common
attempted to make a 5-foot (152 cm) Cassegrain using a blank with a hole
cast in it. His work clearly revealed the problems arising from the relatively
high expansion coefficient of normal plate (crown) glass (ca. 80×10^{-7}) and
the far lower thermal conductivity than that of speculum. The project was
abandoned before 1900, but the mirror was used and reworked later by Fecker
in 1933.

The most notable reflector of the pre-1900 period used the Calver 36-
inch (91 cm) mirror, mounted as a Newton telescope by Common for Ed-
ward Crossley's private observatory near Halifax, England. This telescope
was presented to the Lick Observatory in 1895 and re-figured and remounted
by H. Grubb. The remarkable feature of this Calver mirror was its fast relative aperture of *f/5.8*, an immense advance on the Melbourne reflector of
f/7.5, which itself was a big advance on the values around f/10 - f/9 of Herschel, Rosse and Lassell. From Eq. (5.1), the asphericity of the Calver mirror
compared with similar sized mirrors of Rosse was a factor of about 4 times
higher.

The Crossley reflector, because of its high light efficiency at f/5.8 in the
Newton focus, was really the *first modern reflector* and introduced modern
astrophysics and cosmology above all through the work of Keeler. Keeler's
photographs with this telescope were the first to reveal large numbers of small
or distant nebulae [5.2] [5.16]. The Mayall slitless spectrograph made excellent
use of the Newton focus of the Crossley reflector for UV spectroscopy applied
to faint nebulae. A photograph of the prime focus of this historic telescope,
which is still in operation, is given by Dimitroff and Baker [5.17]. Figure 5.10
shows the Crossley reflector at Lick with its new mounting. The rich harvest
of astronomical results with this telescope, in spite of its modest size even
when made in 1879, demonstrates the advantage of a dynamic and visionary
astronomer (in this case James Keeler) in exploiting to the full the potential
of a telescope. A beautiful account of Keeler's life and work, evoking the
extraordinary productivity of this period in the United States, is given by
Osterbrock [5.18].

Fig. 5.10. The 36-inch (91 cm) Crossley reflector, remounted at Lick in 1900 (courtesy Mary Lea Shane Archives of the Lick Observatory, through D. E. Osterbrock)

The turn of the century introduced the work of one of the greatest of all telescope builders, *George Ritchey*. Excellent accounts of Ritchey's career and achievements, above all due to the remarkable symbiosis with the organisational genius of *G.E. Hale*, are given by King [5.2], Riekher [5.3] and, again in an admirably complete account, recently published, by Osterbrock [5.19]. Hale was instrumental in getting the 40-inch Yerkes refractor built but fully recognised the inherently greater potential of the reflector. Inspired by Draper, Ritchey had made a $23\frac{1}{2}$-inch (60 cm) mirror with *f/3.9* which he set up at Yerkes in 1901 after his appointment by Hale. Ritchey was greatly interested in photography and recognised that higher "speeds" were essential in telescopes intended for nebular photography. In exactly 40 years, the speed (f/no) of large primaries had progressed from $N_1 = 9.2$ (Lassell, 1861) to 7.5

(Melbourne reflector, Grubb, 1869), to 5.8 (Crossley reflector, Calver, 1879), to 3.9 (Ritchey, 1901). This revolution was just as important as silvered glass as the basis for the final triumph of the reflector in the twentieth century. It would not have been possible without Foucault's invention of the knife-edge test. Figure 5.11 shows the 60 cm telescope at Yerkes. We said above, the Crossley reflector was the first modern telescope. This is true in the sense of its impact on astrophysics with f/5.8 in the Newton focus. But Ritchey's 60 cm at Yerkes was, in fact, *the first telescope to possess all the following features characterising a modern telescope,* albeit with a modest size:

Fig. 5.11. Ritchey's 60 cm telescope at Yerkes, 1901 (courtesy Deutsches Museum, Munich)

 – Glass mirrors (silvered)
 – A high speed primary (f/3.9) for photography (at a Newton focus)
 – A Cassegrain focus for spectroscopy with a fixed spectrograph
 – An open frame, completely ventilated tube
 – An equatorial mounting (German type with counterweight)

Figure 5.12 shows another view of the 60 cm telescope taken with Ritchey himself observing at the Newton focus. This photograph was discovered by chance in a private house near the Yerkes Observatory late in 1998. D.E. Osterbrock recognized that the figure was Ritchey, had the photograph cleaned and improved and kindly gave it to me for the second edition of this book, for which my grateful thanks. The brimless hat was typical for astronomers observing in the hard, winter conditions of that time (between 1901 and 1904).

Riekher [5.3] shows a reproduction of a famous photograph by Ritchey with a 4-hour exposure of the M31 nebula in Andromeda. Ritchey stopped the primary down to about $\frac{3}{4}$ aperture to reduce the field coma, an experiment which led to his interest in, and understanding, later, of the aplanatic telescope. This photograph showed far more detail than ever revealed before, even with the 40-inch refractor used visually with an aperture 2.2 times as large. (For comparison, the 3.5 m NTT, which went into operation 88 years later, would achieve the same intensity at far higher resolution with a few seconds exposure on a modern CCD detector).

With such results, the way was open for Hale and Ritchey to produce large telescopes of modern form. Hale had procured a *60-inch (1.52 m)* plate glass blank of about 20 cm thickness from the St. Gobain glassworks in Paris. Ritchey's optical work for this telescope, both in figuring techniques and testing (see RTO II, Chap. 1), represented a milestone. His method of parabolising was conceived to avoid working the edge of the primary at all. In Eq. (5.2), we showed how the aspherising function can be chosen at will by slightly varying the curvature of the reference sphere c_s relative to the vertex curvature c_p of the desired paraboloid. If δz in (5.2) is set to zero for the mirror edge y_m, this 2-dimensional function for the difference is a minimum. In three dimensions, however, this does not represent minimum removal of material and some material must be removed right up to the edge. Ritchey chose to give the reference sphere and the parabola the same *slopes* at the edge. Differentiating z_s and z_p and inserting $c_p - c_s \simeq \delta r/r^2$, where r is the mean radius, gives

$$(\delta r)_{ES} = \frac{D^2}{16 f'} \tag{5.3}$$

for the necessary increase of the reference sphere radius r_s compared with $r_p = 2f'$ for the equal slope (ES) case. Ritchey's mirror had a relative aperture of f/5.0, so $\delta r/r = 0.125\%$, a negligible effect on the focal length for a normal telescope. This technique of aspherisation gives maximum removal of

Fig. 5.12. Ritchey observing at the Newton focus of the 60 cm Yerkes reflecting telescope between 1901 and 1904 (courtesy Yerkes Obervatory, through D.E. Osterbrock)

material at the centre of the mirror, a far better technique than that used by Rosse.

Ritchey's test methods will be discussed in RTO II, Chap. 1. The Foucault knife-edge test was systematically applied by zonal masking and measuring the radius of different zones, the difference between centre and edge zones corresponding to Eq. (5.3). Here we shall mention just one other aspect as an illustration of the care taken. The primary was not only tested in the normal way, at its centre of curvature, but also – as a null test – at its focus in autocollimation and double-pass with a plane mirror of similar size. A plane mirror of the necessary high quality and in a diameter sufficient for a 1.5 m primary was an undertaking in its own right.

Similar care was taken with the support of the mirror. Both for the 60-inch and subsequent 100-inch telescopes, Ritchey systematically applied the astatic support concept of Lassell.

Because of the greater size of the primary, Ritchey no doubt felt it prudent to relax the f/no of the primary to f/5.0 compared with f/3.9 for his 24-inch telescope. The 60-inch telescope had a remarkably versatile optical concept,

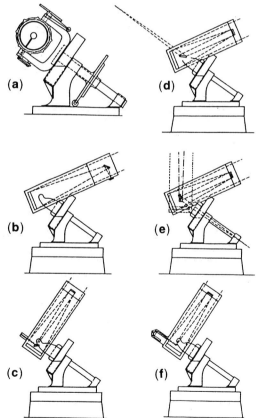

Fig. 5.13. Ritchey's 60-inch Mt. Wilson reflector: optical arrangement (courtesy Rolf Riekher)

as shown in Fig. 5.13. It was the first telescope to offer Newton, Cassegrain and coudé foci, the coudé focus being at the lower end of the hollow polar axis of a fork-type mounting. The Newton focus (b) has $f' = 25$ feet; the Cassegrain arrangement for photography (c) $f' = 100$ feet; the Cassegrain arrangement for spectroscopy (f) $f' = 80$ feet, and the coudé arrangement (e) $f' = 150$ feet. The corresponding relative apertures are f/5.0, f/20, f/16 and f/30, values which have set the style for most of this century. Figure 5.14 shows the telescope, completed in 1908. Ritchey recognised the fundamental importance of avoiding temperature changes in the mirror and strict rules were established to achieve this by control of the conditions in the dome, based on experience with the adjacent Snow solar telescope. It was judged that images were as good as those obtained in the optical shop. The judgment on photographs at the Newton focus after 11 hours exposure was that very round images were obtained whose diameter was not greater than *1.03 arcsec*. This set a new standard of optical quality, above all for a telescope of this size and speed. At last, a versatile, manoeuvrable telescope of very high quality was available with a size only surpassed (still) by Rosse's 6-foot telescope of 1845. The 60-inch telescope at Mt. Wilson spelled the definitive death-knell for the large refractor. The reflector was superior to the 40-inch refractor even for visual planetary observation, apart from its potential in photography at both Newton and Cassegrain foci. Together with W. Herschel's 20-foot focus telescope, Ritchey's 60-inch was arguably the greatest relative advance in astronomical observing potential ever achieved. None of the weaknesses of the Melbourne reflector prevented this potential form being fully realised. Ritchey himself was involved in the observing success and maintained close contact with the function of the telescope. With numerous aspects of modernisation, it is still doing good work today, 83 years after its "first light".

Anticipating the success of the 60-inch telescope while it was still to be built, Hale ordered in 1906 a similar blank of *100-inch* (2.54 m) diameter which was delivered in 1908. Because of the high glass mass (4.5 t), the melted glass was poured from 3 separate pots leading to bubble concentrations at the joints. Ritchey at first refused to work the blank for fear the bubbles could provoke a breakage. Three further castings failed or were too thin, so Ritchey agreed to work the original blank after a statement by Day that the layer of bubbles strengthened rather than weakened it. Optical work started in 1910 and took over five years. Test procedures were similar to those used for the 60-inch, except that a quantitative Hartmann test was added. The aspect ratio of the blank (diameter to edge thickness) was 8.0 and the relative aperture of the finished mirror f/5.1. The telescope has an English cradle-type mount (Fig. 5.15). This does not allow access to the pole but offered advantages of symmetry and rigidity. The optical forms available with Newton, Cassegrain and coudé foci were very similar to the 60-inch (f/5.1, f/16 and f/30).

The first test on the sky in 1917 by Hale and Adams [5.3], observing Jupiter, apparently condemned the telescope as a failure, giving an extremely

Fig. 5.14. Ritchey's 60-inch Mt. Wilson reflector, 1908 (courtesy Donald Osterbrock and the Observatories of the Carnegie Institute of Washington)

poor image. This was due to lack of experience with the thermal inertia of a $4\frac{1}{2}$ ton primary cast in classical plate glass with relatively high expansion coefficient. The problem was caused by preparatory work during daytime with sunlight and insufficient cooling time for the mirror at night. Just before dawn, an excellent image of the star Vega was observed. Classical plate glass blanks gave an immediate measure of the thermal equilibrium. Later blanks in low- or zero-expansion materials were not sensitive; but we shall see that the consequences for *dome seeing* could be just as disastrous.

The light grasp of the 100-inch Mt. Wilson (Hooker) telescope was almost 3 times that of the 60-inch. Together, these two telescopes transformed astrophysics and cosmology. The observations by Hubble and Humason in 1924 of M31 in Andromeda, and other galaxies, resolved the outer stars sufficiently well to enable Cepheid variables to be observed and their distances determined. They proved definitively the "island universe" theory of William Herschel that spiral nebulae were external galaxies similar to our own Milky Way system. This led in 1929 to Hubble's redshift law of the expansion of the universe and the Big Bang theory of cosmology.

Recently, the 100-inch telescope has been taken out of service whereas the 60-inch is still working. The pioneer work and greatest achievement of Ritchey was really in the 60-inch, the optics of the 100-inch being the extension of a brilliant concept and professional work of the highest level. Osterbrock [5.19] gives a fascinating account of the collapse of Ritchey's relations with the Mt. Wilson management during and after the completion of the 100-inch and his subsequent, tragic estrangement from the American astronomical establishment.

As after Herschel in 1789 and Rosse, Lassell and Foucault 60 years later, a period of consolidation followed the building of the 100-inch telescope before larger sizes were attempted. Notable telescopes were the 72-inch (1.83 m) Victoria reflector (primary f/5.0), the optics being made by Brashear, completed in 1919; the 69-inch (1.75 m) Perkins reflector, 1932 (Fecker, primary f/4.3); the 74-inch (1.88 m) Dunlap reflector (Grubb-Parsons, primary f/4.9), and the 82-inch (2.08 m) McDonald reflector (Lundin, primary f/4.0). Beautiful and powerful though these telescopes were, their optics represented no significant advance over Ritchey's achievements. The most notable feature was the use of *Pyrex* (low expansion glass – see RTO II, Chap. 3) blanks for the primaries of the Dunlap and McDonald telescopes, completed in 1933 and 1939 respectively. The McDonald telescope had the fastest large primary (f/4.0) yet made; but this was a fairly modest advance over the 100-inch f/5.1, bearing in mind Ritchey's 24-inch, f/3.9 finished in 1901! Although of smaller size, two telescopes by Carl Zeiss Jena were noteworthy, a 1 m, f/3.0 telescope for Bergedorf in 1911, and a 1.22 m with Newton f/6.9 and Cassegrain f/19.7 in 1915. The latter (subsequently moved to the Crimea) was the largest telescope in Europe for 30 years, but, in contrast to the very

Fig. 5.15. The 100-inch Hooker telescope at Mt. Wilson (1917) (courtesy Deutsches Museum, Munich, and acknowledgement to the Observatories of the Carnegie Institute of Washington)

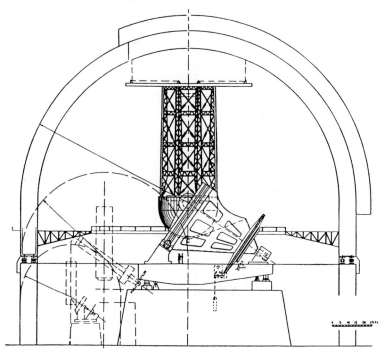

Fig. 5.16. Pease's concept for a 300-inch (7.5 m) telescope in 1921 (reproduced from Dimitroff and Baker [5.17], courtesy Churchill Livingstone, Edinburgh)

short telescope for Bergedorf, was an outdated optical concept (too long) compared with Ritchey's telescopes.

Ritchey's dream was to build still larger telescopes than the 100 inch. He already understood two central problems: the manufacturing and, above all, thermal problems of larger massive (solid) glass blanks; and the problems of field coma with faster parabolic primaries. In 1921 Pease [5.20] made proposals for a 300-inch (7.5 m) telescope (Fig. 5.16). Ritchey moved in 1924 to Paris [5.19] and worked on lightweighted blanks, taking up the idea in glass proposed originally by Rosse for metal. He produced two 30-inch blanks [5.3] and proposed to use such mirrors (Fig. 5.17) for a vertical siderostat telescope (suggested by Foucault in 1869) using a 6 m plane mirror to feed a 5 m primary. His contact with Chrétien had convinced him that the aplanatic form of the Cassegrain, which became known as the Ritchey-Chrétien (RC) form, was the optimum for such a giant telescope. In 1927, while working in France [5.3] [5.19], he successfully made the first RC telescope, with the modest aperture of 0.5 m.

Returning to the USA in 1930, he concentrated on the manufacture of a larger RC telescope of 1 m aperture, discussed in Chap. 3. The data, as quoted by Bahner [5.21], reflected Ritchey's deep perception of the photographic

Fig. 5.17. George Willis Ritchey in Paris, 1927, with a built-up cellular mirror disk (courtesy D. E. Osterbrock, photograph by James Stokley)

requirements of the time with f/4.0 to f/6.8 ($m_2 = -1.7$) giving an aplanatic field of $1.5°$ diameter. The value of R_A was about 0.4, so we have from Eq. (3.114) combined with (2.72)

$$(b_{s1})_{RC} = -1 - \frac{2R_A}{(1 - R_A)m_2^2} = -1.46 \,,$$

a very high asphericity because of the low value of m_2. This extreme form of RC, giving relatively high obstruction and increased difficulty of figuring and testing, was not chosen without good reason: Ritchey wished to achieve the highest speed possible in the RC focus for photography. Such an eccentric primary at f/4.0 was, at that time, at the limit of even his technological possibilities. The telescope was finished in 1934 and represented yet another brilliant technical achievement, effectively the last in his remarkable career before he retired in 1936 [5.19]. An excellent photograph of Ritchey with this primary is shown by Ingalls [5.22].

The consequences of Ritchey's advanced design were not fully understood at the time and the telescope did not produce notable photographic results until much later [5.19]. In 1971 the original optics were replaced by RC optics of essentially the same design, but made from ULE fused quartz to lower the thermal sensitivity, and with a substantially larger secondary to reduce the vignetting in the field which was, above all, a serious disadvantage for photographic photometry[3]. The centering tolerances of the Ritchey layout are appreciably more critical than for normal Cassegrain telescopes, which has sometimes led to the assumption that this is a general negative property of RC telescopes. This is not a correct interpretation, as we may see from Eq. (3.364) for the lateral coma (mm) produced by transverse decenter δ (mm) of an RC telescope:

$$(Coma_t)_{\delta,RC} = -\frac{3}{16}\frac{m_2}{N^2}\left[(m_2^2 - 1) + \left(\frac{1}{1 - R_A}\right)\right]\delta$$

The first term in the square bracket is that corresponding to a classical Cassegrain (parabolic primary) and the second term is the RC supplement. Converting to arcsec with

$$(Coma_t)_{arcsec} = (Coma_t)_{lat}\frac{206\,265}{f'}\ ,$$

we have in Table 5.1 the decentering coma values in arcsec corresponding to $\delta = 1$ mm for the case of the 60-inch Mt. Wilson, 80-foot focal length Cassegrain focus ($m_2 = -3.2$) and the case of Ritchey's RC telescope. We see from this table that the increase in coma due to the optical geometry of the Ritchey 40-inch *as a classical telescope* is a factor 2.16 whereas the increase due to its (extreme) RC form is a further factor of 1.88. As illustrated in Chap. 3, the latter factor is only about 1.05 for modern RC telescopes with much higher m_2 and lower R_A.

---

[3] The original Ritchey optics were lent to the Vienna Observatory in 1971 and were subsequently (1978) used in a 1 m telescope named after Doppler.

I am deeply grateful to Dr. D. Osterbrock of Lick Observatory and Dr. R. Walker of the US Naval Observatory for the above information concerning the *motivation* for replacing the original Ritchey optics and the *nature* of the change. The thermal sensitivity of the original optics was probably mainly due to Ritchey's use of a "low-expansion glass" from St. Gobain with expansion "about one-half that of crown glass". This was a useful improvement in expansion compared with the 60-inch and 100-inch plate-glass blanks, but was inferior to the Pyrex of the time. The strong vignetting of Ritchey's secondary was probably due to his desire to reduce the central obstruction to a minimum and the fact that photographic photometry only later developed its full potential.

Table 5.1. Comparison of the lateral decentering coma with $\delta = 1$ mm of the 60-inch Mt. Wilson telescope and Ritchey's 40-inch RC telescope for the US Naval Observatory

Case	Transverse decentering coma (arcsec) $\delta = 1$ mm
60-inch Mt. Wilson 80-foot focus classical Cassegrain ($m_2 = -3.2$; f/16)	0.183
Ritchey's 40-inch RC ($m_2 = -1.7$; $R_A = 0.4$; f/6.8) a) As RC b) As a classical Cassegrain	0.744 0.395

The manufacturing achievement of Ritchey with the 40-inch RC telescope should be judged from the fact that this was the first time in 300 years that a primary had been manufactured with a form other than the parabola prescribed by Descartes, apart from his own experimental RC telescope of half the aperture completed in France in 1927. Ritchey had to rely largely on zonal testing at the centre of curvature since an autocollimation test against a plane mirror at the focus was no longer an automatic null test.

In the 1930's, two *Schwarzschild* telescopes were manufactured in the United States, a 24-inch for the University of Indiana and a 12-inch for Brown University [5.17]. The problems of obstruction and length made the Schwarzschild form less attractive than the RC form.

Apart from Ritchey's and Pease's ideas for giant telescopes, Hale himself was determined to advance further than the 100-inch Mt. Wilson telescope. Although he was attracted to the visionary plans of Pease for a 300-inch telescope, he finally settled for a *200-inch* [5.2] [5.3]. This was a wise decision, as it already represented a step of a factor of two in diameter. The whole history of the reflector showed that even this represented a high risk. The fundamental issue, as always, was the nature of the primary mirror. This decision (on its structure) was taken in 1928, the site on Mt. Palomar was decided in 1934. We shall confine ourselves here to a brief account of the *optical* aspects of this remarkable telescope. Apart from excellent accounts by King [5.2] and Riekher [5.3], an overall review was given by Bowen [5.23]. General accounts are given by Woodbury [5.24] and Wright [5.25].

Hale and his colleagues, reinforced by the experience of Ritchey, realised that a massive glass blank of the 100-inch type would not only weigh about 40 tons but would lead to insoluble thermal problems – a nine year cooling period of a massive Pyrex blank was estimated! Many materials were considered, the first choice being fused quartz and the second Pyrex. Pyrex was made

by Corning for ovenwear and had normally an expansion about one third of
that of plate glass, while quartz was some 6 times as favourable as Pyrex (see
RTO II, Chap. 3). Between 1928 and 1931, Elihu Thomson of G.E.C. pursued
the development of fused quartz and made two 60-inch blanks. Because of
the high cost of the programme and the limited size achieved, fused quartz
was abandoned in favour of Pyrex. It was not until 20 years later that the
solution of welding together segments of fused quartz led to the use of this
material in large blanks.

Pease had made a design for a lightweighted blank with ribs and cylindrical bores for the support system. This reduced the weight of the 200-inch
(5 m) blank by a factor of 2 to about 20 t, the worked face being only 12 cm
thick (Fig. 5.18). The contract was given to Corning in 1931. A form of Pyrex
was used with 80.5% SiO_2 content [5.3] giving an expansion coefficient only
about 3 to 4 times that of fused quartz [5.2] instead of the normal 5 to 6 times.
Several intermediate blanks were cast, including a 120-inch finally used for the
Lick 120-inch reflector. The first 200-inch blank was cast in March 1934, but

Fig. 5.18. The primary of the 200-inch Mt. Palomar telescope in testing position
with John A. Anderson (left) and Marcus H. Brown (courtesy Palomar/Caltech)

was considered unacceptable because mould cores broke loose and remained in the blank. The cores were then cooled and fixed with chrome-nickel steel attachments for the second casting in December 1934. This was successful, as was the cooling and annealing over an 8-month period. In March 1936, this second blank was transported by rail to Los Angeles. Hale died in 1939 and did not see the completion of the project. But the successful completion of the blank was the most fundamental and critical step, justifying his confidence in the entire project.

For the optical work, starting in 1936, much of Ritchey's technology was taken over. The basic test in achieving the parabolic form was again a zonal test of the differences of radii of curvature. Aspherising was a much greater task than with Ritchey's 100-inch primary because the relative aperture was $f/3.3$, easily the fastest large mirror ever figured at that time. The difference between sphere and parabola according to Eq. (5.1) is 0.136 mm. This was too large to be removed by polishing alone; alternate fine grinding and polishing for testing was performed, a slow and laborious process directed by Anderson, Brown and Hendrix. The second world war caused a further delay and the mirror was not finished until 1947.

Apart from the introduction of low expansion borosilicate glass (Pyrex), the 1930's saw another major revolution in the technology of the reflecting telescope: vacuum evaporation of aluminium to replace chemical silvering. Apart from much improved reflectivity in the UV, the protection by translucent oxide gave a coating much more robust against tarnish (see RTO II, Chap. 6). A special coating plant for Al was built in Palomar. The primary was first tested in the telescope by Anderson in December 1947. Lengthy adjustments of the complex support system followed. This support system comprised 36 supports, combining both axial and lateral functions, inset into the cylindrical bores. The mirror had been left with a "turned-up edge", a trend often encouraged by opticians to avoid the dreaded "turned-down edge", requiring repolishing of the whole surface for its correction. It was hoped that the turned-up edge would compensate in function in the cell. This was not the case. In May 1949 the primary was removed and the edge zone retouched by hand, only 9 hours work but spread over 6 months because of careful testing. In November 1949 it was declared finished and the telescope started operation in 1950.

Figure 5.19 shows one of R.W. Porter's famous drawings of the Palomar telescope. The horseshoe equatorial mounting was a direct modification of the 100-inch cradle mount to permit access to the polar region. The aperture was large enough, for the first time, to include a prime focus cage for the observer, rather than a Newton focus. As discussed in Chap. 4, the PF field was limited to about 12 mm diameter by field coma and this was extended by factors up to about 12 by the Ross correctors giving final apertures from f/3.6 to f/6.0. The Cassegrain system was f/16, giving $m_2 = -4.85$, quite typical of present day telescopes. The coudé focus (f/30) is reached either by one plane

LIGHT PATH TO PRIME FOCUS f 3.3
=CASSEGRAIN = f 16
COUDE = f 30

APPROXIMATE SCALE

R. W. PORTER. '38.

THE TWO HVNDRED INCH TELESCOPE

Fig. 5.19. The Palomar 200-inch (5 m) telescope as drawn by R.W. Porter (courtesy Palomar/Caltech)

mirror of variable rotation angle sending the beam down the polar axis; or via three plane mirrors if the telescope is observing near the pole [5.3]. An important feature for the optical quality was the so-called Serrurier truss, a tube design due to M. Serrurier which compensated lateral decentering coma by equal lateral sags of the tube at the secondary and primary. This is discussed further in RTO II, Chap. 3 in connection with active optics. The success was measured by the result that, for all positions of the tube, the focus of the primary did not vary from its mean position by more than 25 μm [5.2].

In 1950, one of the first photographs of a galaxy using the Ross field corrector was published [5.26]. It showed NGC 147 clearly resolved, the smallest star image being slightly more than 0.5 arcsec in diameter, confirming the excellent quality of the primary and the potential of the telescope. The 100-inch could only resolve this "nebula" with red-sensitive plates. Riekher [5.3] quotes the workshop tested quality of the Palomar primary as *68% of the geometrical optical energy within 0.5 arcsec and 95% within 1.0 arcsec*, which agrees well with the above result, bearing in mind the addition of the Ross corrector.

From the point of view of telescope optics, the following features of the 200-inch were particularly notable compared with its 100-inch predecessor:

- The primary material (low expansion Pyrex)
- The primary structure (lightweighted)
- The primary relative aperture (f/3.3 instead of f/5.1)
- The primary support (36 supports performing axial and radial functions – see RTO II, Chap. 3)
- The Al reflecting coat
- The use of field correctors in the PF
- A very large dome and building including all facilities

The next section will show what a remarkable advance it was, bearing in mind that all of these features were determined in the 1930's.

5.3 Reflectors after the 200-inch Palomar Telescope up to about 1980

As always, after a major advance, there followed a period of consolidation, prolonged in this case by the aftermath of the second world war.

One of the earliest projects was also the biggest of all: the *USSR 6 m* telescope in the Caucasus mountains (Mt. Pasthukhov, Zelenchuk). The optical concept essentially goes back to Maksutov in 1952, who proposed a 6 m parabolic primary working at f/4.0. Maksutov considered a steeper primary would give unacceptable problems of aspherising. Now Eq. (5.1) shows that the aspherisation required is a linear function of aperture and an inverse cube function of the f/no; therefore that required for an f/4 mirror of 6 m aperture is only about two thirds of that for an f/3.3 mirror of 5 m aperture. Viewed like this, since the successful results of the Palomar 5 m telescope were known in 1952, the optical layout at f/4 must be seen as cautious. More important was the nature of the 6 m blank. Like Couder in France [5.27], Maksutov had been experimenting with *metal* mirrors since the thirties [5.3]. Speculum (bronze) is too heavy to be of modern interest, but aluminium, stainless steel and beryllium, among other possibilities, are extremely interesting because of their better *thermal conductivity* than glass. These possibilities are discussed in detail in RTO II, Chap. 3. Maksutov pleaded for a metal mirror for the 6 m. However, the choice finally was a low expansion glass similar to Pyrex. A meniscus form was chosen with a thickness of 650 mm, without lightweighting. The glass volume was thus three times higher than for the Palomar telescope with correspondingly longer cooling period. The blank manufacture was performed between 1963 and 1968, the weight being 42.7 tons. The optical work was finished in 1974. Careful test results, both with an interferometer and by Hartmann testing, gave 62% geometrical energy concentration in 0.5 arcsec and 91% in 1.0 arcsec, values almost as good as Palomar. The first tests in the telescope were performed in 1975. Since the demands on optical quality had been made more severe for other telescopes of this period compared with the Palomar 5 m (a specification from the thirties), a second borosilicate blank was cast and worked, to higher quality. This was inserted in the telescope in 1979. However, the thermal inertia of a massive glass blank of this size is so large that this set the practical limit of quality achievable. It was intended to replace this second borosilicate blank by a blank in *glass ceramic* (Russian "Astro-Sitall"), a zero expansion material which had become available about 1970 in the western world (see RTO II, Chap. 3). Apparently, it has not proved possible to cast such a huge glass mass in this material without breakage, which is confirmed by the experience of Schott in Germany (RTO II, Chap. 3). The solution would be a much thinner blank controlled actively (RTO II, Chap. 3), but the 6 m telescope is a concept of an earlier period. I visited the telescope in 1984 and was given information on the quality in a very open and friendly spirit. The three aspects limiting the optical quality are apparently:

– The thermal inertia of the mirror
– The thermal conditions in the dome
– The site, for which the seeing was less good than hoped

The average image quality was quoted as better than $2\frac{1}{2}$ arcsec and the best about $1\frac{1}{2}$ arcsec.

From an *optical* point of view, the 6 m telescope (Fig. 5.20) is less advanced than the 5 m Palomar telescope. But from the point of view of *mechanics* and *controls* it was epoch-making. This was above all due to B. K. Ioannisiani, the most determined proponent of the *Alt-Az* mounting [5.3]. The control system to solve the 2-axis tracking problem was the work of N. Michelson and O. Melnikov. This bold decision was a great success and a strange contrast

Fig. 5.20. The Russian 6 m telescope at the Zelenchuk Observatory in the Caucasus (courtesy "Ciel et Espace", Paris, through Serge Brunier)

with the western world where, in spite of a fully developed industrial base in electronics, the courage to take the logical step back to the compact and gravity-symmetrical Alt-Az mount was lacking for a whole generation. For this reason alone, the 6 m will go into telescope history as a great telescope. Furthermore, we must remember that the somewhat mediocre optical quality is compensated by the huge size.

Another, smaller telescope of the post-Palomar epoch should be mentioned: the *3 m Lick reflector*. This was commenced in 1946 on the basis of the availability of a 3 m Pyrex blank originally cast in the Palomar 5 m blank development programme (see §5.2). Originally, it had been intended for an autocollimation test flat, but this was abandoned. The thickness of this blank was such that a relative aperture no faster than f/5 could be made, giving a basic optical geometry and tube format no shorter than those of the Mt. Wilson 60-inch and 100-inch telescopes. The telescope is just big enough for a PF cage. A notable aspect of the optics of this telescope is that the first really complete scientific Hartmann test of a large reflector in function was performed by Mayall and Vasilevskis [5.28] (see RTO II, Chap. 2), following the forerunner work of Bowen [5.29] on the 200-inch Palomar telescope. In their most favourable test at the PF, the final quality of the primary was given by the geometrical energy concentrations 70%, 95% and 97% in 0.34, 0.68 and 1.35 arcsec diameter respectively. This set a manufacturing quality standard of $d_{80} \simeq 0.40$ arcsec for the 80% energy concentration which was applied for a whole class of succeeding telescopes.

The optical geometry of a whole generation of telescopes was effectively determined by a notable paper by Bowen in 1967 [5.30]. He considered the optical layout of a telescope at a time when the photographic plate, as the basic detector for nearly 100 years, was being rapidly replaced by photomultipliers for single objects or image-intensifier tubes of various sorts for small fields. The long refractors with f/15 to f/20 were admirably suited to visual observation if supplemented by a relatively cheap battery of eyepieces. The slow photographic plates at the beginning of the century forced the revolution in the speed of primaries because exposure times of galaxies and nebulae for speeds slower than about f/6 were unacceptably long. This dominated Ritchey's thinking and led to his first RC telescope with f/6.8 following his f/5 primaries for the 60- and 100-inch classical telescopes. Bowen's analysis rationalised the known fact that, while the *rate* of accumulation of quanta of a photographic plate depended only on the *aperture*, the *limiting magnitude*, as determined by the information capacity of the photographic plate, is primarily a function of the *focal length*. For the "unbaked" plates of the time, with an efficiency less than 1%, Bowen concluded that the focal ratio for *direct imaging at the Cassegrain focus should be f/8 - f/10*. For spectroscopy, which is less critical in the sense that the sky background is so diluted as to become negligible, such a focal ratio was favourable to pixel matching and echelle spectrographs. These f/nos were also suitable for photomultipliers. At the prime focus, he considered the focal length for direct imaging should be about 6–10 m, giving in 100-inch to 200-inch telescopes relative apertures between f/1 and f/4.

Out of this analysis emerged the concept of the *Bowen-type telescopes* with an optical geometry of about

$$f/3, f/8, f/30$$

in the prime, Cassegrain and coudé foci respectively. The magnification $m_2 = -8/3$ gave reasonable obstruction ratios for normal image positions of the Cassegrain image behind the primary. From Eq. (2.86)

$$R_A = \frac{\bar{b} - 1}{m_2 - 1} ,$$

with $\bar{b} = b/f_1'$, b being positive and f_1' negative (see Fig. 2.12). With $\bar{b}$ typically about -0.13, we have $R_A \simeq 0.31$, giving with normal fields an obstruction from the secondary of the order of 0.35. A PF of f/3 was only 10% steeper than the 200-inch Palomar primary, but few opticians had experience of working large mirrors as steep as this and there was a general reluctance in the 1960's to go much steeper than Palomar.

The Bowen geometry was a big driver for the *RC telescope* since classical photography with large plates at the Cassegrain focus was considered an essential feature, requiring fields of $0.5°$ diameter or more. In view of the success of the Ross correctors at the PF of the Palomar 200-inch, interest in PF field correctors was strong and converged out of the work described in Chap. 4 on a general requirement of a $1°$ diameter field for classical photography. The photographic plate experienced a certain resurgence of interest with the introduction of "baking" plates in nitrogen or forming gas to drive out water and increase the quantum efficiency from less than 1% to possibly 4-5% at maximum. A number of optical forms emerged for a range of telescopes all having apertures in the class 3.5–4 m, thereby large enough to permit a PF cage. The solutions chosen ranged from a classical form with a parabolic primary through a strict RC form to various quasi-RC forms, usually with primaries somewhat more eccentric than that prescribed by the RC form. The motivations were essentially as follows:

a) *Classical telescope with parabolic primary*

The principal motivation was the use of the naked primary without corrector for direct imaging at the PF. The limitation of field coma is, of course, very severe at f/3. From (3.87) we have for the field coma of the parabolic primary

$$(S_{II})_1 = -\left(\frac{y_1}{f_1'}\right)^3 \frac{f_1'}{2} u_{pr1} ,$$

which can be written in the form

$$(S_{II})_1 = -\left(\frac{1}{2N_1}\right)^3 \frac{f_1'}{2} u_{pr1}$$

Converting into angular aberration from (3.198) gives with $n' = -1$ and u_{pr1} expressed in arcsec

$$(\delta u_p')_{Coma_t} = -\frac{3}{16}\frac{1}{N_1^2}(u_{pr1})_{arcsec} \text{ arcsec} \tag{5.4}$$

With the Bowen prescription of $N_1 = -3.0$, Eq. (5.4) gives a field coma of 1 arcsec for a field diameter of *96 arcsec*. For a primary with $D = 3.5$ m working at this relative aperture, the corresponding linear field diameter is only *4.9 mm*. Apart from the fact that a coma limit of 1 arcsec is generous by today's standards, such a linear field is very small for practical use and demands very accurate centering of a detector of similar size.

A second motivation often quoted is that the classical telescope is easier to manufacture. This matter is dealt with in RTO II, Chap. 1. With modern methods of manufacture there is a negligible advantage of a parabolic primary over an RC for the primary and only a modest advantage for the secondary: this argument is no longer valid.

So far as correctors are concerned, as was shown in Chap. 4, there is no advantage of the classical telescope, either in the prime or Cassegrain foci.

The only significant advantage is for the coudé with its high N, typically f/30 or longer. It should be recalled that the classical and RC solutions converge for an afocal system. It follows that the classical telescope is nearly aplanatic for coudé foci with high m_2 values, whereas RC solutions have considerable field coma. However, this would not normally be decisive as coudé systems work with fixed spectroscopic equipment and small fields.

b) *RC solution*

This has the obvious advantage of being aplanatic in the Cassegrain focus, giving good field correction over about 0.5° diameter if field curvature is ignored and astigmatism is the limitation.

The over-correction of the primary is a help in the design of PF correctors (see Chap. 4), although the advantage is modest for modern systems with high values of m_2.

The coudé is less favourable for field coma, but this may not have much weight in the decision (see a) above).

c) *Quasi-RC solutions*

These were discussed in Chap. 4 and consist in optimizing the Cassegrain focus by departing from the RC form when using a corrector. Normally, the axial performance (spherical aberration) is negligibly affected by the corrector.

Such solutions given optimum performance at the Cassegrain focus with corrector, e.g. with a Gascoigne aspheric plate and field flattener. Without corrector, there is some field coma.

Since the primary asphericity is normally somewhat increased, these solutions are also more favourable for PF correctors.

The coudé is slightly worse than the RC solution, but the difference is small.

The telescopes listed in Table 5.2 show examples of all these solutions. The successful manufacture of a number of RC or quasi-RC telescopes of modern design to high optical standards must be seen as a major consolidation of telescope optical technology following the Palomar 5 m, even though all these telescopes were smaller. Equally notable, probably even more so, was the application of *zero-expansion* or *nearly zero-expansion* glasses to the optics (RTO II, Chap. 3). In the 1950's and 1960's, the technology of fusing quartz segments (boules) was perfected, the first major blank being the 4 m blank for the Kitt Peak Mayall telescope, made by the General Electric Company. Normal fused quartz has an expansion coefficient of about $5 \cdot 10^{-7}$, about one sixth that of borosilicate glass (Pyrex). Glass ceramic was invented in the 1950's at Corning, but since then Corning has concentrated on fused quartz blanks, above all in the form of ULE quartz (Ultra-Low Expansion) with an expansion coefficient of about $1 \cdot 10^{-7}$, only about 1% of that of plate glass. Glass ceramic has a coefficient effectively zero, but is slightly temperature dependent. Originally made by Owens Illinois under the name Cervit, it is now made above all by Schott in Germany under the name Zerodur, also in Russia (Sitall) and China. These materials are discussed in detail in RTO II, Chap. 3.

Table 5.2 is very instructive in revealing the trends of the time. The most significant advances in the optics were the switch to very low or zero expansion glasses for the primaries and the successful manufacture of RC or quasi-RC mirror forms. The blanks are all massive as compared with the lightweighted type for Palomar. This was no longer significant for blank distortion from thermal effects, but has often proved disadvantageous from the effects on the local air due to the high thermal inertia of the primaries. The ability to manufacture RC systems successfully was essentially due to null (or compensation) testing introduced by Dall in 1947 [5.46] [5.47], although the basic idea was already published by Couder in 1927 [5.48] but its significance was not then understood. This will be treated in detail in RTO II, Chap. 1.

Also of significance for the optics is the mounting. The reluctance in the western world to follow the *Alt-Az* development initiated in the 1950's and 1960's for the Russian 6 m telescope has been mentioned above. The retention of equatorial mounts increased the volume and hence the air mass in the domes, thereby giving greater problems with dome seeing. The 4.2 m William Herschel Telescope (WHT) was the second large telescope to have an *Alt-Az* mount. It also pushed the f/no of the primary to f/2.5. This advance from f/3.3 of Palomar took 50 years.

The relative conservatism of this period was to some extent explained by a legitimate concentration of effort on electronic detectors. The culmination has been the CCD (Charge Coupled Device) which has increased the efficiency compared with unbaked plates by almost 2 orders of magnitude. This has

Table 5.2. Principal optical characteristics of major telescopes following the Palomar 5 m telescope and with a conventional optical concept

Telescope	Diameter (m)	Completion date	Blank material	Blank nature	f/no primary	Optical form	References	Mount
Palomar	5.0	1949	Borosilicate (Pyrex)	Lightweight	3.3	Classical	[5.2][5.3][5.31]	Eq.
Lick	3.0	1959	Borosilicate (Pyrex)	Lightweight	5.0	Classical	[5.3][5.31]	Eq.
INT {Herstmonceux / La Palma}	2.5	{1967 / 1983}	{Borosilicate (Pyrex) / Glass ceramic (Zerodur)}	{Massive³ / Massive}	3.0	Classical	{[5.3][5.32] / [5.33][5.34]}	Eq.
Zelenchuk/Caucasus (USSR)	6.0	1976	Borosilicate	{Massive (meniscus)}	4.0	Classical	[5.3]	Alt-Az
Kitt Peak (USA)	4.0	1973	Fused quartz	Massive	2.8	RC	[5.3][5.35]	Eq.
AAT (Australia)	3.9	1974	Glass ceramic (Cervit)	Massive	3.3	RC	[5.3][5.36]	Eq.
Cerro Tololo (Chile)	4.0	1976	Glass ceramic (Cervit)	Massive	2.8	RC	[5.3][5.37]	Eq.
ESO La Silla (Chile)	3.6	1976	Fused quartz	Massive	3.0	Quasi-RC	[5.3][5.38][5.39]	Eq.
CFHT (Hawaii)	3.6	1979	Glass ceramic (Cervit)	Massive	3.8	Classical	[5.3][5.40]	Eq.
UKIRT (Hawaii)	3.8	1979	Glass ceramic (Cervit)	{Massive (thin: edge AR=13)}	2.5	Classical	[5.3][5.41][5.42]	Eq.
MPIA (Calar Alto)	3.5	1985	Glass ceramic (Zerodur)	Massive	3.5	RC	[5.3][5.43]	Eq.
WHT (La Palma)	4.2	1988	Glass ceramic (Zerodur)	Massive	2.5	Classical	[5.3][5.44][5.45]	Alt-Az

³ The original blank of the Isaac Newton Telescope (INT) primary was made by Corning in 1936 for the Michigan Observatory in the course of work initiated for the Palomar 200-inch telescope, but the proposed telescope was never built. The blank was presented to the INT project in 1949 [5.33]. I am extremely grateful to Dr. R. Bingham for information about this blank and confirmation that it was *massive*. He also kindly drew my attention to the publication [5.34], the most complete on the original INT.

meant the photon efficiency of a 50 cm telescope equipped with a CCD can rival the Palomar 5 m with classical plates!

Site selection also made a major advance. The sites in Chile established 1 arcsec as the criterion of "good" seeing. As a result, the specification of all the Bowen-type telescopes was $d_{80} \approx 0.5$ arcsec at the Cassegrain focus. Polishing and test techniques were developed to deliver this quality even with primaries in the range f/3.0 - f/2.5.

Apart from the WHT, because of its more advanced *Alt-Az* mount, none of the 6 standard Bowen-type telescopes listed in Table 5.2 has any major feature, either optical or mechanical, which distinguishes it from the others. However, such telescopes allowed the Europeans to recover their position, in both telescope technology and observation, which they had lost to the USA at the turn of the century. As arbitrary examples, two of these telescopes are shown here, the 4.0 m Kitt Peak telescope (Fig. 5.21; see also Pasachoff [5.49]) and the 3.6 m ESO telescope (Fig. 5.22). The building-dome concept of all these telescopes was also similar, following closely the Palomar 5 m tradition (see Fig. 5.23).

The Bowen telescopes [5.30] were originally conceived for photography with plates at the Cassegrain (RC) focus. In fact, although most of them had provision for such photography, very little of it has been done at these foci because photography at the PF at about f/3 was some 6 to 10 times more rapid for extended objects. PF correctors have therefore had much more significance for these telescopes than Cassegrain correctors. As we discussed in detail in Chap. 4, PF correctors are rendered easier by the overcorrection of the RC primary; but modern RC designs with high values of m_2 have only small departures from the parabolic form, so this advantage is no longer very significant. This situation has led to statements, e.g. by Learner [5.50], that the RC telescope had run its course by 1980 after a relatively brief 50 year period of partial supremacy over the classical Cassegrain. This is a view I personally do not share, since the RC form is no more difficult to manufacture with modern techniques and the field quality has technical advantages in active telescopes (see RTO II, Chap. 3), apart from the astronomical significance of better correction in the field for rapidly increasing electronic detector sizes.

The 4.2 m WHT (Fig. 5.24) followed this trend in the Anglo-Saxon world and has the classical Cassegrain form. It was the last, and most notable telescope to be built by the famous firm of Grubb-Parsons, which ceased operations after its completion about 1985, thus ending a tradition going back to William Parsons (Lord Rosse) and Thomas Grubb.

A number of large Schmidt telescopes were also built in this period. The Palomar Schmidt with 1.22 m aperture of the corrector was completed in 1948 and was a notable advance in size. The mirror has a diameter of 1.80 m and works at f/2.5. In 1960 the 1 m Schmidt telescope for the Bjurakan

Fig. 5.21. The 4.0 m Kitt Peak telescope (courtesy National Optical Astronomy Observatories, Tucson)

Fig. 5.22. The 3.6 m ESO telescope (courtesy ESO)

Fig. 5.23. The building of the 3.6 m ESO telescope with the smaller building of the 1.4 m Coudé Auxiliary Telescope (CAT) on La Silla (2400 m) (courtesy ESO)

Observatory in the USSR was completed with a mirror working at f/2.1. The nominal fields of these telescopes were 5° to 6° diameter.

The largest Schmidt telescope yet built emerged from the difficult postwar period in the (then) GDR. The concept was laid down by Kienle [5.3] in 1949 for a "Universal Telescope" combining the properties of Schmidt and classical telescope conceptions, since the means for building separate instruments were not available at that time. Figure 5.25 shows the original concept. The Schmidt corrector plate has a diameter of 1.34 m free aperture, the largest yet made. The glass is UBK7 and the thickness 40 mm. The spherical primary has a focal length of 4 m giving an f/3 Schmidt telescopewith a field of 5° diameter. The Schmidt plate can be removed allowing, in the original concept, an f/2 PF with a doublet corrector (never realised) for the spherical primary, an f/11 Nasmyth focus and an f/46 coudé. The telescope became

Fig. 5.24. The 4.2 m William Herschel Telescope (WHT) with its *Alt-Az* mounting on La Palma (Roque de los Muchachos 2400 m) (courtesy Royal Greenwich Observatory, through Richard Bingham and Peter Andrews)

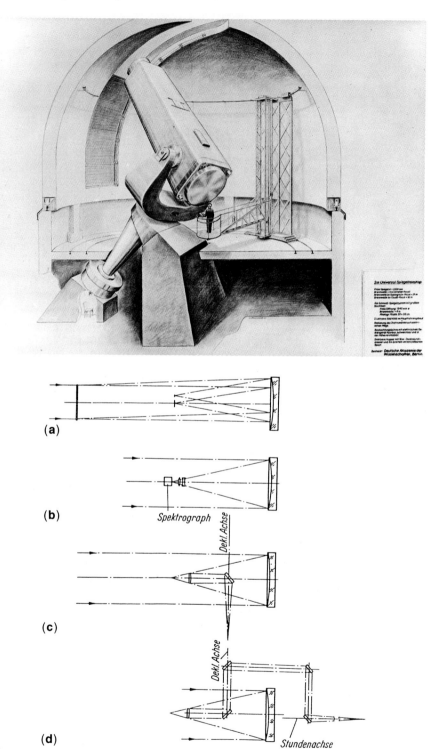

Fig. 5.25. The original optical layout of the "Universal Telescope" of the Karl-Schwarzschild Observatory at Tautenburg, Germany (courtesy Rolf Riekher)

operational in 1960 and has been very successful within the climatic limitations of its site. Since the spherical form of the primary was determined by the Schmidt concept, the two Cassegrain foci are of the "spherical primary" (SP) type treated in §3.2.6.3. This solution has very large field coma, but is acceptable for use with spectrographs provided the slits are accurately placed on the optical axis.

Complex mechanics was required to keep the telescope in adjustment in spite of the radical changeovers operated. This subject will be dealt with in connection with active telescopes (RTO II, Chap. 3). Such a universal concept was a legitimate design in its time but is unlikely to be repeated, although active control would make its realisation easier.

The optical data of Table 5.2 and otherwise in this chapter is intended to give the reader a guide to the larger and most notable telescopes. The literature includes many lists of telescopes, many of them in older books such as Danjon-Couder [5.1] and Dimitroff-Baker [5.17]. An excellent general source of up-to-date information is Riekher [5.3]. A very useful listing of basic data is given by Gochermann and Schmidt-Kaler [5.51]. The most complete listings by far, above all for the optical properties, are by Bahner [5.52] and Wolf [5.53], also Kuiper and Middlehurst [5.31]. Bahner lists all existing reflectors of aperture ≥ 0.9 m for normal telescopes, ≥ 0.5 m for Schmidt telescopes and ≥ 0.4 m for meniscus telescopes. The follow-up work by Wolf lists all normal telescopes ≥ 1.2 m and wide-field telescopes ≥ 0.6 m. Another very useful listing from the optical viewpoint is given by Schielicke [5.54].

The MPIA 3.5 m telescope, which went into operation in 1985 after 15 years of development and manufacturing time [5.43], was the last of the equatorially-mounted Bowen-type telescopes. With its Palomar type mounting, it was also the most massive and most expensive. In spite of the beauty of the optics and mechanics of this telescope, it was by that time generally agreed that telescopes of this size could be built more cheaply and effectively with more modern mounts (usually *Alt-Az*) and new optical and building concepts made possible by modern electronics. These modern developments are the subject of RTO II, Chap. 3. In fact, a copy of the MPIA 3.5 m telescope was built by Carl Zeiss for the proposed observatory in Iraq. For obvious reasons, this instrument has never been installed.

The development of the reflecting telescope in the period covered by this chapter is beautifully illustrated by the work of Fehrenbach [5.55]. This work only covers developments in France and has no pretensions to scientific or technical comprehensiveness; instead, it deals with the background and personalities involved, including the period Ritchey spent in France, in a personal and most revealing way.

A. List of mathematical symbols

Chapter 2

Table A.1. List of symbols for Chapter 2

Symbol	Meaning	Where defined
A	Aperture stop	Fig. 2.6
a_1, $a_2 \cdots$	Coefficients for the definition of a centered surface	Eq. (2.13)
a_E, b_E, c_E	Positions of the Ramsden disk in a refracting telescope	Fig. 2.8
b	Distance from the pole of M_1 to the final image in a 2-mirror telescope	Fig. 2.11, Fig. 2.12, Eq. (2.65)
$\bar{b}$	Normalized value of b	Eq. (2.87)
C	Beam compression factor	Fig. 2.8, Eq. (2.39)
c	Curvature of a surface $= 1/r$ (see surface number ν)	Eq. (2.13)
c_l	Velocity of light in wavefront propagation	Fig. 2.3, Eq. (2.11)
D	Effective diameter (aperture) of a system; diameter of the entrance pupil	§ 2.2.6
$(D_{AX})_2$	Axial beam diameter of M_2 in a 2-mirror telescope	Eq. (2.93)
$(D_{TOT})_2$	Full diameter of M_2 in a 2-mirror telescope	Eq. (2.93)
d	Axial distance between a given surface of a system and the next surface (see surface number ν)	Fig. 2.11, Fig. 2.12, Eq. (2.37)
E, E'	Entrance, exit pupil of a system	Fig. 2.6
F, F'	Object, image focal point of a system	Fig. 2.1, Fig. 2.2
f, f'	Object-side, image-side focal length of a system	Fig. 2.1, Fig. 2.2

Table A.2. List of symbols for Chapter 2 (continued)

Symbol	Meaning	Where defined
f_1', f_2'	Image-side focal lengths of M_1, M_2 in a 2-mirror telescope	Eqs. (2.54), (2.55), (2.60)
H	The Lagrange Invariant	Fig. 2.5, Eq. (2.27)
H (suffix)	Upper ray	Fig. 2.1, Fig. 2.2
I, I'	Axial object, image positions of a system	Fig. 2.2
I_H, I_H'	Object, image positions associated with an upper ray	Fig. 2.2, Fig. 2.5
I_L, I_L'	Object, image positions associated with a lower ray	Fig. 2.8
I_1, I_2'	Primary and secondary axial image points in a 2-mirror telescope	Fig. 2.11, Fig. 2.12
i, i'	Ray incident angles to a surface before, after refraction or reflection (see surface number ν)	Eq. (2.15), Eq. (2.36)
K	Optical power of a system or element	Eqs. (2.22), (2.51), (2.53)
L	Back focal distance in a 2-mirror telescope (distance from the secondary to the image)	Fig. 2.11, Fig. 2.12 Eq. (2.61), Eq. (2.72)
L (suffix)	Lower ray	Fig. 2.8
LTP	Light Transmission Power (Throughput)	Eq. (2.47)
l (suffix)	The last surface of a system before the image	Eq. (2.38)
M	Mirror (see surface number ν)	Fig. 2.11, Fig. 2.12
M_1, M_2	The primary and secondary mirrors of a 2-mirror telescope	Fig. 2.11, Fig. 2.12

Table A.3. List of symbols for Chapter 2 (continued)

Symbol	Meaning	Where defined
m	Magnification of a system	Eqs. (2.1), (2.2), (2.23), Fig. 2.5
m_2	Magnification of the secondary mirror of a 2-mirror telescope	Eq. (2.55)
N	Relative aperture, f/no	Eq. (2.104)
$n,\ n'$	Refractive index in the object, image space of a refracting or reflecting surface (see surface number ν)	Fig. 2.2 Eq. (2.11), Eq. (2.12)
$P,\ P'$	Object-side, image-side principal points of a system	Fig. 2.1, Fig. 2.2
$P_H,\ P'_H$	Object-side, image-side upper ray points defining with $P,\ P'$ the principal planes of a system	Fig. 2.1, Fig. 2.2
$P_L,\ P'_L$	Object-side, image-side lower ray points defining with $P,\ P'$ the principal planes of a system	Fig. 2.2
P	The distance between the primary and secondary image in a 2-mirror telescope	Fig. 2.11, Fig. 2.12, Eq. (2.81), Eq. (2.84)
p (suffix)	Denotes a field ray, or property thereof, which cuts the axis at P or P'	Fig. 2.5, Fig. 2.6
pr (suffix)	Denotes a principal ray, or property thereof, which cuts the axis at E and E'	Fig. 2.6, Fig. 2.8
R_A	Axial obstruction ratio of a 2-mirror telescope	Eq. (2.58), Eq. (2.72)
r	Radius of curvature of a surface $= 1/c$ (see surface number ν)	Eq. (2.13), Eq. (2.36)

Table A.4. List of symbols for Chapter 2 (continued)

Symbol	Meaning	Where defined
$r_1,\ r_2\cdots$	Defined light rays	Figs. 2.1, 2.2, 2.4, 2.6, 2.8
$\mathcal{S}$	Optical system	Fig. 2.1, Fig. 2.6
S	Scale of a telescope (arcsec/mm)	Eq. (2.102)
$\overline{S}$	Inverse scale of a telescope (mm/arcsec)	Eq. (2.103)
$s,\ s'$	Axial distance from an object, image to an imaging surface or a principal plane (see surface number ν)	Figs. 2.2, 2.5, 2.6, 2.11, 2.12, Eq. (2.4)
T	True telephoto effect of a 2-mirror telescope relative to the length L	Eq. (2.57)
T_p	Telephoto effect of a 2-mirror telescope relative to the primary focal length f_1'	Eq. (2.56)
t	Time (light propagation)	Eq. (2.11)
$u,\ u'$	Ray angles to the axis before, after refraction or reflection (see surface number ν)	Fig. 2.5, Eq. (2.26)
$W,\ W'$	Wavefront (or wavefront aberration) before, after refraction or reflection	Fig. 2.3
$x,\ y,\ z$	Cartesian coordinate system (right-hand set): z is the axial distance, y the height in the principal section	§ 2.2.1, § 2.2.3, Eq. (2.13)
$x',\ y',\ z'$	Coordinates after refraction or reflection	§ 2.2.1, § 2.2.3, Eq. (2.13)

Table A.5. List of symbols for Chapter 2 (concluded)

Symbol	Meaning	Where defined
$\sim$	Parameters for the folded Cassegrain case with M_2 plane ($m_2 = -1$)	Fig. 2.15, Eqs. (2.97) – (2.100)
η, η'	Height in the object, image plane	Fig. 2.2, Fig. 2.5, Eq. (2.1), Eq. (2.23)
ν (subscript)	Counter number for system surfaces and paraxial parameters, e.g. r_ν, y_ν, s_ν, d_ν, n_ν, i_ν, u_ν	Eqs. (2.36) – (2.38)

Chapter 3

Table A.6. Additional symbols for Chapter 3

Symbol	Meaning	Where defined
A, $\overline{A}$	Snell invariant for the paraxial aperture, principal ray (see surface number ν)	Eq. (3.19), Eq. (3.20)
A	Area of a circular aperture for diffraction phenomena	Eq. (3.442)
A'	Real area of a pupil, including vignetting	Eq. (3.466)
A, A_0	Common area, whole pupil area for a sheared pupil	Eq. (3.499)
A, B	Constants of a linear dispersion equation	Eq. (3.316)
A, B	Real, imaginary parts of the autocorrelation function	Eq. (3.503)
A_S, B_S, C_S, D_S ...	Supplementary aspherizing coefficients in the general definition of a surface	Eq. (3.77)

Table A.7. Additional symbols for Chapter 3 (continued)

Symbol	Meaning	Where defined
A_δ, $B_{\delta 1}$, $B_{\delta 2}$	Terms in equations for deriving decentering coma	Eqs. (3.355) – (3.357)
$A_{\alpha\beta}$	Normalization constants for deriving Zernike polynomials	§ 3.9, Eq. (3.426)
$\overline{A_1^*}$	Quantity calculated for the secondary as though it were a primary in a Schiefspiegler	Eq. (3.344)
A (suffix)	Pertaining to third order astigmatism for the Strehl Intensity Ratio	Eq. (3.471)
a, b	Semi-axes of an ellipse	Eq. (3.4)
a	Aspheric plate profile constant (defining its spherical aberration)	Eq. (3.221)
a, b	Upper and lower points of a Schmidt plate	Fig. 3.27
(a), (b)	Stop positions in a Maksutov telescope	Fig. 3.36
a_0, a_I	Normalized amplitude of the object, image function (OTF)	Eq. (3.485), Eq. (3.486)
$Afoc$ (suffix)	Pertaining to the afocal form of a 2-mirror telescope	Eq. (3.98)
$Aplan$ (suffix)	Pertaining to the aplanatic form of a 2-mirror telescope	Eq. (3.106)
b_s	Schwarzschild (conic) constant	Eq. (3.10)
$\overline{b}_s$	Laux definition of conic parameter	Eq. (3.79)
B (suffix)	Pertaining to Bouwers' achromatic meniscus	Eq. (3.311)
Bou (suffix)	Pertaining to Bouwers-type meniscus	Eq. (3.278)
BF (suffix)	Best focus	Eq. (3.185)

Table A.8. Additional symbols for Chapter 3 (continued)

Symbol	Meaning	Where defined
C	A constant in the diffraction integral	Eq. (3.434)
C_1, C_2	Primary longitudinal, lateral chromatic aberration	Eq. (3.222), Eq. (3.223)
C_0, C_I	Contrast for the object, image of a given spatial frequency	Eq. (3.483)
C (suffix)	Pertaining to chromatic aberrations	Eq. (3.222), Eq. (3.223)
C (suffix)	Pertaining to a concentric surface	Fig. 3.36
C (suffix)	Pertaining to third order coma for the Strehl Intensity Ratio	Eq. (3.471)
$Coma_t$, $Coma_s$	Tangential, sagittal coma	Fig. 3.18, Eq. (3.196), Eq. (3.197)
CFP (also suffix)	Coma-free point	§ 3.7.2.3
cl (suffix)	Pertaining to a classical 2-mirror telescope	Eq. (3.93)
DK (also suffix)	Dall-Kirkham form of a 2-mirror Cassegrain telescope	§ 3.2.6.3(c)
d	Thickness of an aspheric or plane-parallel plate or filter; or of a meniscus or lens element of a corrector	§ 3.6.2.3
d_{80}	Diameter of the circle containing 80% of the geometrical energy in the point image (PSF)	Table 3.15
d (suffix)	Pertaining to defocus for the Strehl Intensity Ratio	Eq. (3.471)
dec (suffix)	Pertaining to Gaussian effects of transverse despace, i.e. lateral decenter	Eq. (3.422), Eq. (3.423)

Table A.9. Additional symbols for Chapter 3 (continued)

Symbol	Meaning	Where defined
$\tilde{E}$	Total energy incident on a circular aperture for diffraction phenomena	Eq. (3.442)
EFC	Effective field curvature	Table 3.3, Eq. (3.164)
ExP (also suffix)	Exit pupil in the theory of the CFP	Eq. (3.381)
F	Pupil function in autocorrelation theory of the OTF	Eq. (3.494)
F^*	Complex conjugate of F	Eq. (3.494)
$F(\varepsilon)$	Functions of ε (obscuration factor)	Eq. (3.482)
Fi (suffix)	Pertaining to a plane-parallel plate or filter	Eq. (3.253)
FFL (suffix)	Pertaining to a field-flattening lens	Eq. (3.259)
f (suffix)	Pertaining to *field* aberration in a decentered telescope	Eq. (3.354)
GF (suffix)	Pertaining to the Gaussian focus	Eq. (3.184)
g (suffix)	Pertaining to the approximation of geometrical optics for the MTF calculation	Eq. (3.507), Eq. (3.508)
HE	Pupil position parameter (see surface number ν)	Eq. (3.19)
$H\partial E$	Stop shift parameter	Eq. (3.22)
I	Light intensity (Fresnel law)	Eq. (3.269)
I	Intensity at point Q' (diffraction PSF)	Eq. (3.435)
I_0	Intensity at point Q'_0 on the principal ray (diffraction PSF)	Eq. (3.437)

Table A.10. Additional symbols for Chapter 3 (continued)

Symbol	Meaning	Where defined
I_0	Intensity at point Q_0 on the principal ray in the absence of aberrations (Strehl Intensity Ratio)	Eq. (3.458)
I (suffix)	Pertaining to the image (Fourier transform theory)	Eq. (3.504)
i	Fourier transform of intensity function I	Eq. (3.505)
J_0, J_1	Bessel functions of the zero and first orders	Eq. (3.440), Eq. (3.441)
k	A constant $(2\pi/\lambda)$ in the diffraction integral	Eq. (3.434)
$_{(l+n)}k_{(m+n),n}$	Coefficients of the Hamilton Characteristic Function	Eq. (3.14), Eq. (3.16)
k_{pl}	Dimensionless profile constant for an aspheric (Schmidt) plate	Eq. (3.238)
k_Z	"Ripple" amplitude (for Strehl Intensity Ratio)	Eq. (3.479)
L_w	Fraction of the total energy enclosed within the diffraction angle w	Eq. (3.448)
$\underline{L}$	The complex OTF	Eq. (3.487)
LSF	Line Spread Function	Eq. (3.493)
l, m, n	Integers involved in the definition of the Characteristic Function	Eq. (3.16)
l, m, n, s	Integers involved in the definition of Zernike radial polynomials	Eq. (3.431)
M_F	Folding flat mirror at an intermediate image; or a Newton-type flat	Fig. 3.72(b), Eq. (3.76)
MTF	Modulation Transfer Function	Eq. (3.484)
M (suffix)	Pertaining to a Maksutov solution	Fig. 3.36

Table A.11. Additional symbols for Chapter 3 (continued)

Symbol	Meaning	Where defined
m_{pl}	Magnification of an aspheric plate imaged back to object space	Eq. (3.221)
m_{min}	Minimum magnification in order to reveal diffraction resolution to the eye	Eq. (3.451)
$\underline{m}$	Parameter for MTF calculations	Fig. 3.112, Eq. (3.506)
m_+ (suffix)	Pertaining to the right-hand part of an achromatic meniscus	Fig. 3.48, Eq. (3.307)
men (suffix)	Pertaining to a meniscus corrector	Eq. (3.270)
N_H	Order number in the Characteristic Function	Table 3.1
N_s	Aperture number of a surface used afocally with object at infinity	Eq. (3.82)
n	Degree of polynomial V (Zernike polynomials)	Eq. (3.427)
n_Z	Number of complete "ripple" zones (Strehl Intensity Ratio)	Eq. (3.479)
$\tilde{n}$	Total number of powered surfaces in an $\tilde{n}$-mirror system (see surface number ν)	§ 3.6.5.2
O'	Axial image point for diffraction phenomena	Fig. 3.99
OTF	Optical Transfer Function	§ 3.10.7
O (suffix)	Pertaining to the object (Fourier transform theory)	Eq. (3.504)
opt (suffix)	Optimum image surface	§ 3.6.4.1 Fig. 3.46
P_c	Petzval sum (see surface number ν)	Eq. (3.19)

Table A.12. Additional symbols for Chapter 3 (continued)

Symbol	Meaning	Where defined
$\underline{P}$	Phase Function (OTF)	Eq. (3.502)
PF	Prime focus case	§ 3.2.6.2
PSF	Point Spread Function	§ 3.10.7
P (suffix)	Pertaining to a plane surface	Fig. 3.36
P (suffix)	Pertaining to the condition for the CFP to be at the exit pupil in an aplanatic 2-mirror telescope	Eq. (3.382), Eq. (3.383)
PH (suffix)	Pertaining to the plateholder	Eq. (3.262)
Pl	Aspheric plate	Fig. 3.42
p	Pupil magnification for a surface of an all-reflecting system (see surface number ν)	§ 3.6.5, Eq. (3.317)
p, q	Differences of the first two direction cosines of the incident and diffracted waves	Eq. (3.434)
p, q, r	Cartesian coordinate system of Q referred to the principal ray (Strehl Intensity Ratio)	§ 3.10.5
$\bar{p}$	Phase shift with transfer of a sinusoidal wave	Fig. 3.107
ptv	Peak-to-valley wavefront aberration	Table 3.27
pl (suffix)	Pertaining to an aspheric plate	Eq. (3.219)
Q_0'	A Gaussian image point (image "centre" on the principal ray) in the image plane for diffraction phenomena	Fig. 3.99
Q'	A point near Q_0' in the diffraction PSF	Fig. 3.99
Q_0, Q	Simplified forms (without primes) of Q_0', Q'	§ 3.10.5, Eq. (3.458)

Table A.13. Additional symbols for Chapter 3 (continued)

Symbol	Meaning	Where defined
Q (suffix)	Pertaining to the PSF of image point Q (Fourier transform theory)	Eq. (3.504)
q	Number of a Seidel aberration ($I \ldots V$)	Eq. (3.18)
R	Distance from the exit pupil to Q_0 (radius of the reference sphere)	§ 3.10.5
$R(\rho)$	A function only of ρ and equal to $V(\rho, 0)$ (Zernike polynomial)	Eq. (3.427)
$\underline{R}$	Real part of the OTF, i.e. the MTF	Eq. (3.487)
RC	Ritchey–Chrétien form of a 2-mirror Cassegrain telescope	§ 3.2.6.3(b)
r_c	Optimum radius of curvature of the field	Fig. 3.4
rms	Root-mean-square wavefront aberration	Table 3.27
rot (suffix)	Pertaining to angular decenter of the secondary in a 2-mirror telescope	Eq. (3.374)
S_q	Seidel surface contribution to Seidel aberration q	Eq. (3.18)
S_q^+	Seidel term including stop shift	Eq. (3.22)
S_q^0	Basic (power) part of a Seidel term	Eq. (3.24)
S_q^*	Aspheric part of a Seidel term	Eq. (3.24)
δS_I^*	Contribution of an aspheric plate to the third order spherical aberration	Eq. (3.219)
SP	Spherical Primary form of a 2-mirror Cassegrain telescope	§ 3.2.6.3(d)
S (suffix)	Pertaining to a laterally decentered 2-mirror telescope treated as a Schiefspiegler	Eq. (3.352)

Table A.14. Additional symbols for Chapter 3 (continued)

Symbol	Meaning	Where defined
S_1, S_2 (suffix)	Pertaining to third, fifth order spherical aberration for the Strehl Intensity Ratio	Eq. (3.471)
s_{pr1}	Distance of the entrance pupil from the first surface	Eq. (3.25)
s_{pl}	Stop shift of an aspheric plate relative to the primary	Eq. (3.219)
s_{pr1}^*	Quantity calculated for the secondary as though it were a primary in a Schiefspiegler	Eq. (3.344)
s, t	Spatial frequency of the object or image function in the ξ, η directions (MTF)	Eq. (3.485)
sol (suffix)	Pertaining to a solid Schmidt	§ 3.6.4.1
T (suffix)	Pertaining to wavefront tilt for the Strehl Intensity Ratio	Eq. (3.471)
$TC, 0$ (suffix)	Pertaining to a 2-mirror telescope of general or SP form with zero lateral (translation) decentering coma	Eq. (3.386), Eq. (3.389)
t, s, m	Tangential, sagittal and mean astigmatic surfaces, foci or sections	Fig. 3.20, Fig. 3.21
tot (suffix)	Pertaining to the total aberration (coma) defining neutral points (decenter)	Eq. (3.377)
U, U'	Angles of finite rays to the axis	Fig. 3.2, Eq. (3.83)
$U(Q')$	The complex amplitude at Q' (diffraction theory)	Eq. (3.434)
$\delta u_p'$	Angular aberration referred to the principal plane	§ 3.1
$\Delta\left(\frac{u}{n}\right)$	Aplanatic parameter (see surface number ν)	§ 3.2.2, Eq. (3.19)

Table A.15. Additional symbols for Chapter 3 (continued)

Symbol	Meaning	Where defined
$\delta \overline{u}'_y$	Normalized diffraction angle (kqy_m) in the y-direction	Eq. (3.435), Eq. (3.436)
$\tilde{u}, \tilde{v}$	Normalized ("optical") coordinates of Q referred to the principal ray (diffraction theory)	§ 3.10.5
V_α, V_β	Typical orthogonal polynomials of a set (Zernike polynomials)	§ 3.9, Eq. (3.426)
W'_3	Third order wavefront aberration	Eq. (3.21)
$\overline{W}'$	Full wavefront aberration taking account of $\cos n\phi$ term	Eq. (3.192)
$\overline{W}^2$	Mean of the square of the wavefront aberration	§ 3.9, (footnote 8)
$(\overline{W})^2$	Square of the mean wavefront aberration	§ 3.9, (footnote 8)
W_Q	The wavefront aberration associated with Q	Eq. (3.459)
$\overline{W}^n_Q$	Average value of the nth power of W_Q	Eq. (3.461)
ΔW_Q	Root-mean-square (rms) wavefront aberration $(W_Q)_{rms}$	Eq. (3.470)
w	Angular separation in the image plane of the point Q' from Q'_0 (diffraction theory)	§ 3.10.3
$\overline{w}$	Normalized form of the diffraction angle w	Eq. (3.444)
w_0	Radius (angular) of the first dark ring in the Airy disk	Eq. (3.446)
w_R	Angular resolution according to Rayleigh	Eq. (3.449)

Table A.16. Additional symbols for Chapter 3 (continued)

Symbol	Meaning	Where defined
x, y	Coordinates in the pupil for diffraction phenomena	Fig. 3.99
x_m, y_m	Slit dimensions for diffraction at a slit	Fig. 3.99
Y, Y'	Heights of finite rays from the axis	Eq. (3.84)
y_m	Normalized aperture parameter	Eq. (3.21)
Z (suffix)	Pertaining to "ripple"	Eq. (3.479)

Table A.17. Additional symbols for Chapter 3 (continued)

Symbol	Meaning	Where defined
α, β (suffix)	Pertaining to two glasses of an achromatic Schmidt plate	§ 3.6.2.6, Eq. (3.267)
δ	Lateral (transverse) decenter of the secondary in a 2-mirror telescope	§ 3.7.2.1
δ (suffix)	Pertaining to a laterally decentered 2-mirror telescope treated as a Schiefspiegler	Eq. (3.352)
$\delta_{\alpha\beta}$	Kronecker symbol (Zernike polynomials)	§ 3.9, Eq. (3.426)
ε	Eccentricity of a conic	Eq. (3.6)
ε	Obscuration (obstruction) factor for diffraction phenomena at an annular aperture	§ 3.10.4, Eq. (3.453)

Table A.18. Additional symbols for Chapter 3 (continued)

Symbol	Meaning	Where defined
ζ	Parameter for Seidel spherical aberration of the primary mirror of a 2-mirror telescope	Eq. (3.30)
ζ^0, ζ^*	Basic (power), aspheric component of ζ	Eq. (3.30)
ζ', η', ξ'	Cartesian coordinate system in the image plane for diffraction phenomena	Fig. 3.99
$\delta\eta_R$	Linear resolution according to Rayleigh	Eq. (3.450)
η'_m	Normalized field parameter	Eq. (3.21)
η_0, ξ_0	Object coordinates in OTF theory	§ 3.10.7
η, ξ	Image coordinates in OTF theory	§ 3.10.7
λ	Wavelength (spectral)	§ 3.5
λ_Z	Normalized "ripple" wavelength	Eq. (3.479)
$\overline{\lambda}$	Spatial wavelength in OTF theory ($1/s$ or $1/t$)	Fig. 3.107
μ	Reciprocal of magnification m (see surface number ν)	Eq. (3.332)
μ_{pr}	Reciprocal of pupil magnification p (see surface number ν)	Eq. (3.332)
μ^*_{pr1}	Quantity calculated for the secondary as though it were a primary in a Schiefspiegler	Eq. (3.344)
ν_A	Abbe number for an optical glass	§ 3.6.2.6 Eq. (3.244), Eq. (3.252), Fig. 3.32
ξ	Parameter for Seidel spherical aberration of the secondary mirror of a 2-mirror telescope	Eq. (3.40)

Table A.19. Additional symbols for Chapter 3 (concluded)

Symbol	Meaning	Where defined
ξ^0, ξ^*	Basic (power), aspheric component of ξ	Eq. (3.40)
ξ'	See coordinate system ζ', η', ξ' above	
ξ'	See coordinate system η', ξ' above	
ρ	Normalized aperture radius	§ 3.2.1
ρ	A point in a circular pupil for diffraction phenomena	§ 3.10.3
ρ_m	Radius of a circular pupil for diffraction phenomena	§ 3.10.3, Eq. (3.442)
σ	Normalized field radius	§ 3.2.1
τ	Aspheric form parameter (see surface number ν)	§ 3.2.2, Eq. (3.19)
τ	Magic number of Pythagoras and Fibonacci	Fig. 3.5, Eq. (3.129)
ϕ	Azimuth angle containing an aperture ray and a principal ray in the image forming wavefront	§ 3.2.1
ϕ, ψ	Azimuth angles in the pupil, image coordinate systems for diffraction phenomena	§ 3.10.3
$\otimes$	Convolution symbol (Fourier transform relations)	Eq. (3.504) et seq.

Chapter 4

Table A.20. Additional symbols for Chapter 4

Symbol	Meaning	Where defined
ADC	Atmospheric Dispersion Corrector	§ 4.4
b	Barometric pressure (mbar)	Eq. (4.91)
cor (suffix)	Pertaining to the properties of a lens corrector ("central" contributions)	§ 4.3.1.1, Eq. (4.27)
E	Pupil position parameter in a normalized telescope system with corrector	Eq. (4.4)
FR (also suffix)	Focal reducer	§ 4.5
FE (also suffix)	Focal extender	§ 4.5
FF (suffix)	Pertaining to a field flattener	Eq. (4.15)
g	Distance of an aspheric plate field corrector from the image	Eq. (4.2), Fig. 4.2, Fig. 4.5
gh (suffix)	Pertaining to ghost images from a plane-parallel plate	Eq. (4.1), Fig. 4.1
h	Height of the equivalent homogeneous atmosphere	Eq. (4.93)
L (suffix)	Pertaining to a thin corrector lens	Eq. (4.52)
L_1, L_2 (suffix)	Pertaining to the first, second thin lens of a 2-lens corrector	Eq. (4.73)
MFC (also suffix)	Mean field curvature, i.e. the EFC in the presence of a field corrector	Eq. (4.13)

Table A.21. Additional symbols for Chapter 4 (continued)

Symbol	Meaning	Where defined
n'	Refractive index of a single corrector lens	Eq. (4.71)
n'_1, n'_2	Refractive indices of doublet corrector lenses with two different glasses	Eq. (4.81)
n_a	Effective refractive index of the atmosphere	Eq. (4.92)
P (suffix)	Pertaining to an aspheric plate or an effective plate formed by an aspheric surface on a corrector lens	Eq. (4.54), Eq. (4.60)
PP (suffix)	Pertaining to a plane-parallel plate	Eq. (4.16)
$1pl$ (suffix)	Pertaining to a 1-plate corrector	Eq. (4.21) et seq.
$2pl$ (suffix)	Pertaining to a 2-plate corrector	Eq. (4.21) et seq.
R	Radius of the Earth	Eq. (4.93)
$real$ (suffix)	Pertaining to real values of aperture and field compared with the normalized case	§ 4.2.2.2 (end of section)
S_ν	Simplified notation for the spherical aberration contribution $(\delta S_I^*)_\nu$ for aspheric plate number ν	Eq. (4.5)
ΔS_{II}, ΔS_{III}	Changes in telescope contributions to S_{II}, S_{III} due to free aspheric constants combined with a lens corrector	Eq. (4.75)
Tel (suffix)	Pertaining to the contribution of the telescope without its corrector	Eq. (4.67)
t	Absolute temperature ($^\circ$K)	Eq. (4.91)
U	A simplifying parameter defined as a function of f', constant for a given telescope geometry	Eq. (4.44)

Table A.22. Additional symbols for Chapter 4 (concluded)

Symbol	Meaning	Where defined
V	A simplifying parameter defined as a function of f', d_1, L, constant for a given telescope geometry	Eq. (4.44)
V	Dispersion vector produced by a prism (ADC)	Fig. 4.25
X_L, Y_L	Shape, magnification parameters for the Seidel aberrations of a thin corrector lens	Eq. (4.52)
Z	Zenith distance (angle)	Eq. (4.91)
α	Rotation angle of ADC prisms	Fig. 4.25
δ_{gh}	Ghost image displacement in the image plane (plane-parallel plate)	Eq. (4.1), Fig. 4.1
ϕ_{ref}	Angular atmospheric refraction	Eq. (4.91)
$\delta\phi_{ref1,2}$	Atmospheric dispersion between wavelengths 1 and 2	Eq. (4.94)
* (superscript)	Pertaining to the total aberration induced by a corrector lens with an aspheric surface	Eq. (4.55)

Chapter 5

Table A.23. Additional symbols for Chapter 5

Symbol	Meaning	Where defined
c_p	Curvature (vertex) of a parabola (figuring)	Eq. (5.2)
c_s	Curvature of a sphere (figuring)	Eq. (5.2)
ES (suffix)	"Equal slope" case (figuring)	Eq. (5.3)

B. Portrait gallery

This Portrait Gallery is intended to give the basic biographical information of 26 great scientific and technical contributors to the historical development of *reflecting telescope optics*. Such a choice is inevitably personal and to some extent arbitrary. I have chosen the personalities listed below because of their contributions to the *form*, the *theory* and the *optical quality* of the reflecting telescope.

Marin Mersenne
Born: 1588, Oizé, France
Died: 1648, Paris

Mersenne invented the afocal forms of the 2-mirror telescope, fundamental to modern theory. He worked closely with Descartes. Recent historical research (see under "Cassegrain" overleaf) has revealed that he also invented other basic forms of reflecting telescope. (Courtesy Deutsches Museum, Munich)

René Descartes
Born: 1596, La Haye, France
Died: 1650, Stockholm

Descartes used his analytical geometry to lay down the mathematical principles for the *axial* monochromatic correction of images in both mirror and lens systems. This established all forms of the *classical* reflector (parabolic primary). (Courtesy Deutsches Museum, Munich)

No portrait of Cassegrain is known. The biographical details given in the original edition, taken from Robert "Dictionnaire des Noms Propres", were incorrect, due apparently to a confusion between Cassegrain and J. D. Cassini. New historical research by A. Baranne and F. Launay has revealed at last definitive information on the indentity of Cassegrain.*

Laurent Cassegrain

Born: 1628–1629, in or near Chartres, France
Died: 1693, Chaudon, near Nogent-le-Roi (Eure-et-Loir)

Cassegrain invented the 2-mirror telescope form which has become the definitive *focal* form for modern telescopes. Mersenne had already invented the equivalent *afocal* form, and, though this has remained largely unknown, also the focal forms of both Gregory and Cassegrain.*

James Gregory

Born: 1638, Aberdeen, Scotland
Died: 1675, Edinburgh

Gregory is rightly considered the inventor of the 2-mirror telescope in focal (Gregory) form. Mersenne had already invented the equivalent *afocal* form. (Courtesy Royal Astronomical Society, through Peter Hingley)

---

* A. Baranne, F. Launay, 1997, "Cassegrain: un célèbre inconnu de l'astronomie instrumentale", J. Opt. *28*, 158–172.

    I am delighted and proud that questions I addressed to my French colleagues, above all André Baranne, concerning the identity of Cassegrain for this portrait gallery have resulted in this superb piece of historical research, mainly regarding Cassegrain but also with largely unknown information concerning the remarkable work of Mersenne. If a birth certificate of Cassegrain can be found (the search is continuing, I believe), the basic biographical information will be complete.

Isaac Newton
Born: 1643, Woolsthorpe, England
Died: 1727, Kensington

Newton invented the Newton form and made the first usable reflector. But his greatest contribution was his theory of chromatism, albeit incorrectly denying the possibility of *a*chromatism. (Courtesy Deutsches Museum, Munich)

William Herschel
Born: 1738, Hannover, Germany
Died: 1822, Slough

Herschel realised the "front view" form, but his greatest contribution was in advancing the technology to increase size and optical quality. (Courtesy Deutsches Museum, Munich)

Carl Friedrich Gauss
Born: 1777, Braunschweig, Germany
Died: 1855, Göttingen

Gauss gave a complete mathematical exposition of the first order (Gaussian) theory of optical imaging. This was essential for the development of aberration theory. (Courtesy Deutsches Museum, Munich)

Joseph von Fraunhofer
Born: 1787, Straubing, Germany
Died: 1826, Munich

Fraunhofer's main contributions were to the development of the refractor. But his advances in optical glass technology, testing and figuring, as well as the application of ray tracing and diffraction theory, all helped the advance of the reflector to its many modern forms. (Courtesy Deutsches Museum, Munich)

William Lassell
Born: 1799, Bolton, England
Died: 1880, Maidenhead

Lassell was above all a great maker of telescopes. His greatest contribution to the advance of the reflector was his invention of the astatic lever for mirror supports, a principle fundamental to improved image quality in large telescopes. He was the first to apply the equatorial mount to a large reflecting telescope. (Courtesy National Museums and Galleries on Merseyside from an original Daguerreotype of 1845 owned by the Liverpool Astronomical Society, through Peter Hingley and Alan Bowden)

William Parsons
Earl of Rosse
Born: 1800, York, England
Died: 1867, Dublin

Rosse, like Lassell, was principally a maker of telescopes and his contributions were mainly concerned with technological advance. He invented the lightweight mirror blank and successfully applied the "whiffle-tree" support principle of Thomas Grubb to his biggest telescope. (Courtesy Royal Astronomical Society, through Peter Hingley)

William Rowan Hamilton
Born: 1805, Dublin, Ireland
Died: 1865, Dunsink

Hamilton's great contribution to telescope optics was his fundamental work on the nature of the aberration function of a centered optical system, the basis for the theoretical work of Gauss, Petzval and von Seidel. (Courtesy Deutsches Museum, Munich)

Joseph Petzval
Born: 1807, Szepesbéla, Hungary
Died: 1891, Vienna

Petzval was probably the first to develop third order aberration theory and was a great designer of early photographic optics. His field-curvature theorem (Petzval curvature) is fundamental to telescope optics and optical systems in general. (Courtesy Deutsches Museum, Munich)

James Nasmyth
Born: 1808, Edinburgh, Scotland
Died: 1890, London

Nasmyth was another great maker of reflectors. He invented the Nasmyth form and, in applying it, constructed one of the first successful Cassegrain telescopes of appreciable size. (Courtesy Deutsches Museum, Munich)

Léon Foucault
Born: 1819, Paris, France
Died: 1868, Paris

Foucault was largely responsible for the triumph of silvered glass mirrors over speculum for reflectors. Equally important was his invention of the "knife-edge" test of mirror quality, a great advance in image quality improvement. He also developed the coelostat form of telescope. (Courtesy Deutsches Museum, Munich)

Ludwig von Seidel
Born: 1821, Zweibrücken, Germany
Died: 1896, Munich

Seidel was the first to formulate and publish an explicit practical form for calculating the five monochromatic third order (Seidel) aberrations. This theory was later applied systematically to reflecting telescopes by Schwarzschild. (Courtesy Deutsches Museum, Munich)

Ernst Abbe
Born: 1840, Eisenach, Germany
Died: 1905, Jena

Abbe was one of the great contributors to the development of optical design and aberration theory. Particularly important in telescope optics is the Abbe sine condition, defining freedom from coma. (Courtesy Deutsches Museum, Munich)

George Ritchey
Born: 1864, Meigs County, Ohio, USA
Died: 1945, Azusa, California

Ritchey was a practical genius rather than a great theoretician, but he developed deep understanding of the theoretical requirements of reflecting telescopes. His encouragement of, and association with, Chrétien produced the Ritchey-Chrétien form, the modern form of 2-mirror aplanatic telescope. His advances in figuring and test technology were also a major contribution. (Courtesy U.S. Naval Observatory, through Brenda Corbin)

Karl Schwarzschild
Born: 1873, Frankfurt/Main, Germany
Died: 1916, Potsdam

In 1905 Schwarzschild formulated the complete third order aberration theory of 1- and 2-mirror telescopes. Furthermore, his formulation can be extended to any system and forms the basis of all modern reflecting telescope optics. He also developed a practical "Eikonal" theory giving the total aberration at a given field point. (Courtesy Martin Schwarzschild)

Bernhard Schmidt
Born: 1879, Naissaar Island, Estonia
Died: 1935, Hamburg

Schmidt was the first to realise the full significance of stop shift from the primary if combined with an aspheric refracting plate. This has led not only to the famous Schmidt wide-field telescope, but also to a whole range of modern telescope forms using aspheric plates. (Courtesy Hamburg Observatory, through Christian de Vegt)

Henri Chrétien
Born: 1879, Paris, France
Died: 1956, Washington

Chrétien was a great optical designer who invented, among other optical systems, the principle of Cinemascope. In association with Ritchey, he set up in 1910 the design of the aplanatic Ritchey-Chrétien telescope, apparently before learning of Schwarzschild's work. (Courtesy Observatoire de la Côte d'Azur, through Françoise le Guet Tully)

Frits Zernike
Born: 1888, Amsterdam, Holland
Died: 1966, Naarden

Zernike made major contributions to diffraction theory and received the Nobel prize in physics for his invention of phase-contrast microscopy. In telescope optics, his greatest contribution was the derivation of the Zernike circle polynomials, of fundamental importance in optical design and specification because of their orthogonal properties. (Courtesy Rijksuniversiteit Groningen, through H. J. Frankena, TU Delft)

Maurice Paul[1]
Born: 1890, Fontainebleau, France
Died: 1981, Cervens, Haute Savoie

Paul has been largely forgotten, even in France, but he wrote in 1935 a classical paper on modern reflector optics. A pupil of Chrétien, he proposed the fundamental form of 3-mirror anastigmat and critically analysed the scope of aspheric plates and lenses for field correctors. (Courtesy Claude Paul, through André Baranne)

---

[1] The biographic details of Maurice Paul were the most difficult to obtain of all the personalities in this portrait gallery. I am deeply grateful to André Baranne (Marseille) for a veritable detective work; also to Françoise le Guet Tully (Nice), Marie-José Vin (OHP) and Françoise Chavel (Ecole Supérieure d'Optique, Paris) for valuable information.

Albert Bouwers
Born: 1893, Dalen, Holland
Died: 1972

Bouwers invented the "concentric meniscus" wide-field telescope, the most fundamental of all forms. In the 1940s and 1950s, various design variations by Bouwers and others produced a new family of telescope designs. (Courtesy Delft Instruments NV, formerly de Oude Delft NV, through H. J. Frankena, TU Delft)

Dimitri Maksutov
Born: 1896, Odessa, Ukraine
Died: 1964, Leningrad

Maksutov invented independently of Bouwers the meniscus telescope that bears his name. Above all for amateurs, but also for smaller professional telescopes, the Maksutov form has been widely used. Maksutov ranked as a general expert on telescope optics. His experiments with metal mirrors were particularly notable. (Courtesy Rolf Riekher)

André Couder
Born: 1897, Alençon, France
Died: 1979, Paris

Couder contributed in a major way to nearly every aspect of telescope optics. He invented the 2-mirror Couder anastigmat, a development of the Schwarzschild telescope. He also invented "Null Testing" and advocated metal mirrors from practical experiments. The "Couder law" is the fundamental law of primary mirror supports. (Courtesy Charles Fehrenbach)

James Baker
Born: 1914, Louisville, Kentucky, USA

Baker is the doyen of living optical designers of telescope optics. His contributions are vast both in scope and significance. Among the greatest designs are the Baker-Schmidt-Cassegrains; the Paul-Baker 3-mirror flat-field telescope; a 3-mirror, 2-axis concept; the "reflector corrector", and the Baker-Nunn Super-Schmidt. He is one of the giants in the history of reflecting telescope optics. (Courtesy James Baker)

Group picture taken at the International Solar Union (ISU) meeting at Mount Wilson in 1910 (names Schwarzschild and Chrétien underlined)

This historic picture was sent to me for the second edition of this book by D.E. Osterbrock, to whom I express my grateful thanks. The positive copy was in Karl Schwarzschild's possession, who himself wrote in the names of nearly all those on the photograph. It includes the majority of the famous names in astronomy at that time, although – strangely – the great names of the host institute, Hale, Adams and Ritchey are not present, though Ellerman, Gale and Pease are included. Karl Schwarzschild and Henri Chrétien are shown (names underlined and written in capitals). Other historic figures connected with optics are Fabry, Hartmann, Turner and Babcock.

Schwarzschild and Chrétien met each other here for the first time and Chrétien learnt of Schwarzschild's comprehensive theory of 1905 [3.1]. However, there is strong historical evidence, given by Osterbrock [5.19], that Chrétien had already independently developed his own theory for the specific Ritchey-Chrétien form of reflecting telescope, *before* this meeting in 1910 took place. His delay in publication till 1922 [3.14] was due to his reluctance to publish his theory before an experimental telescope had confirmed it, but Ritchey only realised this in 1927 in Paris (see also §3.2.6.3b and §5.2).

Courtesy Mary Lea Shane Archives of the Lick Observatory, through D.E. Osterbrock

References

General

1. Bahner, K., 1967, "Teleskope" in Handbuch der Physik, Vol. XXIX, Springer Verlag, Heidelberg, 227–342
2. Born, M., Wolf, E., 1987, "Principles of Optics", 6th ed., Pergamon Press, Oxford
3. Czapski, S., Eppenstein, O., 1924, "Grundzüge der Theorie der optischen Instrumente", 3. Aufl., J.A. Barth, Leipzig
4. Danjon, A., Couder, A., 1935, "Lunettes et Télescopes", reissued 1983, Blanchard, Paris
5. Gascoigne, S.C.B., 1968, "Some Recent Advances in the Optics of Large Telescopes", Quart. J. Roy. Astron. Soc., 9, 98
6. Herzberger, M., 1958, "Modern Geometrical Optics", Interscience Publishers, New York
7. Hopkins, H.H., 1950, "Wave Theory of Aberrations", Oxford
8. King, H.C., 1955, "The History of the Telescope", Griffin, London
9. Korsch, D., 1991, "Reflective Optics", Academic Press, San Diego
10. Laux, U., 1993, "Astrooptik", Verlag Sterne u. Weltraum, München, Taschenbuch 11
11. Maréchal, A., Françon, M., 1960, "Diffraction: Structure des Images", Vol. 2, Editions de la Rev. d'Opt., Paris
12. Michelson, N.N., 1976, "Optical Telescopes: Theory and Design", published by "NAUKA" (Mathematical and Physical Section), Moscow. (In Russian)
13. Osterbrock, D.E., 1993, "Pauper and Prince: Ritchey, Hale and Big American Telescopes", Univ. of Arizona Press, Tucson and London
14. Riekher, R., 1990, "Fernrohre und ihre Meister", 2. Aufl., Verlag Technik GmbH, Berlin
15. Rutten, H.G.J., van Venrooij, M.A.M., 1988, "Telescope Optics", Willmann-Bell, Richmond, VA, USA
16. Schroeder, D.J., 1987, "Astronomical Optics", 2nd ed. 2000, Academic Press, San Diego
17. Welford, W.T., 1974, "Aberrations of the Symmetrical Optical System", Academic Press, London; 2nd ed., 1986, "Aberrations of Optical Systems", Adam Hilger, Bristol
18. Wetherell, W., 1980, "Applied Optics and Optical Engineering", Vol. 8, Academic Press, New York, Chap. 6, 171–315

Chapter 1

1.1　King, H.C., 1955, "The History of the Telescope", Griffin, London
1.2　Riekher, R., 1957, "Fernrohre und ihre Meister", 1. Aufl., Berlin; 1990, 2. Aufl., Verlag Technik GmbH, Berlin
1.3　Danjon, A., Couder, A., 1935, "Lunettes et Télescopes", reissued 1983, Blanchard, Paris
1.4　Grant, R., 1852, "History of Physical Astronomy", H.G. Bohn, London, 514
1.5　Martin, L.C., 1950, "Technical Optics", Vol. II, Pitman, London, 303
1.6　Schwarzschild, K., 1905, Abh. der Königl. Ges. der Wissenschaften zu Göttingen, Bd. IV (2)
1.7　Czapski, S., Eppenstein, O., 1924, "Grundzüge der Theorie der optischen Instrumente", 3. Aufl., J.A. Barth, Leipzig, 454
1.8　Wolter, H., 1952, Ann. Phys., **10**, 94 and 286
1.9　Herschel, W., 1800, Phil. Trans. Royal Society, **90**, 49
1.10　Hockney, D., 2001, "Secret Knowledge", Thames & Hudson, London

Chapter 2

2.1　della Porta, Giambattista, 1589, "Magia Naturalis", Naples, Book 17 of the revised edition from the original of 1558
2.2　Riekher, R., 1990, "Fernrohre und ihre Meister", 2. Aufl., Verlag Technik GmbH, Berlin, 16
2.3　Welford, W.T., 1974, "Aberrations of the Symmetrical Optical System", Academic Press, London; 2nd ed., 1986, "Aberrations of Optical Systems", Adam Hilger, Bristol
2.4　Gauss, C.F., 1841, "Dioptrische Untersuchungen", Göttinger Abh., **1**
2.5　Born, M., Wolf, E., 1964, "Principles of Optics", 2nd ed., Pergamon Press, Oxford
2.6　Schroeder, D.J., 1987, "Astronomical Optics", 2nd ed. 2000, Academic Press, San Diego
2.7　Herzberger, M., 1958, "Modern Geometrical Optics", Interscience Publishers, New York
2.8　Hansen, G., 1950, Optik, **6**, 337
2.9　Mütze, K. et al., 1961, "ABC der Optik", Brockhaus Verlag, Leipzig, 477
2.10　Zimmer, H.G., 1970, "Geometrical Optics", Springer Verlag, Heidelberg, 5
2.11　Bahner, K., 1967, "Teleskope" in Handbuch der Physik, Vol. XXIX, Springer Verlag, Heidelberg, 227–342
2.12　Conrady, A.E., 1929, "Applied Optics and Optical Design", Part 1, OUP, Oxford. Reprinted in USA, 1943. Reprinted 1960, Dover, New York
2.13　Hopkins, H.H., 1950, "Wave Theory of Aberrations", OUP, Oxford

Chapter 3

3.1　Schwarzschild, K., 1905, Göttinger Abh, Neue Folge, Band IV, No. 1, "Untersuchungen zur geometrischen Optik", I, II, III
3.2　Hamilton, W.R., 1833, Report Brit. Assoc., **3**, 360
3.3　Hopkins, H.H., 1950, "Wave Theory of Aberrations", Oxford

3.4 Seidel, L., 1856, Astron. Nachr., **43**, 289

3.5 Bahner, K., 1967, "Teleskope" in Handbuch der Physik, Vol. XXIX, Springer Verlag, Heidelberg, 227–342

3.6 Welford, W.T., 1974, "Aberrations of the Symmetrical Optical System", Academic Press, London; 2nd ed., 1986, "Aberrations of Optical Systems", Adam Hilger, Bristol

3.7 Herzberger, M., 1958, "Modern Geometrical Optics", Interscience Publishers, New York, (a) 82, 457

3.8 Laux, U., 1993, "Astrooptik", Verlag Sterne u. Weltraum, München, Taschenbuch 11, (a) 150, (b) 78, 101

3.9 Abbe E., 1873, Schultzes Arch. f. mikr. Anat., **9**, 413–468

3.10 Clausius, R., 1864, Pogg. Ann., **121**, 1–44

3.11 Czapski, S., Eppenstein, O, 1924, "Grundzüge der Theorie der optischen Instrumente", 3. Aufl., J. A. Barth, Leipzig, (a) 229, (b) 242, (c) 513, (d) 295

3.12 Rutten, H.G.J., van Venrooij, M.A.M., 1988, "Telescope Optics", Willmann-Bell, Richmond, VA, USA, (a) 65, (b) 272, (c) 96, (d) 103, (e) 98, (f) 109, (g) Chap. 13, (h) 132, (i) 133-135, (j) Chap. 12

3.13 Wilson, R.N., 1994, "Karl Schwarzschild Lecture of the German Astronomical Society", Bochum, 1993, Reviews in Modern Astronomy, **7**, 1

3.14 Chrétien, H., 1922, Rev. d'Opt., **1**, 13, 19

3.15 Theissing, H., Zinke, O., 1948, Optik, **3** (5/6), 451

3.16 Krautter, M., 1986, SPIE Vol. **655**, 127

3.17 Wolter, H., 1952, Ann. der Physik, 6. Folge, **10**, 94, 286

3.18 Kirkham, A. R., 1951, Sci. Amer., **185**, 118

3.19 Wilson, P.I., 1995, J. theor. Biol., **177**, 315

3.20 Bravais, L., Bravais, A., 1839, Ann. Sci. Nat. Bot., (Paris), **12**, 5, 65

3.21 Chrétien, H., 1959, "Calcul des Combinaisons Optiques", Ecole Supérieure d'Optiques, Paris, 385

3.22 Schroeder, D.J., 1987, "Astronomical Optics", Academic Press, San Diego, (a) 151, (b) 114, (c) 276, (d) 110, (e) 113, (f) 189, (g) 192, (h) Chap. 11

3.23 Dimitroff, G.Z., Baker, J.G., 1945, "Telescopes and Accessories", Blakiston, New York, (a) 294, (b) 102, (c) 107, (d) 105, (e) App. VII, 295

3.24 Danjon, A., Couder, A., 1935, "Lunettes et Téléscopes", reprinted 1979 by Blanchard, Paris, (a) 192–195 and Fig. 83, (b) 206

3.25 Couder, A., 1926, C.R. Acad. Sci. Paris, **183** (II), 1276

3.26 Maréchal, A., Françon, M., 1960, "Diffraction: Structure des Images", Vol. 2, Editions de la Rev. d'Opt., Paris, (a) 152, (b) 105, (c) 108, (d) 116, (e) 127, (f) Chap. 2, (g) 136, (h) 25, 40, (i) 34, (j) 39

3.27 Nijboer, B.R.A., 1942, Thesis, Univ. of Groningen; 1947, Physica, **10**, 679, and **13**, 605

3.28 Burch, C.R., 1942, MNRAS, **102**, 159

3.29 Linfoot, E.H., 1955, "Recent Advances in Optics", Oxford

3.30 König, A., Köhler, H., 1959, "Die Fernrohre und Entfernungsmesser", 3. Aufl., Springer Verlag, Heidelberg, (a) 152, (b) 157, (c) 158, (d) 159

3.31 Schmidt, B., 1931, Centr. Ztg. f. Opt. u. Mech., **52**, 25

3.32 Bouwers, A., 1946, "Achievements in Optics", Elsevier, Amsterdam, 25

3.33 Harrington, R.G., 1952, PASP, **64**, 275

3.34 Richter, N., 1961, Sterne, **37**, 89

3.35 Piazzi-Smyth, C., 1874, Brit. J. of Phot.-Alman., 43

3.36 Baker, J.G., 1940, Proc. Am. Phil. Soc., **82**, 323; Harvard Reprint No. 198

3.37 Herzberger, M., 1956, Internat. Projectionist, **31**, 7; 1957, JOSA, **47**, 584

3.38 Maksutov, D.D., 1944, JOSA, **34**, 270

3.39 Riekher, R., 1990, "Fernrohre u. ihre Meister", 2. Aufl., Verlag Technik, Berlin, (a) 335, (b) 332, (c) 334, (d) 342, (e) 335, (f) 345, (g) 232 et seq., (h) 318

3.40 Wright, F.B., 1935, PASP, **47**, 300

3.41 Väisälä, Y., 1936, Astron. Nachr., **259**, 197

3.42 Väisälä, Y., 1950, Turku Inform. No. 6

3.43 Baker, J.G., 1940, Proc. Amer. Phil. Soc., **82**, 339

3.44 Slevogt, H., 1942, Z. f. Instrumentenkunde, **62**, 312

3.45 Linfoot, E.H., 1943, MNRAS, **103**, 210

3.46 Köhler, H., 1949, Astron. Nachr., **278**, 1

3.47 Linfoot, E.H., 1944, MNRAS, **104**, 48

3.48 Mackintosh, A., 1986, "Advanced Telescope Making Techniques", Vol. 1: Optics, Willmann-Bell, Richmond, VA, USA

3.49 Wright, F.B., 1953, "Amateur Telescope Making III", A.G. Ingalls ed., Scientific American, New York, 574

3.50 Hawkins, D.G., Linfoot, E.H., 1945, MNRAS, **105**, 334

3.51 Gabor, D., 1941, Brit. Pat. No. 544 694

3.52 Maxwell, J., 1972, "Catadioptric Imaging Systems", American Elsevier, New York

3.53 Penning, K., 1941, German Pat. No. P 82 128 IX a/42h

3.54 Baker, J.G., 1945, U.S. Patent No. 2 458 132

3.55 Whipple, F.L., 1947, Harvard Meteor Program, Cambridge, Mass.; Harvard Reprint Series II, No. 19

3.56 Bradford, W.R., 1956, The Observatory, **76**, 172

3.57 Davis, J., 1963, Quart. J. Roy. Astron. Soc., **4**, 74

3.58 Henize, K.G., 1957, Sky and Telescope, **16**, 108

3.59 Wynne, C.G., 1947, MNRAS, **107**, 356

3.60 Wynne, C.G., 1947, Nature, **160**, 91

3.61 Bennett, H.F., 1945, U.S. Patent No. 2 571 657

3.62 Sonnefeld, A., 1936, DRP No. 697 003

3.63 Richter, R., Slevogt, H., 1941, German Patent Application No. Z 26592 IXa 42h

3.64 Houghton, J.L., 1944, U.S. Patent No. 2 350 112

3.65 Mangin, A., 1876, Mémorial de l'officier du génie **25.(2)10**, 211-289

3.66 Rosin, S., Amon, M., 1967, Appl. Opt., **6**, 963

3.67 Silvertooth, E.W., 1968, Bull. Inst. Phys. and Phys. Soc., **19**, 257

3.68 Hamilton, W.F., 1814, Engl. Pat. No. 3781

3.69 Schupmann, L., 1899, "Die Medial-Fernrohre", B. G. Teubner, Leipzig

3.70 Kerber, A., 1893, Centr. Ztg. f. Optik u. Mechanik, **14**, 145

3.71 Daley, J., 1984, "Amateur Construction of Schupmann Medial Telescopes", privately printed (see [3.12(h)])

3.72 Delabre, B., 1990, Internal ESO communication

3.73 Korsch, D., 1972, Appl. Optics, **11**(12), 2986

3.74 Paul, M., 1935, Rev. d'Opt., **14**, 169

3.75 Baker, J.G., 1969, IEEE Trans. Aerosp. Electron. Syst., **5**, 261

3.76 McGraw, J., et al., 1982, Proc. SPIE Conf. **331**, 137

3.77 Willstrop, R.V., 1984, MNRAS, **210**, 597

3.78 Epps, H.W., Takeda, M., 1983, Ann. Tokyo Astron. Obs., 2nd series, **19**, 401

3.79 Loveday, N., 1981, Sky and Telescope, **61** (6), 545

3.80 Zimmer, H.G., 1970, "Geometrical Optics", Springer Verlag, Heidelberg, (a) 47

3.81 Meschkowski, H., 1965, "Meyers Handbuch über die Mathematik", Bibl. Institut, Mannheim, D57, D83, D95

3.82 Robb, P.N., 1978, Appl. Opt., **17** (17), 2677

3.83 Shack, R.V., Meinel, A.B., 1966, JOSA, **56**, 545

3.84 Rumsey, N.J., 1970, Optical Instruments and Techniques, ed. J. Home Dickson, Oriel Press, Newcastle, 516

3.85 Rumsey, N.J., 1969, U.S. Patent No. 3 460 886

3.86 Vigroux, L., 1993, Proc. IAU Symposium No. 161 on "Astronomy from Wide-Field Imaging", Potsdam, Aug. 1993; Kluver, Dordrecht, 73

3.87 Korsch, D., 1991, Paper presented at the JPL Workshop on the "Next Generation Space Telescope", Pasadena, March 1991

3.88 Korsch, D., 1986, Opt. Engin., **25**(9), 1034

3.89 Strong, J., 1933, Phys. Rev., **43**, 498

3.90 Jacobson, M.R., et al., 1992, GEMINI Project Investigation of Low Emissivity Durable Coatings, Optical Data Associates, Tucson

3.91 Schmidt-Kaler, Th., 1994, Astron. Nachr., **315**(4), 323

3.92 Ardeberg, A., et al., 1992, Proc. ESO Conf. on "Progress in Telescope and Instrument Technologies", ESO, Garching, 159

3.93 Ardeberg, A., et al., 1993, Proc. Conf. on "Metal Mirrors for Astronomical Telescopes", University College, London, 1992

3.94 Wilson, R.N., Delabre, B., 1995, Astron. and Astrophys., **294**(1), 322

3.95 Wilson, R.N., Delabre, B., Franza, F., 1994, Proc. SPIE Conf. on "Advanced Technology Optical Telescopes V", Vol. **2199**, 1052

3.96 Baranne, A., Lemaître, G., 1986, Proc. "Workshop on Large Telescopes", Hamburg, Mitt. der Astron. Gesellschaft No. 67, Hamburg, 1986, 226

3.97 Baranne, A., Lemaître, G., 1987, C.R. Acad. Sci. Paris, **305**, Série II, 445

3.98 Wilson, R.N., et al., 1991, J. Mod. Optics, **38**(2), 219

3.99 Robb, P.N., 1979, JOSA, **69**(10), 1439

3.100 Sasian, J.M., 1990, Opt. Engin., **29**(10), 1181

3.101 Schafer, D.R., 1979, "Instrumentation in Astronomy III", SPIE Vol. **172**, 19

3.102 Shectman, S.A., 1983, Proc. SPIE **444**, 106

3.103 Kutter, A., 1953, "Der Schiefspiegler", Verlag F. Weichardt, Biberach

3.104 Kutter, A., 1964, "Bauanleitung für den Kosmosschiefspiegler", Franckh'sche Verlagshandlung, Stuttgart

3.105 Czerny, M., Turner, A.F., 1930, Zeits. f. Physik, **61** , 792

3.106 Wilson, R.N., Opitz, A., 1976, FRG Patent No. 24 26 325

3.107 Buchroeder, R.A., 1969, Sky and Telescope, **38** (6), 418

3.108 Kutter, A., 1975, Sky and Telescope, **49** (1) and (2), 46 and 115

3.109 Conrady, A.E., 1919, MNRAS, **79**, 384

3.110 Maréchal, A., 1950, Rev. d'Opt., **29**, 1

3.111 Slevogt, H., 1963, Optik, **20**, 488

3.112 Baranne, A., 1966, J. Observateurs, **49**, 75

3.113 Meinel, A.B., Meinel, M.P., 1984, Opt. Eng., **23** (6), 801

3.114 Delabre, B., 1991, ESO private communication

3.115 Sand, R., 1991, Conference of the Deutsche Ges. d. angew. Opt., and private communication

3.116 Robb, P.N., Mertz, L., 1979, SPIE Proc. "Instrumentation in Astronomy III", Vol. **172**, 15

3.117 Schafer, D.R., 1978, Appl. Opt., **17**(7), 1072

3.118 Burch, C.R., 1947, Proc. Phys. Soc. London, **59**, 41

3.119 Zernike, F., 1934, Physica, **1**, 689

3.120 Born, M., Wolf, E., 1964, "Principles of Optics", 2nd ed., Pergamon Press, Oxford, (a) 464, 767, (b) 384, 392, (c) 416, (d) 459, (e) 480, (f) 484, (g) 381

3.121 Bhatia, A.B., Wolf, E., 1954, Proc. Cambridge Phil. Soc., **50**, 40

3.122 Dierickx, P., 1989, ESO Internal report for the VLT

3.123 Flügge, S., 1979, "Mathematische Methoden der Physik I", Springer Verlag, 107, 160

3.124 Malacara, D., 1978, "Optical Shop Testing", John Wiley, New York, 489

3.125 Sumita, H., 1969, Jap. J. Appl. Phys. **8**, 1027

3.126 Zernike, F., 1938, Physica, **5**, 785

3.127 Hopkins, H.H., 1951, Proc. Roy. Soc. A, **208**, 263

3.128 Hopkins, H.H., 1953, Proc. Roy. Soc. A, **217**, 408

3.129 Airy, G.B., 1835, Trans. Camb. Phil. Soc., **5**, 283

3.130 Rayleigh, Lord, 1881, Phil. Mag. (5), **11**, 214

3.131 Rayleigh, Lord, 1879, Phil. Mag. (5), **8**, 261

3.132 Steward, G.C., 1928, "The Symmetrical Optical System", Cambr. Univ. Press, 89

3.133 Jacquinot, P., Roizen-Dossier, B., 1964, "Progress in Optics", Vol. 3, North Holland, Amsterdam, 29

3.134 Lansraux, G., 1953, Rev. d'Opt., **32**, 475

3.135 Strehl, K., 1902, Z. f. Instrumentenkunde, **22**, 213

3.136 Duffieux, P.M., 1946, "L'Intégrale de Fourier et ses applications à l'optique", published by the author. 2nd ed., 1970, published by Masson, Paris

3.137 Wetherell, W., 1980, "Applied Optics and Optical Engineering", Vol. 8, Academic Press, New York, 171–315

3.138 Goodman, J., 1968, "Introduction to Fourier Optics", McGraw-Hill, New York

3.139 O'Neill, E.L., 1956, JOSA, **46** (4), 285, 1096

3.140 Lloyd, J.M., 1975, "Thermal Imaging Systems", Plenum Press, New York, 102

3.141 Rayleigh, Lord, 1903, Scientific Papers, **3**, Cambridge University Press

3.142 Hopkins, H.H., 1955, Proc. Roy. Soc. A, **231**, 98

3.143 Black, G., Linfoot, E.H., 1957, Proc. Roy. Soc. A, **239**, 522

3.144 De, M., 1955, Proc. Roy. Soc. A, **233**, 96

3.145 Hopkins, H.H., 1956, Proc. Phys. Soc. B, **LXIX**, 562

3.146 O'Neill, E.L., 1963, "Introduction to Statistical Optics", Addison-Wesley, Reading, Mass., 88

3.147 Ditchburn, R.W., 1976, "Light", Vol. 1, Academic Press, London, 188

3.148 Babinet, J., 1837, C. R. Acad. Sci. Paris, **4**, 638

3.149 Scheiner, J., Hirayama, S., 1894, Abh. d. Königl. Akad. Wissensch., Berlin, Anhang I

3.150 Cox, R.E., 1960, Sky and Telescope, **20** (3), 166

3.151 Petzval, J., 1843, "Bericht über die Ergebnisse einiger dioptrischer Untersuchungen", Verlag C.A. Hartleben, Pesth. Reprinted 1975, Akadémiai Kiadó, Budapest (ISBN 96305 02690)

3.152 von Rohr, M., 1899, "Theorie und Geschichte des photographischen Objektivs", J. Springer, Berlin, 246–271

3.153 Klügel, G.S., 1778, "Analytische Dioptrik in zwey Theilen", J.F. Junius, Leipzig

3.154 Buchdahl, H.A., 1954, "Optical Aberration Coefficients", Oxford Univ. Press

3.155 Focke, J., 1965, "Higher Order Aberration Theory", Progress in Optics, Vol. IV, Ed. E. Wolf, North Holland Publ. Co., Amsterdam, 1–36

3.156 Smith, T., 1921–22, Trans. Opt. Soc. London, **23**, 311

3.157 de Meijère, J.L.F., Velzel, C.H.F., 1989, JOSA A, **6** (10), 1609

3.158 Lurie, R.J., 1975, JOSA, **65** (3), 261

3.159 Turco, E., 1979, Sky and Telescope, **58** (5), 473

3.160 Busack, H.J., 2000, FRG Patent (Offenlegungsschrift) DE 19847702 A1

3.161 Busack, H.J., 2002, FRG Patent (Offenlegungsschrift) DE 10036309 A1

3.162 Rakich, A., 2001, M.Sc. Thesis "A Complete Survey of Three-Mirror Anastigmatic Reflecting Telescope Systems with One Aspheric Surface", Univ. of Canterbury, Christchurch, New Zealand

3.163 Rakich, A., Rumsey, N.J., 2002, JOSA A, **19** (7), 1398

3.164 Aldis, H.L., 1900, Photographic Journal (London), **24**, 291

3.165 Rakich, A., 2001, private communication

3.166 Stevick, D., 1993, "The Stevick-Paul Off-Axis Reflecting Telescope", Amateur Telescope Making (ATM) Journal No. 3, Spring/Summer 1993

3.167 Conrady, A.E., 1960, "Applied Optics and Optical Design", Part 2, Dover, New York, 628–631

Chapter 4

4.1 Sampson, R.A., 1913, Phil. Trans. Roy. Soc. (London), **213**, 27

4.2 Sampson, R.A., 1913, MNRAS, **73**, 524

4.3 Violette, H., 1922, Rev. d'Opt., **1**(9), 397

4.4 Paul, M., 1935, Rev. d'Opt., **14**(5), 169

4.5 Wynne, C.G., 1972, Progress in Optics X, ed. E. Wolf, North Holland Publ. Co., Amsterdam – London, 137

4.6 Meinel, A.B., 1953, Astrophys. J., **118**, 335

4.7 Gascoigne, S.C.B., 1965, The Observatory, **85**, 79

4.8 Gascoigne, S.C.B., 1968, Quart. J. Roy. Astron. Soc., **9**, 98

4.9 Gascoigne, S.C.B., 1973, Appl. Opt., **12**(7), 1419

4.10 Burch, C.R., 1942, MNRAS, **102**, 159

4.11 Schulte, D.H., 1966, Appl. Opt., **5**(2), 313

4.12 Köhler, H., 1967, ESO Bulletin No. 2, 13

4.13 Köhler, H., 1968, Appl. Opt., **7**, 241

4.14 Wynne, C.G., 1968, Astrophys. J., **152**, 675

4.15 Wilson, R.N., 1971, ESO Conf. "Large Telescope Design", ESO/CERN, Geneva, 131

4.16 Cao, C., Wilson, R.N., 1984, Astron. Astrophys., **133**, 37

4.17 Schulte, D.H., 1966, Appl. Opt., **5**(2), 309

4.18 Bowen, I.S., Vaughan, A.H., 1973, Appl. Opt., **12**(7), 1430

4.19 Brown, R.A., Ford, H.C. (eds), 1990, "Report of the HST Strategy Panel: A Strategy for Recovery"

4.20 Ross, F.E., 1933, Astrophys. J., **77**, 243

4.21 Ross, F.E., 1935, Astrophys. J., **81**, 156

4.22 Hopkins, H.H., 1950, "Wave Theory of Aberrations", O.U.P., London

4.23 Wynne, C.G., 1949, Proc. Phys. Soc., **B62**, 772

4.24 Wynne, C.G., 1965, Appl. Opt., **9**(4), 1185

4.25 Bahner, K., 1967, "Teleskope" in "Handbuch der Physik", Vol. XXIX, 227–342

4.26 Rosin, S., 1961, JOSA, **51**(3), 331

4.27 Baker, J.G., 1953, "Amateur Telescope Making", Book 3, Scientific American, Inc., New York, 7 et seq.

4.28 Wynne, C.G., 1967, Appl. Opt., **6**(7), 1227

4.29 Bourdet, M., Cayrel, R., 1971, INAG Report 71-10, Meudon

4.30 Faulde, M., Wilson, R.N., 1973, Astron. Astrophys., **26**, 11

4.31 Baranne, A., 1966, Proc. IAU Symposium No. 27, "The Construction of Large Telescopes", Academic Press, 22

4.32 Baranne, A., 1966, Publ. de l'Observatoire de Haute Provence, **8**(22), 75

4.33 Richardson, E.H., Harmer, C.F.W., Grundmann, W.A., 1982, "Better but bigger prime focus corrector lenses for RC telescopes", Dominion Astrophys. Obs. (Preprint)

4.34 Richardson, E.H., Harmer, C.F.W., Grundmann, W.A., 1984, MNRAS, **206**, 47

4.35 Bahner, K., 1975, Mitt. Astron. Gesellschaft, **36**, 57

4.36 Epps, H.W., Angel, J.R.P., Anderson, E., 1984, Proc. IAU Colloq. No. 79 "Very Large Telescopes, their Instrumentation and Programs", ESO, Garching, 519

4.37 Angel, J.R.P., Woolf, N.J., Epps, H.W., 1982, SPIE Proc. Vol. 332, 134

4.38 Epps, H.W., Takeda, M., 1984, Japanese J. Optics, **13**(5), 400

4.39 Richardson, E.H., Morbey, C.L., 1984, IAU Colloq. No. 79 "Very Large Telescopes, their Instrumentation and Programs", ESO, Garching, 549

4.40 Brodie, J.P., Epps, H.W., 1986, Report of Space Sciences Lab., Univ. of California, Berkeley, on "Six alternative PF field correctors for the 120-inch Shane telescope"

4.41 Bowen, I.S., 1961, PASP, **73**, 114

4.42 Wynne, C.G., 1973, MNRAS, **163**, 357

4.43 Rosin, S., 1964, Appl. Opt., **3**(1), 151

4.44 Harmer, C.F.W., Wynne, C.G., 1976, MNRAS, **177**, 25P

4.45 Wilson, R.N., 1968, Appl. Opt., **7**, 253

4.46 Epps, H.W., 1989, MMT Project Report for the Carnegie Institution

4.47 Epps, H.W., 1989, Magellan Project Report No. 9

4.48 Epps, H.W., 1987, "MMT Project Design" in SPIE Proceedings Vol. 766, 140

4.49 Refsdal, I.N., 1968, Appl. Opt., **7**, 1645

4.50 Rosin, S., 1966, Appl. Opt., **5**(4), 675

4.51 Su, D.-q., Zhou, B.-f., Yu, X.-m., 1990, Science in China (Series A), **33**(4), 454

4.52 Wallner, E.P., Wetherell, W.B., 1980, Proc. Conf. "Optical and Infrared Telescopes for the 1990s", Vol. II, KPNO, Tucson, 717

4.53 Airy, G.B., 1869, MNRAS, **29**, 333

4.54 Russell, H.N., Dugan, R.S., Stewart, J.Q., 1945, "Astronomy", Vol. I, App. VI

4.55 Barlow, C.W.C., Bryan, G.H., 1944, "Elementary Mathematical Astronomy", Univ. Tutorial Press, London, 108

4.56 Von Hoerner, S., Schaifers, K., 1967, "Meyers Handbuch über das Weltall", 163

4.57 Beddoes, D.R., Dainty, J.C., Morgan, B.L., Scaddon, R.J., 1976, JOSA, **66**, 1247

4.58 Breckenridge, J.B., McAlister, H.A., Robinson, W.G., 1979, Appl. Opt., **18**, 1034

4.59 Wynne, C.G., 1984, The Observatory, **104**, 140

4.60 Wynne, C.G., Worswick, S.P., 1986, MNRAS, **220**, 657

4.61 Wynne, C.G., 1986, The Observatory, **106**, 163

4.62 Wynne, C.G., Worswick, S.P., 1988, MNRAS, **230**, 457

4.63 Bingham, R.G., 1988, Proc. Conf. "Very Large Telescopes and their Instrumentation", Vol. II, ESO, Garching, 1167

4.64 Su, D.-q., 1986, Astron. Astrophys., **156**, 381

4.65 Wang, Y.-n., Su, D.-q., 1990, Astron. Astrophys., **232**, 589

4.66 Rutten, H., van Venrooij, M., 1988, "Telescope Optics", Willmann-Bell, Inc., Richmond VA, 149

4.67 Hartshorn, C.R., 1953, "Amateur Telescope Making", Book 3, 277

4.68 Meinel, A.B., 1956, Astrophys. J., **124**, 652

4.69 Jörsäter, S., 1991, NOT News No. 4, July 1991, 9

4.70 Boulesteix, J., Courtès, G., Laval, A., Monnet, A., 1974, Proc. ESO/SRC/ CERN Conf. "Research Programmes for New Large Telescopes"

4.71 Geyer, E.H., Nelles, B., 1984, Proc. IAU Colloq. No. 79 "Very Large Telescopes, their Instrumentation and Programs", ESO, Garching, 575

4.72 Meinel, A.B., Meinel, M.P., Wang, Y.-n., 1985, Appl. Opt., **24**(17), 2751

4.73 MacFarlane, M., 1982, "Report of the Optical Conference on the 7.6 m Telescope", Univ. of Texas Publ. in Astron. No. 22, 157 and 181 et seq.

4.74 Rosin, S., 1968, Appl. Opt., **7**, 1483

4.75 Delabre, B., 1996, Internal Memo ESO INS-96-0080, 7 May, 1996

4.76 Delabre, B., 2002, private communication November, 2002

4.77 Seifert, W., Mitsch, W., Nicklas, H., Rupprecht, G., 1994, SPIE Conf. Proc. Vol. 2198 "Instrumentation in Astronomy VIII", Eds. D.L. Crawford, E.R. Crain, 213

4.78 Avila, G., Rupprecht, G., Beckers, J., 1997, Lund Univ., ESO, SPIE Conf. Proc. SPIE Vol. 2871 "Optical Telescopes of Today and Tomorrow", Ed. A. Arderberg, 1135

Chapter 5

5.1 Danjon, A., Couder, A., 1935, "Lunettes et Télescopes", reprinted 1983 (Blanchard, Paris), 677 et seq.

5.2 King, H.C., 1955, "The History of the Telescope", Griffin, London

5.3 Riekher, R., 1990, "Fernrohre und ihre Meister", 2. Aufl., Verlag Technik GmbH, Berlin

5.4 Schwarzschild, K., 1905, "Untersuchungen zur geometrischen Optik II", Abh. der königlichen Ges. der Wissensch. zu Göttingen, Math.-Phys. Klasse, Neue Folge Bd. IV (2)

5.5 Rosse, Earl of, 1861, Phil. Trans., **151**, 681

5.6 Grant, R., 1852, "History of Physical Astronomy", Bohn, London, 569

5.7 Lassell, W., 1842, Mem. Roy. Astron. Soc. XII, 265

5.8 Lassell, W., 1867, Mem. Roy. Astron. Soc. XXXVI, Plate XI (Frontispiece)

5.9 Robinson, T.R., 1869, Phil. Trans., **159**, 132

5.10 Foucault, J.B.L., 1857, C. R. Acad. Sci. Paris, **44**, 339

5.11 Tobin, W., 1987, Vistas in Astronomy, **30**, 153

5.12 Tobin, W., 1987, Sky and Telescope, **74** (4), 358

5.13 Hyde, W.L., 1987, Optics News, Jan. 1987, 6

5.14 Ritchey, G.W., 1904, Smithsonian Contrib. to Knowledge, **34**, 47

5.15 Foucault, J.B.L., 1859, Ann. de l'Obs. de Paris, **5**, 197

5.16 Hale, G.E., 1908, "The Study of Stellar Evolution", Univ. of Chicago Press, Chicago, 45

5.17 Dimitroff, G.Z., Baker, J.G., 1945, "Telescopes and Accessories", Blakiston, New York, 150

5.18 Osterbrock, D. E., 1984, "James E. Keeler: Pioneer American Astrophysicist", CUP, Cambridge,UK

5.19 Osterbrock, D. E., 1993, "Pauper and Prince: Ritchey, Hale and Big American Telescopes", Univ. of Arizona Press, Tucson & London

5.20 Pease, F.G., 1935, JOSA, **25**, 156

5.21 Bahner, K., 1967, "Teleskope" in Handbuch der Physik, Vol. XXIX, Springer Verlag, 266

5.22 Ingalls, A.G., 1945, "Amateur Telescope Making", Vol. 1, Munn and Co., New York, 396

5.23 Bowen, I.S., 1960, in "Stars and Stellar Systems", Vol. 1 "Telescopes", Ed. Kuiper and Middlehurst, Univ. Chicago Press, 1

5.24 Woodbury, D.O., 1939, "The Glass Giant of Palomar", Dodd, Mead & Co., New York

5.25 Wright, H., 1953, "The Great Palomar Telescope", Faber & Faber, London

5.26 Sky and Telescope, 1950, Vol. IX, February Issue

5.27 Couder, A., 1931, Bulletin Astron., 2me. Série, Tome VII, Fasc. VII, 283

5.28 Mayall, N.U., Vasilevskis, S., 1960, Astron. J., **65**, 304

5.29 Bowen, I.S., 1950, PASP, **62**, 91

5.30 Bowen, I.S., 1967, Q. J. RAS, **8** (1), 9

5.31 Kuiper, G.P., Middlehurst, B.M. (Eds), 1960, "Stars and Stellar Systems", Vol. 1 "Telescopes", Univ. Chicago Press, 239

5.32 Brown, P.L., 1967, Sky and Telescope, **34**, 356

5.33 Smith, F. G., Dudley, J., 1982, J. Hist. of Astron., **13**, Part 1, No. 36, 1

5.34 (Anon)(Special Article), 1965, Engineering (6 August 1965), 164

5.35 Crawford, D.L., 1971, Proc. ESO Conf. "Large Telescope Design", Ed. R.M. West, ESO Geneva, 23

5.36 Bowen, E.G., Gascoigne, S.C.B., Wehner, H., 1968, Proc. ASA, **1**(3), 74

5.37 Blanco, V.M, 1968, Sky and Telescope, **35** (2), 72

5.38 Laustsen, S., Madsen, C., West, R.M., 1987, "Exploring the Southern Sky", Springer-Verlag, Plates 214 and 215

5.39 Wilson, R.N., 1974, Technical Report No. 2, ESO Geneva

5.40 Odgers, G.J., Richardson, E.H., Grundmann, W.A., 1977, Proc. of ESO Conf. "Optical Telescopes of the Future", Eds. F. Pacini, W. Richter, R. Wilson, ESO Geneva, 79

5.41 Carpenter, G.C., Ring, J., Long, J.F., 1977, as [40], 47

5.42 Brown, D.S., Humphries, C.M., 1977, as [40], 55

5.43 Bahner, K., 1986, Sterne u. Weltraum, **25** (6), 310

5.44 Pope, J.D., 1977, as [40], 67

5.45 Ridpath, I., 1990, Sky and Telescope, **80** (2), 136

5.46 Dall, H.E., 1947, J. Brit. Astron. Ass. **57** (5), 201

5.47 Dall, H.E., 1953, "Amateur Telescope Making", Book 3, Scientific American Inc., New York, 149

5.48 Couder, A., 1927, Rev. d'Opt., **6**, 49

5.49 Pasachoff, J. M., 1978, "Astronomy Now", W. B. Saunders, Philadelphia, 42

5.50 Learner, R., 1982, "Astronomy through the Telescope", Evans Bros., London, 148

5.51 Gochermann, J., Schmidt-Kaler, T., 1989, Sterne u. Weltraum, **28** (1), 30

5.52 Bahner, K., 1965, "Astronomical Instruments" in Landolt-Börnstein, Neue Serie, Ed. K.H. Hellwege, Vol. VI/1, Springer-Verlag, 6

5.53 Wolf, R., 1982, "Astronomical Instruments" in Landolt-Börnstein, Ed. K.H. Hellwege, Vol. VI/2a, Springer-Verlag, 6

5.54 Schielicke, R., 1982, Die Sterne, **58** (2), 93

5.55 Fehrenbach, Ch., 1990, "Des hommes, des télescopes, des étoiles", Editions du CNRS, Paris

List of Figures

Chapter 1

1.1 Zucchi's attempt at a Herschel-type front-view reflecting
telescope, 1616 2

1.2 Spherical aberration of a spherical concave telescope mirror.
"Paraxial" rays are nominally at a negligible height
from the axis... 3

1.3 Two of Mersenne's designs for reflecting telescopes, adapted
from "L'Harmonie Universelle", 1636, and King [1.1], compared
with Galileo-type and Kepler-type refracting telescopes 4

1.4 (a) Facsimile of the Gregory telescope from "Optica Promota",
1663 (after Danjon and Couder [1.3]).
(b) Raypath of the Gregory form (after King [1.1]) 7

1.5 The Newton reflecting telescope, 1668 8

1.6 The Cassegrain reflecting telescope, 1672, (a) as drawn by
de Bercé. (b) Raypath of the Cassegrain form (after King [1.1]).. 10

1.7 John Hadley's 6-inch, f/10.3 Newton reflector, 1721
(courtesy Royal Astronomical Society, through Peter Hingley) ... 12

1.8 The "Ramsden disk" (exit pupil) explained by Ramsden in 1775
(after King [1.1]) 15

1.9 Sir William Herschel (1738–1822) painted by L.T. Abbot in 1785
(courtesy Deutsches Museum, Munich) 16

1.10 William Herschel's "large" 20-foot focus telescope,
aperture 18.8 inches (f/12.8), completed in 1784, (reproduced
from an engraving of 1794, courtesy Science History Publications
Ltd., Cambridge, England) 17

1.11 William Herschel's largest telescope: 4 feet in aperture,
40-foot focus (f/10), completed in 1789
(courtesy Deutsches Museum, Munich) 19

Chapter 2

2.1 The ideal optical system: the principal planes
and unit magnification between them 23

2.2 Geometrical construction of ideal image formation 24

2.3   Geometrical wavefronts and rays 27
2.4   The relationship between the focal lengths 29
2.5   Derivation of the Lagrange Invariant 31
2.6   Aperture stop, entrance and exit pupils 32
2.7   Telecentric aperture stop 34
2.8   Gaussian optics of a conventional refracting telescope with ocular
       (afocal in both object and image spaces)................... 36
2.9   Image principal plane in the defocused telescope of Fig. 2.8,
       producing a real image at I_2' 39
2.10   Prime focus forms of reflecting telescope 41
2.11   Gaussian optics of a Gregory telescope 42
2.12   Gaussian optics of a Cassegrain telescope 43
2.13   Exit pupil position E' in the Cassegrain form
       with the entrance pupil E at the primary 51
2.14   Entrance pupil position E in the Cassegrain form
       with the exit pupil E' at the secondary 52
2.15   Limit case of a Cassegrain telescope with a plane secondary mirror 53

Chapter 3

3.1   Wavefront, longitudinal and lateral aberration............... 58
3.2   The Abbe sine condition.................................... 86
3.3   Normal representation of spot-diagrams 88
3.4   (a) Spot-diagrams for a classical Cassegrain telescope
       with the geometry of the ESO 3.5 m NTT (f/11; $m_2 = -5$)
       for an optimum field curvature $r_c = -1955$ mm
       (concave to the incident light) 99
3.4   (b) Spot-diagrams for an RC aplanatic telescope
       with the geometry of the ESO 3.5 m NTT (f/11; $m_2 = -5$)
       for an optimum field curvature $r_c = -1881$ mm............... 100
3.5   The function $f(m_2) = \frac{(m_2+1)(m_2-1)^2}{m_2^3}$ for DK telescopes.
       The left-hand curve refers to Cassegrain solutions,
       the right-hand curve to Gregory solutions, if the image is real ... 104
3.6   Spot-diagrams for a DK Cassegrain telescope with the geometry
       of the ESO 3.5 m NTT (f/11; $m_2 = -5$) for a flat field.
       Compare with Fig. 3.4 where the field is 10 times larger 106
3.7   Spot-diagrams for an SP Cassegrain telescope with the geometry
       of the ESO 3.5 m NTT (f/11; $m_2 = -5$), for a flat field.
       Compare with Fig. 3.6 with field $4\frac{1}{2}$ times larger and Fig. 3.4
       with field 45 times larger 110
3.8   Karl Schwarzschild's first impractical telescope solution fulfilling
       four Seidel conditions [3.1] 113
3.9   Karl Schwarzschild's original aplanatic telescope (1905)
       [3.1][3.13] 114

3.10 Spot-diagrams for the Schwarzschild telescope 1905 [3.1]
for an aperture of 1 m with f/3.0 118

3.11 Geometrical construction from the sine condition of the form
of an RC telescope compared with a classical Cassegrain
(from Danjon and Couder [3.24(a)]) 119

3.12 The Couder (aplanatic) anastigmatic telescope (1926) [3.25] 124

3.13 Spot-diagrams for the Couder telescope (1926) [3.25]
for an aperture of 1 m with f/3.0 125

3.14 Third order spherical aberration as wavefront aberration 129

3.15 Third order spherical aberration: longitudinal and lateral forms . 130

3.16 Third order coma as wavefront aberration 132

3.17 Third order coma: lateral aberration form 132

3.18 Third order coma: the "coma patch" 133

3.19 Third order astigmatism: wavefront aberration reversal
in the t- and s-sections due to the $\cos 2\phi$ term 136

3.20 Third order astigmatism: astigmatic surfaces and lines 136

3.21 Tangential and radial astigmatic lines at the t-focus
and s-focus respectively 137

3.22 Distortion: (a) barrel, (b) pincushion 139

3.23 Stop-shift effect for a single third order aspheric plate shifted
from the pupil 141

3.24 Heights of the paraxial aperture and paraxial principal rays
as they pass through a Cassegrain telescope 143

3.25 Fundamental form of a wide-field telescope
without correction of spherical aberration 149

3.26 The Bouwers concentric telescope (1941) 151

3.27 The Schmidt telescope (1931) 151

3.28 Profile function $(\rho_{pl}^4 - k_{pl}\rho_{pl}^2)$ for Schmidt corrector plates
with various values of the form profile parameter k_{pl}.
The glass plate is formed by considering the area under the curves
to be filled with glass down to an abscissa tangential to the curve
in question. To the resulting axial thickness, the constant
thickness $(d_{pl})_0$ is added to give the necessary minimum plate
thickness. (After Bahner [3.5]) 154

3.29 Spot-diagrams for the ESO 1 m, f/3.0 Schmidt telescope
with the original singlet corrector plate. Optimum curved field
of radius 3050 mm and $\pm 3.20°$ field for 24 cm $\times$ 24 cm plates ... 162

3.30 The dispersion function for a typical optical glass $\left(\frac{1}{\nu_a} = \frac{n_1' - n_0'}{n_0' - 1} \right)$. 163

3.31 The effect of achromatisation: the dispersion function is rotated
to minimize its slope, giving desired zero points λ_1 and λ_2 164

3.32 The optical glass diagram (from the Schott Catalogue,
courtesy Hans F. Morian and the Schott Glaswerke, Mainz) 165

3.33   Spot-diagrams for the ESO 1 m, f/3.0 Schmidt telescope with
       the achromatic (doublet) corrector plate (glasses UBK7 and
       LLF6). Optimum curved field of radius 3050.5 mm and ±3.20°
       field for 24 cm × 24 cm plates (format identical with Fig. 3.29,
       but scale five times larger) 166
3.34   The basic form of the Maksutov telescope (1944) 167
3.35   Spot-diagrams for the "short" Maksutov telescope of Table 3.11
       with aperture 400mm and f/3.0 171
3.36   Effect of stop shift on transverse chromatic aberration C_2
       in a Maksutov meniscus: the "short" version (b) causes refraction
       and dispersion of the principal ray at the first surface,
       but the effect is largely (though not entirely) compensated
       at the second surface 172
3.37   Spot-diagrams for a "short" Maksutov telescope
       with aperture 400mm and f/3.0 optimized
       with an achromatic field flattener (Table 3.12) 173
3.38   The solid Schmidt camera in the direct form (a)
       and folded form (b), with effective focal length f'/n' 175
3.39   The semi-solid Schmidt camera with effective focal length f'/n' . . 176
3.40   The Wright-Väisälä telescope (1935) 177
3.41   Spot-diagrams for a Wright-Väisälä telescope
       of aperture 400 mm and f/4.0 179
3.42   Schmidt-Cassegrain systems proposed by Baker (1940) 181
3.43   Monocentric (concentric) Schmidt-Cassegrain
       proposed by Linfoot (1944) 183
3.44   Spot-diagrams for a Linfoot monocentric Schmidt-Cassegrain
       with spherical mirrors and a singlet (non-achromatic)
       corrector plate (400 mm, f/3.0 - f/6.0) 187
3.45   Typical modern aplanatic Schmidt-Cassegrain
       for advanced amateur use with aperture 400 mm, f/2 - f/10 188
3.46   Spot-diagrams for the aplanatic Schmidt-Cassegrain system
       of Fig. 3.45 with 400 mm, f/2 - f/10,
       and an achromatic corrector plate 192
3.47   Spot-diagrams for a Slevogt aplanatic Schmidt-Cassegrain
       with 400 mm, f/2.0 - f/3.25, and an achromatic corrector plate.
       The field is flat 194
3.48   Two-glass concentric (monocentric)
       Bouwers-Cassegrain telescope 195
3.49   Spot-diagrams for an achromatic, monocentric
       Bouwers meniscus-Cassegrain telescope as in Fig. 3.48, but with
       a singlet field flattener added. The geometry is lightly modified
       to 400 mm, f/3.11 - f/6.0, and the stop is shifted
       to the meniscus 197

3.50   Classical Bouwers telescope with additional weak lens
        at the stop... 198
3.51   Spot-diagrams of a classical Bouwers telescope (prime focus)
        with weak achromatising positive lens in the pupil (400 mm, f/3).
        The angular field is large (±3°) 199
3.52   Spot-diagrams of a classical Bouwers-Cassegrain with weak
        achromatising lens (400 mm, f/3.0 - f/6.0), as shown in Fig. 3.50 200
3.53   Maksutov-Cassegrain in "long" and "short" versions
        with secondary separated from the meniscus. Example with
        aperture 400 mm, f/3.5 - f/10.71 202
3.54   Spot-diagrams for the "short" version of the Maksutov-Cassegrain
        of Fig. 3.53 and Table 3.14 for an optimum curved field 203
3.55   A Maksutov-Cassegrain with secondary combined
        with the meniscus. Aperture 400 mm, f/3.5 - f/15.20 204
3.56   Spot-diagrams of the Maksutov-Cassegrain of Fig. 3.55......... 205
3.57   Hawkins-Linfoot Schmidt-Bouwers telescope with f/1.2
        in the prime focus 208
3.58   Spot-diagrams for the Hawkins-Linfoot monocentric Cassegrain
        of Fig. 3.57, aperture 400 mm, f/3.0 - f/6.0 210
3.59   Baker Super-Schmidt, with $f' = 200$ mm, effective aperture ratio
        f/0.82 and field ± 26° 211
3.60   Baker-Nunn camera, designed for satellite tracking,
        aperture 508 mm, f/1.0..................................... 211
3.61   Double-meniscus system due to Wynne [3.59] (schematic)....... 212
3.62   Double-meniscus system due to Wynne with strongly
        asymmetric meniscus thicknesses [3.59] (schematic) 213
3.63   Buchroeder design of a Houghton-type corrector
        in Schmidt geometry (200 mm, f/3) given by Rutten
        and van Venrooij [3.12(g)] 214
3.64   Lurie design of a Houghton-type corrector
        in Wright-Väisälä camera geometry (200 mm, f/4)
        given by Rutten and van Venrooij [3.12(g)] 215
3.65   Spot-diagrams for a modified Lurie-Houghton design
        with aperture 400 mm at f/3.5 and geometry like the
        Wright-Väisälä system of Fig. 3.40 216
3.66   Original Mangin system for searchlight projection 217
3.67   Dialyte telescope due to Plössl (1850) 219
3.68   Brachymedial due to Hamilton (1814) 220
3.69   Brachymedial due to Schupmann (1899) 220
3.70   The Medial telescope due to Schupmann...................... 221
3.71   Compact system using Mangin secondary
        and Brachymedial geometry due to Delabre 222
3.72   Two 3-mirror anastigmatic, flat-field solutions proposed
        by Korsch (1972): (a) single-axis system, (b) 2-axis system 225

3.73 3-mirror system due to Paul (1935) 227

3.74 The Willstrop Mersenne-Schmidt telescope
      with f/1.6 and a 4° diameter field (1984) 230

3.75 Baker 3-mirror, 2-axis anastigmatic telescope (1945) 231

3.76 Dual-purpose Newton telescope due to Loveday (1981) 232

3.77 3-mirror system proposed by Robb (1978) 239

3.78 3-mirror system given by Laux (1993) for a fast,
      flat-field 2.5m wide-field survey telescope with f/2.18 primary
      and f/4.0 final image, with a field diameter of 2.0° to 2.5° 239

3.79 3-mirror, 4-reflection telescope proposed by Korsch (1991)
      for a future large space telescope 240

3.80 *First solution* of a 2-axis system with 4 powered mirrors
      (spherical primary and secondary) and a folding flat (f/1.5
      and f/7.29), proposed by Wilson and Delabre (1993, 1995) 243

3.81 Spot-diagrams of the first, 2-axis solution of Table 3.19
      and Fig. 3.80: (a) axis to ± 9 arcmin with circle 0.20 arcsec;
      (b) ± 12 arcmin to ± 18 arcmin with circle 1.00 arcsec 245

3.82 First solution, 2-axis system as in Fig. 3.80,
      but with two identical "Nasmyth-type" foci 246

3.83 *Second*, 2-axis solution with 4 powered mirrors (spherical
      primary and secondary) and a folding flat (f/1.5 and f/6.01),
      proposed by Wilson and Delabre (1993, 1995) 247

3.84 Spot-diagrams of the second, 2-axis solution of Fig. 3.83:
      (a) axis to ±9 arcmin with circle 0.2 arcsec
      (b) ± 12 arcmin to ± 18 arcmin with circle 1.00 arcsec 248

3.85 Single-axis, 4-mirror system with
      f/1.2 - f/2.657
      giving a field diameter of 1.50°. The primary is spherical 249

3.86 Spot-diagrams for the fast, wide-field, 4-mirror design of Fig. 3.85.
      The circle diameter is 1 arcsec 249

3.87 Single-axis, 4-mirror concept for a fast, wide-field telescope
      with improved field curvature 250

3.88 Single-axis, 4-mirror system using an afocal feeder
      and a spherical primary 250

3.89 A 2-axis system with 5 powered mirrors capable of a fast
      output beam (faster than f/3.0) and a flat field. The primary
      and secondary mirrors are spherical as in Fig. 3.83 251

3.90 A 2-axis solution with 4 powered mirrors proposed
      by Sasian (1990). Either M_1 or M_2 is spherical, M_3 is toroidal .. 252

3.91 2-axis form of the system of Fig. 3.85 proposed by Baranne
      and Lemaître (1986), the mirror pair M_3M_4 forming a corrector
      and focal transfer system with a magnification of –1
      in the TEMOS concept, giving f/2.0 - f/4.5 - f/4.5 253

3.92   Double-Cassegrain 4-mirror telescope with intermediate image
after M_2, proposed by Korsch (1986) 254

3.93   The Kutter Schiefspiegler [3.103] [3.104] showing 3 solutions
(after Rutten and van Venrooij [3.12(j)]) 256

3.94   The basis of coma compensation
in a Czerny-Turner monochromator 257

3.95   Schiefspiegler achieved by off-axis sections of a centered,
2-mirror telescope 261

3.96   Schiefspiegler interpretation of lateral decentering
in a Cassegrain telescope 261

3.97   Strict case of lateral decenter in a 2-mirror telescope 263

3.98   Schiefspiegler with spherical primary and insensitive
to lateral decenter [3.114] 277

3.99   Exit pupil (x, y, z) and image plane (η, ξ, ζ) coordinate systems . . 294

3.100 Fraunhofer diffraction at a rectangular aperture showing
the function $I = [\frac{\sin(\delta \overline{u}')}{\delta \overline{u}'}]^2$ (after Born-Wolf [3.120(b)]) 296

3.101 Fraunhofer diffraction pattern of a rectangular aperture
8 mm × 7 mm, magnification 50×, $\lambda = 579$ nm. The centre
was deliberately overexposed to reveal the secondary maxima
(after Born-Wolf [3.120(b)] and Lipson, Taylor and Thompson,
courtesy Brian Thompson) 297

3.102 Fraunhofer diffraction at a circular aperture showing
the function $I = [\frac{2J_1(\overline{w})}{\overline{w}}]^2$ (after Born-Wolf [3.120(b)]) 299

3.103 Fraunhofer diffraction at a circular aperture 6 mm in diameter,
magnification 50×, $\lambda = 579$ nm. The central maximum
has been overexposed to reveal the weak subsidiary maxima.
(After Born-Wolf [3.120(b)] and Lipson, Taylor and Thompson,
courtesy Brian Thompson) 300

3.104 Energy encircled in the radius w of the pattern
due to Fraunhofer diffraction at a circular aperture. The fraction
of energy is given by $L_w = 1 - J_0^2(\overline{w}) - J_1^2(\overline{w})$ with $\overline{w} = k\rho_m w$.
(After Born-Wolf [3.120(b)]) 301

3.105 The diffraction PSF at an annular aperture showing the effect
of increasing the central obscuration factor ε
(after Born-Wolf [3.120(c)] and G.C. Steward [3.132]) 303

3.106 Idealized sinusoidal ripple showing 3 complete wavelengths
of ripple in the pupil radius 309

3.107 Transfer of a sinusoidal wave with reduced contrast
through an optical system
(a) without phase shift, (b) with phase shift $\overline{p}$ 313

3.108 Shearing of a circular pupil corresponding to the calculation
of the OTF from the autocorrelation function
for shears of λRs, λRt 316

3.109 One-dimensional pupil shear to demonstrate
the low bandpass filter function of an optical system 317
3.110 The MTF for a circular pupil free from aberrations
and obstruction, corresponding to the diffraction PSF
with incoherent illumination 318
3.111 MTF for a circular pupil free of aberration
with central obstruction ε (after Lloyd [3.140]) 319
3.112 The MTF for a non-obstructed circular aperture
with incoherent illumination in the presence
of pure defocus aberration (after Hopkins [3.142]) 321
3.113 Normal form of supporting spider for secondary mirrors
shown here without central obstruction of the secondary 323
3.114 Typical astronomical photograph of a star field where diffraction
spikes appear on the bright star images. The galaxy is NGC 253,
photographed with the 2.2 m telescope at La Silla
with 40 m exposure. (Courtesy ESO) 323

Chapter 4

4.1   Ghost images through 2 reflections at a plane parallel plate 326
4.2   Transformation of a real corrector plate
to a virtual plate in object space 328
4.3   Spot-diagrams for the PF Gascoigne plate-field flattener
corrector (with filter) of the ESO 3.6 m telescope on La Silla.
The Schwarzschild constant of the primary is $b_{s1} = -1.1567$
for a quasi-RC solution 332
4.4   Schematic appearance of aspherics
on PF plate correctors: (a) Gascoigne plate (singlet) corrector
for RC hyperbolic primaries; (b) 2-plate corrector for strongly
hyperbolic primaries correcting $\sum S_I = \sum S_{II} = \sum S_{III} = 0$;
(c) 2-plate corrector for parabolic or RC primaries correcting
$\sum S_I = \sum S_{II} = 0$; (d) 3-plate corrector for parabolic
or RC primaries correcting $\sum S_I = \sum S_{II} = \sum S_{III} = 0$ 338
4.5   Conjugate virtual plate in object space for a real aspheric plate
at distance g in front of the Cassegrain focus 341
4.6   Spot-diagrams for a quasi-RC telescope
with Gascoigne plate corrector and field flattener
based on the geometry of the ESO 3.6 m telescope (f/3 - f/8) . . . 346
4.7   The 3-lens Ross corrector for the Mt. Palomar 5 m, f/3.3
parabolic primary (schematic, after Wynne [4.5] [4.24]) 353
4.8   Reflector-corrector due to Baker [4.27] 357
4.9   Wynne design for a 4-lens corrector of the Palomar 5 m, f/3.34
paraboloid. The cross shows the focus of the naked primary
(after Wynne [4.28]) 359

4.10   Spot-diagrams for the Wynne design of Fig. 4.9: (a) on axis,
       (b) 6 arcmin off-axis, (c) 9 arcmin
       off-axis, (d) 12.5 arcmin off-axis.
       The circle has a diameter of 0.5 arcsec.
       (After Wynne [4.28]) 360
4.11   3-lens corrector for paraboloids using one aspheric
       by Faulde and Wilson [4.30]. 360
4.12   Spot-diagrams for the 3-lens corrector of Fig. 4.11.
       The circle represents 0.5 arcsec. (a) Basic focus 361
4.12   Spot-diagrams for the 3-lens corrector of Fig. 4.11.
       The circle represents 0.5 arcsec. (b) Focus shift +0.05 mm,
       (c) focus shift −0.05 mm 362
4.13   3-lens corrector by Wynne [4.14] [4.5] for the Kitt Peak 3.8 m,
       f/2.8 - f/8 RC telescope. All three lenses are of UBK7.
       (After Wynne) 364
4.14   Spot-diagrams for two correctors for the ESO 3.6 m
       quasi-RC telescope: (a) the basic plate system of Köhler
       with field flattener, recalculated without vignetting;
       (b) the Wynne-type lens corrector also recalculated
       without vignetting. (After Cao and Wilson [4.16]) 365
4.15   3-lens prime focus corrector designed by Richardson et al. [4.34]
       for the then proposed 7.6 m, f/2 primary of University of Texas.
       (After Richardson et al.) 367
4.16   3-lens PF corrector for the 10 m Keck primary
       (after Epps et al. [4.36]) 368
4.17   Spot-diagrams for the system of Fig. 4.16 for two of the four
       spectral bands given (after Epps et al. [4.36]) 369
4.18   Spot-diagrams for the new Delabre prime focus ($f/3.0$) corrector
       for the ESO 3.6 m telescope using two glasses, Schott FK5
       and fused silica (1996) 371
4.19   Spot-diagrams for the singlet lens corrector of the ESO 3.6 m
       quasi-RC telescope (after Wilson [4.15]). Circle 0.18 arcsec 383
4.20   Three-element corrector for an f/2.00 to f/5.28 classical Cassegrain
       designed by Epps et al. [4.36] for a 300-inch telescope,
       giving f/6.00 with the corrector 385
4.21   Spot-diagrams for the design of Fig. 4.20 for a field of 1° diameter
       (Epps et al. [4.36]) 386
4.22   Spot-diagrams for the 2.2 m MPIA telescope as *quasi*-RC using
       a 2-lens corrector of one glass (quartz) (after Wilson [4.15]).
       Circle 0.47 arcsec 387
4.23   Spot-diagrams for the 2.2 m MPIA telescope as *strict* RC using
       a 2-lens corrector of 2 different glasses (PK50 and BaF$_3$) (after
       Wilson [4.15]) Circle 0.48 arcsec 388

4.24 Spot-diagrams for a doublet corrector using a single glass
      designed by Su, Zhou and Yu [4.51] for the Chinese 2.16 m,
      strict RC telescope with f/3 to f/9 389
4.25 Spot-diagrams for a three-element corrector designed by Epps
      et al. [4.36] for an f/1.80 – f/4.50 Cassegrain system
      using a significantly hyperbolic primary $(b_{s1} = -1.152\,310\,5)$ 391
4.26 Dispersion variation by opposite rotation of two prism pairs..... 394
4.27 Performance of the ADC designed by Wynne and Worswick
      for the William Herschel 4.2 m telescope with $Z = 70°$.
      Curve A shows the uncorrected atmospheric dispersion; curve B
      the correction achieved with the glasses used (UBK7 and LLF6);
      curve C what could be achieved with FK50 and
      Calcium Fluoride. (After Wynne and Worswick [4.60]) 395
4.28 Configurations of prisms for maximum dispersion setting,
      in cases 1 and 2 (angles exaggerated) for ADC
      in the prime focus (PF). (After Wynne and Worswick [4.62]) 396
4.29 (a) Spot-diagrams, reproduced from Wynne and Worswick [4.62],
      for their PF ADC for the 4.2 m, f/2.5 William Herschel Telescope.
      (a) Zero dispersion setting 397
4.29 (b) and (c) Spot-diagrams, reproduced from Wynne
      and Worswick [4.62], for their PF ADC for the 4.2 m, f/2.5
      William Herschel Telescope. (b) At ± 45°
      and (c) maximum dispersion setting ± 90° 398
4.30 Section through the complete ADC-corrector system for the
      4.2 m, f/2.5 PF of the WHT, reproduced from Bingham [4.63]... 399
4.31 "Lensm" design of Su [4.64] for an ADC integrated into a doublet
      corrector for a 5 m, f/2 to f/4.5 strict RC (Cassegrain) focus.
      The glasses are from the Chinese glass catalogue............... 399
4.32 Two types of lensm corrector designed by Wang and Su [4.65]
      for the PF of a 7.5 m, f/2 paraboloid........................ 400
4.33 Points in the field for the calculation of spot-diagrams,
      from Wang and Su [4.65] 400
4.34 Spot-diagrams for lensm corrector type I for a field diameter
      of 45 arcmin with a 7.5 m, f/2 paraboloid. Rotation angles of
      the lensms are 0°, 0°, i.e. zero dispersion. Circle diameter = 1
      arcsec. Reproduced from Wang and Su [4.65] 401
4.35 Spot-diagrams for lensm corrector type II for a field diameter
      of 45 arcmin with a 7.5 m, f/2 paraboloid. Rotation angles
      of the lensms are 0°, 0°, i.e. zero dispersion.
      Circle diameter = 1 arcsec. Reproduced from Wang and Su [4.65] 401
4.36 Optical design of the LADC – schematic
      (after Avila, Rupprecht and Beckers [4.78]) 402

4.37 Basic design (schematic) for an FR *without* intermediate image
      for a field of 0.9° diameter at the Cassegrain (RC) focus
      of the 3.5 m MPIA, f/3 to f/8 telescope [4.15]. $m_{FR} = 1/2.67$ ... 407
4.38 Spot-diagrams for the FR system of Fig. 4.37 [4.15].
      The circle is 0.98 arcsec = 50μm.......................... 408
4.39 Basic design (schematic) for an FR *with* intermediate image
      for a field of 0.9° diameter at the Cassegrain (RC) focus
      of the 3.5 m MPIA f/3 to f/8 telescope [4.15]. $m_{FR} = 1/2.67$ ... 409
4.40 Spot-diagrams for the FR system of Fig. 4.39 [4.15].
      The circle is 0.98 arcsec 410
4.41 Basic focal reducer geometry using a Schmidt-based
      mirror system for the 3.5 m MPIA telescope.
      Reduction is f/8 to f/1.7 ($m_{FR} = 1/4.71$) [4.15]............... 410
4.42 Spot-diagrams for a focal reducer f/8 to f/1.7 designed
      for the 3.5 m MPIA telescope for a field diameter of 1°.
      Doublet corrector (corrected intermediate image), one field lens
      and a Hawkins-Linfoot camera. Circle = 0.50 arcsec = 14 μm.
      Image radius 1031 mm [4.15] 412
4.43 Spot-diagrams for a focal reducer f/8 to f/1.7 designed
      for the 3.5 m MPIA telescope for a field diameter of 1°.
      Doublet corrector (*un*corrected intermediate image), 2 field lenses
      and Baker-type camera with 2 menisci and one plate.
      Circle = 0.50 arcsec = 14 μm. Plot field for
      axial spot-diagrams = 10 μm. Image radius = 1052 mm [4.15] ... 413
4.44 Focal reducer designed by Meinel et al. [4.72]
      for the Texas 7.6 m telescope project. The f/13.5 Nasmyth focus
      is converted to f/3.0 over a field of 8 arcmin diameter 415
4.45 INCA (Inverted Cassegrain) focal reducer
      proposed by MacFarlane [4.73] 415
4.46 FR concept using a field mirror proposed by MacFarlane [4.73] .. 416

Chapter 5

5.1 90 cm lightweighted, built-up blank made by Lord Rosse in 1839
     (courtesy Rolf Riekher)..................................... 420
5.2 Whiffle-tree support system in 4 stages designed by
     Thomas Grubb for the Rosse 6-foot reflector completed in 1845
     (courtesy Rolf Riekher)..................................... 421
5.3 Lord Rosse's 6-foot (1.82 m) telescope completed in 1845
     (courtesy Deutsches Museum, Munich) 422
5.4 Original single astatic counterweight described by Lassell in 1842
     [5.7] (reproduced from Danjon and Couder [5.1]) 423
5.5 Lassell's 1.22 m telescope set up in 1861 in Malta
     (reproduced from the original plate of [5.8]) 424

5.6   James Nasmyth's 20-inch Cassegrain-Nasmyth telescope
       about 1845 (reproduced from King [5.2]) 425
5.7   The 4-foot (1.22 m) Melbourne reflector erected in 1869
       (reproduced from King [5.2]) 427
5.8   The casting of the first 48-inch speculum blank for the
       Melbourne reflector (courtesy Royal Astronomical Society,
       through Peter Hingley, original engraving reproduced
       in the Strand Magazine, London, Vol. XII, 1896, p. 372) 428
5.9   Foucault's largest (80 cm) silver-on-glass reflector,
       completed in 1862 (reproduced from King [5.2]) 430
5.10   The 36-inch (91 cm) Crossley reflector,
       remounted at Lick in 1900 (courtesy Mary Lea Shane Archives
       of the Lick Observatory, through D. E. Osterbrock) 433
5.11   Ritchey's 60 cm telescope at Yerkes, 1901
       (courtesy Deutsches Museum, Munich) 434
5.12   Ritchey observing at the Newton focus of the 60 cm Yerkes
       reflecting telescope between 1901 and 1904
       (courtesy Yerkes Obervatory, through D.E. Osterbrock) 436
5.13   Ritchey's 60-inch Mt. Wilson reflector: optical arrangement
       (courtesy Rolf Riekher)..................................... 437
5.14   Ritchey's 60-inch Mt. Wilson reflector, 1908
       (courtesy Donald Osterbrock and the Observatories
       of the Carnegie Institute of Washington) 439
5.15   The 100-inch Hooker telescope at Mt. Wilson (1917)
       (courtesy Deutsches Museum, Munich, and acknowledgement
       to the Observatories of the Carnegie Institute of Washington) ... 441
5.16   Pease's concept for a 300-inch (7.5 m) telescope in 1921
       (reproduced from Dimitroff and Baker [5.17],
       courtesy Churchill Livingstone, Edinburgh) 442
5.17   George Willis Ritchey in Paris, 1927, with a built-up
       cellular mirror disk (courtesy D. E. Osterbrock,
       photograph by James Stokley) 443
5.18   The primary of the 200-inch Mt. Palomar telescope in testing
       position with John A. Anderson (left) and Marcus H. Brown
       (courtesy Palomar/Caltech).................................. 446
5.19   The Palomar 200-inch (5 m) telescope as drawn by R.W. Porter
       (courtesy Palomar/Caltech).................................. 448
5.20   The Russian 6 m telescope at the Zelenchuk Observatory
       in the Caucasus (courtesy "Ciel et Espace", Paris,
       through Serge Brunier)..................................... 451
5.21   The 4.0 m Kitt Peak telescope
       (courtesy National Optical Astronomy Observatories, Tucson) ... 458
5.22   The 3.6 m ESO telescope (courtesy ESO)..................... 459

5.23   The building of the 3.6 m ESO telescope with the smaller
       building of the 1.4 m Coudé Auxiliary Telescope (CAT)
       on La Silla (2400 m) (courtesy ESO) 460
5.24   The 4.2 m William Herschel Telescope (WHT) with its *Alt-Az*
       mounting on La Palma (Roque de los Muchachos 2400 m)
       (courtesy Royal Greenwich Observatory,
       through Richard Bingham and Peter Andrews) 461
5.25   The original optical layout of the "Universal Telescope"
       of the Karl-Schwarzschild Observatory at Tautenburg, Germany
       (courtesy Rolf Riekher)................................... 462

List of Tables

Chapter 2

2.1 Gaussian optics of a prime focus reflecting telescope
with a single powered mirror: sign of the paraxial parameters . . . 41

2.2 Gregory and Cassegrain telescope forms:
sign of paraxial ray trace quantities
(*denotes sign inversion between Gregory and Cassegrain) 44

2.3 Signs of derived quantities from the paraxial ray trace
for the Gregory and Cassegrain forms 53

Chapter 3

3.1 Aberration types from the Characteristic Function 61

3.2 Paraxial values for deriving the Seidel coefficients (Table 3.3)
for some basic telescope systems 67

3.3 Seidel coefficients for some basic telescope systems.
The asterisk denotes the aspheric contribution 68

3.4 Third order aberrations for a 1-mirror telescope
(concave primary) 80

3.5 Third order aberrations and associated relations
for a 2-mirror telescope in *focal* form 81

3.6 Third order aberrations and associated relations
for a 2-mirror telescope in *afocal* form 82

3.7 Schwarzschild's data for the aplanatic telescope of Fig. 3.9 115

3.8 Comparison of the essential parameters in the evolution
of the aplanatic telescope (from [3.13]) 116

3.9 Constructional data of the Couder
anastigmatic telescope (1926) [3.25] 126

3.10 Angular spherical aberration, coma and astigmatism
for three telescope cases of Table 3.3, with an f/10 image beam
and a semi-field angle u_{pr1} of 30 arcmin 140

3.11 Data for the "short" Maksutov telescope
giving the results of Fig. 3.35 with $D = 400$ mm, f/3.0 170

3.12 Data for the "short" Maksutov system ($D = 400$ mm and f/3.0)
      optimized with an additional achromatic field flattener,
      giving the results of Fig. 3.37 172
3.13 Optical data of Baker Schmidt-Cassegrain Type B
      with f/3.0 and $f' = 1$ 180
3.14 Design data for the Maksutov-Cassegrain of Fig. 3.53. Aperture
      400 mm, f/3.5 - f/10.71 201
3.15 d_{80} values (80% encircled energy diameters in μm)
      for the "short" system of Fig. 3.53 and Table 3.14 202
3.16 Design data for the Maksutov-Cassegrain of Fig. 3.55 204
3.17 Surface contributions for the aplanatic Maksutov-Cassegrain
      of Table 3.14 207
3.18 Data of the system of Fig. 3.72(a), adapted from Korsch [3.73] . . 226
3.19 Optical design data of the first, 2-axis solution
      with 4 powered mirrors and flat of Fig. 3.80 with primary f/1.5
      and final image f/7.29 244
3.20 Angular tangential coma produced by transverse decentering
      of the secondary in the 2-mirror telescopes of Table 3.2.
      $|m_2| = 4$; $|R_A| = 0.225$. The relative aperture at the final image
      is $N = 10$. The decenter is $|\delta/f'| = 10^{-4}$ 267
3.21 Angular despace spherical aberration at best focus (BF)
      and angular despace field coma for 2-mirror telescopes defined
      as in Table 3.2: $|m_2| = 4$, $|R_A| = 0.225$, $|N| = 10$
      and the despace is $|dd_1/f'| = 10^{-3}$, or $dd_1/f_1' = -4 \cdot 10^{-3}$,
      ten times the decenter δ of Table 3.20. The semi-field angle
      for the coma is $u_{pr1} = 15$ arcmin 284
3.22 The Zernike radial polynomials $R_n^m(\rho)$ up to degree 8
      (after Born-Wolf [3.120(a)]) 291
3.23 Zernike polynomials resolved in the x, y directions. This table
      gives all terms up to R_5^5 and subsequent terms up to $n = 10$
      with $n + m \leq 10$ (after Dierickx [3.122]) 292
3.24 The first five maxima of the rectangular aperture function
      $I = [\frac{\sin(\delta\overline{u}')}{\delta\overline{u}'}]^2$ (after Born-Wolf [3.120(b)]) 296
3.25 The first 3 subsidiary minima and maxima
      of the function $[\frac{2J_1(\overline{w})}{\overline{w}}]^2$ (after Born-Wolf [3.120(b)]) 300
3.26 Coefficients of various aberrations giving a Strehl Intensity Ratio
      of 0.8 (incoherent light, point source, circular unvignetted pupil) . 309
3.27 The effect of the obscuration factor ε on the wavefront variance
      and the ratio (ptv/rms) with $\varepsilon = 0$ 311

Chapter 4

4.1   Refractive index values n_a as a function of wavelength λ, taken
      from von Hoerner and Schaifers [4.56], for 760 Torr and $0°C$ 393

Chapter 5

5.1 Comparison of the lateral decentering coma with $\delta = 1$ mm of the
60-inch Mt. Wilson telescope and Ritchey's 40-inch RC telescope
for the US Naval Observatory 445

5.2 Principal optical characteristics of major telescopes following the
Palomar 5 m telescope and with a conventional optical concept .. 456

Appendix

A.1   List of symbols for Chapter 2 467
A.2   List of symbols for Chapter 2 (continued) 468
A.3   List of symbols for Chapter 2 (continued) 469
A.4   List of symbols for Chapter 2 (continued) 470
A.5   List of symbols for Chapter 2 (concluded) 471
A.6   Additional symbols for Chapter 3 471
A.7   Additional symbols for Chapter 3 (continued) 472
A.8   Additional symbols for Chapter 3 (continued) 473
A.9   Additional symbols for Chapter 3 (continued) 474
A.10 Additional symbols for Chapter 3 (continued) 475
A.11 Additional symbols for Chapter 3 (continued) 476
A.12 Additional symbols for Chapter 3 (continued) 477
A.13 Additional symbols for Chapter 3 (continued) 478
A.14 Additional symbols for Chapter 3 (continued) 479
A.15 Additional symbols for Chapter 3 (continued) 480
A.16 Additional symbols for Chapter 3 (continued) 481
A.17 Additional symbols for Chapter 3 (continued) 481
A.18 Additional symbols for Chapter 3 (continued) 482
A.19 Additional symbols for Chapter 3 (concluded) 483
A.20 Additional symbols for Chapter 4 484
A.21 Additional symbols for Chapter 4 (continued) 485
A.22 Additional symbols for Chapter 4 (concluded) 486
A.23 Additional symbols for Chapter 5 486

Name Index

Abbe, E. 85, 86, 89
Adams, W.S. 438
Airy, G.B. 259, 299, 392
Aldis, H.L. 63, 232
Amon, M. 218
Anderson, E. 368
Anderson, J. 446, 447
Andrews, P. 461
Angel, J.R.P. 368
Ardeberg, A. 249
Avila, G. 403
Ayscough, J. 14

Babinet, J. 322
Bahner, K. 45, 63, 65, 70, 72, 77, 101,
    144, 146, 154, 157, 161, 163, 167, 176,
    178, 209, 266, 279, 281, 293, 354, 360,
    442, 463
Baker, J.G. 117, 161, 176, 178, 180–
    183, 191, 209, 228, 229, 231, 239, 241,
    248, 322, 357, 432, 442, 463
Baranne, A. 249, 250, 253, 266, 366
Barlow, P. 5, 404
Bass, G. 14
Baxandall, D. 13
Beckers, J. 403
Beddoes, D. 393
Bennett, H.F. 212
Bertola, F. 2
Bhatia, A.B. 288
Bingham, R. 399, 456, 461
Bird, J. 14
Black, G. 320
Born, M. 27, 291, 293–301, 303–305,
    315, 320, 322
Boulesteix, J. 414
Bouwers, A. 150, 152, 165, 167, 169,
    172, 174, 193, 195, 196, 198, 206, 208

Bowen, I.S. 345, 373, 445, 452–454
Bradford, W.R. 209
Bradley, J. 11
Brashear, J. 432, 440
Breckenridge, J.B. 393, 394
Brodie, J.P. 369
Brown, M. 446, 447
Bruns, H. 86
Buchdahl, H.A 87
Buchroeder, R.A. 214, 260
Burch, C.R. 140, 232, 327, 341, 377
Busack, H.J. 222, 223

Calver, G. 432, 434
Cao, C. 340, 364–367
Caravaggi, C. 2
Caravaggio, M.M. da 2
Cassegrain, L. 5, 9, 10, 13, 88, 90, 91,
    426
Cassini, J.D. 392
Chrétien, H. 88, 94, 111, 117, 148, 174,
    325, 327, 372, 373, 442
Clairaut, A. 14
Clark, A. 431
Clausius, R. 85
Common, A.A. 432
Comstock, G.C. 392
Conrady, A.E. 45, 129, 167, 266
Cook, L.G. 278
Couder, A. 1, 7, 116, 119, 122, 123,
    419, 423, 450, 455, 463
Cox, J. 8
Cox, R.E. 323
Crossley, E. 432
Czapski, S. 86, 167, 217
Czerny, M. 256

Daley, J. 222

Dall, H.E.   88, 101, 455
Dallmeyer, J.H.   5
Danjon, A.   1, 7, 119, 419, 423, 463
Davis, J.   209
Day, A.L.   438
de Bercé   9, 10
de Meijère, J.L.F.   87
De, M.   320
Delabre, B.   222, 243, 247, 248,
    370–372, 416
della Porta, G.   22
Descartes, R.   3–6, 8, 10, 13, 88, 419,
    445
Dierickx, P.   290, 292
Dimitroff, G.Z.   117, 178, 322, 432, 442,
    463
Ditchburn, R.W.   322
D'Odorico, S.   2
Dollond, J.   14
Dollond, P.   14
Doppler, C.   444
Draper, H.   429, 432, 433
Duffieux, P.M.   312

Eichens, F.W.   432
Ellery, R.L.J.   431
Elsässer, H.   360
Eppenstein, O.   86, 167, 217
Epps, H.W.   230, 368–370, 385, 386,
    389–392, 396
Euler, L.   14

Faulde, M.   359–361, 367, 400
Fecker, J.W.   432, 440
Fehrenbach, C.   463
Fermat, P. de   26, 27, 86
Fibonacci, L.   103
Focke, J.   87
Forster, J.   255, 256
Foucault, L.   429, 430, 432, 434, 437,
    440, 442
Fourier, J.   312
Françon, M.   293, 304, 306, 308,
    313–315, 319, 321
Fraunhofer, J.   5, 14, 293, 419
Fresnel, A.   165, 293
Fritsch, K.   255, 256

Gabor, D.   208

Galileo, G.   1, 2, 10, 15
Gallert, F.   223
Gascoigne, S.C.B.   327, 330, 334, 336,
    341, 342, 345, 348, 350, 355, 426
Gauss, C.F.   23, 28, 233
Gautier, P.   432
Geyer, E.H.   414
Gochermann, J.   463
Goodman, J.   313
Grant, R.   1
Gregory, D.   14
Gregory, J.   5, 6, 8–10, 14, 88, 90
Grubb, H.   431, 432
Grubb, T.   420, 421, 424, 426, 434, 457
Grundmann, W.A.   366
Guinand, P.   14

Hadley, J.   11–13
Hale, G.E.   433, 435, 438, 445, 447
Hall, C.M.   14
Hamilton, W.F.   220
Hamilton, W.R.   59, 60, 62, 63, 85, 255,
    288
Hansen, G.   38
Harmer, C.F.W.   366, 382, 384
Hartshorn, C.R.   405
Hawkins, D.G.   208
Helmholtz, H.   30
Hendrix, D.O.   176, 447
Henize, K.G.   209
Henneberg, P.   366
Herschel, J.   426, 429
Herschel, W.   2, 15–20, 38, 90, 320,
    419–422, 426, 431, 432, 438, 440
Herzberger, M.   29, 86, 87, 233
Hirayama, S.   322
Hockney, D.   2
Hopkins, H.H.   45, 60, 64, 141, 146,
    167, 297, 320, 321, 333, 349, 351
Houghton, J.L.   213–217
Hubble, E.   440
Humason, M.L.   440
Huygens, C.   9, 11, 167

Imperiali, B.   2
Ingalls, A.G.   444
Ioannisiani, B.K.   450

Jacquinot, P.   304

Jörsäter, S.   414

Keeler, J.E.   432
Kepler, J.   1, 3, 5, 6, 10
Kerber, A.   221
Kienle, H.   460
King, H.C.   1, 4, 7, 10, 15, 419, 425, 427, 429, 430, 433, 445
Kirkham, A.R.   88, 101
Klingenstierna, S.   14
Klügel, G.S.   63
Köhler, H.   146, 176, 178, 208, 209, 213, 340, 364, 365, 383, 384
König, A.   146, 176, 178, 208, 209, 220
Korsch, D.   223–227, 230, 231, 233, 238, 240, 242, 246, 248, 252, 254
Krautter, M.   101
Kuiper, G.P.   463
Kutter, A.   255, 256, 259, 260, 320

Lagrange, L. de   30, 31
Lansraux, G.   304
Lassell, W.   419, 420, 422–424, 426, 431–433, 437, 440
Laux, U.   84, 88, 239
Le Sueur, A.   431
Learner, R.   457
Lemaître, G.   249, 250, 253
Leonard, A.S.   260
Liebig, J.   429
Lihotzky, E.   85, 86, 120
Linfoot, E.H.   140, 160, 161, 183, 191, 208, 320
Lipson, H.   295, 297, 300
Lloyd, J.M.   319
Loveday, N.   231, 232, 238
Lundin, C.A.R.   440
Lurie, R.J.   214–217

MacFarlane, M.   414–417
Mackintosh, A.   198, 207, 260
Maksutov, D.D.   150, 165, 167, 169, 170, 172, 174, 193, 195, 196, 198, 206, 212, 449, 450
Malacara, D.   293
Malus, E.L.   86
Mangin, A.   217, 404
Maréchal, A.   266, 293, 304, 306, 308, 313–315, 319, 321

Marsili, C.   2
Martin, A.   432
Maskelyne, N.   14
Maxwell, J.   208, 209, 212, 218, 219
Mayall, N.U.   432, 452
McGraw, J.   230
Meinel, A.B.   238, 272, 327, 337, 340, 414, 415
Meinel, M.P.   272, 414, 415
Mellish   429
Melnikov, O.   450
Mersenne, M.   4–6, 9, 10, 44, 45, 93
Mertz, L.   278
Meschkowski, H.   289
Messier, C.   18
Michelson, N.   450
Middlehurst, B.M.   463
Morbey, C.L.   369

Nasmyth, J.   424–426
Nelles, B.   414
Newton, I.   1, 8–10, 13, 14, 89
Nijboer, B.R.A.   130, 133, 322

O'Neill, E.L.   319–321
Opitz, A.   259
Osterbrock, D.E.   432, 433, 440, 443, 444

Parseval, M.A.   315, 319
Pasachoff, J.M.   457
Paul, M.   227, 230, 241, 242, 278, 325, 327, 329, 333, 334, 337, 351–359, 363, 367, 381
Pease, F.G.   442, 445, 446
Penning, K.   208
Petzval, J.   5, 63, 87
Piazzi Smyth, C.   159
Plössl, G.S.   219
Porter, R.W.   447, 448
Pound, J.   11
Pythagoras   103

Rakich, A.   63, 232, 233, 278
Ramsden, J.   15, 37, 302
Rayleigh, J.W. Strutt, Lord   13, 301–303, 308, 320
Reeves, R.   8
Refsdal, I.N.   390

Richardson, E.H.   366, 367, 369
Richter, R.   213
Riekher, R.   1, 2, 176, 178, 208, 209,
    213, 219, 255, 266, 419–422, 433, 435,
    437, 445, 449, 462, 463
Ritchey, G.W.   88, 89, 94, 95, 111, 266,
    354, 431, 433–440, 442–445, 447, 452,
    463
Robb, P.N.   238, 239, 248, 278
Robinson, T.R.   426
Roizen-Dossier, B.   304
Rosin, S.   218, 357, 363, 382, 384, 390,
    414
Ross, F.E.   348–353, 358, 359, 363, 381,
    447, 453
Rosse, W. Parsons, Lord   419–422,
    426, 431, 432, 437, 438, 440, 442, 457
Rumsey, N.J.   232, 233, 238–240, 278
Rupprecht, G.   403
Russell, H.N.   392
Rutten, H.   88, 105, 186, 196, 198, 206,
    207, 209, 213–215, 217–219, 222, 231,
    256, 259, 260, 405

Sagredo, G.   2
Sampson, R.A.   325, 348, 350, 351, 353,
    354, 372, 373
Sand, R.   277, 278
Sasian, J.M.   252, 253, 416
Schafer, D.R.   254, 278
Schaifers, K.   392, 393
Scheiner, J.   322
Schielicke, R.   463
Schmidt, B.   148–151, 174, 176, 213
Schmidt-Kaler, T.   463
Schroeder, D.J.   27, 43, 44, 46, 117,
    167, 174–176, 178, 183, 186, 223, 227,
    257, 266, 279, 286, 293, 294, 304, 308,
    310, 312, 313
Schulte, D.H.   340, 345, 347, 356, 384
Schupmann, L.   219–221, 248
Schwarzschild, K.   3, 5, 59, 77, 84, 86,
    87, 89, 94, 111, 113–117, 119, 122, 123,
    126, 140, 148, 150, 151, 174, 176, 325,
    327, 372, 373, 415, 419, 445
Seidel, L.   5, 59, 61–65, 67, 68, 85, 89,
    120, 128, 148
Serrurier, M.   449
Shack, R.V.   238

Shapley, H.   404
Shectman, S.A.   254
Short, J.   13
Silvertooth, E.W.   218
Slevogt, H.   178, 183, 186, 191, 213, 266
Smith, R.   30
Smith, T.   87
Snell, W.   27
Sonnefeld, A.   213
Southall, J.P.C.   167
Staeble, F.   85, 86, 120
Steinheil, C.A.   429
Stevick, D.   278
Steward, G.C.   303, 319
Strehl, K.   301, 304
Strong, J.   241
Su, D.-q.   389, 390, 399–401
Sumita, H.   293

Takeda, M.   230
Taylor, C.A.   295, 297, 300
Theissing, H.   101
Thompson, B.J.   295, 297, 300
Thompson, E.   446
Tobin, W.   429
Turner, A.F.   256

Väisälä, Y.   176–178
van Venrooij, M.   88, 105, 186, 196,
    198, 206, 207, 209, 213–215, 217–219,
    222, 231, 256, 259, 260, 405
Varnish   429
Vasilevskis, S.   452
Vaughan, A.H.   345
Velzel, C.H.F.   87
Violette, H.   325, 372, 373, 380, 381
von Hoerner, S.   392, 393
von Rohr, M.   63

Walker, R.   444
Wallner, E.P.   392, 393, 395
Wang, Y.-n.   400, 401
Welford, W.T.   23, 24, 29, 35, 63–65,
    85, 128, 141, 146, 163, 293, 304, 306,
    308, 313–315, 318, 322, 333
Wetherell, W.B.   313, 318, 320, 321,
    392, 393, 395
Willstrop, R.V.   230, 241, 242

Wilson, R.N.   111, 243, 247, 259, 340, 359–361, 364–367, 383, 384, 387–389, 400, 406, 416
With, G.H.   432
Wolf, E.   27, 288, 290, 291, 293–301, 303–305, 315, 320, 322
Wolf, R.   463
Wolter, H.   101
Woodbury, D.O.   445
Worswick, S.P.   395–398, 401, 403
Wright, F.B.   176–178, 206
Wright, H.   445

Wynne, C.G.   212, 213, 326, 337, 340, 348, 351–354, 356–361, 363–367, 373, 375, 381–385, 389, 393–398, 400–403

Young, T.   320
Yu, X.-m.   389, 390

Zernike, F.   62, 86, 288, 297
Zhou, B.-f.   389, 390
Zimmer, H.G.   234
Zinke, O.   101
Zucchi, N.   2, 6, 8

Subject Index

Abbe number   155, 157, 186, 195, 221, 405

Abbe sine condition   37, 85, 86, 89, 117, 119–121
- graphical demonstration   119

Aberration   3, 29, 33, 41, 57, 65, 70, 89, 101, 141, 146, 158, 299, 314, 321
- angular   *see* lateral
- astigmatism   *see* Astigmatism
- axisymmetric   134
- "central"   349, 351, 374
- chromatic   **6**, **8**, 9, **10**, 14, 107, 124, **146**, 148, 152, 160, 165, 167, 182, 183, 186, 191, 193, 206, 208, 230, 260, 326, 331, 339, 347, 351, 353, 363, 364, 376, 379, 381, 383, 384, 406, 409, 411
- chromatic (lateral)   170, 196, 345, 384, 389, 390, 395, 407
- chromatic (longitudinal)   146, 163, 193, 195, 212, 214, 218, 349, 394, 396
- chromatic variation   146, 148, 394
- classical   288
- coefficient   64, 65, 139, 233, 307, 310, *see also* Seidel
- coma   *see* Coma
- decentering   261, 269, 394
- decentering astigmatism   269
- decentering coma   *see* Coma
- defocus, pure   61, 140, 321
- despace   *see* Despace
- distortion   *see* Distortion
- family   61, 63, 290, 292
- field   93, 101, **111**, 140, 142–144, **148**, 150, 159, 202, 255, **325**, 348, 404, 405
- field curvature   *see* Field curvature
- field dependence   62

- fifth order   60, 85, 174
- first order (Gaussian)   60, 61, 69, 136, 139, 146
- first order chromatic   146, 349
- formulae   78
- function   59, 60, 269, 270, 288, 314
- Gaussian   *see* first order
- geometrical   27
- higher order   **57**, 62, **82**, **85**, 117, 145, 160, 161, 169, 255, 278, 333, 339, 356, 359, 381
- higher order chromatic   339, 345, 347, 348, 359, 363–366, 368, 384, 407, 409
- higher order field   152, 351
- higher order theory   85, 86
- large (relative to diffraction limit)   312, 320, 321
- lateral (angular)   13, **57**, **58**, 64, 65, 85, 87, 90, 121, 122, **130**, 131, 133, **134**, **138**, 139, 156, 161, 190, 283
- lateral chromatic   *see* chromatic (lateral)
- longitudinal   57, 58, 65, 130
- monochromatic   146–148, 193, 339
- near diffraction limit   304–312
- order   57, 58, 60, 61, 85, 87, 288
- peak-to-valley (ptv)   310
- primary chromatic   169
- primary longitudinal chromatic   150, 358
- pupil   33, 411, 416
- quasi-telecentric   405
- root-mean-square (rms)   290, 310
- Seidel   61, 64, 89, 114, 145, 327, 329, *see also* third order
- small   304–312, 320, 321

– spherical   *see* Spherical aberration
– spherochromatism   *see* Spherochromatism
– symmetrical   310, 314
– theory   5, 41, 42, 48, 50, 55, 57, 69, 109, 111, 128, 396
– third order (Seidel) monochromatic   **59–63**, 65, 66, **70**, **80–82**, 85, 90, **128**, 141, 144, 146, 158, 177, 185, 201, 238, 270, 280
– third order chromatic   146
– third order theory   126, 128
– total   306
– transverse   *see* lateral
– transverse chromatic   94, 124, 160, **172**, 175, **201**, 202, 207, 212, 222, 352, 358
– type   61, 255
– wavefront   60, **64–66**, 69, 88, **127–129**, **131, 132, 134–136**, 138–140, 142, 145, 153, 155, 158, 167, 283, 314, 328
Aberration coefficient   *see* Seidel
Aberration function   *see* Aberration
Absorption   165, 314, 359, 394
Access (image)   115
Accuracy
– third order   186
Achromat   *see* Corrector
– thin doublet for FR or FE   404
Achromatic doublet (objective)   8, 14, 36, 144, 163, 164
– manufacture   14
Achromatisation   163, 164, 339, 359, 404, 407
Achromatism   146, 169, 259, 350, 351, 368, 369, 373, 380, 384, 432
Active control   62, **69**, 111, 242, **245**, 252, 255, 272, **277**, 293, 449, 450, 457, 463
Active correction   *see* Active control
Active optics   *see* Active control
ADC   *see* Atmospheric Dispersion Correction
Afocal case   **32**, 33, 36, 40, 42, **54**, 55, 78–80, **82**, 92, 93, 96, 98, 102, 104, 105, 107–109, 272, 282, 283, 285
Afocal corrector   *see* Corrector

Afocal feed system   55, 238, 246, 250
Afocal mode   54
Afocal supplementary lens system   55
Afocal system   48, 55, 221
– definition   31
Air   30, 31, 36, 39, 43
– local   301, 455
Airy disk   259, 299, 300, 320, 405
Alignment   17, 86, 269
– Schmidt   161
Aluminium (reflecting coat)   204, 219, 241, 449
– vacuum evaporation   204, 241, 447
Amateur   109, 186, 188, 191, 193, 217, 222, 231, 256, 304, 320, 405
Analysis   *see also* Theory
– chemical   259
– third order   69, 88, 90, 180
Anastigmat   *see* Reflector
Anastigmatism   227, 242, 259, 260, 278, 352
Angle
– azimuth   60, 133
Angular field   *see* Field angle
Aperture   21, 31, 60, 66, 71, 82, 85, 128, 269
– annular   302, 310, 320
– circular   298, 301, 317
– double slit   320
– excentric   260
– finite   86
– rectangular   293, 294, 298, 317, 322
– relative   11, 47, 267, 318, 438, 449, 452–454
– slit   295
– unobstructed   303, 321
Aperture diaphragm   41
Aperture number   84
Aperture ratio   111, 188
Aperture stop   32, 33, 36, 51, 52
– telecentric   34
Aplanatic   *see also* Reflector
– supplement   96, 265
Aplanatic Cassegrain (RC)   67, 68, 94, 263
Aplanatic Gregory   67, 68
Aplanatic modification   119

Aplanatic solution (general)   95, 101,
  186
Aplanatism   96, 116, 121–123, 201,
  202, 204, 207, 215, 218, 221
Apodisation   293, 304, 315
Approximation
– Gaussian optics   27
– Seidel   57–59, 63–65, 69, 82, 96, 127
– third order   *see* Seidel
Array detectors   94
Aspheric   *see also* Reflector
– form   3, 5, 6, 11, 13, 64, 70, 98, 228,
  229, 243, 244, 251, 262, 270
– general form   82–84
– higher order   83
– mirror form relaxation   377
– off-axis   260
– steep   178
– surface   3, 8, 82–84, 241, 363,
  366–370, 381, 385, 390, 411, 415
– term   83
– third order   84
Aspheric contribution   68
Aspheric parameter   70
Aspheric plate   140–142, 145, 185
– third order   141
Asphericity   **59, 82–84**, 89, 91, 103,
  105, 107, 116, 164, 180, 184, 190, 193,
  212, 329, 330, 333, 338, 345, 347, 351,
  353, 361, 366, 372, 373, 380, 384, 385,
  420, 432, 443
– three mirror solution   223
– total   339
– very high   178
Aspherising   13, 208, 366, 435, 447, 449
Astigmatic
– lines   101, 136, 138, 190
– patch   137
– radial line   137
– sagittal line   136, 137
– sagittal surface   136
– tangential line   136, 137
– tangential surface   135, 136
Astigmatic types   62
Astigmatism
– angular   138, 190
– chromatic difference   147, 345, 347
– definition   135–138

– field   2, 14, **62**, **65**, **69**, 76, 79, 90,
  92–94, 97, 98, 101, 105, 108, 109,
  114–117, 126, **135**, 136, 137, 139–141,
  143, **148, 149**, 170, 177, 178, 182–184,
  186, 190, 191, 204, 212, 215, 217, 221,
  227, 229, 242–244, 256, 257, **259**, 260,
  290, 292, 312, 327, 330, 331, 333–337,
  339, 341, 343, 348, 350, 354–359, 364,
  373, 375, 376, 378, 379, 383, 384, 396,
  405–407, 411, 431, 454
– lateral   136
– longitudinal   136
– quadratic field dependence   269
– sagittal   115, 116
– tangential   115, 116
– third order   136, 152, 306, 308, 309,
  311, 320
Astrometric reflector   *see* Telescope
Astronomical observation
– visual   38
Atmosphere   147, 392
– equivalent homogeneous   392
Atmospheric dispersion   392–396, 400
Atmospheric Dispersion Corrector
  (ADC)   392–404
– Cassegrain focus   396, 399, 402–404
– diffraction limited   393
– ESO linear (LADC)   402–404
– oiled contact   396
– prime-focus   396–400
– prime-focus, 3- or 4-lens   396
– prisms   396, 399
– RC (Cassegrain) focus   399
– Risley variable dispersion prism
  393
– two plane-parallel plates   392
– with existing corrector   396
– zero-deviation   392–394
Atmospheric refraction   392, 393
Atmospheric turbulence   *see* Seeing
Autocollimation   96, 437
Axial imagery   *see* Imagery
Axial obstruction ratio   *see* Obstruction
Axis   3, 23, 24, 60, 61, 88, 101, 148, 150,
  152, 161, 255, 263
– altitude   242, 246, 251, 424
– aspheric   261

– optical system (definition)   23, 28,
  255, 463
– second   225, 227, 230, 231, 242, 243,
  247
– single   241, 242
Axis of symmetry   23, 59, 255
Azimuth angle   *see* Angle, azimuth

Back focal distance   35, 73
Back-reflection   219, *see also* Reflector,
  Mangin
Baffle   144, 246, 302
– front   180
– stray-light   *see* Baffling
Baffling   115, 124, 152, 180, 230, 239,
  240, 245, 255, 278
– by secondary   227
– stray-light   8, 180, 227, 230
Bandwidth
– spectral   150
Barlow lens   404, 405
Beam
– afocal   228
– converging   158, 333, 393, 394
– diverging   158
– telecentric   268, 269
Beam compression   4, 37, 257
– factor   37, 38
– laws   38
– ratio   37, 38, 48, 55, 257
Beam compressor   4, 78, 231
– Mersenne afocal anastigmatic   78,
  227, 229, 231, 352
– two-mirror, afocal   *see* Mersenne
  afocal anastigmatic
Bending   405
– lenses   170, 221
– meniscus   204, 206, 218
– optical   218
– photographic plates   160
Best focus   *see* Focus, best
Binocular   38
Blank (mirror)
– aluminium   450
– beryllium   450
– borosilicate   450, 456, *see also* Pyrex
– fused quartz   444, 446, 455, 456
– lightweighted   442, 446, 449, 450,
  455, 456

– lightweighted, built-up   420
– massive   420, 442, 445, 450, 455, 456
– meniscus   450, 456
– 100-inch   438, 447
– Palomar 5 m   446, 452
– Pyrex   445, 446, 456
– 6 m   450, 456
– 60-inch   435, 446
– speculum   11, 14, 18, 419, 420, 424,
  429, 432
– stainless steel   450
– thin   450, 456
– 3 m Pyrex (Lick)   452
– 200-inch   *see* Palomar
Bore
– cylindrical   447
Bowen camera   111, 414
Boyden station (South Africa)   183
Buchroeder design   *see* Reflector
Building   *see also* Dome and Air, local
– ESO 1.4 m Coudé Auxiliary
  Telescope   460
– ESO 3.6 m telescope   460

Cage
– prime focus   447, 452, 453
Calculation data
– input   98
Camera   *see also* Reflector
– Baker flat-field   183
– Baker-Nunn   209, 211, 214
– photographic   21, 26
– pinhole   22
– Schmidt   *see* Reflector
– Schmidt semi-solid   174, 176
– Schmidt solid   174–176
– Schmidt solid (direct form)   175
– Schmidt solid (folded form)   175
– spectrograph   114, 174, 175
Camera objective
– triplet   144
Camera obscura   21, 22
Carl Zeiss (optical works)   366, 406
Cartesian
– oval   3
– system   24, 34, 77
– theory   5, 11
Case   *see also* Reflector
– afocal   *see* Afocal case

540　　Subject Index

- aplanatic   94, 97
- Cassegrain   *see* Reflector and Corrector
- focal   85, 93, 283, 286
- prime focus   *see* Reflector and Corrector
Cassegrain reflector   *see* Reflector
Cassegrain system with concave secondary   114
CAT (ESO)   *see* Telescope
Catoptric system   15, 88
CCD (Charge Coupled Device)   94, 98, 239, 326, 455
CCD array   239, 245
Cement
- optical   165
Centered optical system   57, 59, 158, 206
- definition   23
- ideal   23
Centering   11, 206
- tolerances   111
Central obscuration factor   303
Cervit (glass ceramic)   455, 456
Characteristic Function   59, 61–66, 85, 120, 127, 255, 288, 290
Chromatic limitation   152
Chromatic variation   147, 148, 152, 160, 333, 339, 394
- astigmatism   147, 407
- coma   147, 339, 345, 347, 358, 359, 365, 407, 409
- first order (Gaussian) terms   147
- focus (longitudinal colour)   147
- higher order terms   152
- spherical aberration   147
- wavefront tilt (lateral colour)   147
Chromatism
- primary   14
Circle (equation)   *see* Equation, circle
Classical Cassegrain   *see* Reflector
Classical Gregory   *see* Reflector
Coat
- aluminium   219, 241, 447, 449
- cleaning   241
- durable   241
- multi-dielectric   244, 255
- silver   219, 241, 255, 426, 429

Coefficients
- chromatic   147, 349
- third order aberration   63, 90, 128
- third order aspheric   84
Coherence   297
- partial   297
Colour   *see also* Chromatic variation
- lateral   147
- longitudinal   146
Coma
- angular tangential   343
- chromatic difference   *see* colour
- coefficient   134
- colour (chromatic variation)   147, 339, 345, 347, 358, 359, 365, 407, 409
- compensation   257
- decentering   261–272, 274–277, 287, 288, 444, 445, 449
- definition   131–134
- field   2, 5, 14, 19, **62**, **65**, **69**, 75, 76, 78, **85, 86**, 89–92, **95**, 97, 101, 103, 105, 107–109, **117**, 121, **131**, 134, 135, 138–141, 143, **148, 149**, 161, 176, 182, 206, 217, 218, 222, **227**, 229, 243, 244, **255–260**, 262–265, 268, 270, 272, 273, 275, 276, 287, 290, 292, 314, 327, 329, 333–335, 339, 341–343, 345, 348, 350, 354, 356, 357, 359, 373, 376–384, 396, 405–407, 409, 421, 435, 442, **453, 454**, 463
- field (linear dependence)   85, 86, 269
- field-uniform (decentering)   *see* decentering
- fifth order   361
- figure (pattern)   *see* patch
- inward   134, 202
- outward   202, 204
- patch   101, 133, 134, 139
- rotation decentering   270, 271, 274, 276
- sagittal   134
- Staeble–Lihotzky condition   85, 86, 120
- stop-independent   92
- tangential   19, 133
- third order   62, 132, 133, 152, 242, 306, 309, 311

– third order (with optimum tilt) 309, 311
– total   271
– translation (lateral, transverse) decentering   *see* decentering
– triangular   62
Coma limits
– acceptable   269
Coma-free point   271–275, 277
Combination   *see also* Reflector
– aspheric plate-meniscus   208
– thin-lens   39, 184
– two-mirror   42–55, 71–110
Committee (telescope design)   431
Companar   *see* Reflector
Compression ratio   *see* Beam compression, ratio
Computer drawing   98
Computer program (optical design)
– ACCOS V   207, 286
– ZEMAX-EE   98
Computers   85–87, 98, 117, 161
Concentric zones (ripple)   308–311
Concentricity   150, 183, 196, 206
Condensor
– two-mirror   259
Condition
– aplanatic   115, 189, 390
– Bouwers   169
– coma-free   259
– effective flat field   116
– flat field (Petzval)   94, 115, 116, 225, 228, 239, 241, 245, 251
– Maksutov   169
Conditions   *see also* Air, local
– thermal (in dome)   450
Conic (Schwarzschild) constant   59, 64, 82, 84, 226, 243, 332, 361, 363, 369, 376, 379, 420
Conic form   84
Conic parameter   84
Conic section   3, 13, 28, 58, 59, 82–84, 96
– higher order variations from   269
– strict   58, 59, 84, 269
Conjugate plane   24, 31
Conjugate point   23, 26
Constant

– aspheric (plate)   146, 151, 328, 347
– dimensionless profile   154
Construction (primary)
– monolithic   242
– segmented   242, 249, 368
Contrast
– imagery   256, 313, 316, 319, 323
– loss   320
– negative   321
– reduction   320
Contribution
– "central"   *see* Aberration, "central"
Control
– telescope (pointing and tracking) 450
Convex lens   *see* Lens
Convolution   319
Coordinates
– Cartesian   24, 117, 306
– "optical"   304
– polar (secondary mirror)   117
– system (diffraction)   294
Correction
– chromatic   150, 219
– field   144, 326
– fifth order   191
– monochromatic   152, 407
Corrector   97, 123, 160, 161, 208, 253, 325–391, 454
– achromatic   208
– achromatised doublet meniscus   384
– afocal   214, 215, 219, 325, 348–350, 363, 372
– afocal doublet   352–355, 372, 373, 381, 382
– afocal doublet plus plate   358
– afocal doublet with single glass   373, 382, 384
– aspheric   356
– aspheric (Schmidt)   150
– aspheric plate   325, 327–349, 353, 375
– aspheric plate plus field flattener 384
– Atmospheric Dispersion (ADC) 368, 385, 392–404
– Baker concave mirror pair   250
– Baker reflector-corrector   357

– Baranne three-lens   366
– bigger   367
– bigger, with aspheric   366
– Cassegrain focus   331, 340–348, 363–391, 454, 457
– COSTAR 2-mirror   348
– Delabre 4-lens, 2-glass PF   370–372
– Dialyte   219
– double meniscus   217, 221
– double-pass Mangin   222
– doublet lens   211, 213, 215, 348–358, 363, 372, 373, 376, 378–382, 384, 385, 399, 460
– doublet plus aspheric   356
– doublet with ADC   399
– doublet with hyperboloid   357, 363
– doublet with single glass   389, 390
– doublet with two different glasses 376
– doublet, fused silica   390
– Epps   368–370, 385, 386, 390, 392, 396
– Faulde-Wilson three-lens   361
– field   119, 145, 146, 287, 325–404
– for paraboloids   358, 359, 366–368
– for RC primaries   359, 361, 363
– four-lens all spherical   358–361, 366
– Gascoigne   327, 330, 334, 336, 341, 342, 345, 348, 350, 355
– Gascoigne PF aspheric plate 330–332, 355, 356
– Gascoigne plate (Cassegrain) 342–348
– Gregory focus   340
– Hawkins-Linfoot aspheric achromatic 209, 211
– Houghton-type lens   211, 213–217
– Köhler (Cassegrain)   365
– lens   325, 339, 340, 348–391
– "lensm"   399–402
– lens-mirror   221
– Meinel   327, 337
– meniscus   150, 353, see also Reflector, Bouwers and Maksutov
– modern RC   363–367
– multi-element   339
– multiple plate   329, 347, 356
– multiple plates at Cassegrain   347

– multiple thin lenses   352
– nearly afocal doublet (Harmer-Wynne)   385
– nearly afocal spaced doublet (Sampson)   372
– Paul   325, 327, 329, 333, 351–359, 363
– Paul-Baker reflecting-type   369
– plate   327–348
– position (relative to stop)   329, 340, 341, 374
– powered   406
– prime focus (PF)   97, 138, **327–340**, 345, **348–373**, 408, 411, 416, 417, **449**, 453, 454, 457, 460
– quasi-afocal (with FR)   404, 405, 409
– refracting   151, 326, 369
– Ritchey–Chrétien (RC)   342, 367, 373, 381, 390
– Ross   348–353, 358, 359, 363, 381, 447, 449, 453
– Ross three-lens   353, 358, 447, 453
– Ross-type doublet   357, 363, 364
– Sampson   325, 348
– Schmidt   151, 161, 165, 214, 457
– Schmidt achromatic   163–165
– Schulte   345, 356, 384
– Schwarzschild-Bowen pair   250
– secondary focus   see Cassegrain focus
– secondary focus with aspheric plates 340
– secondary focus with lenses   372, 383
– separated doublet, single glass   349, 380, 381
– separated doublet, two glasses   351
– simple meniscus   172
– single glass   165, 214, 353
– single lens   353, 354, 364, 375, 376, 381, 383, 384
– single thick meniscus   384, 390
– single-plate   329, 336, 337, 340–348
– size   368
– Su   389, 390, 399–404
– supplementary (plate)   140
– system   147, 255, 373

– thin   354, 377, 379, 380
– thin afocal   350, 351, 354
– thin afocal doublet   349, 350, 354–356, 358, 376, 382
– thin doublet with one glass type   379–382
– thin doublet with two different glasses   378, 379, 381
– thin singlet lens   376, 378
– three aspheric plate   329, 337–340, 347, 364–366, 385
– three quartz lenses   369
– three separated thin lenses   381, 382, 385
– three weak aspheric lenses   366
– three-element (Epps et al.)   385, 386, 391
– three-element (meniscus)   213
– three-lens   340, 348, 352, 356, 358, 359, 361, 363–365, 367–369, 384, 385, 390
– three-lens plus one aspheric   360–362
– three-lens with aspherics   366
– three-lens, one glass (Wynne)   364
– three-plate   337–340, 347, 348
– three-plate plus field flattener   340, 364, 365
– tilt   161
– two aspheric plates   333–337
– two fused quartz lenses with quite large separation (Su et al.)   390
– two plates at Cassegrain   347
– two separated lenses   357, 358, 380, 389, 390
– two separated thin afocal doublets   357, 358
– two-element (with FR)   409
– two-glass   390, 411
– two-glass doublet   357, 381
– two-glass meniscus   390
– two-lens afocal   see thin afocal doublet
– two-lens solutions   384, 385
– two-lenses with one glass type   387
– two-lenses with two different glasses   388
– two-plate   333–338, 348

– weak achromatic (Hawkins-Linfoot)   208
– Wilson   359–362, 365–367, 383, 384, 387–389
– Wynne   **326**, 337, **348**, 351, 352, 354, 356–360, **363–367**, 373, 375, 381–382, 384, 390, **393–398**, 400
– Wynne four-lens   359, 360, 363, 366
– Wynne three-lens, all spherical   364–366
Corrector (Cassegrain focus)
– aspheric plate   340–348
– lens   372–391
Corrector (PF)
– aspheric plate   327–340
– lens   348–372
Corrector plate   101, 160, 188, 191, 193, 196
– achromatic (doublet)   161–166, 191, 192, 194
– aspheric   186, 208, 340
– bending   161
– for astigmatism of an RC telescope   342
– Gascoigne (for a quasi-RC system)   344–348
– Gascoigne (for a strict RC system)   341–345
– Gascoigne (for PF)   331, 333, 338
– non-achromatic (singlet)   152, 165, 186, 187, 191, 214
– sag   161
– Schmidt   see non-achromatic
– Schmidt-type   357
– singlet (Paul)   330
Cosine effect   245
Cost advantage (spherical primary)   242
Coudé Auxiliary Telescope (ESO CAT)   see Telescope
Crystals   146, see also Glass
Curvature   28
– vertex   59, 83
Czerny-Turner
– condition   257
– monochromator   257

Dall–Kirkham (DK)   see Reflector

Decentering   54, 86, 266, 267, 269, 272, 287, *see also* Aberration
– angular   270
– lateral (transverse)   161, 261–270, 275, 276, 282, 444
– rotational   270, 271
– transverse   *see* lateral
Definitionshelligkeit (Strehl Intensity Ratio)   304
Defocus   61, 158, 292, 306, 308–312, 320
– effect   135
– pure   61, 320
Deformation
– aspheric   157
Derived quantity   *see* Quantity, derived
Design   *see* Reflector
Despace
– angular field coma   287
– axial   279–287
– field coma   283–287
– Gaussian terms   279, 280
– pointing change   288
– sensitivity   280, 287
– spherical aberration   280–284, 287
– transverse   287, 288
– zero sensitivity (spherical aberration)   283
Detector   **21**, 40, 123, 124, 169, 190, 230, 241, 243, **302**, **319**, 368, 405, 409, 411, **452**, 454
– CCD   94, 98, 326, 435, 455
– CCD-array   94, 98, 239, 245, 326, 331, 414, 417
– electronic   326, 455, 457
– faceplate   369
– image-intensifier tube   452
– non-linear   138
– photomultiplier   452
Diagram
– see-saw (Burch)   377
Diameter
– field   32, 36, 454
– free   54
Diaphragm   141, *see also* Stop
Diffraction   293–324
– annular aperture   302–304

– circular aperture   298–302
– cross   322, 324
– effects of spider   322, 324
– Fraunhofer   293, 294, 296, 297, 299–301
– Fraunhofer integral   294
– Fresnel   293
– image   295, 297, 299, 300, 305
– integral   298, 314, 318
– limit   301, 312
– rings   300, 304, 320
– small aberrations   304–312
– spike   322, 323
Diffraction limited   19, 215, 229, 302, 312
Diffraction pattern   295, 297, 300, 303
Diffraction point-spread-function (PSF)   293, 298, 302–304, 310, 312, 318, 319
Diffraction theory   26, 293–324
– with aberrations   293, 304–322
Disk
– Airy   *see* Airy disk
– of least confusion   130, 131, 156, 283, 308, 351, 353
– Ramsden   *see* Ramsden disk
Dispersion
– abnormal   209
– atmospheric   147, 368, 385, 392, 393, 395, 396, 400, *see also* ADC
– chromatic   **8**, 14, **146**, **148**, 155, 157, **164**, 172, 193, 195, 196, 208, 214, 351, 359, 376, 400
– curve   163
– glasses for ADC   394, 397
– mismatch (ADC)   395
– properties   163
– vector (ADC)   394
Displacement theorem   305, 312
Distinct vision
– minimum distance of   39
Distortion   62, 65, 66, 138, 141, 148, 160, 175, 180, 227, 352, 373, 380
– barrel   138, 139
– chromatic difference   345
– pincushion   138, 139
Dome
– air mass   455
– concept   457

Focal reducer (FR)   31, 119, 325, 404–417
– Baker-type   411, 413, 414
– Bouwers-Maksutov-type   411
– Cassegrain pair (Wilson-Delabre) 416
– flat-field anastigmat (MacFarlane) 415
– four-mirror (Meinel)   414
– Hawkins-Linfoot-type   411, 412, 414
– ideal   406
– INCA (Inverted Cassegrain) 414–416
– inverse (FE)   325
– Konica camera objective (Stockholm) 414
– Leitz Summicron photographic objective (Meinel)   414
– lens solutions   408–410
– lens system with field lens   409
– mirror solutions   407, 409–417
– Nasmyth (MacFarlane)   414
– non-spectral mode (spectrograph) 408, 417
– power   409
– replacing a PF facility   404, 409
– Schmidt camera   416, 417
– Schmidt-Cassegrain   416
– small-field doublet   404, 406
– Stockholm   414
– supplementary, Schmidt-based   409, 410
– thin lens theory   404, 406
– wide-field   406–417
– with dispersive means (spectrograph) 409
– with intermediate image   409–417
– with Tessar-type collimator   414
– with uncorrected intermediate image 409
– without intermediate image   407, 408
Focus
– best   129, 131, 156, 283, 343
– best astigmatic   138
– Cassegrain   114, **119**, **325–327**, 335, 341, 347, 372, 373, **394**, **409**, 414, **435**, 438, 447, **453**, 454, 457

– Cassegrain (RC)   407, 408, 414, 443, 454, 457
– compensation   130
– coudé   438, 447, 453–455, 460
– diffraction   304
– error (aberration)   62, 292
– final   101
– fixed (Nasmyth)   424
– Gaussian   128, 129, 131, 135, 153, 156, 304, 321
– Herschel (front-view)   19
– marginal   3, 130, 308
– mean astigmatic   137
– Nasmyth   424, 460
– Nasmyth-type   242, 245–247
– Newton   see Newton focus
– optimum   312
– paraxial   3, 29, 129, 130, 308
– prime   see Prime focus
– Ritchey–Chrétien (RC)   see Cassegrain (RC)
– sagittal   136
– tangential   136
Focus shift   153, 154, 158, 293, 308
– lateral   69, 131, 140
– longitudinal   69, 140, 153
– transverse (image height)   see lateral
Focusing   279
Folded Cassegrain   78
Folding flat   see Flat mirror, folding
Form   see also Reflector
– afocal   see Afocal case
– angular (aberration)   131
– aplanatic   96, 282
– aspheric   3, 5, 6, 11, 13, 64, 70, 98, 151, 152, 228, 229, 243, 244, 281, 283, 330
– brachy   278
– concentric   165, 184
– normalized   92
– parabolic   96
– paraxial (Snell's law)   29
– stop-shifted   150
Formulae
– aberration   65, 80–82, 186
– analytical   69–82, 233
– Cassini   392

– conversion (aberration forms)   128, 139, 283
– decentering (telescope)   261
– G-sum   167
– Korsch 3-mirror   223, 225
– Maksutov   193
– paraxial (reflectors)   41–55, 186
– recursion   55, 70, 233, 234, 236, 238, 257
– recursive forms   234
– stop-shift   66, 69, 89–91, 170, 180
– thick lens   169
Formulation
– explicit (Seidel)   65
– general, afocal   82, 92
– Hamilton   62, 63
– Zernike   288
Foucault "knife-edge test"   429, 432, 434, 437
Fourier theory   312, 315
Fourier transformation   130, 315, 319, 321, 322
Free diameter   54
Freedom
– degrees of (for corrector lenses)   376
Frequency (spatial)   312
– limit   316, 320
– low   320
– normalized atmospheric limit   320
Fresnel law (of reflection)   165
Function
– aberration   59–63, 288–292, 315
– aspheric   91
– aspherising   435
– autocorrelation   315, 316, 318
– axisymmetrical   60, 61, 290
– Bessel   298, 301
– circular aperture   299–301
– dispersion   163, 164, 220
– line spread   315, 317
– normalizing   69
– orthogonal or non-orthogonal   288
– phase   318
– profile   154
– pupil   314–316, 318, 320
– rectangular aperture   294–296
– sinusoidal spatial intensity   316
– slit   295, 297, 301

– variance (Strehl Intensity Ratio)   307
– wavefront aberration   127
Fused quartz   444–446, 455, 456

Galaxies
– external   422
Gauss-error   see Spherochromatism
Gaussian (paraxial) region   28, 36, 57, 61, 86
Gaussian brackets   233
Gaussian condition   60
Gaussian optics   21, 23, 29, 30, 36, 41, 42, 55, 57, 60, 86, 119
Gaussian optics approximation   27–29
Gaussian parameters   50
Gaussian properties   40
Gaussian relationships   41
Gaussian terms   61, 290–292
Geometrical optics   23, 24, 26, 27, 59, 87, 293, 295, 312, 322
Geometry
– Baker 2-axis   231
– Brachymedial   222
– Cassegrain   94, 98, 109, 111, 327
– Korsch 2-axis   225, 248
– normalized   104
– Schmidt   174, 230
– single-axis   243
– three-mirror   225
– two-axis   230, 231, 243
Ghost image   see Image, ghost
Ghost reflections   326
Glass   14, 146, 148, 154, 155, 165, 166, 193, 195, 196, 206, 215, 218, 351
– Astro-Sitall (ceramic)   450, 455
– block   174, 176
– borosilicate (Pyrex)   440, 444–447, 449, 450, 455, 456
– bubbles   438
– ceramic   450, 455, 456
– Cervit (Owens Illinois)   455, 456
– choice   196
– classical main sequence   221
– common   14, 405
– crown   196, 333, 339
– diagram (optical glass)   193
– expensive   193, 405
– filter   326

– flint   14, 186, 209
– lead   165
– low-expansion   444, 450
– low-expansion borosilicate (Pyrex)
  see borosilicate (Pyrex)
– mass   438, 450
– near zero-expansion   455
– normal   163, 164, 174, 219–221
– optical   163, 193, 195, 217, 419
– pairs (ADC)   393
– plate   431, 432, 435, 440, 446, 455
– plate (crown)   432
– quartz   *see* Quartz
– real (ADC)   393
– Schott UBK7   157, 163, 460
– silvered   426, 429, 430, 432, 434
– single type   214, 339, 348, 349, 358,
  360, 363, 376, 378, 380, 385
– special   146, 196, 209, 218, 220
– two types   363, 368, 373, 380, 392,
  394, 404
– volume   450
– zero-expansion   455
– Zerodur (Schott)   455, 456
Glass catalogue
– Chinese   399
– Schott   193
Glasses
– different (PF correctors)   364
Glassworks
– Corning   446, 455, 456
– General Electric Company   446, 455
– Owens Illinois   455
– Schott   193, 450, 455
– St. Gobain   435, 438, 444
Graticule   34
Grating   257
Grazing incidence   5, 101
Gregory   *see* Reflector

Hamilton theory   59–63, 255, 288
Herschel condition   421
Higher order (aberration) theory
  82–84
Hubble Space Telescope (HST)   *see*
  Telescope
Hubble's redshift law   440
Hyperbola   9, 59, 84, 91, 96, 102, 108,
  226, 238, 242, 335, 336, 348, 368, 384

Image   24
– "axial" (Schiefspiegler)   256–260,
  278
– comatic   101
– diffraction   293–321
– Gaussian   *see* paraxial
– Gaussian lateral shift (despace)   287
– geometrical quality   87
– ghost   165, 326, 327, 331, 345, 359,
  366, 384, 414
– motion   319
– paraxial   37, 130
– position (final)   47–49, 94, 255, 273,
  274
– primary   54, 97, 326, 327
– quality   82
– real   21, 33, 103, 104
– round geometrical   101
– secondary   97
– shift (despace)   280, 283, 284
– stability   275
– uncorrected intermediate (FR)   409
– virtual   103, 112
Image formation   24, 28, 130
– ideal   24, 28
Image height   71
Image shift
– despace   280
– transverse (despace)   287
Image space   24
Imagery
– axial   3, 11
– direct   175, 206, 241
– field   3
Imaging   21
– direct   408, 409, 452
Incidence
– angle of   27
Incident beam   *see also* Czerny-Turner
– compression   257
– expansion   257
Incoherent illumination   297, 301, 308,
  309, 313–315, 318, 320, 321
Index   *see* Refractive index
Infra-red observation   51, 97
Instrument (auxiliary)   21, 111, 402,
  408
Intensity   294, 295

– distribution   295, 299, 300, 302
Interferometry
– Fabry-Perot   414
Invariance   64, 289
Isoplanatism   314

Lagrange Invariant   30–32, 38, 39, 48,
    64, 70, 141, 331, 349, 374
Lateral (transverse) decentering   261,
    263
– Schiefspiegler interpretation   261
– strict case   263
Layout (telescope)
– initial   77, 188, 189
Lead   *see* Glass
Length   180, 186, 191, 215, 218, 445
– constructional   165, 178, 190, 191
– effective   185
– favourable   178
– optical   124, 215
– physical   114, 161
– reduced   182
– total   201
Lens   3, 5, 6, 148, 160, 208, 219, 326
– additional corrector   196, 198
– aspherised   209, 356
– bending   167, 349, 351, 353, 354, 356,
    358, 365, 376, 381, 382, 384
– field   22, 331, 405, 408, 414, 416
– field-flattening   94, 123, 159, 160,
    190, 331, 366, *see also* Field flattener
– finite thickness   351, 356, 409
– liquid   404
– meniscus (PF corrector)   384
– negative   190, 193
– negative Mangin   222
– negative, field-flattening (RC
    corrector)   383
– objective   6
– power   144, 146, 148, 349, 351, 358,
    365, 378, 381
– powered   146
– quartz (correctors)   363
– shape   168, 170, 351
– singlet, field (FR)   409
– spectacle quality   9, 11
– thick   168, 169
– thickness   *see* finite thickness
– thin   41, 65, 146, 159, 167, 168, 195,
    348, 351, 352, 354, 356, 373–375, 381
– thin positive (biconvex)   41, 65
– tilted plano-convex   260
– triplet group corrector   209
Lens corrector   *see also* Corrector
– Houghton-type   213–215, 217
– Lurie-Houghton   214–216
– Richter-Slevogt   213
Lens forms   101
Lens formula   25, 43
Lens system
– afocal supplementary   55
Lenses
– off-axis sections (ADC)   399
– separated and powered   359
– thin, separated   40, 219, 221, 380,
    390
Light
– incident   24
– wave nature   26
Light ray   *see also* Ray
– definition   26, 28, 57
Light transmission power   38, 39, 48,
    50
Light-gathering power   11, 15, 18, 255
Limit case (afocal telescope)   98
Line spread function (LSF)   315, 317
Linearity   29, 35
Losses
– absorption   38
– reflection   38
LSF   *see* Line spread function
Luminous wire
– incoherent   297

Magic number (Fibonacci)   103, 273
Magnification   **22**, **24–26**, **37**, **38**, 54,
    55, 58, 146, 224, 242, 259, **302**, 313,
    378, **404–407**, 409, 415, 422
– afocal system   48
– angular   55
– Barlow   405, 406
– empty   302
– field angle   55
– inverse   234
– secondary   **43**, 73, **113**, **115**, 142,
    185, 190, 191, 247, 256, 286, 287, 354,
    378, 453

– transverse   30
– unit   23, 24, 31
Magnification factor (lens)   351
Magnification laws   38, 55, 279
Magnifying power   20
Magnitude
– limiting   452
Maintenance (telescope optics)   128, 269
Major planets (Schiefspiegler advantage)   256
Maksutov circulars   198
Malus' theorem   86
Mangin-type systems   *see* Reflector
Manufacture   1, 2, **9, 10**, 11, 13, 46, 88, **101**, 105, 107, 111, 140, **165**, 186, 191, **207**, 253, 256, 260, 279, 340, 345, 347, 361, 426, 431, 432, **442**, **445**, 454
– errors   319
– specification   98
Mask (testing)   421
– zonal   421, 437
Material (mirror blank)   *see also* Blank
– low-expansion   440
– zero-expansion   440
Matrix spot-diagram
– single-column   88
– single-focus field-wavelength   87, 98
– through-focus   88, 98
Maximum (diffraction PSF)
– central   295, 297, 299, 303–305
– secondary   295, 297, 299, 300, 303–304
Mechanics
– telescope   18, 450
Medium (optical)   24, 26, 27, 29, 33, 36, 145, 148, 172
– image   24, 39
– object   24, 39
Meniscus   165, 168, 170, 193, 196, 202, 217, 222
– achromatic   218
– back surface (as secondary)   202
– Bouwers   196
– concave   168
– concentric   150, 168–170, 172, 195, 209, 213, 396
– concentric Bouwers   152, 167, 193

– concentric shells   209
– contribution   193
– double-pass   217
– Maksutov   169, 172, 193, 195, 206, 218
– Mangin   219
– quasi-concentric   213
– simple   196
– solution   165
– strongly asymmetric thickness   213
– thick   168
– thick concentric   167
– two-glass achromatic Bouwers concentric   193, 218
Mersenne afocal telescope   *see* reflector
Metric (length)   353, 374
Microscope   85
Miniature camera format   201, 204
Minimum (diffraction PSF)
– secondary   297, 300, 301, 303
Mirror   3, 5, 6, 219, 409, *see also* Blank (mirror)
– aspheric   3, 105, 145, 174, 176, 177, 180, 223, 240, 252, 257, 325, 377, 380
– built-up cellular   443
– centered concave   71, 89
– concave   2, 3, 6, 13
– confocal, paraboloidal   92
– cylindrical Nasmyth   252, 416
– deformed plane   325
– field   410, 416
– field, concave   416
– flat (2-axis solutions)   242, 253, 416
– large   241
– Mangin   213, 217, 218, 220, 221
– metal   450
– Newton, plane   9, 89
– plane   53, 71, 89, 142, 176, 185, 242, 255, 267, 278, 345, 416, 437, 442, 445
– powered   89, 142
– primary   *see* Primary mirror
– Schmidt-type spherical   242
– secondary   *see* Secondary mirror
– segmented   249
– silver-on-glass   426, 429, 430, 432, 435

– speculum   11, 14, 18, 255, 419, 422, 424, 450
– spherical   3, 9, **67**, **68**, **89**, 97, **148**, 153, 168, 183–187, 214, 215, 229, 241, **242**, 256, 257, 415
– support   18, 89, 245
– tertiary   228, 238, 240
– toroidal   252, 253, 260
Mirrors
– four, powered   230, 241–244, 247, 254
– three aspheric   251
– two aspheric (Schwarzschild)   114
– two spherical (concentric Schmidt-Cassegrain)   183
– two tilted (Schiefspiegler)   255
Modal concept
– aberration function   306
Mode
– natural vibration   62, 293
Modulation Transfer Function (MTF)   278, 313, 314, 317–321
– degradation (aberrations)   321
– diffraction   316–321
Monocentric (system)   191, 193
Monocentricity   184
Monomial functions   293
Mounting   89
– altazimuth (alt-az)   18, 242, 245, 423, 450, 452, 455–457, 461, 463
– English cradle-type   438, 447
– equatorial   423, 426, 435, 455, 456
– fork-type   438
– horseshoe equatorial (Palomar type)   447, 463
MTF   *see* Modulation Transfer Function

Natural (vibration) mode   62, 86, 293
Nebula
– M31 in Andromeda   435, 440
– spiral   422
Negative (test glass)
– concave   101
Neutral point (coma)   271
Neutral point (pointing)   271, 272
Newton focus   8, 70, 89, 325, 423, 424, 426, 432, 434, 435, 438, 447
Nodal point   26, 33

Normalization   **66**, **70**, 71, 77, **90**, 92, 105, 107–109, **115**, 116, 122, 123, **126**, 128, 134, **144**, 153, 177, 180, 182, 185, 190, 193, 225, 235, 238, 276, 283, **287**, 289, 290, 295, 307, 308, 314–317, 320, 321, 328–347, 349, 350, 353, 374, 379
NTT (ESO)   *see* ESO Telescopes
Null test (system)   96, 361, 437, 445

Object
– extended   312, 429, 457
Object space   24, 26, 33, 54, 142
Objective
– achromatic (doublet)   8, 14, 17, 21, 36–38, 40, 144, 163, 164, 218, 405
– camera   22
– convex lens (singlet)   2, 6, 14, 17
– lens   218–221, 404
– photographic   5, 21, 40, 54, 55, 114, 147, 148, 408, 414
– singlet (Dialyte)   219
– telescope   22, 36–40
– thin   36, 39, 41
– triplet   14
Objective prism (ADC)   396, 397, 399
Oblate spheroid   59, 107, 117, 178, 182, 274, 275, 283
– convex   108
Obscuration factor   302–304, 310, 311
Observation
– astrometric   138
– astronomical programme   431
– bird-watching   38
– Cepheid variables   440
– galaxy M31 in Andromeda   435, 440
– infra-red   51, 97, 272
– Jupiter (test object)   438
– Milky Way   440
– Moon and major planets   256
– photographic   186, 431
– planetary   320, 438
– spectroscopic   431
– spiral structure   422
– Sun   39
– Vega (test object)   440
– visual   186, 190, 204, 256, 438, 452
Observatory
– Armagh   183, 424
– Bjurakan   460

– Dublin   183
– ESO La Silla   102, 456, 460
– Harvard College   183
– Karl Schwarzschild (Tautenburg)
  157, 462
– Lick   369, 444
– Michigan   456
– Mt. Palomar   157, 445, 446, 448
– Mt. Wilson   440
– Royal Greenwich   461
– St. Andrews University   183
– Tautenburg   *see* Karl Schwarzschild
– Urania, Berlin   221
– US Naval, Washington   54, 95, 266,
  444, 445
– Zelenchuk   451
Obstruction   115, 116, 191, **223**, **247**,
  251, 252, **276**, **278**, 384, **406**, 409,
  411, 443, 445, **453**
– central (axial)   230, **256**, 259, 260,
  277, **278**, 287, **302**, 303, 316, 319,
  320, 323, 352, 414, 444
– detector   411
– field supplement   126, 230
– front-view   8
– linear   124
– ratio (axial)   **46**, 54, **77**, 91, 113, 115,
  116, 160, **180**, 184–186, 209, **229**, 239,
  243, 273, 276, **319**, 373, **453**
Ocular   *see* Eyepiece (ocular)
Oil
– optical immersion   394
Oiled contact   394
Opera glasses   40
Operation (telescope)   89
Optical design   24, 63, 70, 77, 144, 170
Optical design program   83, 98, 155,
  161, 170, 269, 288, 322
– ACCOS V   207, 286
– ZEMAX-EE   98
Optical figuring   18
Optical geometry   *see also* Layout,
  telescope
– Bowen   452
Optical glass   *see* Glass, optical
Optical length   *see* Length
Optical path   26, 27, 35, 117, 161
– length   27, 86, 160, 293

– total   85
Optical power   41, 52, 89, 148, 151, 255
– zero (plane secondary)   54, 327
Optical system
– centered   23, 26, 28, 32, 66
– ideal   23, 24
– non-centered   23, 255
Optical theory
– Gaussian   21
– third order   *see* Theory
Optical Transfer Function (OTF)   304,
  312–317, 319, 322
– multiplicative form   319
– theory   312
Optical works
– Carl Zeiss   366, 406, 440, 463
– Grubb-Parsons   457
Optics
– Gaussian   *see* Gaussian optics
– geometrical   23, 24, 26, 27, 59, 87
– large   255
– physical   26, 293–324
– spectrograph   111
– telescope   26, 402
Optikzentrum, Bochum   241
Optimization   170, 174, 191, 238, 288,
  325
– computer   98, 155
Order (aberration)   *see* Aberration,
  order
Orthogonality   289, 290
OTF   *see* Optical Transfer Function
Output data (optical design)   98
Overcorrection (spherical aberration)
  159
Overhang method   13

Parabola   **3**, 5, 11, 13, 18, **28**, **58**, **59**,
  67, **68**, **83**, 84, **89**, **90**, 96, 102, 103,
  129, 178, 182, 189, 218, 229, 244, 255,
  336, **348**, 351–354, 357, 367, 372, 390,
  400, 401, **420**, 421, **435**, 447, 457
Parabolic primary mirror   67, 68, 83,
  89
Parabolising   13, 421, 435
Paraboloid   *see also* Parabola
– confocal (Mersenne)   227, *see*
  also Reflector
Parameter

– aplanatic   64
– aspheric   64, 237, 344
– basic system (constructional)   70,
  234, 238
– dependent   279, 284
– derived   77
– independent   279, 284
– natural   233, 234
– normalized   77
– paraxial   41, 55
– profile   154
– ray tracing   46
Paraxial   29
– calculation   129
– quantity   35
– ray   29, 42, 51
– region   28, 29, 33, 86
Paraxial aperture ray   55, 65, 66, 70,
  142, 143, 234, 262
Paraxial data   66, 67
Paraxial equations
– linear nature   65
Paraxial parameter   41, 55
Paraxial principal ray   55, 65, 66, 70,
  73, 74, 142, 143, 224, 234, 262
Paraxial ray-trace   35
Paraxial relations (2-mirror)   42–55,
  73, 77–79, 264, 269, 285
Parseval's theorem   315, 319
Petzval curvature   70, 191, 331, see
  also Field curvature
Petzval sum   64, 147, 159, 180, 184,
  191, 228, 373, 378, 416
– chromatic variation   147
Phase error   27, 57
Phase shift   65, 313, 314
Photographic camera   see Camera
Photographic emulsion   111
Photographic grain   214
Photographic law   429
Photographic system   33
Photography   21, 190, 206, 406, 432,
  435, 438, 443, 457
– classical   453
– nebular   433
Photometry
– photographic   444
Physical length   see Length, physical

Pitch lap (polisher)   9, 18
Plane
– tangential   see Section
Plane mirror   see Mirror, plane
Plane-parallel plates   see Plate,
  plane-parallel
Plant growth   103
Plate   144, 152, 153, 155, 160, 327–348
– achromatic   124, 161–166, 182, 186,
  191, 211, 339
– aspheric   **144, 145, 148**, 165, **174**,
  177, 180, 182, 184, 193, 213, 214, 227,
  254, 326–348, 356, 411
– aspheric refracting corrector
  (Schmidt)   151, see also Schmidt
  corrector
– aspheric singlet   183
– aspheric surface   152
– asphericity   157, 178, 185, 186, 191,
  208, 337, 343
– bending   see photographic, bending
– conjugate, virtual   see virtual
– corrector   142, 143, 152, 161, 165,
  176–191, 208, 327–348
– deformation   178
– diagram (Burch)   140
– distance (from primary)   191
– doublet corrector   161, see
  also achromatic
– filter   159
– form   190
– full-size aspheric (object space)
  143–145, 340
– Gascoigne (Cassegrain)   342–347
– Gascoigne (PF)   331–333, 337, 342
– Gascoigne (with field flattener)
  331–332, 454
– glass, in air   146
– larger (corrector, PF)   337
– non-existent   191
– photographic   115, 138, 160, 161,
  165, 166, 326, 373, 383, 452
– photographic (baked)   363, 453
– photographic (unbaked)   452, 455
– photographic, bending   160, 383
– plane-parallel   154, 157, 158, 172,
  326, 333, 392, 394, 396

– power (strength)   144, 337, 339, 342,
    347, 381
– profile   145, 156, 157
– profile constant   146
– real aspheric (corrector)   328, 340,
    341
– refracting   141, 142
– Schmidt corrector   **151**, **154**,
    157–159, **164**, 165, 175, 209, 213, **330**,
    333, 337, **342**, 347, 460
– shift   144
– single-glass set   339
– singlet corrector (field)   144,
    327–336, 340–348
– singlet corrector (pupil)   144, 151,
    157, 161, 186, 190
– spacing   143–144, 337, *see also*
    power
– strength   *see* power
– strong   186, 339
– theory   140–146, 193, 227, 229
– thickness   154, 158, 326, 333
– triplet corrector   143–144, 337–340,
    348, 361–367
– two separated   327, 334, 348
– virtual   146, 327, 328, 340, 341
Plate theory
– aspheric   140–146, 193, 227, 229
Plateholder   115, 160
Point source   21, 26, 293, 297, 298, 308,
    309
Point-spread-function (PSF)   293, 298,
    302, 303, 315, 317–319
– geometrical (optical)   88, 322
Pointing (tracking)   21, 61, 147, 272,
    393
Points
– coma-free (neutral)   271
– pointing-free (neutral)   271
Polarisation   293
Polishing   420, 457, *see also* Figuring
    techniques
– repolishing (speculum)   431
Polynomial
– aberration   291, 306, 309
– aspheric surface   59, 83
– degree   289
– Hermite   293

– Jacobi   290, 293
– Laguerre   293
– Legendre   290, 293
– orthogonal   288, 293
– radial   289–290
– Tschebyscheff   293
– Zernike   62, 86, 288–293
Power
– aspheric   144
– cylindrical   260
– dispersive   193, *see also* Dispersion
– finite (FE and FR)   404, 407
– high individual (separated lenses)
    359, 385
– individual (achromatic Schmidt
    plates)   164
– lens   25, 40, 366
– optical   26, 30, 71, 144, 146, 170, 242,
    404
– residual optical   43, 164
– total   40, 146, 168, 404, 406
Power of penetration   20
Powered mirror
– single   40, 41
– third   223
– two   42
Primary image   54, 97
– real   46, 109, 112
Primary mirror   6, 8, 11, 13, 40, 50–52,
    66, **70, 71**, 74, 77, 90–96, 102, 103,
    105, 108, 113, 114, 117, 123, 124, 126,
    141, 142, 144, 145, **148**, **152**, 153, 159,
    161, 176, 180, 183, 191, 244, 257, 258,
    261, 262, 266–268, 272, 274, 284, 285,
    327, 329, 331, 339, 373, 374, 420, 426,
    445
– aberration coefficients for   72
– asphericity   161, 337, 372, 454
– Cassegrain   71, 186
– classical   90, 96, 119
– concave   107, 112, 149
– convex   111, 113
– Dall–Kirkham (DK)   102
– eccentricity   331, 336, 345, 354, 364,
    390, 453
– elliptical   372
– ESO 3.6 m, f/3.0 quasi-RC   340
– fast   366, 435

– field curvature   159, 331
– fixed spherical   278
– hyperbolic   97, 116, 327, 330, 334, 335, 337, 338, 353–355, 357, 358, 390, 391
– Kitt Peak 3.8 m, f/2.8 RC   340
– lightweighted structure   *see* Blank (mirror)
– modern RC   367
– naked   453
– Newton paraboloidal   *see* parabolic
– non-parabolic (with correctors)   330
– oblate-spheroidal   178, 357
– overcorrection (RC)   457
– Palomar 200-inch   446, 449, 453
– parabolic   101, 120, 151, 178, 186, 229, 231, **242**, 254, 264, **327**, 329, 333–338, 340, **347–358**, 375, 382, 385, 389, 390, **399**, **419**, 442, 445, 449, **453, 454**
– perforated   229, 230
– quasi-Ritchey–Chrétien   337, 338, 344, 381, 454–456
– Ritchey–Chrétien (RC)   **95**, 97, **119, 330, 331**, 336–338, 354–356, 359, **363, 364**, 368, **454–456**
– Schmidt   151, 159
– segmented   242, 249, 368
– single concave   71, 80
– spherical   67, 68, 88, **89**, 97, **105**, 107, 124, **128, 139, 149, 151, 152**, 178, 186, 188, 191, **213**, 217, 240, **242**, 243, 247–255, 266, 276, 330, 334, 338, 357, 416, 460
– Spherical Primary (SP)   105
– steep   152, 190, 242, 247, 416
– strongly hyperbolic   336–338
– structure (blank)   445, *see also* Blank (mirror)
Prime focus (PF)   **40**, 42, 51, **89**, 90, **97**, 101, 119, 198, 217, 271–273, 275, **325, 327**, 329, 331–340, 345, **347–372, 406**, 414–417, 447, 452–454, 457
Prime focus field corrector   *see* Corrector, prime focus
Principal plane   23, 26, 29, 31, 32, 36, 39, 40, 42, 86, 119

– image   23, 24, 39, 42
– object   23, 24
Principal point   23, 26, 41, 86
Principal ray   28, 32–34, 36, 172
Principal section   28
Prism   8, 394
– curved face   399
– objective   396, 399
– pair   396
Problem
– flexure   *see* Flexure
– handling (FR)   406
– manufacturing (massive blanks)   442
– mechanical   18, 419, 422
– tarnishing   11, 18, 429, 431, 432, 447
– thermal   440, 442, 445, 450, 455
– 2-axis tracking   450
Procedure (third order theory)
– non-normalized   70
Profile
– corrector plate   152
Program
– optical design   *see* Optical design program
Projection
– solar   22, 39
Proof plate
– interference   180
Protection
– dust and wind   429
PSF   *see* Point-spread-function
Pupil   **28**, 87, 133, 137, 142, **148, 152**, 157, 160, 217, 218, 222, **242**, 244, 245, 250, 252, 258, 259, **289**, 306, 307, 314, **322**, 327, 409, **411**, 414
– circular   298, 302, 304, 309, 310, 317, 318
– division   87
– entrance   **32–34, 36–38**, 41, **51, 52**, 66, 70, 87, 145, **223**, 231, 242, 244, **272**, 273, 275, 284
– exit   5, 6, 9, **15, 32, 33, 37**, 38, 40, 41, **51, 52**, 55, 133, 135, 227, 231, **242, 272–275**, 294, 304, 316, 404, 407, 408, 414
– eye   38, 40

- free from aberrations and obstruction 318
- full circular (unvignetted)   298, 304, 309, 310
- masking   323
- non-axial symmetrical obstruction 322
- position   *see* Pupil position
- shearing   316, 317
- spider obstruction   322–324
- telecentric   34, 404
- theory of   55
- transfer   416
- transferred   242, 246, 248, 250–254, 409, 411
Pupil position   33, 64, 66, 95, 144, 148, 273
Pupil shift   66, 230, 258, *see also* Stop shift
Pupil transfer
- Schmidt-type   *see* Pupil, transferred
Pyrex   445, 446, 456, *see also* Blank (mirror)

Quadratic (aberration term)   290, 292, *see also* Aberration, family
Quantity
- derived   53, 279
- normalized   77
Quantum efficiency   453, 455
Quartz
- fused   444–446, 455, 456
- fused segments (boules)   446, 455
- ULE fused   444, 455

Radiation   26
Ramsden disk   15, 37, 38, 40, 302
Ray   *see* Light ray
- aperture   142, 143, 268
- definition   26, 28, 57, 293
- finite   85, 87, 117
- marginal   64, 130, 133, 136, 158, 322
- paraxial   **3, 29**, 55, 64, 65, **70**, 85, 130, **136, 142, 143**, 158, 224, **233**, 234, 262, 339, 340, 351
- paraxial aperture   65, 66, 70
- paraxial principal   64–66, 70, 73, 74, 224, 235, 258

- principal   27, **32, 33**, 34, 71, 131, 133–136, **142, 143, 148**, 150, **170, 172**, 175, **235**, 268, 326, 339, 347, 361, 409
- skew   28
- wavefront normal   57
Ray tracing   14, 29, 41, 57, 65, 82, 85–87, 155, 161, 234, 238, 286, 288
- paraxial   35, 43, 52, 53, 64, 69, 71, 90, 233
Rayleigh criterion   129, 132, 301, 302, 306, 316
Rectangular aperture   293–298
Rectangular grid (for spot-diagrams) 87
Recursion formulae   55, 70, 233, 236, 238, 257, 261, 263, 268, 269, 280, 284
Redundant digits   98
Reference sphere   61, 129, 131, 421, 435
Reflecting telescope   *see* Reflector
Reflection   27, 34, 165
Reflection equation   35, 50
Reflectivity   11, 14, 19, 219, 241, 244, 255, 279, 422, 429, 432, 447
- low   11, 419, 432
Reflector   1–6, 8, **10, 11**, 13, 14, 17, 18, 20, 36, 98, 111, **148**, 247, 302, **419, 431, 432**, 433, 463, *see also* Telescope
- advances   89
- afocal   37, 82, 281
- afocal (Mersenne)   **4, 5**, 6, 9, **48**, 55, **78, 91–93**, 96, 103, 117, **227, 230**, 241, **242**, 267, 268, 285
- afocal Dall–Kirkham (DK)   105, 267
- afocal Spherical Primary (SP)   108, 109, 268, 274
- anastigmatic   93, 184, 191, 225–227, 259
- aplanatic (2-mirror)   **5, 89**, 93, **94–98**, 101, **111**, 114–122, 126, 265, 266, 272, 273, 284, 286, 287, 373, **442, 443, 454**
- aplanatic Gregory   95, 97, 101, 267, 273, 274, 277, 284, 342
- aplanatic Maksutov-Cassegrain 198, 201, 207

- aplanatic Ritchey–Chrétien (RC)
  94, 98, 100, 101, 119, 122, 265, 442,
  454, 456
- aplanatic Schmidt-Cassegrain
  (Baker-type)   186–192
- Ardeberg et al. (Nordic) 25 m   249
- Arecibo (fixed primary)   254, 278
- Baker 3-mirror, 2-axis anastigmatic
  231, 247, 248, 250
- Baker reflector-corrector   357
- Baker Schmidt-Cassegrain   178,
  180–185, 191
- Baker Super-Schmidt   209, 211, 213
- Baker-Nunn Super-Schmidt   209
- Baker-type FR   see Focal reducer
- Baranne-Lemaître 4-mirror, single
  axis   249
- Baranne-Lemaître 4-mirror, two axis
  253
- basic forms   40
- Bouwers (concentric)   148, 150–152,
  165, 167, 168, 174, 193, 206, 208
- Bouwers achromatic concentric PF
  or Cassegrain   186, 193–196, 208
- Bouwers concentric (with additional
  weak lens)   196, 198–200
- Bouwers-Maksutov-type FR   see
  Focal reducer
- Buchroeder Houghton-type   214
- Burch anastigmatic   278
- Busack improved Medial-type   222,
  223
- Cassegrain   4, **5, 9, 10**, 13, 15, 40,
  42, 43, 46–53, 55, 88, **90**, 93–97, 101,
  102, 107, 108, 111, 114, 126, 134, **142,
  143**, 147, 152, 183–185, 204, 217, 227,
  240, 241, 252, 256, 261, 265, 267, 268,
  270, 272–276, 282, 287, 288, **302**, 322,
  326, **340, 341**, 347, 348, **373**, 374,
  382, 407, 415, 419, 423, **426**, 431, 432,
  444
- Cassegrain afocal   see afocal
  (Mersenne)
- Cassegrain Dall–Kirkham (DK)   see
  Dall–Kirkham
- Cassegrain Ritchey–Chrétien (RC)
  see Ritchey–Chrétien (RC)

- Cassegrain Spherical Primary (SP)
  see Spherical Primary
- centered   255
- centered, two-mirror (off-axis)   260
- Chrétien (original RC)   116, 117
- classical (2-mirror)   **5, 88**, 92, 93,
  95–98, 101, 107–109, 111, **134**, 152,
  265, 272, 282, 286, 444, **452–454**,
  456, 460
- classical afocal Mersenne   see afocal
  (Mersenne)
- classical Cassegrain   **67, 68**, 88,
  90, 91, 92, 94, 98, **99**, 101, 102, 104,
  105, 119, 120, 122, 139, **140**, 254,
  264, 265, 266, 267, **271**, 275–277,
  282, **284**, 286, 347, 372, 373, 375–381,
  384–386, 389, 391, 404, 405, 444, 457,
  see also Cassegrain
- classical Gregory   67, 68, 88, 90–92,
  267, 271, 284, see also Gregory
- Companar   209
- compound (2-mirror)   **4**, 5, **6, 9**,
  11, 42, 43, 46, 50, 54, 69, **71**, 77, **78**,
  80–82, 88, **90**, 95, 98, 101, **109, 111**,
  114, 124, 127, 138, 144, 145, 150, 174
- concentric (monocentric) Cassegrain
  208, see also Hawkins-Linfoot
- Couder (anastigmat)   111, 116, 119,
  122–126, 152, 159
- Dall–Kirkham (DK)   **67, 68**, 88,
  101–109, 266, 267, 271–273, 275–277,
  282–284, 372
- Dall–Kirkham (DK) Gregory   104,
  267, 282, 284
- decentered   255, see also Schiefspiegler
- Delabre 4-mirror, single axis   248,
  249
- Delabre Brachymedial   222
- design (third order)   89
- double or multiple meniscus   212
- double-pass Cassegrain   240, 254
- Epps and Takeda 3-mirror   230
- fast, wide-field 4-mirror   see Delabre
  4-mirror
- first modern (Crossley)   432
- five-mirror 2-axis (Wilson-Delabre)
  251

– folded (flat) Cassegrain   54, 78, 89, 103
– folded Gregory   78
– Forster and Fritsch Brachy Cassegrain   255
– four-mirror   223, 235, 236, 238, 240–250, 252, 254
– four-mirror, single-axis Cassegrain-Gregory   250, 251
– Gregorian equivalent of Paul-Baker (Baker)   231
– Gregory   **4, 6, 7, 9**, 13, 17, 40, **42, 43, 46–53, 55**, 88, **90**, 94–97, **101–103**, 108, 109, 112, 114, 126, **266**, 267, 268, 270, 273–276, **340**, 341, **372**, 414, 419
– Gregory afocal   78, see also afocal (Mersenne)
– Gregory Dall–Kirkham (DK)   see Dall–Kirkham
– Gregory limit case   103
– Gregory Spherical Primary (SPG)   284, 287
– Hamilton Brachymedial   219, 220
– Harmer-Wynne   382, 384
– Hawkins-Linfoot monocentric Cassegrain   208–210
– Hawkins-Linfoot Schmidt-Bouwers PF   208, 209
– Hawkins-Linfoot-type (FR)   see Focal reducer
– Herschel (front-view)   2, 6, 11, 18, 40, 41, 90, 255
– Houghton afocal doublet corrector (Richter-Slevogt)   213–217
– hyperbolic primary (high eccentricity)   336, 337
– inverted Cassegrain (INCA-type) FR see Focal reducer
– Korsch 3-mirror, 4-reflection   240, 241
– Korsch 3-mirror, single-axis   67, 68, 223–227, 238, 239, 278
– Korsch 3-mirror, two-axis   223–225, 231, 238, 246, 247
– Korsch double Cassegrain (4-mirror)   254

– Kutter anastigmatic   see Schiefspiegler
– Kutter catadioptric   see Schiefspiegler
– Kutter coma-free   see Schiefspiegler
– Kutter Schiefspiegler   255–259, 278, 320, see also Schiefspiegler
– laterally decentered 2-mirror   264
– Laux 3-mirror   239
– lens telescopes with catadioptric correction   219
– Leonard Schiefspiegler (Solano and Yolo)   260
– Linfoot monocentric (concentric) Schmidt-Cassegrain   183–187, 191, 193, 196, 198, 208
– "long" Maksutov   170, 201, 202
– Loveday   231, 232, 238, 240
– Lurie (Houghton)   214, 215
– Maksutov   150, 165–175, 178, 196, 206, 207, 212, 214, 217
– Maksutov-Cassegrain   193, 198, 201, 204, 205, see also "long" and "short" Maksutov
– Maksutov-Cassegrain, secondary on meniscus   204–206
– Maksutov-Newton   215
– Mangin   217–219
– Mangin (achromatic)   218
– Mangin prime focus   218
– Mangin-type, Delabre   222
– Mangin-type, Rosin and Amon   218
– Mangin-type, Silvertooth   218
– meniscus, double (Wynne)   212, 213
– meniscus, multiple   212
– meniscus-type   150, 193, see also Bouwers
– Mersenne (afocal)   see afocal (Mersenne)
– Mersenne (grazing incidence)   5, 101, see also Wolter
– Mersenne afocal feeder   242
– Mersenne-Schmidt   230, see also Willstrop
– mirror forms   3
– modern   419
– modern aplanatic Schmidt-Cassegrain   186–191

– modern RC   116, 117, 444, 457
– monocentric Schmidt-Cassegrain
  (Linfoot)   *see* Linfoot
– multi-mirror (centered)   223
– Newton   **8–13**, 17, 18, **40, 41**, **89**,
  134, 204, 215, **218**, 231, 240, **329**, 348,
  419, 423, 426
– non-aplanatic (with decentering
  coma)   287
– normal (focal)   78, 80, 81
– normal Cassegrain (with ADC)   393
– normalized   *see* Normalization
– off-axis (Schiefspiegler)   255
– original Ritchey–Chrétien (RC)
  116, 117
– Paul 3-mirror   227–230, 242, 325,
  352
– Paul-Baker 3-mirror   228–232, 239,
  241, 278
– Plössl Dialyte   219–221
– plate-meniscus systems   207
– prime focus form   41, 70, 80, 88, 208,
  240, 329
– quasi-classical Cassegrain   376, 379,
  384, 385, 390
– quasi-concentric (monocentric)
  Bouwers-Cassegrain   193
– quasi-Ritchey–Chrétien   331, **332**,
  335, 340, **344–348**, 364–366, 372,
  379, 381, 383, 384, **387, 389**, 390, 453,
  454, 455, 456
– Rakich–Rumsey 3-mirror solutions
  63, 232, 233, 278, 279
– RC modification (of classical
  Cassegrain)   119
– Richter-Slevogt (afocal doublet
  corrector)   *see* Houghton
– Ritchey–Chrétien (RC)   **67, 68**,
  88, **94**, 95, 97, 98, 101, 105, 109, 111,
  119–122, 139, **140**, 178, 255, **266**,
  267, 272–274, 277, 282, 284, 286, 287,
  325, **331**, 335, 336, 340, **341–344**,
  347, 348, 354–356, **363**, 367, **372,
  373**, 376, 378, **381**, 383, **389**, 390,
  404, 405, **442, 444, 445, 452, 453**,
  455–457
– Robb 3-mirror   238–239, 278

– Rosin (with doublet field corrector)
  382, 384
– Rumsey 3-mirror   238
– Sampson (Mangin secondary
  corrector)   372
– Sand excentric 3-mirror   277
– Sasian two-axis, 4-mirror   252
– Schafer 4-mirror Schiefspiegler   278
– Schiefspiegler   *see* Schiefspiegler
– Schmidt   119, 124, 145, **151, 152**,
  157–160, **161, 163**, 165, 174–176, 178,
  180, 182, 183–186, **190**, 196, 209,
  214, 230, 241, **242**, 252, 357, 409,
  416, **457, 460**
– Schmidt achromatic   161, 163
– Schmidt, basic principles   151
– Schmidt, solid or semi-solid   174
– Schmidt-Cassegrain   178–191, 193,
  405
– Schmidt-type with afocal doublet
  see Houghton
– Schupmann Brachymedial   219–222,
  248
– Schupmann Medial   219–222, 248
– Schwarzschild   445
– Schwarzschild (aplanatic)   94, 111,
  114–119, 122–124, 126, 178
– Schwarzschild (impractical)   113,
  126, 279, 372
– Schwarzschild-Couder types   273
– Shack and Meinel 3-mirror   238
– "short" Maksutov   170–173, 198,
  201–203
– single-axis, 4-mirror   242, 249–250,
  254, 255
– single-mirror   46, **69, 71, 80, 88–90**,
  92, 93, 111, **128**, 134, **138–140**, 144,
  145, 150, 226
– Slevogt (aplanatic Schmidt-
  Cassegrain)   183, 186, 191, 194
– small-field (Mangin)   217
– Solano Schiefspiegler (Leonard)   260
– Sonnefeld (Mangin with corrector)
  213, 219
– Spherical Primary (SP)   101,
  107–110, 122, 191, 266, 267, 274–277,
  284, 287, 463

- Spherical Primary (SP) Gregory 105, 267, 284
- Spherical Secondary   see Dall–Kirkham (DK)
- Stevick–Paul Schiefspiegler   278
- strict aplanatic   341
- strict classical Cassegrain   375–376, 385
- strict Ritchey–Chrétien (RC)   345, 347, 377–381, 388, 453
- TC,0 (insensitive to lateral decenter) 275–277
- TC,0 (off-axis)   276
- TC,0 (with spherical primary)   276
- three-mirror (centered)   223–231, 238–242, 345
- three-mirror excentric (uniaxial) (Sand)   278
- total afocal (zero power system) 352
- Tri-Schiefspiegler (three-mirror) 255, 260, see also Schiefspiegler
- two-axis Baker   231, 248, 250
- two-axis Baranne-Lemaître   253
- two-axis Korsch   225, 241, 247, 252
- two-axis Sasian   252
- two-axis Wilson-Delabre   243–251
- two-glass concentric (monocentric) Bouwers-Cassegrain   195–196, 208
- two-mirror   **42–55, 71–82, 90–126**, 226, 233, 238, **263, 265–268**, 271, 275, 277, **279**, 281, 283, 284, 327, 340, 380, **382**
- two-mirror afocal   see afocal (Mersenne)
- two-mirror and plate   178, 223, see also Schmidt-Cassegrain
- two-mirror non-aplanatic (with decentering coma)   287
- two-mirror Schiefspiegler   255–261, see also Schiefspiegler
- wide-field   21, 55, 140, 143, 145, 146, 148–255
- Willstrop 3-mirror   230, 238, 239, 242, 243, 246
- Wilson 4-mirror, single-axis with Bowen camera   250
- Wilson-Delabre 4-mirror, two-axis 243–251
- Wilson-Delabre 5-mirror, two-axis 251
- Wolter   5, 101, see also Mersenne
- Wright-Väisälä   176–179, 215–217, 357
- Wynne three-meniscus   213
- Wynne two-meniscus   212
- Wynne two-meniscus (asymmetric) 212, 213
- Yolo Schiefspiegler (Leonard)   260
Reflector-corrector (Baker)   357
Refracting telescope   see Refractor
Refraction   8, 14, 27, 34, 172
- angle of   27
- atmospheric   392, 393
- Snell's law of   27
Refraction equation   35
Refractive index   14, 24, 27, 34, 43, 124, 131, 145, 148, 155, 158, 161, **164, 165**, 174, 209, 376, 381, 394
Refractor   1, 4–6, 8, 9, 11, 14, 17, 21, 22, 146, 191, 219, 221, 222, **419**, 431, see also Telescope
- afocal   39, 55
- conventional   36, 54
- conventional visual   78
- defocused   39, 40, 42
- Galileo   4–6, 9, 40, 42
- Galileo-type (afocal)   405
- Kepler   4, 5, 37, 40, 42
- large   431, 433, 438, 452
- long   10, 13, 452
- normal equivalent to Schupmann Medial   221
Relative aperture   114, 160
Remittance   22
Resolution   15, 19, 26, 38, 301, 306, 316, 320, 435
- angular   302
- limit   301, 320
- linear   302, 317
Resolving power   301, 303, see also Resolution
Reticle   6
Ripple   308–311
- idealized sinusoidal   308, 309

Ritchey–Chrétien (RC)   *see* Reflector
Rotation
– secondary (angular decenter)   270,
  271

Sagitta (sag)   41, 84
Sagittal section   *see* Section
Satellite tracking (Baker-Nunn camera)
  211
Scale   26, 40, 46, 54, 127, 175, 267, 288,
  326
– inverse   54
– linear   138, 139
Scaling   128
Scaling factor (correctors)   374
Scaling laws   126, 356, 358, 367
Schiefspiegler   255–263, 268, 273, 277,
  304, 320
– analysis   270
– anastigmatic (Kutter)   257, 259, 260
– basic   259
– catadioptric (Kutter)   260
– coma-free (Kutter)   256–258, 260
– excentric (off-axis) section   260, 261
– formula (decentering coma)
  261–266
– four-mirror   278
– lateral decenter, strict case   263
– perfect   260
– Solano   260
– spherical primary (TC,0 form)   277
– term (decenter)   265
– three-mirror (Sand)   277, 279
– with off-axis sections   260, 261,
  276–279
– with three or more mirrors   277, 278
– Yolo   260
Schmidt plate   *see* Plate
Schmidt telescope   *see* Reflector
Schwarzschild (conic) constant   **59**,
  64, 82, **84**, 226, 243, 332, 361, 363,
  369, **420**
Schwarzschild formulation (two-mirror
  telescope)   77
Schwarzschild theorem   89, 98, 101,
  111, 114, 176, 242, 325
Schwarzschild theorem generalisation
  89, 145, 241, 242, 248, 251
Searchlights   217

Secondary mirror   **11**, 13, 50–55, **73**,
  91, 93–97, 107, 111–114, 117, 142, 147,
  150, 183, 189, 191, 207, 230, 256, **262**,
  268–271, 273–276, 280, 281, 284, 286,
  373, 406
– aplanatic form   96
– aspheric   189, 228, 384
– Cassegrain (convex) hyperbolic   **5**,
  9, 15, **91**, 92, 101, 107, 111, 117, **120**,
  287, 426
– changed asphericity   385
– circular (obstruction)   302
– combined with (coated on) meniscus
  204
– concave   113, 114, 124, 273
– concave (Schwarzschild impractical)
  113, 114
– convex (virtual image)   112
– decentered   263, 287
– dummy flat   185, 327, 328, 350, 355
– elliptical (convex)   231
– Gregory (concave) elliptical   8, 9,
  91, 101, 111, 278, 287
– high magnification   287
– hyperbolic concave   231
– magnification   *see* Magnification,
  secondary
– Mangin-type   219, 222, 372
– oblate spheroid   117, 122, 191
– parabolic   92, 242, 285
– plane   185, 267, 327, 328, 350, 355
– plane (folding)   53, 78
– power   54, 115
– Ritchey–Chrétien (RC)   95, 119, 120
– rotation (angular decenter)   270,
  271, 288
– separate (from meniscus)   206
– spherical   **101**, 103, **180**, 182, 191,
  227, **243**, **247**, 251–255, 272, 382,
  384, 416
– spider   322–324
– steep   6, 94
– vertex (Paul telescope)   227
– virtual   126
– weakly curved   93
Secondary spectrum   *see* Spectrum,
  secondary
Section

– sagittal   133, 135, 314
– tangential   132, 133, 135, 255, 313, 314
Seeing
– atmospheric   19, 38, 139, 301, 312, 320, 422, 450, 457
– dome   440, 455
– limit   414
Segment
– aspheric   242
– fused quartz   446
Segmented construction (primary mirrors)   107, 242
Seidel
– aberrations   127
– approximation   57–59, 63–65, 96, 120, 121
– coefficients   63–68, 139
– monochromatic conditions   111, 241
Seidel sum   64
Self-achromatic (2-meniscus system)   212
Sensitivity (despace)   see Despace
Separation   143
– aspheric plates   143, 333, 337, 356
– thin lenses   40, 356
Series
– aspheric surface (Schwarzschild)   59
– Fibonacci   103
Serrurier truss   449
Shape   see Lens
Shape factor   168, 170, 206, 351, 353
Shearing
– circular pupil   316
Shellac (speculum protection)   429
Shift of image (transverse despace)   288
Short constructional length   190
Sign convention   24, 35, 41, 43, 52, 53, 55, 64, 65, 77, 233, 234
– Cartesian   24, 31, 63, 70, 233
– strict   224
Signal/noise ratio   302
Silica   see also Quartz
– fused   345, 347, 385, 390, 446
Silver (reflecting coat)   255, 426, 432, 447
Silvering

– chemical   429, 432, 447
Sine condition (Abbe)   see Abbe
Single surface   34
Singlet objective   219, 220
Singularity   235, 236, see also Recursion formulae
Sitall (glass ceramic)   450, 455
Site   457, 463
Site selection   457
Skew ray   28
Sky
– background   452
– limited coverage (mounting)   422
Slit   294, 295, 298
– long   295
– two, at right-angles   322
Snell Invariant   64
Snell's law   27, 29, 34, 57
Solar projection   22, 39
Solution   see also Reflector
– effective flat field   117
– quasi-RC   331, 453, 454, 456
– strict-RC   331, 453, 454, 456
– third order limitation   278
– three reflection   240
Solution matrix
– well-conditioned   243
Source
– extended   293, 297
– point   297
Space
– image   24–26, 34, 36, 331
– object   24–26, 34, 36, 145, 146, 148, 327, 328, 340, 341
Space telescope
– Hubble   see Telescope
– Next Generation Large   240
Spatial frequency   313–315, 321
Speckle camera   393
Spectral band
– narrow   147
– wide   147
Spectral line (resolution)   301
Spectral range (FR using lenses)   407, 409
Spectrograph   175, 250, 260, 463
– design   408
– echelle   452

– fixed   435, 454
– Mayall slitless   432
– multi-fibre (multi-object)   98, 326,
   363, 370, 390
– optics   111, 217
Spectroscopy   241, 435, 452
– multi-object   98, 326, 363, 370, 390
– with fixed equipment   435, 454
Spectrum   8, 147, 175
– normal secondary   163
– primary   163, 220
– secondary   14, **146**, **163**, 174, 209,
   218–221, 348, 352, 358, 359, 364, 368,
   373, 376, 379, **381**, **389**, 390, 400, 406,
   411, 419
Speculum metal   11, 14, 18, 419, 420,
   422, 426, 429, 432
– casting large mirrors in   18, 419, 429
– crystallisation   419
– tarnishing   *see* Problem, tarnishing
Speed (photographic)   48, 175, 267,
   373, 443
Sphere   6, 11, 13, 18, 28, 59, 84, 86, 421,
   447
Spherical aberration   **3, 5**, 6, 8, **14**,
   62, **65**, **69**, 73, 76, 88, 91, 97, 102, 117,
   128, 130, 131, 134, 140, 142, **143,
   144**, **148–150**, 152, 156, 158, 159,
   164, 168, 176, 182, 206, 217, **242**, 244,
   248, 250, 255, 264, 265, 267, 268, 275,
   280, 282, **284**, 290, **292**, 293, 299,
   310, 312, 327, 329, 334, 335, 341–343,
   345, **348**, 349, 351–354, 356–358, 361,
   363, 364, 367, 372, 373, 376, 382, 384,
   396, 405, 406, 454
– definition   3, 128, 130, 131
– fifth order   165, 193, 198, 202, 207,
   208, 212, 215, 218, 306, 308
– lateral   131
– longitudinal   153, 167
– third order   **129, 130**, 144, 145, 150,
   152, 153, 158, **167**, 168, 169, 217, **283**,
   306, 308, 309, 311, 320
– third order (with optimum focus)
   309, 311
Spherical mirror   *see* Mirror, spherical
Spherical primary   *see* Primary
   mirror, spherical

Spherical Primary (SP) 2-mirror system
   see Reflector
Spherochromatism   **147**, 150, **152**,
   155–157, 158, **161**, **163**, 165, 182,
   183, 186, 191, 196, 202, 206, 213–215,
   218, 339, 348, 411
Spider
– diffraction effects   322, 324
Spot-diagram   **87, 88, 98–101**, 105,
   106, 109, 110, 117, 118, 124, 125, **161**,
   165, 166, 170, 171, 173, 174, 178, 179,
   186, 187, 191, 192, 194, 196, 198–205,
   209, 210, 214–216, 243, 245, 248, 249,
   252, 253, 259, 260, 322, 332, 333, 340,
   346, 359–365, 368, 369, 383, 385–391,
   396–401, 405, 407–415
– single-column matrix   88
– single-focus field-wavelength matrix
   88, 98
– through-focus matrix   88, 98
Stop   89–92, 113, 124, **141, 142, 148,
   149, 151**, 160, **165, 172**, 182, 184,
   208, **211**, 212, 255, **256**, 262, 270, 285,
   349, 357, 374, 378, *see also* Aperture
   stop, Field stop
– at primary   105, 108, 148, 374
– central   351, 408
– concentric   170
– front   170–172, 196, 357
– quasi-telecentric (FE or FR)   405
– shifted   183, 218
Stop position   66, 89, 90, 151, 170, 180,
   201, 228, 230, 244, 268, 329
– different   69
– front   170–172, 196, 357
– meniscus   175
– original (zero shift)   141
– symmetrical   148
– uncritical   170, 227
Stop shift   **66, 69**, 71, 87, **141, 145,
   148**, 160, 164, 170, **172**, 201, **227**, 257
– formulae   66, 77, 89, 90, 170
– normalized   77
Stray light   152
Strehl criterion   *see* Strehl Intensity
   Ratio
Strehl Intensity Ratio   129, 130, 301,
   304, 305, 307–309, 320, 321

Structure
- spiral   422
Sun (observation)   39
Superposition
- linear (of Seidel coefficients)   64
Supplementary feeder system   55, 227, 229, 230, 242, 243
Support   *see also* Mirror
- active (push-pull)   245
- astatic lever   423, 437
- astatic principle   420, 423
- primary mirror   420, 423, 437, 446, 449
- whiffle-tree   420, 421, 426
Support (mirror)   *see* Mirror and Support
Support system
- 200-inch primary   447
Surface
- aspheric   **59, 82–84,** 87, 148, 176, 178, 218, 349, 353, 354, 356, 359, 361, 366, 367, 370, 381, 401, 402
- centered reflecting   55
- chemically silvered   217, 429, 430, 432, 447
- concave (for aspheric testing)   361
- contributions (Seidel)   207
- first of 2-mirror system   71
- glass   165
- $\tilde{n}$-mirror   233
- reflecting   11, 34, 57, 233, 234
- refracting   11, 34, 57, 142
- second of 2-mirror system   73
- spherical   **11,** 18, **58,** 70, 165, 198, 207, 215, 218, 222, 255, 278, 359, 369, 370, 389, 399, 411
- toroidal   253, 260
Symmetry   59, 150, 255, *see also* Axis
- axial   58, 59
- concentric   148–151, 184, 198
System   *see also* Reflector
- aberration-free   320
- afocal Cassegrain feed   55
- afocal Gregory feed   55
- catoptric   238
- centered   278
- concentric (obstruction problem)   209

- Czerny-Turner (monochromator)   257
- evaluation   82, 87
- Milky Way   440
- mirror (for optical power)   146, 325
- $\tilde{n}$-mirror   70, 233
- normalized   65, 70, 92, *see also* Normalization
- push-pull (active mirror support)   245
- reflecting (catoptric)   98
- reflecting and refracting (catadioptric)   98
- Schmidt equivalent (Paul)   227
- Schmidt-Cassegrain (for FR)   414
- supplementary feeder   55
- telecentric   *see* Telecentric system
- telephoto   *see* Telephoto system
- tilted   257
- Wilson-Opitz 2-mirror condensor   259

Tangential section   *see* Section
Tarnishing   *see* Problem, tarnishing
Telecentric   *see also* Pupil
- output (corrector)   370, 390
Telecentric system   33, 37
Telecentricity (FR)   407
Telecommunications (off-axis system)   277
Telephoto effect   **5,** 6, 9, 33, **40, 43,** 46, 54, 93, 98, 109, **143,** 186, 190, 240, **326,** 426
- true   46
Telephoto objective   40
Telephoto system
- lens   5, 40, 404
- mirror   5, 42
Telescope   21, 22, 26, *see also* Reflector or Refractor
- aberration theory   59, 66
- afocal   32, 39, 48, 82, 454
- afocal feeder   55, 109, 227, 238, 242, 246, 250
- alt-az   242
- Anglo-Australian 3.9 m (AAT)   366, 367, 396, 456
- aplanatic   89, 94, 109, 111, 435
- Astrometric Reflector (USNO)   54

- basic function   21
- Bergedorf 1.22 m (later Crimea) 440
- Bergedorf 1 m   440
- Bjurakan 1-m Schmidt   457
- Bowen-type   453, 457
- Brown University 12-inch (Schwarzschild)   445
- centering   11, 261
- Cerro Tololo 4.0 m   456
- CFHT (Hawaii) 3.6 m   456
- Chinese 2.16 m RC   389, 390
- classical   419, 453, see also Reflector
- Columbus 2×8 m   390
- conventional refractor   36
- Crossley 36-inch   432–434
- defocused   39, 40
- Doppler 1 m RC   444
- Dorpat refractor   419
- Dunlap 74-inch   440
- equatorially-mounted Bowen-type 456, 463
- ESO   see ESO Telescopes
- fast   111, 114
- flexure   18, see also Support
- 40-inch Ritchey–Chrétien (RC) (USNO)   95, 266, 444, 445
- Foucault 33 cm   429
- Foucault 80 cm   429, 430
- Gaussian layout   42, 49, 186, 188, 189, 286
- giant (25 m) USSR   107
- glass optics   429, 431
- ground-based   312
- Grubb 15-inch Newtonian-Cassegrain 424
- Herschel   15, 419
- Herschel 20-foot focus   17–19, 438
- Herschel 4-foot aperture   see 40-foot focus
- Herschel 40-foot focus   18, 19, 419
- Herschel tilted single mirror   421
- history   1, 419
- Hubble Space Telescope (HST)   140, 141, 143, 144, 293, 348
- 100-inch Hooker (Mt. Wilson)   see Mt. Wilson 100-inch Hooker
- Isaac Newton 2.5-m   359, 456

- Keck 10 m   242, 368, 369
- Kitt Peak 3.8 m, f/2.8 RC (concept) 363, 364, 389
- Kitt Peak Mayall 4 m   455–458
- Lassell 9-inch   423
- Lassell's largest (4-foot)   422–424, 426
- layout   42, 49, 186, 188, 189, 286
- Lick 120-inch (3 m) Shane   353, 369, 370, 446, 452, 456
- Lick 36-inch refractor   431
- lists of   456, 463
- LITE (project)   239
- Magellan 315-inch   389, 390
- manufacture   see Manufacture
- McDonald 105-inch   363
- McDonald 82-inch   414, 440
- mechanical problems   18, 419, 422
- Melbourne 48-inch   426, 427, 429, 431, 432, 434, 438
- modern astronomical   54, 109, 426, 434
- MPIA 2.2 m (A), Calar Alto 387–389
- MPIA 2.2 m (B), La Silla   322, 323
- MPIA 3.5 m, Calar Alto   360, 363, 366, 406–410, 412, 413, 456, 463
- Mt. Wilson 100-inch Hooker   437, 438, 440, 441, 445, 449, 452
- Mt. Wilson 60-inch   111, 348, 352, 353, 435–440, 444, 445, 452
- Multi-Mirror Telescope (MMT)   385
- Nasmyth 20 inch   424, 425
- Newton form   419, 422, see also Reflector
- Nordic 25 m (project)   249
- Nordic Optical 2.5 m (NOT)   414
- normalized   65, 70, see also Normalization
- obstruction problem   223, 241
- 1.52 m quasi-RC, Cerro Tololo   345, 347
- optical specifications (quality)   301, 304, 322
- optimization (functional, on site) 431
- Palomar 1.22-m Schmidt   457

- Palomar 200-inch (5 m)   352, 353, 359, 360, 445–457
- Pease 300-inch concept   442, 445
- Perkins 69-inch   440
- photographic   266
- prime focus   40, *see also* Reflector
- reflectivity problem   241
- Ritchey 100-inch optics   *see* Mt. Wilson 100-inch
- Ritchey 24-inch (60 cm)   433–437, 440
- Ritchey 40-inch (RC)   *see* 40-inch RC (USNO)
- Ritchey 60-inch   *see* Mt. Wilson 60-inch
- Rosse 6-foot   419–422, 438
- Russian 6-m, Zelenchuk   449–452, 455, 456
- solar   46, 101, 109
- solar (Snow)   438
- space   240, 301, 312, *see also* Hubble Space Telescope
- specifications   *see* optical specifications
- TEMOS (project)   107, 248, 253
- 300-inch concept (correctors)   385, 390
- two-mirror   42, 71, 88
- UKIRT (Hawaii) 3.8 m   456
- Universal K. Schwarzschild, Tautenburg   157, 460, 462
- University of Indiana 24-inch (Schwarzschild)   445
- University of Texas 7.6 m (project)   366, 367, 414
- vertical siderostat   442
- Victoria 72-inch   440
- visual use of   38, 302, *see also* Eyepiece
- wide-field survey   239
- William Herschel Telescope (WHT) 4.2 m   394, 395, 397–399, 455–457, 461
- Yerkes 40-inch refractor   431, 433, 435, 438
Telescope design parameters   49, 186, 188

Telescope form   *see* Reflector or Refractor
Telescope manufacture   *see* Manufacture
Telescope optics
- active control   *see* Active control
- adjustment   69, 255, 260
Telescopes
- optical characteristics of major   456
Television (OTF theory)   313, 315
Temperature (mirror)
- changes   438
Terms
- aspheric   59, 65, 140, 142, 354
- Gaussian optical   61, 144
- higher order   28, 60, 269, 270, *see also* Aberration
- non-orthogonal   288, 308
- orthogonal   288, 290
- stop-shift   66, 71, 92, 108, 148, 172, 382, 404, 405
Testing   46, 62, **89**, **91**, 95, **101**, 111, 117, 279, 361, 426, 431, **432**, **435**, **437**, 438, **443**, **447**, 457
- autocollimation   437, 445, 452
- autocollimation pinhole   11
- double-pass   437
- flat   437, 452
- Hartmann   438, 450, 452
- interferometer   450
- null (compensation)   117, 455
- problems   91
- procedures (Ritchey)   437, 438, 445
- zonal   421, 445, 447
Theorem
- binomial   13
- displacement   305, 312
- Fermat   27, 86, 117
- general, of plates   140–146, 193, 227, 229
- Malus   86
- Parseval   315, 319
- Schwarzschild   89, 98, 101, 111, 114, 176, 242, 325
- Schwarzschild (generalized)   89, 145, 241, 242, 248, 251
Theory

– analytical (third order) **57**, **69**, 82, **89**, 117, 148, 161, **174**, 226, 228, **269**, 270, 288, 339, 373, **382**
– aspheric plate 140–146, 193, 227, 229, 381
– Big Bang 440
– diffraction 293
– Fourier 312–314
– Hamilton 59–63, 255, 288
– higher order 85
– "island universe" 440
– reflector (Schwarzschild) 77, 94, 114, 325
– Schiefspiegler (for decenter) 261
– thin lens 167–168, 196, 373, 381, 404, see also Lens, thin
Thermal see also Problem, thermal
– effects (dome seeing) 301, 455
Thermal conductivity (mirrors) 432, 450
Thermal sensitivity (mirrors) 18, 440, 444, 450, 455
Thermodynamics 31, 38
Thickness (meniscus) 206, see also Meniscus, Maksutov
– finite 358, 407
Thin lens
– approximation 167–168, 206, 351
Thin objective 36, 39, 41
Third order approximation see Seidel
Third order calculation 64
Third order theory see Theory, analytical
Throughput 38, 141
Tilt
– in one plane 255, 257
– in two dimensions 255
– mirror 90
– small 90
– wavefront 61, 138, 292, 306, 311, 312
Tolerance
– criterion (Rayleigh) 301, 306
– criterion (Strehl) 304–308
Tolerances
– centering 111, 245, 270, 444
– manufacturing 11, 165
– Schmidt 161

Tolerancing 26, 98, 207, 279, 304, 307, 308, 339, 345
Tracing see ray-tracing
Tracking 21
Transfer equation 35
Transformation
– paraxial (virtual corrector plate) 328, 340
Transit instrument (Paul-Baker reflector) 229
Transmission
– UV 411
Transmission (optical glass) 14, 165
Triangular aberration (comatic) 62, 290, 292
Triplet objective see Objective
Tube
– open frame (ventilated) 423, 426, 435
Tube extension 123, 180
Tube length see also Length
– physical 152
Turned-down edge 9

Ultra-violet (aluminium coats) 447
Ultra-violet (transmission for FR) 406
Undercorrection (spherical aberration) 159
Unit circle (Zernike polynomials) 289, 290
Unit magnification 23, 24, 31
Use
– ground-based (specification) 301

Vacuum 30, 31
Variance (Strehl theory) 305, 310, 311, 321
Variation
– chromatic see Chromatic variation
Velocity of light 27
Ventilation 426
– natural 423, 426, 435
Vertex curvature 59, 83
Very Large Telescope (VLT) 269
Vignetting 33, 50, 52, 115, 126, **149**, 150, 160, 193, 204, 243, 306, 316, 338, 340, 364, 365, 393, 414, 444
Visual use see Observation, visual
VLT (ESO) see ESO Telescopes

Wave
- sinusoidal (OTF)   312, 313
Wave theory of light   13, 26, 293
Wavefront   26–28, 30, 57, 60, 61, 65, 71,
  86, 138, 290, 294
- geometrical   26, 27, 293
- higher order effects (decentering)
  270
- imaging   60, 61, 69
- spherical   26, 61
- tilt   61, 138, 292, 306
- variance   305, 310, 311, 321
Wavefront (phase) aberration   57, 58
- coefficient (Seidel)   63–65, 68, 134
- peak-to-peak   134
Wavefront concept
- image formation   26, 57, 58, 86
- Seidel contributions   64
Wavefront error
- rms   290
Wavelength
- central (correction)   155, 156, 159,
  163, 193, 202, 208, 209, 394
- light   293, 323
- secondary   158
- visible light   128
Wedge (prismatic correction)   260

Wind-shake   431
Windloading   426
Window   333, 368
- cold-box   369
Winkeleikonal   86, 117
Wire
- fine (diffraction)   324

Z-form (Czerny-Turner)   257, 258
Zenith   147
Zenith distance (ADC)   392
Zernike (circle) polynomials   62, 288,
  289, 291, 292
Zernike (radial) polynomials   86, 290,
  291
Zero optical power
- plane mirror   270
- plane plate   270
- plane secondary   54, 185, 267, 327,
  328, 350, 355
Zerodur (glass ceramic)   455, 456
Zonal error   150, 152, *see also* Aberration, fifth order
Zonal mask   421, 437
Zone   421, 437
- neutral   157
- neutral (Schmidt plate)   157

Printing: Saladruck, Berlin
Binding: Stein+Lehmann, Berlin